Physics of
Nonmetallic Thin Films

NATO ADVANCED STUDY INSTITUTES SERIES

A series of edited volumes comprising multifaceted studies of contemporary scientific issues by some of the best scientific minds in the world, assembled in cooperation with NATO Scientific Affairs Division.

Series B: Physics

RECENT VOLUMES IN THIS SERIES

Volume 7 — Low-Dimensional Cooperative Phenomena
edited by H. J. Keller

Volume 8 — Optical Properties of Ions in Solids
edited by Baldassare Di Bartolo

Volume 9 — Electronic Structure of Polymers and Molecular Crystals
edited by Jean-Marie André and János Ladik

Volume 10 — Progress in Electro-Optics
edited by Ezio Camatini

Volume 11 — Fluctuations, Instabilities, and Phase Transitions
edited by Tormod Riste

Volume 12 — Spectroscopy of the Excited State
edited by Baldassare Di Bartolo

Volume 13 — Weak and Electromagnetic Interactions at High Energies
(Parts A and B)
edited by Maurice Lévy, Jean-Louis Basdevant,
David Speiser, and Raymond Gastmans

Volume 14 — Physics of Nonmetallic Thin Films
edited by C.H.S. Dupuy and A. Cachard

Volume 15 — Nuclear and Particle Physics at Intermediate Energies
edited by J. B. Warren

Volume 16 — Electronic Structure and Reactivity of Metal Surfaces
edited by E. G. Derouane and A. A. Lucas

Volume 17 — Linear and Nonlinear Electron Transport in Solids
edited by J.T. Devreese and V. van Doren

The series is published by an international board of publishers in conjunction with NATO Scientific Affairs Division

A	Life Sciences	Plenum Publishing Corporation
B	Physics	New York and London
C	Mathematical and Physical Sciences	D. Reidel Publishing Company Dordrecht and Boston
D	Behavioral and Social Sciences	Sijthoff International Publishing Company Leiden
E	Applied Sciences	Noordhoff International Publishing Leiden

Physics of Nonmetallic Thin Films

Edited by
C. H. S. Dupuy and A. Cachard
Université Claude Bernard Lyon 1
Villeurbanne, France

PLENUM PRESS • NEW YORK AND LONDON
Published in cooperation with NATO Scientific Affairs Division

Library of Congress Cataloging in Publication Data

Nato Summer School on Metallic and Nonmetallic Thin Films, 2d, Corsica, 1974.
  Physics of nonmetallic thin films.

  (NATO advanced study institutes series: Series B, Physics; v. 14)
  Includes index.
  1. Nonmetallic materials—Addresses, essays, lectures. 2. Thin films—Addresses, essays, lectures. I. Dupuy, Claude H. S. II. Cachard, A. III. Title. IV. Series.
  QC176.N38 1974   530.4′1   76-8385
  ISBN 0-306-35714-3

Lectures presented at the Second NATO Summer School on
Metallic and Nonmetallic Thin Films held in Corsica, Serra di Ferro,
September 1-5, 1974

© 1976 Plenum Press, New York
A Division of Plenum Publishing Corporation
227 West 17th Street, New York, N.Y. 10011

United Kingdom edition published by Plenum Press, London
A Division of Plenum Publishing Company, Ltd.
Davis House (4th Floor), 8 Scrubs Lane, Harlesden, London, NW10 6SE, England

All rights reserved

No part of this book may be reproduced, stored in a retrieval system, or transmitted, in any form or by any means, electronic, mechanical, photocopying, microfilming, recording, or otherwise, without written permission from the Publisher

Printed in the United States of America

Preface

For several years now the intense development in the field of microelectronics, the interest in coating materials, and activity in integrated optics have produced many advances in the field of thin solid films.

The research activity has become so intensive and so broad that it is necessary to divide the field into metallic and non metallic thin films. A summer school in the area of non metallic thin films appeared to be a very fruitful concept and, hence, in October, 1973, A.S.I.M.S. made a proposal to N.A.T.O to hold this second summer school in Corsica in September 1974.

The basic idea behind this summer school was essentially to stress and synthesize physical properties and structure of non metallic thin films. The main reason for this was the feeling that many laboratories are very specialized and that few engage in both physical and structural analysis of these films.

The program included a large section on physical studies : electrical (transport, interface effects, switching), mechanical and optical. There was also a large section oncharacterization, crystal structure, chemical composition (stoichiometry is always a difficult problem), bonding and electronic structure.

Certainly it is different for every laboratory to pursue all these avenues of research. However, perhaps a good result of this summer school will be a deeper realization among scientists of the connection between physical properties and structure. Collaboration is probably necessary in many of these areas and in this respect I am sure that the summer school was profitable.

I want to acknowledge the many people who helped me organize this summer school. I want to especially acknowledge the members of the scientific committee and the lecturers. I wish to give special mention to the scientific codirectors, Dr. A. Cachard, Université Claude Bernard (Lyon), and Dr. S. Fonash, Penn State University, and Ms. Pivot and Chassagne, treasurer and secretary, respectively.

My special thanks to the staff of the Village de Detente et de Loisirs in Serra di Ferro.

I want to mention especially Mr. A. Paolini, Director and Mrs Pelamourgue, Assistant.

To all the people of Corsica I give my thanks. I am sure that no one will forget the friendly reception of Serra di Ferro and its mayor Mr. J.R. Tomi.

As was the case for the first A.S.I.M.S. summer school in Porto Vecchio, Mr. J. de Rocca Serra, Chef de la Mission Régionale, helped us and I thank him.

During this summer school Dr. J. A. Roger presented and defended his Ph. D. Thesis. It is very pleasant for me to thank him for presenting this first Ph. D. in Corsica.

The help from N.A.T.O allowed the school to be organized and permitted it to take place. The D.G.R.S.T. helped us too, and I thank them.

C.H.S. DUPUY

Contents

General Introduction 1
  A.K. Jonscher

Part I

BASIC NOTIONS

Preparation Methods for Thin Films 9
  D.S. Campbell

Growth Processes 49
  R. Niedermayer

Electronic States in Semiconductors 93
  D.A. Greenwood

Part II

CHARACTERIZATION

Structure Determination of Thin Films 123
  H. Raether

Physico-Chemical Analysis of Thin Films 141
  A. Cachard

Thickness Measurements 163
  D.S. Campbell

Part III

PHYSICAL PROPERTIES

Electronic Transport Properties 189
  R.M. Hill

Ionic Transport in Thin Films 219
  S.J. Fonash

Dielectric Properties of Thin Films: Polarization
      and Effective Polarization 225
  S.J. Fonash

Threshold Switching: A Discussion of Thermal and
      Electronic Issues 253
  H.K. Henisch and C. Popescu

Mechanical Properties of Non-Metallic Thin Films 273
  R.W. Hoffman

Thin Films in Optics 355
  G. Baldini and L. Rigaldi

Radiation Effects in Thin Films 383
  A. Holmes-Siedle

Part IV

APPLICATIONS

Thin Film Applications in Microelectronics 417
  V. Le Goascoz

Application of Thin Non-Metallic Films in Optics 445
  E. Pelletier

Some Applications of Non-Metallic Thin Films 459
  M.H. Francombe

List of Participants 495

Index 501

GENERAL INTRODUCTION

A.K. Jonscher
Chelsea College, University of London
Pulton Place, London SW6 5PR. U.K.

The purpose of this Introduction is to attempt to place the subject of Non-Metallic Thin Films into proper perspective against the more general background of modern physics and technology and also to discuss some problems relevant to research policiy in this field.

The fist question that may be asked is : "What is so significant in this subject that it should have been chosen as the theme of a specialist Summer School ? ". The answer to this, in my opinion, is that thin non-metallic films are commonly found in nature, that they have many useful applications and that their physical properties present certain unique feature which are not normally associated with bulk materials.

From the standpoint of a Summer School, with its definite didactic purpose, the subject of Thin Films is uniquely suited as a vehicle for conveying to the student a wide range of specialist topics - it is a multidisciplinary field which is both challenging and rewarding as a subject of study, provided care is taken to approach it in the correct manner- I shall have something more to say about this later.

It is desirable at the outset to define what is meant by the term "Thin Films". The commonly accepted limitation at the upper end of thickness is the concept of " Thick Films", typically some tens of micrometers in thickness. Thus it would appear that we would be justified to descaribe films up to one micrometer in thickness as "thin", without placing too much significance on the precise figure. However, even 1 μm contains some four thousand atomic layers and it may have many properties of "bulk" material. Figure 1 shows schematically some relevant orders of magnitude.

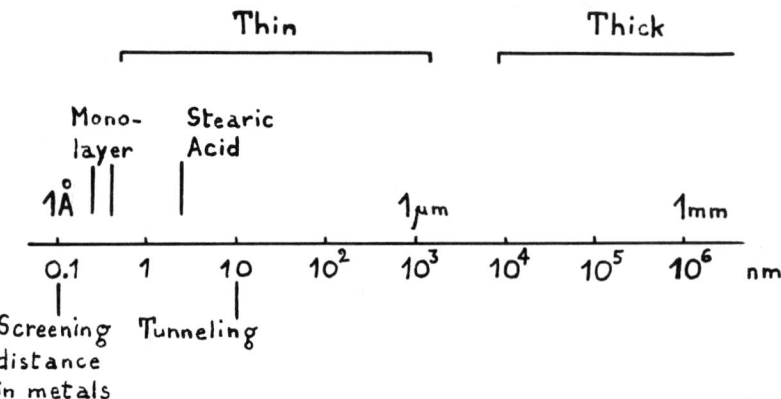

Fig. 1   Typical order of magnitude relating to thin non-metallic films.

Whether the behaviour of a thin film is distinct from that of the corresponding bulk material depends upon the precise context. Thus the structure may be influenced by interfacial processes, e.g. epitaxy or strain due to lattice mismatch and this may carry on for hundreds of atomic layers from the interface. The electrical properties will be determined by the relation between the mean free path and the film thickness - in very disordered non-metallic films the mean free path is effectively the tunnelling distance, of the order of 5 nm and thus most films are bulk-like from this point of view. However, this is not so in metallic or highly crystalline films in which the mean free path may be significantly large than, say, 10 nm and thus film thickness may become a limitation. The electrical effect of the interface may extend to several Debye lengths into the material and this, as we shall see in Dr. Fonash's lectures, may be a serious consideration in non-metallic films, but it is not important in metallic ones where the screening length is of the order of 0.1 nm, or less than one atomic layer.

Not much will be said in the context of the present Summer School of very thin films - of tunnelling thickness - nor of the problem of quantisation of energy levels in the direction normal to the plane of the film. Some of these questions are however implicit, for example, in the operation of MNOS transistors, as discussed by Dr. Le Goascoz.

The presence on metallic surfaces of naturally occuring oxide and other films was known for a long time to influence their optical and electrical properties. Where the films are mechanically robust and chemically inert, as in the case of aluminium, tantalum and silicon, they may act as protective coatings and they have been

GENERAL INTRODUCTION

used extensively as capacitor materials. It is hardly necessary to stress the fact that silicon owes its predominant position as a semiconductor device material to the excellent chemical, dielectric and mechanical properties of its dioxide.

Anodically produced films of oxides on aluminium and tantalum are the basis of a long-established and important capacitor industry.

We shall hear a good deal during this Summer School about the preparation of thin films. There are many specialised methods for doing this but on the whole their common feature is that films prepared in this way are significantly less perfect from the point of view of structure and purity than the best bulk materials. This is an important practical limitation which has to be borne in mind when choosing between samples prepared by film deposition techniques and samples obtained by thinning of bulk-grown material.

In many cases the methods most suitable for the deposition of thin films are not the same as those suitable for the preparation of good bulk samples : anodisation, evaporation, sputtering and chemical vapour deposition are all restricted in the attainable thickness of the deposits. It may even be difficult to grow films of nominally the same structure and composition but of different thickness - a serious limitation when it comes to testing the effect of the thickness on, say, the electrical properties.

It is understandable, therefore, that the problem of reliable characterisation of thin films is central to this branch of science and that it receives due attention in the second part of this Summer School.

Those methods of deposition which employ the condensation of the material from the vapour phase are at the same time most suitable for the formation of heavily disordered layers, including completely amorphous ones, even for those materials which do not occur naturally in the amorphous phase in the bulk, e.g. silicon, germanium and other non-glass-forming semiconductors. For this reason the subject of non-metallic thin films is closely related to the study of amorphous materials which has acquired considerable popularity in recent years, even though the time is ripe to question the continuing justification for it.

The advancement of our understanding of the growth processes of thin films represents one of the most significant developments in modern physics and may be attributable in a large measure to the interest in the science and technology of thin films. Professor Niedermayer's lectures will discuss the various processes governing the growth of thin films at the interface between the solid and the gaseous or liquid phase ouside.

Dr. Geenwood's lectures on the electronic properties of thin films reflect many of the problems encountered in this area of physics. The disordered nature of the films has immediate consequences for the bulk electronic energy structure, a subject which has received a considerable amount of attention in the last ten years or so. In addition, the very proximity of the surface has its own consequences and the study of surface and interface states is of great interest in this context.

Any discussion of the electrical conduction in thin films must distinguish between transport along and across the film. The former is of principal interest in metallic films, while in insulating films we are usually forced to study transverse conduction because of the excessively high impedance. However, conduction in the plane of the films is of great significance in some semiconductor films ; one notable examle is the conduction in the channel of a Metal-Insulator-Semiconductor Field Effect Transistor (Insulated Gate FET), where transport occurs in a very narrow potential trough near the insulator-semiconductor interface. There is no chemical film present here but an effective thin film situation arises nonetheless.

Dr. Hill's contribution to this Summer School is concerned entirely with transport across the film and he is effectively referring to modes of direct current conduction characteristics of heavily disordered solids, i.e. hopping conduction. The last decade has seen some spectacular advances in our understanding of this area of transport and Dr. Hill will outline the most important aspects, stressing the significance of a distribution of localised states in the forbidden gap of material.

The problem of dielctric characterisation of non-metallic films receives its proper attention in this Summer School in the lectures of Dr. Fonash. Dielectric measurements on thin films represent a favoured test method, perhaps next in popularity only to optical measurements and the interpretation of the experimental data is frequently very complicated, since the dielectric properties are determined by the interplay of the barrier characteristics and the bulk region response.

One of the most distinctive applications of non-metallic thin films is as optical filters and anti-reflection coatings. The lectures of Professor Baldini and of Dr. Pelletier deal with this aspect of optical properties and applications and show the degree of sophistication in design and in the execution of optimal structures for particular purposes which are made possible by modern methods of film deposition.

GENERAL INTRODUCTION

Dr. Francombe's and Dr. Le Goascoz's lectures provide further insight into different applications of thin films in technology and both make it clear how important it is to back up any development programme with adequate technological facilities if success is to be achieved.

After this brief look forward to the present Volume I wish to turn, as it were, to a retrospective appraisal of the Summer School.

A survey of the content of the present Volume will reveal what I consider to be the most striking common characteristic of this whole subject – it is its complex nature. This complexity derives from the combination of boundary conditions, from the uncertainty of the exact structural configuration, from the multiplicity of the possible materials, from the complexity of the various deposition and testing techniques, and so on.

An immediate consequence of this complexity is that the instrumentation required for the successful deposition, testing and characterisation of thin film structures has become very expensive and the processes of research and development are correspondingly time- and money - consuming. The study of thin films has cessed to be the "cheap" subject which could be pursued with minimal resources, the proverbial "string and sealing wax" approach so typical of many university laboratories.

This brings me to the point of research policy in relation to thin films. The complexity and the great potential depth of the subject, with its manifold possibilities with regard to the choice of materials and geometrical configurations demands a careful choice of research topics for specific studies. This choice has to be based on certain criteria regarding the potential value of the expected results. I suggest that one question that may be asked, profitably, is whether it is intended to make a scientific or a technological study. Since a scientific study is intended to elucidate fundamental questions regarding the mechanisms or processes in question, it is desirable that the system under investigation should be as simple and well-defined as possible. However, this is precisely what is very difficult to obtain in the majority of thin film systems and this is where the principal problems arise. In order to be scientifically meaningful, the study should proceed from simpler systems to more complex ones, while in practice it is often impossible to devise experiments on simple systems.

By contrast with the scientific approach in which the principal objective is the understanding of the laws of nature, the technological approach is aimed at mastering a set of processes leading to some desired practical end. Technology is mostly concerned with complex systems, arrived at empirically and even accidentally and there is often little hope of elucidating fully the processes involved. However, good technology is, by definition, a technology relevant to some important objective - a device, a process, a system. Once the objective ceases to be important, for example because of obsolescence, because the market for it has changed or even disappeared, because it has been overtaken by other developments, etc., the technology concerned ceases to be good technology.

While good science contributing to our increased understanding of the material universe and good technology aimed at producing desirable goods are both fully justified in their respective diverse pursuits, what is never justified is the pursuit of bad science aimed at obsolete or irrelevant tecnology.

I must confess to a feeling of unease in this respect when looking at much of the published material relating to the general field of thin films. While there is some good science and some good technology being pursued, there is also a great deal of work done at a considerable expense of public money which would not be scientifically justified in its own right while its technological justification has also disappeared long ago.

I hope that this Summer School may help its participants and also the readers of the present Volume to avoid these pitfalls in the future and may give them fresh inspiration for the pursuit of really good science in support of relevant technology. Many examples of this may be found in the lectures that follow this Introduction.

Basic Notions

PREPARATION METHODS FOR THIN FILMS

D.S. Campbell

Department of Electronic and Electrical Engineering
University of Technology
Loughborough, Leicestershire, U.K.

1. INTRODUCTION

It is possible to divide the methods of making thin layers (i.e. layers which are less than 1 μm thick) in several different ways[1][2]. One classification is in terms of the separate groups, (1) chemical methods and (2) physical methods. Physical methods cover deposition techniques which depend on the evaporation or ejection of material from a source, i.e. evaporation or sputtering, whereas chemical methods depend on a specific chemical reaction. This chemical reaction may depend on the electrical separation of ions as in electro-plating and anodisation or it may depend on thermal effects as in vapour phase deposition and thermal growth. However, in all these cases a definite chemical reaction is required to obtain the final film.

When one seeks to classify deposition of films by chemical methods, one finds that it is possible to further sub-divide the methods that are available, into two more classes. The first of these classes is concerned with the chemical formation of the film from the medium, and typical methods involved are electro-plating, chemical reduction plating and vapour phase deposition. A second class, however, is that of formation from the substrate and examples are anodisation, gaseous anodisation and thermal growth.

It must be emphasised that there is often considerable overlap between the physical and chemical classification, and also between the sub-classification of formation from the medium and formation from the substrate.

The methods summarized under the classifications given are, in some cases, capable of producing both films less than and also greater than 1 μm. However, there are certain techniques that are only capable of producing thick films (i.e. greater than 1 μm) and these include methods such as screen printing, as in thick film technology, glazing, electro-phoretic deposition, flame spraying and painting. However, these techniques will not be examined in the context of this chapter. (See Chapman and Anderson[3] for surveys of these techniques)

The aim of this paper therefore is to examine the various methods available for film deposition, be they physical or chemical. At the end of this examination it will be possible to identify the techniques which are suitable for the preparation of non-metallic thin films.[4]

2. PHYSICAL DEPOSITION TECHNIQUES

2.1 Evaporation[5]

2.1.1. Introduction

Evaporation techniques are widely used for the preparation of thin layers. A very large number of materials can be evaporated and if this evaporation is effected in a vacuum system, then the evaporation temperature will be very considerably lowered, the formation of oxides will be considerably reduced, the amount of impurities included in the growing layer will be reduced and finally, straight line propagation will occur from the source to the substrate and this will allow for reproduction of finely defined patterns on the substrate if a mask with the necessary holes in it is placed between the source and the substrate itself. Substrates can be of a wide variety of materials and can be held at a temperature appropriate to the properties of the deposited films that are required. Most materials can be boiled in a suitable crucible, but there are several that can be sublimed before their melting point is reached.

The pressures that are required in a vacuum system to obtain satisfactory deposition, both in terms of the reduction of oxides, reduction in included impurities in the deposition, and the obtaining of a sharply defined pattern on the substrate due to the presence of a mask between the source and the substrate, is less than 10^{-4} torr with an ideal pressure for normal evaporation work being 10^{-5} torr.

Rates of evaporation and condensation can vary over very wide limits, dependent upon the type and temperature of source and the material used, and a typical curve for gold as a function of source temperature is given in Fig. 1. An average figure for growth rate can be reckoned as 10 Å per second, but it is possible to obtain rates as high as 10^{+4} Å/second, using special boats. Very low rates of growth are also possible (e.g. the work of Walton et al on Nucleation studies[6]).

Control of the deposition rate and film thickness can be effected by several systems which are discussed elsewhere[5][7].

2.1.2. Basic Principles

The thermodynamics of the evaporation processes have been considered in detail by various authors (c.f. Glang[5]). The rate of evaporation G from a surface at a temperature T is given by the Langmuir expression:

$$G = p \sqrt{M/2\pi RT} \qquad (1)$$

where p is the vapour pressure of the material and M the molecular weight. R is the gas constant per mole. The relationships between p and T have been published by various authors for a wide variety

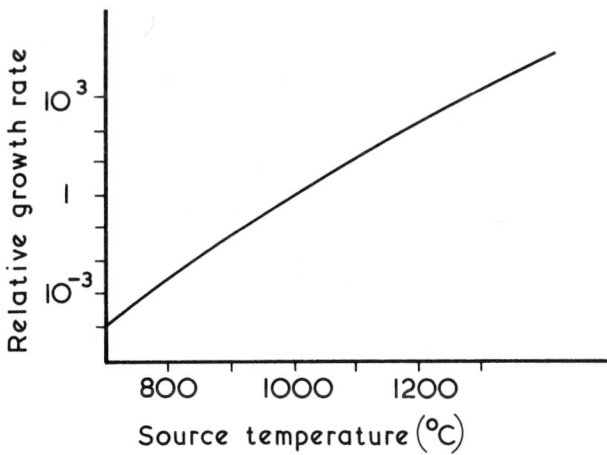

Fig. 1   Relative growth rate of gold as a function of source temperature.

of materials (Honig [8]) and Figure 2 shows a typical curve for gold. A temperature which is normally considered suitable for evaporation is that at which the vapour pressure of the material is equal to 10^{-2} torr. As an example, for gold this corresponds to a source temperature of 1650°C and at this temperature the rate of evaporation from the gold surface is 0.2×10^{-3} gms/cm²/second, which is equivalent to 6×10^{17} atoms/cm²/second.

The average energy of atoms from the evaporating source is given by :

$$\bar{E} = \frac{3}{2} kT \qquad (2)$$

where k is the gas constant per atom. A usual value of $\bar{E}$ is 0.25 eV compared with the average energy for gas atoms at room temperature (the normal temperature of the vacuum in which the evaporation is occurring) of 0.03 eV.

The velocity of the atoms evaporated from the source will be distributed in a roughly Gaussian manner, about the most probable

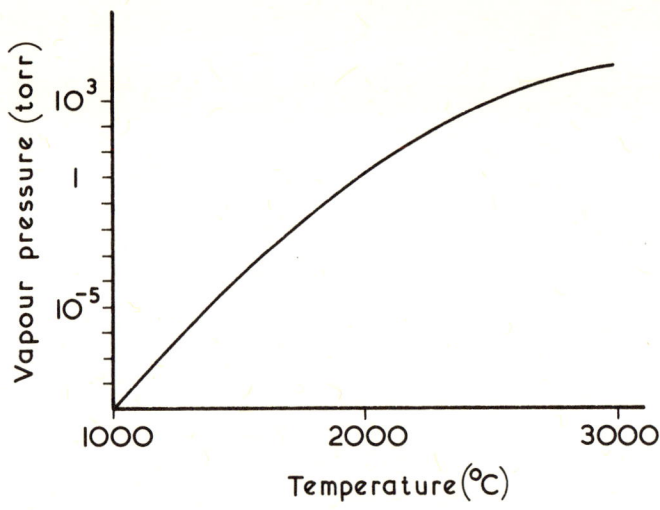

Fig. 2 Vapour pressure of gold as a function of temperature

PREPARATION METHODS FOR THIN FILMS

value of velocity given by :

$$\alpha = \sqrt{2\ kT/m} \qquad (3)$$

where m is the mass of an evaporant atom. The root mean square velocity will be given by :

$$\bar{c} = \sqrt{3\ kT/m} \qquad (4)$$

(a value for $\bar{c}$ for gold at temperature of 1800°K will be 10^5 cms. per second). This value of root mean square velocity will depend on the source geometry and the figure quoted is for an open filament type of evaporation. If, for example, the emission occurs through a small hole in a crucible type of source then $\bar{c}$ will be modified to be equal to :

$$\bar{c} = \sqrt{4\ kT/m} \qquad (5)$$

Such a source with such a characteristic is known as a Knudsen source.

The mean free path of the evaporating material in the residual gas will be a function of the residual gas. A typical figure for a pressure of 10^{-5} torr is 500 cms. and this value will be inversely proportional to the residual pressure.

The rate of arrival n of molecules or atoms of molecular weight M of evaporant at the substrate, is given by :

$$n = \frac{t\rho N}{M} \qquad (6)$$

where t is the thickness deposited in unit time, ρ is the density of the deposition and N is Avogadro's number. This value of n needs to be compared with the arrival rate of residual gas atoms in the vacuum which is given by :

$$n_g = \frac{3}{4}\ p\ \sqrt{N^2/3\ RTM_g} \qquad (7)$$

where M_g is the molecular weight of the residual gas. In normal systems it is found that a ratio of 100 : 1 exists between the number leaving a source/unit area and the number arriving, and this typical figure holds for a source-substrate distance of around 15 cms. Under these circumstances an arrival rate of 10^{15} atoms per cm^2/second is fairly normal, and this corresponds to a growth rate of 10 Å per second. However, the value for the rate of arrival of residual gas atoms at 10^{-5} torr for nitrogen and/or oxygen, is also 10^{15}, so that it can be seen that in order to be sure of obtaining a film with the least number of impurities, either the source-substrate temperature has to be considerably increased so as to raise the rate of evaporation, the source substrate distance must be reduced (with the subsequent unavoidable heating of the substrate that will result), or the residual gas pressure must be reduced below 10^{-5} torr. It is this final solution which is normally used if very pure films are required, and the techniques for obtaining very low pressures in vacuum systems have been widely discussed by many authors in the literature[9]. It is now possible, at least in experimental systems, to work at pressure of 10^{-10} torr or less, but these sorts of pressures are not widely used in commercial production.

2.1.3. Types of Evaporation Sources

In order to evaporate materials in a vacuum, a vapor source is required that will support the evaporant and supply heat of vaporisation while allowing the charge of evaporant to reach a temperature sufficiently high to produce the desired vapor pressure, and hence rate of evaporation, without reacting with the source. In order to avoid the contamination of the evaporant and hence of the growing film, the support material itself must have a negligible vapor and dissociation pressure at the operating temperature. There are, therefore, two types of material that can be used for this, either refractory metals or certain non-metallic materials such as oxides, nitrides etc. The form in which these support material are used depends very much on the evaporant and tables are now available, summarising the best support materials used for the evaporation of the elements (Glang[10]), inorganic compounds (Glang[11]) and alloys (Glang[12]) to which the reader is referred.

A. Wire and Metal Foil Structures

Of the two basic types of material used for sources mentioned, wire and metal foil structures are very widely used for a wide variety of evaporants. The simplest vapor sources are resistance heated wires and metal foils of various types, examples of which are shown in Figure 3. These wires and foils are generally made out of tungsten, molybdenum or tantalum. Wire source are generally made from wire of diameter 0.02" to 0.06" and their use is limited to evaporants which wet the filament upon melting and are then held on by surface tension. However, it is also possible to use wires of the material to be deposited, provided that the wires will sublime. This implies that a vapor pressure of 10^{-2} torr is reached before the melting point of the wire itself. Such a technique has been widely used for the deposition of nickel[13] and also for nickel-chromium,[14] to mention but two, although in the latter case it is important to note that the composition of the depo-

PREPARATION METHODS FOR THIN FILMS

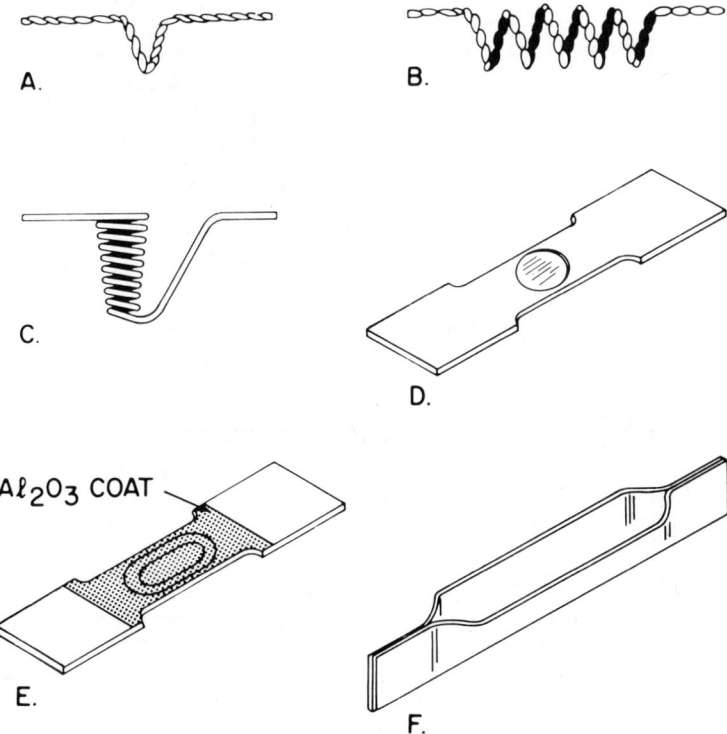

Fig. 3 Wire and metal foil sources. a) Hairpin. b) Wire Helix c) Wire Basket. d) Dimpled foil. e) Dimpled foil with alumina coat. f) Trough type.

sited film will change with time because of the different sublimation rates of nickel and chromium from the wire.

Metal foil structures can take the form of open boats of various shapes, or for films that are apt to de-gas on evaporation, structures which prevent the ejection of solid particles directly from the source. A particularly well-known example of this type of source is that developed by Drumheller[15] for the evaporation of silicon monoxide ;(see Figure 4.)

It is possible to use metal foil structures for the deposition of alloys as well as elements and oxides, provided enough information is available on the relative vapor pressures of the alloy constituents. In this context two alloys have been examined in considerable detail ; nickel-iron (permalloy) and nickel-chromium. The nickel-iorn alloy has been widely used for magnetic memory elements and nickel-chromium for resistors. Both are often eva-

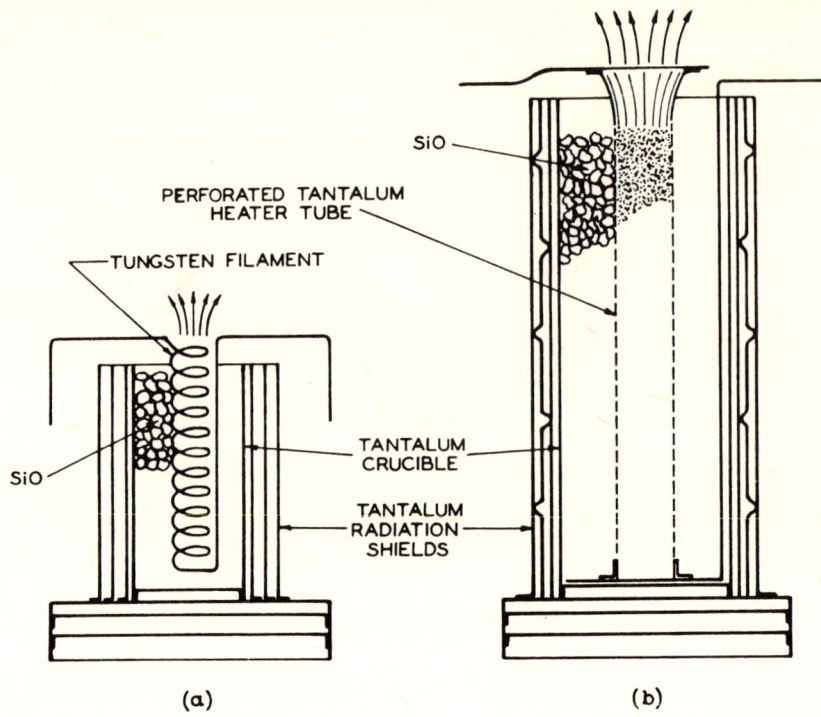

Fig. 4. Chimney evaporation sources (after Drumheller [15])

porated from tungsten boats. However, change of composition with time during the evaporation process, is a considerable disadvantage[16], particularly in the case of Ni-Cr, where the electrical properties are dependent on the Ni-Cr ratio of the final deposition. These problems have been overcome by using evaporation from two separate sources with the rates of evaporation separately controlled, or by the technique known as flash evaporation. Various arrangements for flash evaporation have been derived[17][18] and a typical solution is shown in Figure 5. The principle is that a finely divided powder of the alloy is vibrated on to a very hot tungsten strip where it evaporates immediately on contact with the strip. In these circumstances the composition of the deposited film is the same as the original alloy powder. Care has to be taken, however, to shield the source in order to prevent any loose powder from finding its way into the pumping system of the vacuum chamber.

B. Non-Metallic Structures

Crucibles of non-metallic materials are often used as evaporating sources[20]. These crucibles, which are non-conductive of

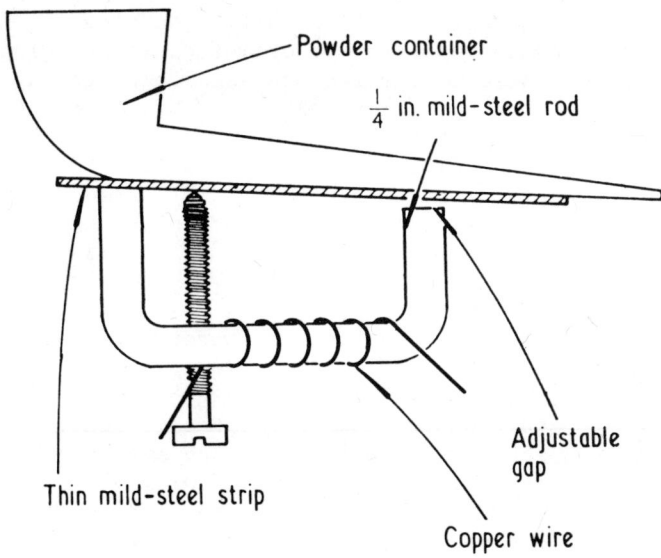

Fig. 5 Flash evaporation source (after Campbell and Hendry[16])

electricity, have to be supported in a suitable metal cradle, the cradle then being heated in the normal way by the direct passage of an electric current. This cradle can take the form of a wire coil directly wound round the crucible or of a foil structure. It should also be noted that another form of heating of such crucibles is that of employing an R.F. coil which is placed around the crucible but not actually touching it. Such a source has been used for depositing aluminium.[19]

Alumina is a widely used crucible material which can be used up to a temperature of 1900°C and which has a fairly good thermal conductivity (0.014 calories per second per degree per cm^2) enabling heat to be transferred easily from the heated metal coil or boat. A better material even than alumina, from the thermal conduction point of view, is beryllia which has a thermal conduction over three times that of alumina. However, there are certain toxic disadvantages in the use of beryllia, which have to be taken care of in any commercial use of such a source. Other oxide materials are available and all of these have been summarised by Glang[20].

Other materials that have been used for crucible sources are boron nitride, carbon either as ordinary graphite, pyrolitic graphite or in the form of vitreous carbon.

C. Electron-Beam Heating[21]

The two main types of source examined so far have been heated either by resistance heating or by induction heating. It is, however, possible to cause vaporisation of materials by using electron bombardment. A stream of electrons is accelerated up to 10 kV and focused on to the evaporant surface. By this means, temperatures exceeding 10,000°C may be obtained, enabling a variety of otherwise non-evaporatable materials to be used, and also, as only a very limited portion of the evaporant is heated, reducing considerably any interaction between the evaporant and the support materials. This technique is widely used for the preparation of very pure films.

A wide variety of electron gun structures have been used for this, and these have been classified by Glang[21], into work-accelerated guns, heavy accelerated guns and bent-beam guns. Two typical electron beam sources are shown in Figure 6. Figure 6 (a) showing a pendant drop configuration in which the metal to be evaporated is in the form of a rod or wire centred within a cathode loop. Evaporation takes place from the molten tip. The molten drop is held on the tip by surface tension, and as result, careful control of the electrical energy supply is required to prevent the drop falling off. The second source shown in Figure 6 (b) is that developed by Unvala and Booker[22], which uses a hearth to contain the evaporant which can be water-cooled. Also, because of the electron beam configuration, the evaporating unit is self-contained.
ned.

D. Reactive Evaporation[23][24]

Evaporation from metal wires or foils, from refractory oxide crucibles, or from electron beam sources, can be used to deposit

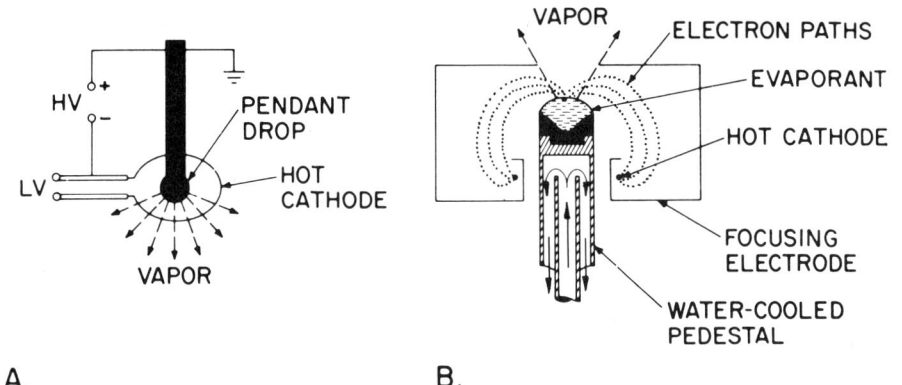

Fig. 6 Electron gun sources. a) Pendant drop. b) Shielded filament.

PREPARATION METHODS FOR THIN FILMS

oxides. However, it should be noted that it is also possible to evaporate metals in relatively high oxygen pressures. Under these circumstances the oxide of the metal being evaporated will then be deposited on the substrate. The oxidation reaction takes place in the main, at the surface of the depositing film, and such techniques have been used successfully for silicon oxide, tantalum oxide aluminium oxide and even $BaTiO_3$.

2.1.4 Summary

The various types of evaporation sources have been briefly described, and the reader is referred to summary tables that are available [5], with regard to the most suitable source for a particular evaporant. The applicability of the basic evaporation principles outlined, to the practical sources used, depends very much on the actual geometry of the source being employed. Crucible type sources with a flat evaporating surface will approximate fairly well to the basic equations noted for evaporation rate etc. However, long thin sources of the type associated with evaporation or sublimation from a wire, will be more complex in behaviour and metal box type structures (Drumheller's silicon monoxide source, Figure 4) and Unvala's electron beam source, Figure 6 (b), will approximate to the Knudsen behaviour of emission from a box with a small exit hole.

2.2. Sputtering[25][26][27]

2.2.1 Introduction

If a surface is bombarded with energetic particle, it is possible to cause ejection of the surface atoms, a process known as sputtering. These ejected atoms can be condensed on to a substrate to form a thin film. Such a process can be realised by forming positive ions of a heavy neutral gas such as Argon and bombarding the surface of the target material by making the surface the cathode in an electrical circuit. Such a method of obtaining a film has various advantages over normal evaporation techniques, in as much as no container contamination will be obtained. It is possible to deposit alloys without worrying about any fractionation of the materials. High melting point materials can be used as easily as low melting point ones and finally, using an R.F. technique, both metals and insulators can be deposited.

2.2.2 Basic Principles[28]

When a charged particle strikes a surface, a variety of interactions are possible. The most important reactions are shown diagramatically in Figure 7 and are (a) the ejection of neutral atoms

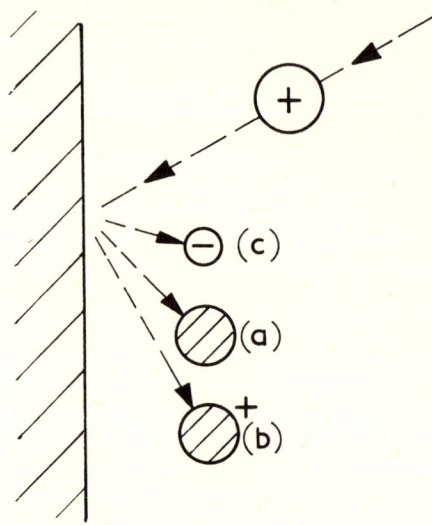

Fig. 7   Ejected species from an ion bombarded surface.

of the surface material, (b) the ejection of a small number of charged atoms of the surface material (usually only about 1% or less of the number of un-charged atoms) and finally, (c) the ejection of free electrons - the number of free electrons usually being greater than 10 for each arriving incident ion. The first major effect of this process is, as has previously been mentioned, that the neutral ejected atoms (Process a) can be collected on suitably placed substrates to form a film. The second most important effect is that the electrons ejected can, if desired, be accelerated away from the target cathode to a suitable placed anode. On their way to the anode they can cause further ionisation of neutral gas in the surrounding   space and the positive ions so formed will then be accelerated towards the cathode target. A self-sustaining system can therefore be obtained and such a situation is known as a glow discharge condition. The pressure at which this self-sustaining system will hold, will depend on the cathode/anode spacing and on the residual pressure and typical figures are that, for a cathode/anode spacing of 15 cms., a pressure of 10^{-2} torr of residual gas will be sufficient. Below 10^{-3} torr the discharge will be eliminated.

PREPARATION METHODS FOR THIN FILMS

To obtain the reactions as shown in Figure 7, the accelerating voltage must be limited. Figure 8 shows a typical graph of sputtering yield, i.e. the number of neutral atoms ejected for one incident ion as a function of voltage. The three types of interactions can be identified on this graph, namely :

(a) Hard sphere ejection, below approximately 10 keV.

(b) A region in which the electron clouds of the incident ion and surface atoms begin to interact.

(c) Finally, a region in which the nuclei interact.

The majority of sputtering situations are concerned with the region of hard sphere ejection and studies on single crystal materials have shown that in this region, ejection from a target is a function of the crystallographic direction. On a polycrystalline target, however, the effect of crystallographic directions is averaged out.

An important feature of ejection under hard sphere conditions, from a polycrystalline target, is that the most probable energy obtained for ejected atoms is much higher than that obtained in the case of evaporation, e.g. 4 eV for Cu bombardment with Hg^+ at 1 keV. Such an average energy in the case of Cu just quoted, corresponds to a root mean square velocity of 4×10^5 cm/second and an equivalent surface temperature of 4×10^4 °K.

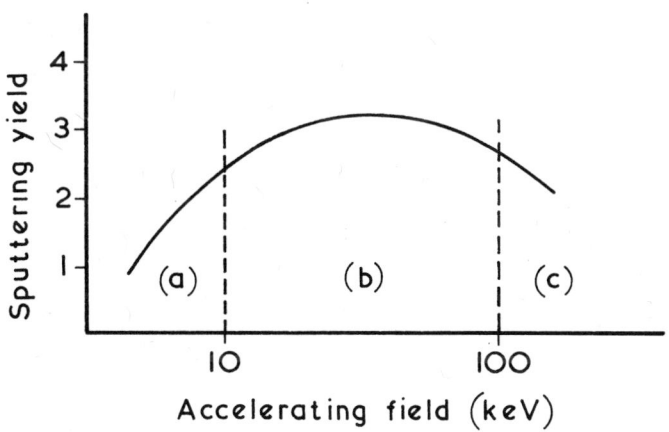

Fig. 8 Idealised sputtering yield / bombardment energy curve

The rate of removal of atoms from the surface under polycrystalline conditions, is a function of the number of ions bombarding the target surface, i.e. a function of current. 1 mA cm^{-2} current corresponds to an ion impingement rate of 5×10^{15} ions/cm^2/second. For sputtering yield figures of around unity, this therefore implies a sputtering rate of 5×10^{15} atoms/cm^2/second. Comparison with evaporation situations that has previously been discussed, implies therefore, that a current of around 100mA/cm^{-2} would be necessary to obtain rates of removal from the target, equivalent to that which can be obtained from a typical evaporation situation (e.g. gold at 1650 °K).

The rate of arrival of atoms at the substrate surface will depend on source-substrate distance as with the case of evaporation, and the typical figures quoted for evaporation can also approximately apply here. The major difference, however, is that if a glow discharge situation is being used, it will not be possible to reduce the impingement rate of residual gas atoms to anything like the level that is possible in evaporation.

2.2.3. Types of Sputtering Systems

A. Glow Discharge

A normal glow discharge sputtering system operates under the minimum conditions in which sufficient secondary electrons are generated to replace those lost to the anode or to the walls of the discharge chamber. Under these conditions, which are typified by an ∿ 5 cm cathode/anode separation and an accelerating field of 1.5 kV at a residual pressure of 10^{-2} torr, material will be sputtered from the metal cathode into the space between the cathode and the anode. Figure 9 shows an experimental arrangement. The position of the substrate to collect this material would seem, at first glance, to be best as close to the cathode as possible. However, such a position would effectively block ions in the discharge from reaching the target cathode so that as a result, substrates placed close to the cathode receive very thin coatings of sputtered materials. This screening effect can be avoided if the substrate is at some distance from the cathode and a useful rule of thumb quoted by Maissel[27] is that the cathode substrate distance should be about twice the length of the Crooke's dark space. This dark space is the region of low luminosity found adjacent to the cathode, and represents the acceleration distance of the electrons from the cathode, which they require to reach an energy at which ionisation of the residual gas can occur. It is also, often found convenient to place the substrate directly on the anode.

Films grown in such an environment are often classified as "dirty" because of the high background pressure of gas, but it should be noted that such a statement needs qualification, since

PREPARATION METHODS FOR THIN FILMS

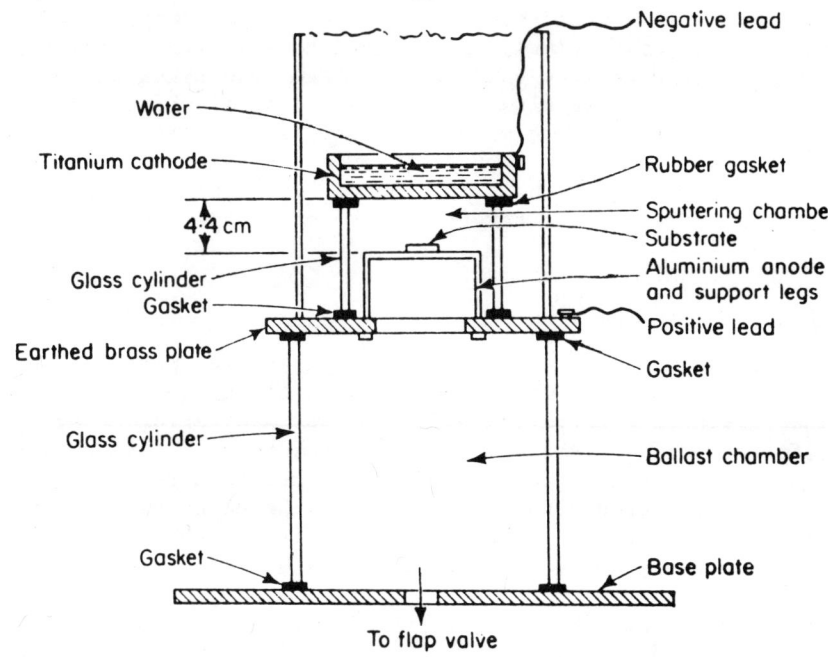

Fig. 9. Glow discharge sputtering system.

the high pressure in a sputtering system is in the main due to an inert gas such as Argon and the partial pressure of reactive gases could be as low as in evaporation systems.

Typical growth figures for substrates placed directly on the anodes in glow discharge systems can be given. For gold sputtered by Argon at a cathode substrate distance of 4.8 cms. with a voltage of 1.5 kV between anode and cathode and a total current of 1.7 ma over a cathode area of 48 cm^2, a growth rate of 0.5 Å/second can be obtained.

B. Reactive Sputtering[29]

If advantage is taken of the residual gas in a glow discharge system such that a large proportion of the residual gas is a reactive species relative to the film being deposited, then a deposition can be obtained of a completely reacted material such as an oxide or nitride. This system has been used for the growth of silicon dioxide to prepare capacitors.

Reactive sputtering need not to be confined to a flow discharge system, but can use any of the sputtering systems which will be discussed (triode sputtering, R.F. supported sputtering, etc.)

The mechanism of reactive sputtering has been considered by various authors, particularly with regard to the site of the oxidation reaction. Extensive studies have been undertaken on the reactive sputtering of tantalum[63][64] in a glow discharge using resistivity, growth rate and mechanical stress observations. It has been concluded that at low oxygen pressure, tantalum metal is sputtered and reacts at the substrate, whereas at high oxygen pressures the reaction occurs at the target and the oxide species is sputtered.

C. Getter Sputtering [30]

Decreases in the partial pressure of reactive gases beyond those usually associated with a normal glow discharge can be effected by utilising the gettering action of the sputtering material to purify the Argon of the discharge system, before it reaches the part of the system where coating of the substrate occurs. This is accomplished by surrounding one or more of the cathodes with an anode can (see Figure 10). Under these conditions the partial pressure of impurities can be reduced to levels as low as 10^{-10} torr.

D. Bias Sputtering[31]

In the systems so far discussed, the substrate has either been floating electrically or kept at anode potential. It is, however, possible to give the film, assuming it is conductive, a small negative bias relative to the anode. The film is then subjected to ion bombardment throughout its growth, a process which effectively cleans the film of absorbed gases which would otherwise become trapped in it as impurities. This technique is known as bias sputtering

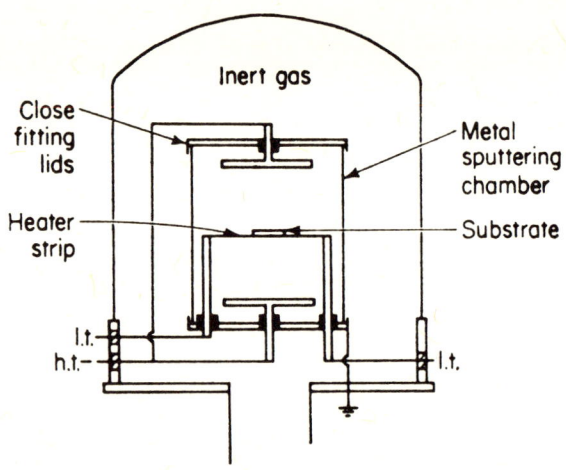

Fig. 10. Getter sputtering apparatus.

PREPARATION METHODS FOR THIN FILMS

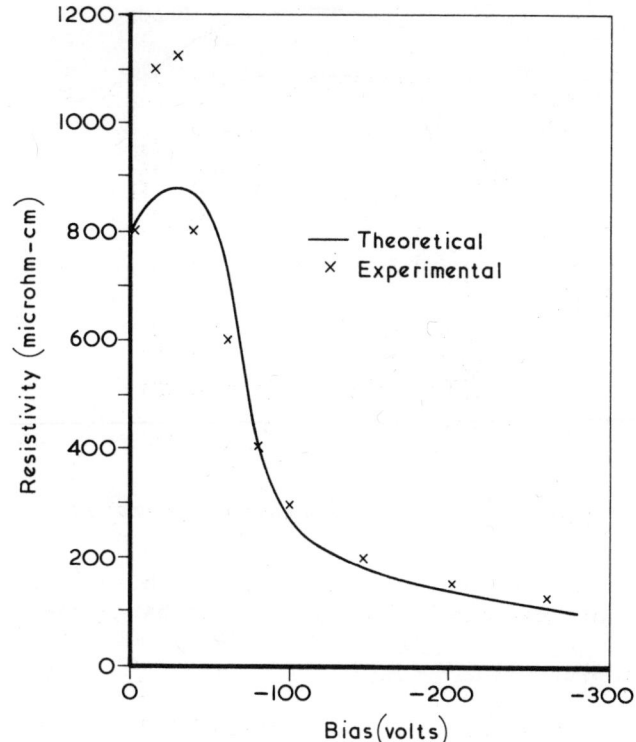

Fig. 11 Effect of bias sputtering voltage on purity of tantalum films (Ref. 31)

Figure 11 shows the effectiveness of this process for the case of a tantalum film deposited at various negative biases. The resistivity is a direct measure of the degree of impurities incorporated in the film and it can be seen that after a small initial rise, probably due to stresses introduced into the film by the high energy ion bombardment, the resistance falls to a very low value for bias voltages of around -200 volts.

E. Triode Sputtering

Another method of reducing the effect of impurities in a glow discharge system is to work at lower pressure than 10^{-2} torr. Such a requirement implies that it is necessary to generate electrons which can be injected into the discharge by other means than the discharge itself. One such system is known as triode sputteting and Figure 12 shows a diagram of the apparatus. A filament at a voltage approximately 50 volts less than the anode is used to inject electrons into the discharge systems and a magnet, external to the vacuum chamber, is used to increase the path of the electrons

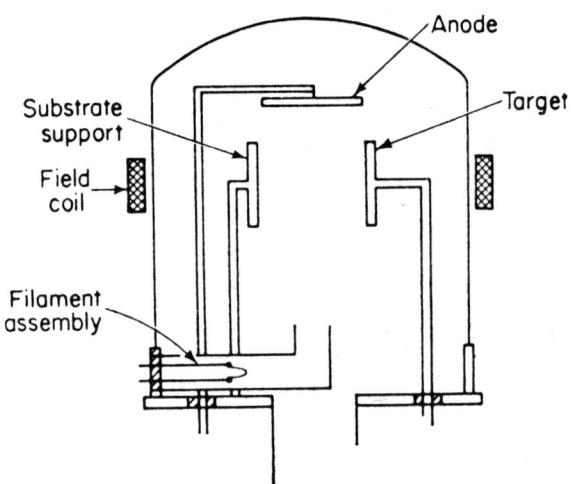

Fig. 12 Triode sputtering apparatus

prior to their collisions with the anode. In this situation, pressures of 10^{-3} or less can be used in the discharge chamber.

F. Ion Beam Sputtering

If instead of using a discharge system to generate the ions either at 10^{-2} or at lower pressure as with triode sputtering, a separate ion source is used, then it is possible to grow films sputtered from targets in a residual pressure as low as 10^{-5} torr or less. A typical apparatus for doing this is shown in Figure 13. This type of apparatus has been developed commercially by Balzers and is capable of growing clean sputtered films at high rates of deposition. It is interesting to note that this technique is very closely allied to that of ion implantation[62], as the ion beam could be any material that can be ionized in the ionisation chamber

G. R.F. Supported Sputtering

A further method for working at lower pressure than 10^{-2} torr is that of running the discharge in an R.F. field. Such an R.F. field, typically at a frequency of the order to 5 MHz, will cause both negative and positive ions to be formed, and the positive ones will then bombard the cathode in a normal manner. A typical apparatus is shown in Figure 14.

H. R.F. Sputtering

The system described so far have involved the use of a contacting cathode as the material to be sputtered, although it has been noted that it is possible be reactive sputtering, to grow

PREPARATION METHODS FOR THIN FILMS

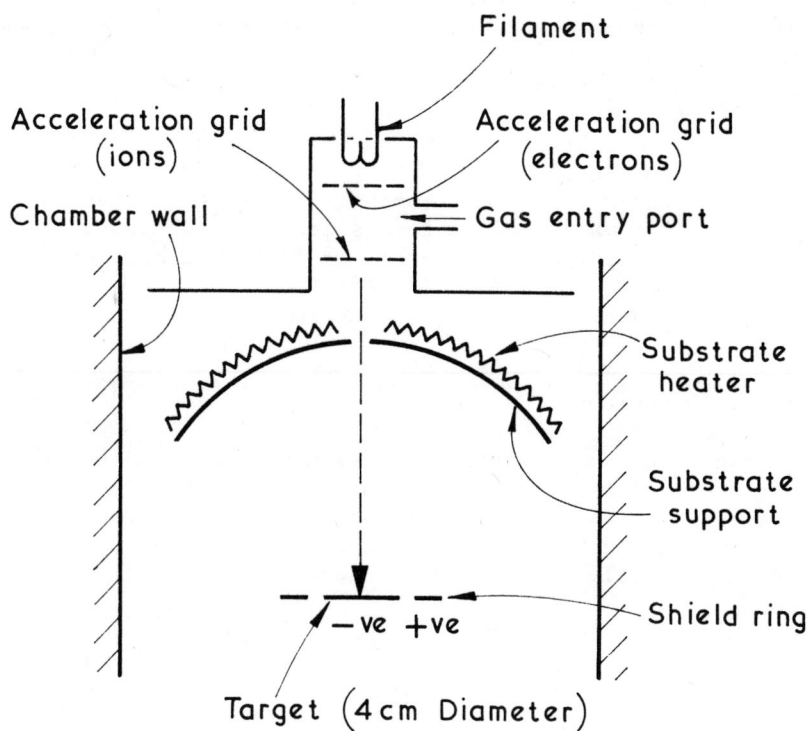

Fig. 13   Ion beam sputtering apparatus

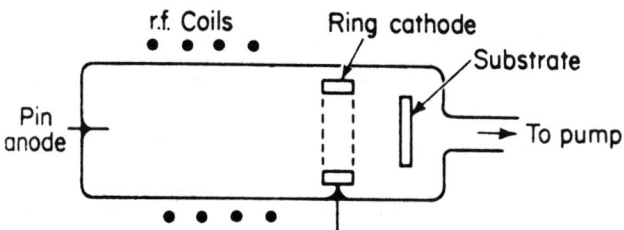

Fig. 14   R.F. supported sputtering apparatus

insulating films on the substrate. It has, however, been found possible to sputter directly from insulators by applying an R.F. potential between the cathode and anode. This process has become know as R.F. sputtering and is not be confused with R.F. supported sputtering. A simple way of viewing the reactions that occur is to note that provided the frequency is high enough(greater than 50 KHz), negative charge accumulated on the insulating target will not be sufficient during the half cycle in which the target is positive, to prevent positive ions bombarding the target during the half cycle in which the target is negative. The actual behaviour of an R.F. sputtered system is in fact more complex when examined in detail and the reader is referred to various authors for further information.(32) The system is important to the preparation of insulators.

I. Plasma Reactions

Plasma reactions of various types have been used for the deposition of thin films. These reactions are difficult to classify and they are sometimes considered under sputtering. They are, however, essentially vapor phase or thermal growth reactions, with the discharge supplying the energy necessary to effect the chemical changes.

A plasma reaction that must be noted is that of gaseous anodisation.(33). Figure 15 shows a typical apparatus. Although a glow discharge is used, this glow discharge is effectively replacing the liquid electrolyte of a conventional wet process. It is important in such a system to use a non-reactive anode in the discharge circuit, otherwise all the voltage in the anodising circuit will be dropped across the oxide formed on the discharge anode. Various metals have been successfully oxidised in this way, including aluminium and tantalum.

A plasma assisted thermal growth has been used by Ligenza[34], who showed that using an R.F. excited discharge and an oxygen pressure of between 0.1 and 1. torr, silicon could be oxidized to a thickness of around 3,500 Å at around 300°C. If the silicon was made the anode of a 50 volt system then negatively charged oxygen ions will bombard the surface and thickness above 3,500 Å could be obtained. However, in this latter case, material will be sputtered from the cathode as well, but the atoms from the cathode can be prevented from landing on the anode by placing a suitable bend in the discharge tube so that the cathode cannot "see" the anode. R.F. plasma oxidation of this type has recently been used with effect, for preparing insulating films on superconductors for Josephson Tunneling Devices[35].

If a normal glow discharge is used and the monomer of an

PREPARATION METHODS FOR THIN FILMS

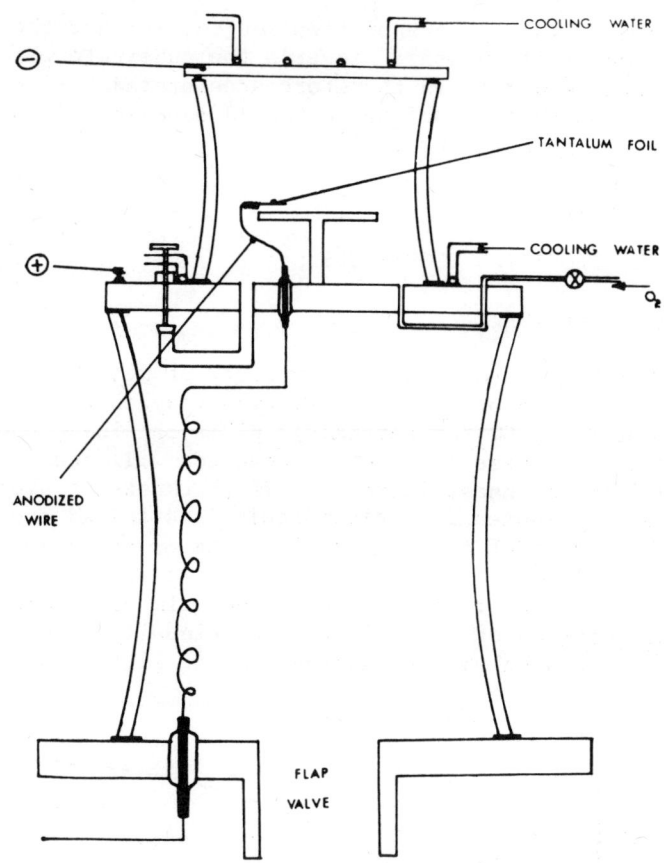

Fig. 15   Gaseous Anodisation apparatus   (Ref. 33)

organic or inorganic material is introduced into the discharge chamber, it has been found that the monomer will be polymerised in the glow discharge so that insulating films can be grown on suitably placed substrates [36]. Over 40 monomers have been examined in this way [37]. Glow discharge conditions can also be used to effect vapor phase reactions such as the deposition of silicon nitride from a gaseous mixture of silicon hydride, ammonia and hydrogen [38]. R.F. excitation of the discharge has also been used and an apparatus has been described by Connel and Gregor[39] that can produce insulating films from styrene at high rates of deposition (20 Å/second).

J. Ion Plating[41][61]

The system of ion plating is essentially a combination of evaporation from a heated wire and the use of a discharge. The anode of the system is also the evaporation source, so that the evaporant is ionised in its passage towards the substrate which is placed on the cathode and it is therefore accelerated before reaching the substrate. By this technique, enhanced adhesion of the metal film to the substrate is obtained.

3. CHEMICAL DEPOSITION TECHNIQUES

3.1 Thermal Growth[42]

Films can be formed on a large variety of metal substrates by heating them in gases of the required type (oxygen for oxides, nitrogen for nitrides, CO for carbides). Films obtained, however, are limited in thickness because the reaction will become very slow as the film thickness increases. If thickness is plotted against time, an exponential relationship is obtained and a typical curve for the growth of alumina on aluminium is shown in Figure 16.

If a non-coherent film is formed, the film will continue to grow because sections of the film will continually flake off from the surface. Such a behaviour, however, is of little use in prepa-

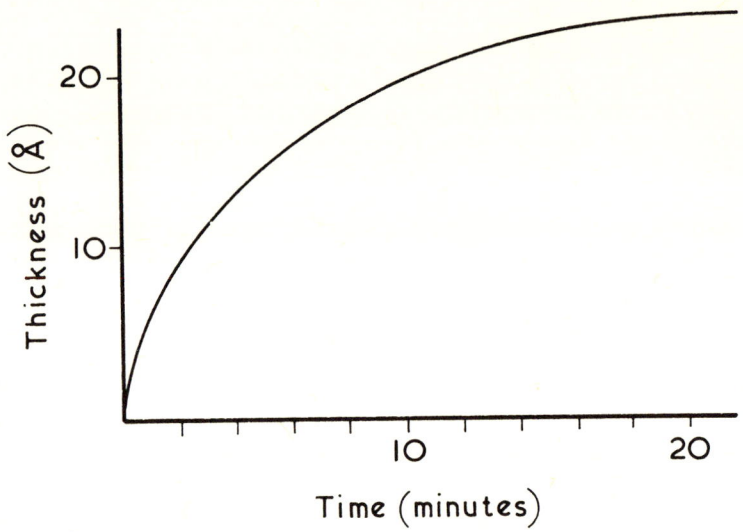

Fig. 16 Thermal growth of Al_2O_3 on Al at 20°C as a function of time.

ring coherent layers of practical importance. Since the mobility
of ions through the oxide is dependent on temperature, the higher
the temperature the greater the thickness that can be obtained.

3.2. Anodisation[43][44]

3.2.1 Introduction

The thickness of oxide obtained in normal thermal growth is
essentially limited by the ability of the ions to migrate through
the film to the film/metal interface. This migration, however, can
be considerably enhanced if the film are grown on a metal substrate in an electrolytic bath. In such a system, the parent metal is
made the anode of an electrolytic cell and a voltage is applied
between the anode and the cathode.

3.2.2 Basic Principles

The reactions that occur at the cathode and anode can be represented by the following equations :

$$M + H_2O \rightarrow MO + 2 H^+ + 2 \varepsilon \qquad \text{(anode)}$$

$$2 \varepsilon + 2 H_2O \rightarrow H_2 \uparrow + 2 OH^- \qquad \text{(cathode)}$$

These equations express the fact that oxide will grow on the metal/
anode surface and hydrogen will be evolved on the cathode. The equations imply the presence of water and anodisation is usually effected in aqueous electrolytes, such as a solution of phosphoric acid.
The acidity of the electrolyte is important in obtaining a coherent
film since if the electrolyte is too acid or too alkaline, the film
can dissolve as it grows and porous oxide structures can result.

Thickness as a function of time for coherent films will depend
on the voltage applied across the electrolytic cell and typical growth
curves are shown in Figure 17. These curves are analogous to the
thermal growth curve previously shown (Figure 16), but in this case
the ultimate thickness is only limited by the applied voltage. The
ultimate thickness is characterised by the "anodisation constant"
that is, the thickness of film that will be obtained after an infinite time, for 1 volt applied. In the case of aluminium this is
13.6 Å, tantalum 16.0 Å, silicon 3.5 Å and niobium 43 Å. In practice, it is usually not possible to keep the metal in the anodisation bath for sufficient time to obtain thicknesses corresponding
to those expected from the anodisation constant. Therefore, the oxide is often somewhat less thick usually between 70-80% of the maximum that can be obtained. The technique of growth illustrated by

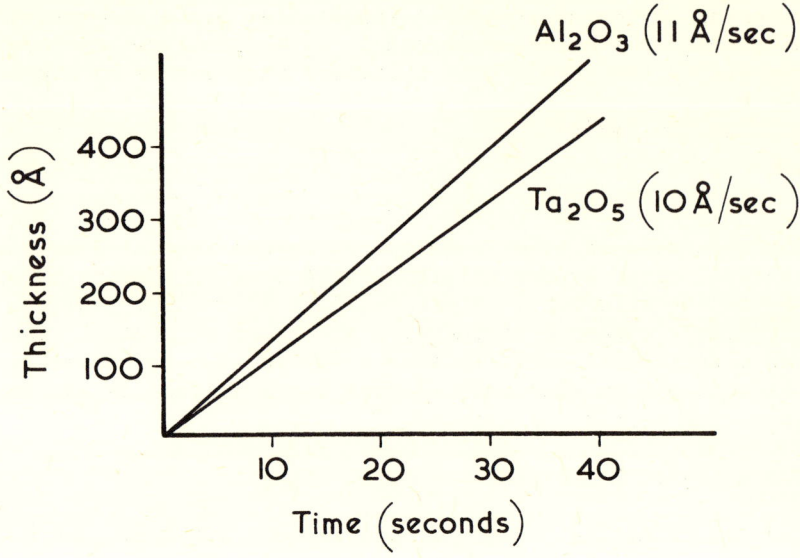

Fig. 17 Constant voltage growth of Ta_2O_5 and Al_2O_3 by anodisation. Thickness/Time for 30 V applied.

Figure 17, assumes that a constant voltage is applied to the cell. The disadvantage of this approach is that in the initial stages of growth, very high current densities are required. One way round this difficulty is to anodise under constant current conditions rather that constant voltage. Growth will then be as illustrated in Figure 18. The limiting condition in these circumstances is the voltage that needs to be applied as the thickness of oxide increases and this will eventually lead to breakdown in the film. In the case of aluminium, this occurs at 1.5 μm, and with tantalum, 1.1 μm, although the limit is essentially a function of the purity of substrate of the electrolyte used.

Growth rates obtained under constant current conditions may be compared with previous figures for evaporation and sputtering. Aluminium can be anodised at a current density of 2 mA/cm^2 to

PREPARATION METHODS FOR THIN FILMS

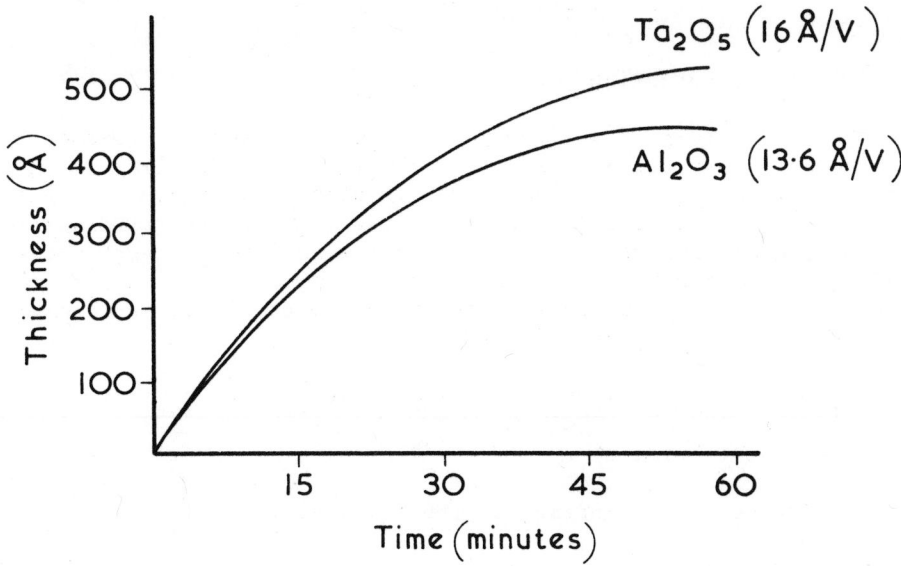

Fig. 18   Constant current growth of Ta_2O_5 and Al_2O_3 by anodisation Thickness/Time for a current density of 2 mamp/cm^2.

give a growth rate of 11 Å/second. This corresponds to the voltage being increased at the rate of 50 volts/minute. In the case of tantalum at 2 ma/cm^2 current density, a growth rate of 10 Å/second is obtained, corresponding to a voltage increase of 40 volts/minute.

3.2.3 Practical Aspects

Anodisation is a widely used technique for obtaining amorphous, highly insulating films. The mobility of ions in the anodising electrolyte is often such that highly convoluted surfaces can be oxidised in a very even manner, as is illustrated by the aluminium and tantalum structures that are used in electrolytic capacitors.[45]

3.2.4 Gaseous Anodisation

Gaseous anodisation has already been referred to under plasma reactions in sputtering (see Figure 15).

3.3 Vapour Phase Growth[36][46][47]

3.3.1 Introduction

The deposition of a film on a surface composed of the same or a different substrate by means of a chemical reaction from the gas phase at the surface, is known as vapour phase growth or vapour phase plating. Usually the surface is hotter than the surroundings so that a heterogenous reaction occurs at the surface. However, other means of activating the chemical reaction may be used, such as a glow discharge (see Sputtering-plasma reactions), electron-beam excitation or ultra violet radiation.

3.3.2 Types of Vapour Phase Reactions

A. Disproportionation

The reaction is typified by the equation :

$$A + AB_2 \rightleftharpoons 2AB$$

where A and B are the two elements. The higher valency state is more stable at lower temperatures so that if a hot gas of AB is passed into a colder region, deposition of A can occur.

Figure 19 shows a typical closed tube system that has been

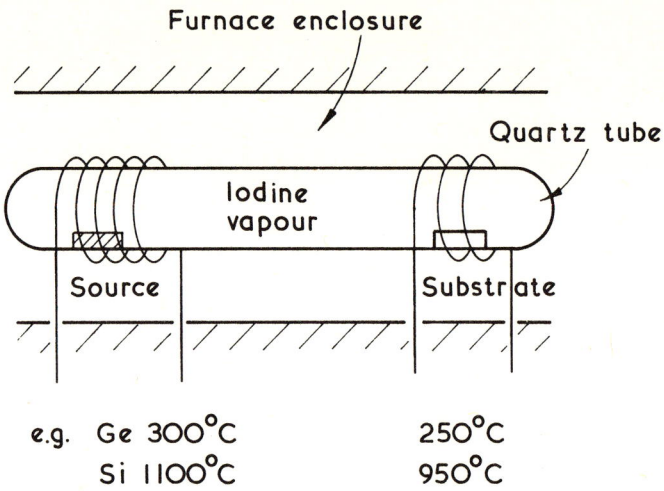

Fig. 19   Disproportionation reaction vessel

employed for the deposition of germanium or silicon. For both these materials two iodides can be formed, and the reaction is used to transport silicon or germanium from the high temperature source zone to the low temperature substrate zone.

Alternatively, a continuous flow open system can be used in which iodine vapour, usually diluted with hydrogen, is continually passed over the source and then the substrate. Growth rates can be high, for example, germanium can be deposited at rates up to 400 Å per second in the above system.

B. Polymerisation(36)

Both organic and inorganic polymers may be prepared from monomer vapour by the use of electron beam, ultra violet irradiation or glow discharge. Insulating films prepared in this manner can have very desirable properties.

Electron beam irradiation has been applied to a large number of materials, including styrene, butadiene etc. Recent workers have described an apparatus for the production of polymer films from evaporating epoxy resin.

Ultra violet irradiation techniques are widely known in photo-resist etching(48). In the etching a relatively thick layer of photo-sensitive material is spread evenly over the surface and then irradiated through a mask. The irradiated areas polymerise to give a material that is insoluble in the solvent used for the un-polymerised film. A similar technique has been used to prepare insulating films. White(49), for example, exposed metal layers in butadiene vapour to irradiate and thus built up thin dielectric films. Various other vapours have also been used (e.g. acrolein) and growth rates of around 1 Å/second are easily obtained.

Glow discharge techniques have already been referred to under Sputtering-plasma reactions - using either a straightforward glow discharge or R.F. excited systems. As noted previously, this latter technique has been used to produce insulating films from a variety of materials including styrene.

C. Oxidation

This is usually undertaken using a halide of the required metal oxide, because of the high vapour pressure of the halide and the ease of removal of the by-products. The reaction used is :

$$2\ AX \xrightarrow{H_2O} A_2O + HX$$

Oxides may be deposited using this technique by the use of steam mixed with the halide and allowed to flow over a hot substrate. To ensure thorough mixing and an even temperature in the reactor, a fluidised bed is often used. Oxides that can easily be prepared in this way are those of aluminium, titanium, tantalum and tin, and high growth rates are possible. (Greater than 100 Å/second).

D. Nitriding

If instead of steam, an atmosphere of ammonia is used, then it is possible to grow layers of nitrides. This reaction has been used effectively for the growth of silicon nitride.

E. Reduction

Metal films can be prepared if hydrogen is substituted for steam in the oxidation reaction previously described. This type of reaction is widely used for the preparation of silicon and germanium. In the silicon case, $SiHCl_3$ or $SiCl_4$ is used and high growth rates are obtained (200 Å/second at 1100°C). The growth kinetics of these reactions have been studied in detail by various workers.

F. Decomposition

Decomposition as represented by the equation :

$$AB \rightarrow A + B$$

can be effected both by heat (pyrolysis) and by glow discharge. Pyrolitic reactions have been widely applied to the preparation of silicon (from SiH_4), nickel from nickel carbonyl and SiO_2 from the decomposition of silicon esters. Rates of growth for nickel can be very high at moderate substrate temperatures (1,000 Å/second at 200 °C).

Glow discharges have also been used to prepare insulating films by decomposition, either directly or with the discharge excited by radio-frequency.[50]

3.3.3. Summary

Various possible reactions for vapour plating have been briefly examined. The apparatus is sometimes quite complex with gas pumping systems or vacuum systems being required, and for pyrolitic work, furnaces may be required to run at high temperatures (2000°C for carbon by the decomposition of toluene and benzene). This high temperature moreover, limits the type of substrate that can be used.

3.4. Electroplating[51][52][53]

3.4.1 Introduction

Electroplating has been known for a considerable time, and many standard textbooks now exist on the subject. The apparatus involved is basically simple, consisting of an anode and cathode immersed in a suitable electrolyte. Metal is deposited on the cathode, and the relationship between the weight of material deposited and the various parameters, can be expressed by the first and second laws of electrolysis. These state :

1. The weight of the deposit is proportional to the amount of electricity passed.

2. The weight of material deposited by the same quantity of electricity is proportional to the electrochemical equivalent E.

Expressed as an equation, the weight deposited per unit area G/A is given by :

$$\frac{G}{A} = JtE\alpha \quad (g\ cm^{-2})$$

where J is the current density and t the time. This equation introduces another term, the current efficiency α, which is the ratio of the experimental to theoretical weight deposited ; it can generally be expected to be between unity and 0.5.

The above equation can be written in a slightly different form to give the rate of deposition. If a thickness l is deposited in time t, then the rate of deposition l/t is given by :

$$\frac{l}{t} = \frac{JE\alpha}{\rho} \quad (cm\ s^{-1})$$

where ρ is the film density.

The rate of deposition values can be very high at high current densities. For example, silver will deposit at 10 Å s^{-1} at a current density of 1 mA cm^{-2}, and this will rise to 1 μm s^{-1} at 1 mA cm^{-2} Such a proportionality to current density holds only if α remains unchanged. This can be expressed in another way by stating that no secondary reactions must occur.

3.4.2 Basic Principles

Of the 70 metallic elements, it is found possible to plate only 33 successfully, and of this latter number, only 14 are deposited commercially. A large variety of baths can be used for the possible elements to improve the adhesion, crystalline structure, current efficiency etc. However, it is not possible to plate elements outside the group of 33, as other reactions (e.g. formation of hydrogen) can more readily occur. This can be illustrated by considering the I-V characteristics of a plating solution. For a simple system the curve will be as shown in Figure 20. Such a curve is obtained with a probe placed near the cathode. The equilibrium potential of the cathode in the solution is indicated by the intercept value on the voltage axis. A negative intercept implies that the cathode will dissolve in the electrolyte at zero voltage (i.e. it will corrode). Equilibrium potentials for the different metals vary from + 1.7 to -1.66 V. Saturation is seen in the curve - at a high cathode voltage, ions cannot get to the cathode fast enough.

If two reactions are possible, each will affect the other. Two I-V curves can now be drawn for the two reactions, and Figure 21. shows this case. In the case of alloy plating this means that the composition of the alloy will depend on the voltage used, as indicated by the dashed line in Figure 21. If one of the reactions is the formation of hydrogen, the curves will now be as in Figure 22.

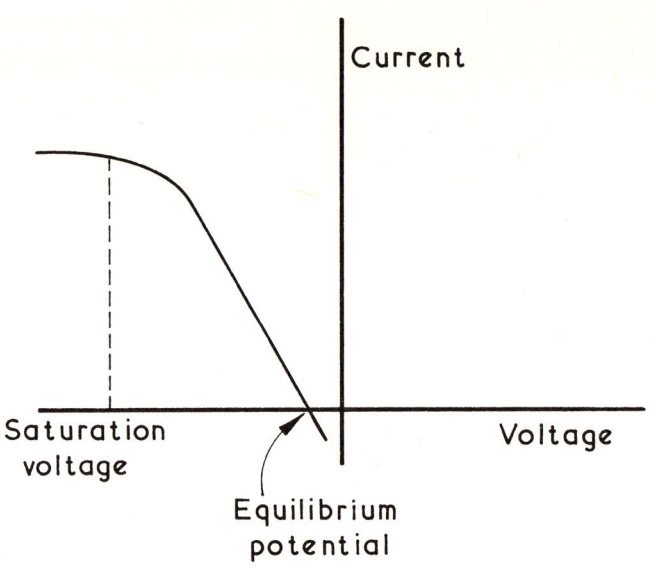

Fig. 20   Ideal i/V curve for electroplating bath.

PREPARATION METHODS FOR THIN FILMS

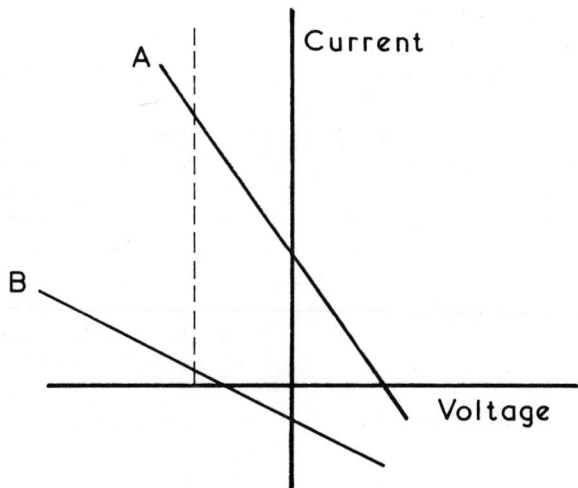

Fig. 21   Ideal i/V curves for two reaction electroplating.

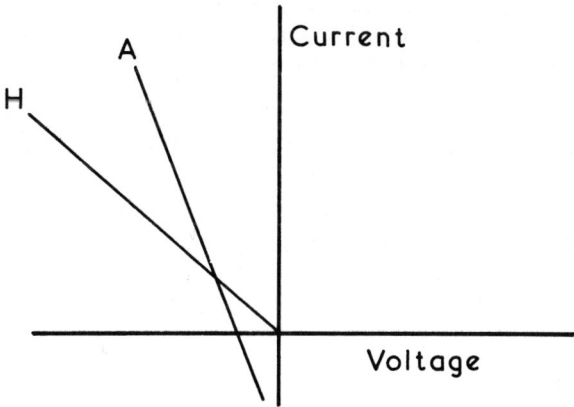

Fig. 22   Ideal i/V curves for one reaction and hydrogen evolution.

The voltage drop in the bath must be reduced to as low a value as possible to reduce waste of power in heating, and this is usually by the addition of conducting salts.

3.4.3. Practical Aspects

In practical deposition systems, care must be taken to control the current density so as to avoid the inclusion of gas bubbles etc. in the film. The effect of solution temperature will not generally be important unless α changes with temperature, (if it does α will generally increase with temperature rise). As the deposition rate can be high, it is possible to use electrochemical deposition for forming thick layers – a process known as electroforming – and for refining. An example of electroforming is the preparation of master disks for gramophone records. In this case, the initial deposit to give a suitable cathode is obtained either by using a colloidal suspension of metal, or by chemical reduction plating.

The many alloys that have been successfully deposited (100 or so) are discussed in detail in Brenner's two volumes on the electroplating(54). It is not possible to deposit every combination because of the characteristics of the separate elements, although it is often possible to slow one of the reactions down by suitable chemical complexing. The effect of complexing is to lower the equilibrium potential to a more negative value, and this results in a crowding together of the i-V characteristics for the separate elements (Figure 21). It is not necessarily the case that the potentials of the separate metals come sufficiently close to permit co-deposition, and it may also be necessary to vary the individual concentrations or the concentration of the complexing agent if it affects both elements. Cyanide is a typical complexing ion for Ag., Cd., Zn., and Cu.

Ternary alloys can be deposited by electroplating ; Brenner lists 15 that can be easily formed.

It is possible to form oxides of elements successfully on the anode of the electrode system. The oxides are deposited from the solution (c.f. anodisation, where it is the oxide of the anode material that is formed). Films of Pb and Mn oxides have been grown in this way, but the method has little importance in thin film technology.

3.5. Chemical Reduction Plating(55)(56)(57)

3.5.1 Introduction

Films of metal can be deposited directly without any electrode potentials being involved, by the chemical reduction of a suitable compound in solution. Such deposition is known as chemical-reduction plating or electroless deposition. Four different types of reaction may be distinguished and these are summarised below.

3.5.2 Practical Aspects

A. Non-Catalytic Reactions

These take place at any surface submersed in the bath. Silver mirrors are usually formed in this way, by the use of a mild reducing agent such as formaldehyde in a solution of silver nitrate. Very thick layers may be built up.

B. Catalytic Reactions

The ability of the metal to deposit on anything can sometimes be a considerable nuisance, and more controlled reactions are often more useful. In these, the metal will deposit only on certain surfaces of other metals and nowhere else. The deposition of nickel, for example, can be achieved by such techniques as the reduction of $NiCl_2$ by sodium hypophosphite, when the metal will grow on a surface of nickel itself, cobalt, iron, and aluminium - the metal acts as the catalyst. (Note : the use of sodium hypophosphite as the reducing agent means that between 5 and 10% of phosphorus will become incorporated in the film). This type of reaction has become so important that a complete book is now available on chemical reduction plating of just nickel.[58] Other metals, particularly the Pt group, can be deposited in this manner.

C. Catalytic Reactions Using Activators

The number of metal surfaces that will catalyze deposition is limited. It is found, however, that it is possible to activate the surfaces of non-catalytic metals so that deposition will take place on these surfaces. The role of the activator is to lower activation energy for the reduction reaction at particular points on the surface so that deposition will occur at these points. Islands of metal will thus grow and spread and eventually, give a continuous film. The best activators to be used for particular metals are listed in standard texts on the subject[55] ; $PdCl_2$ is often used for Cu and Ni. Very little of the activator is required - in the case of $PdCl_2$ a dip in a 0,01% solution, followed by a rinse in water is all that is required.

D. Catalytic Reactions Using Activators and Sensitisers

For non-metallic surfaces, a sensitisation before activation is required. For Ni this takes the form of a dip in a 0.1% solution of $SnCl_2$ followed by a rinse. The activation is then carried out in the normal way. The advantage of such reactions is that it is possible to plate onto glass and other non-conducting surfaces. Also, and this applies in general, it is possible to plate surfaces that are difficult of access, such as the inside of tubes.

3.5.3. Summary

Chemical reduction plating uses very simple apparatus, provided a suitable reaction and if necessary, catalysts, are available. The rates of deposition depend on solution and temperature.

3.6. Solution Deposition[59]

3.6.1 Oxide Films

It is possible to deposit dielectric films on to non-metallic substrates, using organic solution techniques. The substrate to be coated is generally immersed in a solution of a suitable organic material so that a thin layer of material is formed on both sides of the substrate. Uniformity of this liquid film can often be improved by spinning the wetted surface, but after uniformity has been achieved, the substrate is then baked to a temperature of between 200-500°C to convert the liquid layer to a solid, usable structure. For example, it has been found that a colloidal silicon dioxide hydrate can be prepared on a substrate by immersing in a silicate solution to which acids have been added. This colloidal hydrate will then yield a film of SiO_2 on baking.

Film thickness as low as 100 Å or less can be obtained by this technique, and it is possible to easily form multiple layers of different materials. Two materials, SiO_2 and TiO_2 have been extensively prepared in this manner and the films used for a variety of optical purposes.

3.6.2. Hydrophilic Films[60]

Multimonolayers of long chain fatty acids can be built up on a substrate by repeated immersion of the substrate in a liquid, on which long chain fatty acids are floating on the surface. The layers obtained at each immersion are around 20 Å thick and successful films up to 20,000 Å have been prepared. Such films are extremely stable and electrodes may be applied by normal evaporation techniques.

3.6.3. Liquid Phase Epitaxy[65]

Under the heading of Solution Deposition it is worth noting that it is possible to prepare single crystal films from the liquid phase provided a suitable substrate is available. Normal deposition from the liquid phase gives randomly nucleated growth and polycrystalline layers which are often of variable thickness in the initial stages of growth. However, it has been found possible to control the system well enough so that material will grow on a large planar seed in a form that is thin and itself planar enough to be called a layer[65]. This technique is now widely used for the growth of semiconductor films (e.g. GaAs from Ga at temperature down to 700°C) and also for other applications such as magnetic garnets for magnetic bubble systems.

One of the major advances of this type of technique is that it is possible to grow layers at temperatures which are several hundred degree lower than the melting point of the compound and thus, to reduce the crystalline defects obtained in the film.

3.6.4. Growth of Polymer Films

Polymer films of materials such as polypropylene, polystyrene[66] and P.V.C.[67] have been obtained by the simple technique of direct isothermal immersion of a substrate into a suitable solution of the polymer, and also by allowing the evaporation of the solvent from the polymer solution placed on a substrate. The former technique appears to be very promising for obtaining durable and useful films. Detailed studies on P.V.C.[67] have shown that the immersion technique applied to dilute solutions of P.V.C. in either benzene plus acetone or in cyclohexanone, yields films of a limited thickness which is not increased by prolonged immersion (e.g. for a solution of 0.6 gms P.V.C. in 100 cm^3 of cyclohexanone held at 40°C, a polymer film will grow on glass to a limiting thickness of $\sim$ 2000 Å in 15 minutes).

The growth model suggested for P.V.C.[67] allows initially of the P.V.C. chain segments being adsorbed at suitable, unspecified sites on the surface. After the initial adsorption, further growth occurs by adsorption or attachment of new chain segments. However, the outer layer is also capable of being dissolved by the solvent so that a final thickness is reached in which the outer layer growth rate is exactly equal to its solution rate.

4. SUMMARY

Various techniques available for the deposition of thin films have now been examined. Not all of these techniques are applicable to the growth of non-metallic thin films. The ones which are, are summarised in Table 1.

Table 1

Summary of preparation methods
applicable to dielctric films

TECHNIQUE	MAJOR REFERENCES
Evaporation (using all source types)	(4a) (4b) (5)
Sputtering. Reactive.	(4a) (4b) (26) (27)
R.F.	(4a) (4b) (26) (27)
Plasma Reactions Gaseous anodisation	(27) (36) (40)
assisted thermal growth	(27) (36)
polymerisation	(27) (36)
assisted vapour-phase	(27) (36)
Thermal Growth	(4a) (4b) (40)
Anodisation	(4a) (40) (44)
Vapour Phase – Polymerisation (E.B. & U.V.)	(4b) (40) (46)
Oxidation	(4b) (40) (46)
Nitriding	(40)
Solution Deposition – organic	(59)
– hydrophilic	(60)
– Polymer	(67)

REFERENCES

1. D.S. Campbell. In "The use of thin films in physical examinations " (Ed. J. C. Anderson) Academic Press, London, p. 11-25 1966.

2. K.L. Chopra. "Thin film Phenomena" McGraw Hill, N.Y. p. 10-82 1969.

3. B.N. Chapman and J.C. Anderson. "Science and Technology of surface coatings".Academic Press, London. 1974.

4(a) P.J. Harrop and D.S. Campbell. In "Handbook of Thin Film Technology" (Ed. L.I. Maissel and R. Glang) McGraw Hill, N.Y. p. 16.29-16.33. 1970.

4(b) A. Pliskin, D.R. Kerr, J.A. Perri in "Physics of Thin Films" (Ed.G. Hass and R.E. Thun)$\underline{4}$, p. 257-270. 1967.

5. R. Glang. In "Handbook of Thin Film Technology" (Ed. L.I. Maissel and R. Glang). McGraw Hill, N.Y. p. 1.3-1.30. 1970.

6. D. Walton, T.N. Rhodin and R. Rollins. J. Chem. Phys. $\underline{38}$, p. 2695. 1963.

7. D.S. Campbell. Present Volume, 1975.

8. R.E. Honig, R.C.A. Review $\underline{23}$, p. 567. 1962.

9. R. Glang, R.A. Holmwood and J.A. Kurtz. In "Handbook of Thin Film Technology" (Ed.L.I. Maissel and R. Glang) Mc Graw Hill. N.Y. p. 2.1 - 2.142. 1970.

10. As per 5. p. 1.37

11. Ibid. p. 1.86

12. Ibid. p. 1.87 - 1.88

13. K.H. Behrndt. J. Appl. Phys. $\underline{33}$, p. 193. 1962.

14. P. Huijer, W.T. Langedam and J.H. Laby. Philips Tech. Rev. $\underline{24}$ p. 144-149, 1963.

15. C.E. Drumheller. Trans. 7th Nat. Symp. on Vacuum Technology. p. 306-312. 1960.

16. D.S. Campbell and B. Hendry. Brit. J. Appl. Phys. $\underline{16}$ p. 1719-1725. 1965.

17. J.L. Richards. In "The use of thin films in Physical Examinations". (Ed. J.C. Anderson) Academic Press, London. p. 71-86. 1966.

18. As per 5. p. 1.92-1.97

19. I. Ames, L.H. Kaplan and P.A. Roland. Rev. Sci. Inst. 37 p. 1737, 1966.

20. As per 5. p. 1.43-1.50

21. Ibid. p. 1.50-1.54.

22. B.A. Unvala and G.R. Booker, Phil. Mag. 9, p. 691. 1964.

23. R.F. Bunshah. In "Science and Technology of surface coatings" (Ed. B.N. Chapman and J.C. Anderson) Academic Press, London p. 361-368. 1974.

24. As per 5. p. 1.80-1.85.

25. M. Kaminsky. "Atomic and Ionic Impact Phenomena on Metal Surfaces" Springer-Verlag. Berlin. 1965.

26. L.I. Maissel. In "Physics of Thin Films" (Ed. G. Hass and R.E. Thun) Academic Press, N.Y. 3, p. 61-129. 1966.

27. L.I. Maissel. In "Handbook of Thin Film Technology" (Ed. L.I. Maissel and R. Glang) McGraw Hill, N.Y. p. 4.1-4.44. 1970.

28. G.K. Wehner and G.S. Anderson. In "Handbook of Thin Film Technology" p. 3.1-3.38. 1970.

29. As per 27. p. 4.26-4.31

30. Ibid. p. 4.21-4.22

31. L. T. Maissel and P. Schaible, J. Appl. Phys. 36, p. 237. 1965.

32. As per 27. p. 4.31-4.39.

33. N.F. Jackson. J. Mat. Sci. 2, p. 12-17. 1967.

34. J.R. Ligenza. J. Appl. Phys. 36, p. 2703. 1965.

35. J.H. Greiner. J. Appl. Phys. 45, p. 32. 1974.

36. L.V. Gregor. In "Physics of Thin Films" (Ed. G. Hass and R.E. Thun). Academic Press, N.Y. 3, p. 131-164. 1966.

37. A. Bradley and J.P. Hammes. J. Electrochem. Soc. 110, p.15 1963.

38. H.F. Sterling and R.C.G. Swann. Solid State Elec. **8** p. 653. 1965.

39. R.A. Connel and L.V. Gregor. J. Electrochem. Soc. **112** p. 1198 1965.

40. D.S. Campbell. In "Handbook of Thin Film Technology" (Ed. L.I. Maissel and R. Glang) McGraw Hill, N.Y. p. 5.1-5.25. 1970.

41. G.M. Mattox. Electrochem. Tech. **2** p. 295. 1964.

42. As per 40. p. 5.21

43. Ibid. p. 5.17-5.20

44. C.J. Dell'Oca, D.L. Pulfrey and L. Young. In "Physics of Thin Films" (Ed. M.H. Francombe and R.W. Hoffman) Academic Press, N.Y. **6**, p. 1 -- 79 . 1971.

45. D.S. Campbell. Rad. and Elec. Eng. **41**, p. 5-16. 1971.

46. W.M. Feist, S.R. Steele and D.W. Ready. In "Physics of Thin Films" (Ed. G. Hass and R.E Thun) Academic Press. N.Y. **5**, p.237-322. 1969

47. As per 40. p. 5.12 - 5.16

48. D. I. Gaffee. In "Thin film microelectronics" (Ed. L. Holland) Chapman and Hall, London. p. 260-270. 1965.

49. P. White. Elec. Reliab. and Micromin. **2**, p. 161. 1963.

50. L.L. Alt, S.W. Ing Jnr. and K.W. Laendle. J. Electrochem. Soc. **110**, p. 465. 1963.

51. F.A. Lowenheim. "Modern Electroplating" John Wiley. N.Y. 1965.

52. K.R. Lawless. In "Physics of Thin Films" (Ed. G. Hass and R.E. Thun) Academic Press, N.Y. 4. p. 191-225. 1967.

53. As per 40. p. 5.3-5.8

54. A. Brenner. "Electrodeposition of alloys" Vol 1 and 2. Academic Press. N.Y. 1963.

55. W. Goldie. "Metallic Coating of Plastics" Electrochemical Publications Ltd., U.K. **1**, p. 55-152. 1968.

56. As per 40. p. 5.9-5.11

57. M. Schlesinger. In "Science and Technology of Surface Coatings" (Ed. B.N. Chapman and J.C. Anderson). Academic Press, London. p. 176-182. 1974.

58. K.M. Gorbunova and A.A. Nikiforova. "Physical principles of nickel plating". Israel Prog. for Scientific Translations. Jerusalem. 1963.

59. H. Scroeder. In "Physics of Thin Films" (Ed. G. Hass and R.E. Thun) Academic Press, N.Y. $\underline{5}$, p. 87-141. 1969.

60. L. Holt. Nature $\underline{214}$ p. 1105. 1967.

61. R. Carpenter. In "Science & Technology of Surface Coatings" (Ed. B.N. Chapman and J.C. Anderson) Academic Press, London p. 393-403. 1974.

62. M. Martini. In "Science & Technolgy of Surface Coatings" (Ed. B.N. Chapman and J.C. Anderson) Academic Press, London. p. 404-410. 1974.

63. E. Krikorian and R.J. Sneed. In J. Appl. Phys. $\underline{37}$ p. 3674. 1966.

64. E. Hollands and D.S. Campbell. In J. Mat. Sci. $\underline{3}$, p. 544-552. 1968.

65. L.R. Dawson. In "Progress in Solid State Chemistry". (Ed. H. Reiss and J.O. McCaldin). Pergamon Press, Oxford, p. 117-139. 1972.

66. J.A. Koursky, A.G. Walton and E. Baer. J. Polymer Sci. $\underline{B.5}$ p. 177. 1967.

67. A.C. Rastogi and K.L. Chopra. Thin Solid Films $\underline{18}$, p. 187-200. 1973.

GROWTH PROCESSES

R. Niedermayer

Ruhr Universität Bochum, Experimentalphysik 4

463 Bochum Postfach 2148, Germany

1. INTRODUCTION

The growth of thin films has attracted considerable attention during the last years, and several reviews of the field as well as many original contributions have been published. For a recent review, which contains all literature, the article of J.A. Venables and G.L. Price[1] should be consulted. It is not the aim of this series of lectures to repeat this work. I shall try to make clear the fundamental connection of statistical and kinetic considerations, which govern the description of growth processes. Though this is not entirely new, it has been neglected to a certain extent in the literature. The reason for this is the large weight of the difficulties, which have been encountered in the solution of the system of kinetic equations, which have however been overcome now.

The following lectures will consequently be divided into three parts :

1. the connection between material constants and the rates of growth and decay of clusters. This is achieved mainly by thermodynamic considerations, from the results of which the constants of the kinetic equations can be derived and which allow a judgement of the general direction of growth ;

2. methods for the solution of the kinetic equations of film growth. These methods are new to a certain extent and allow the easy solution of large systems of almost linear differential equations by numerical methods ;

3. experimental expectations from these theoretical considerations and comparison with the evaluation of some experiments

2. THERMODYNAMIC CONNECTIONS BETWEEN THE PROPERTIES OF SUBSTRATE AND FILM AND THE RATES OF GROWTH AND DECAY OF CLUSTERS.

2.1. Formulation of the problem.

The formulation of the kintetics of film growth follows essentially the same lines as that of polymerisation of macromolecules from single atoms. If B_i is the symbol for a cluster of i atoms, processes of the kind

$$B_i + B_k \rightleftharpoons B_{i+k} \qquad (1)$$

are interesting. The general expectance is that the main contribution to growth comes from the monomers (k=1)[2,3] and so the subsequent work will be confined to this case. Some observations and theoretical considerations show however, that migration of misaligned large clusters occurs and is important for the growth of nonepitaxial thin films[4].

The description of growth will use the following parameters:

J_{imp} the impingement rate of monomers on the substrate
J_{twin} the rate of twin formation $B_1 + B_1 \rightarrow B_2$
J_i the capture rate of a cluster B_i leading to $B_i + B_1 \rightarrow B_{i+1}$
α the desorption rate of monomers from the substrate
α_i the decay rate of B_i leading to $B_i \rightarrow B_{i-1} + B_1$
i^* the size of the last subcritical cluster, a natural number defined so that $J_i/\alpha_i \leq 1 \leq J_{i+1}/\alpha_{i^*+1}$
n_i the number of clusters B_i per unit area, containing i monomers
N_c the total number of supercritical clusters
N_a the total number of condensed atoms

N_c and N_a can be defined in terms of n_i

$$N_c = \sum_{i=i^*+1}^{\infty} n_i \qquad (2)$$

$$N_a = \sum_{i=i^*+1}^{\infty} i \cdot n_i \qquad (3)$$

Definition (3) states that an atoms will be considered condensed, if it has joined a supercritical cluster. The use of these variables allows a description of the polymerisation process in a set of equations which can be found in Frenkel's book[6] and which have been used by many authors in chemical kinetics.

$$\dot{n}_2 = J_{twin} - (\alpha_2 + J_2) \cdot n_2 + \alpha_3 \cdot n_3 \qquad (4)$$

$$\dot{n}_i = J_{i-1} \cdot n_i - (\alpha_i + J_i) \cdot n_i + \alpha_{i+1} \cdot n_{i+1} \qquad (5)$$

GROWTH PROCESSES

for $3 \leq i \leq i^* + m$, where m is so defined that $\alpha_{i^*+m} \ll J_{i^*+m}$. These equations contain the main formulation of the problem. For the comparison with experiment the numbers N_c and N_a are necessary, which may exactly be derived from (5) by use of the definitions (2,3):

$$\dot{N}_c = J_{i^*} \cdot n_{i^*} - \alpha_{i^*+1} \cdot n_{i^*+1} \tag{6}$$

$$\dot{N}_a = (i^*+1) \cdot J_{i^*} \cdot n_{i^*} - i^* \cdot \alpha_{i^*+1} \cdot n_{i^*+1} + \sum_{i=i^*+1}^{\infty} ((J_i - \alpha_i) \cdot n_i) \tag{7}$$

The total number of monomers on the free area A_f of the unit area substrate can be derived from

$$\dot{N}_1 = A_f \cdot J_{imp} - \alpha \cdot N_1 - 2 \cdot J_{twin} + \alpha_2 \cdot n_2 - \sum_{i=2}^{\infty} ((J_i - \alpha_i) \cdot n_i) \tag{8}$$

Equations (6, 7, 8) can in principle be obtained by integration, if the solutions of equs. (4,5) are known. Direct impigement onto nuclei and coalescence of islands is not yet incorporated in these equations and will be considered separately. The infinite sums in (6, 7, 8) can be approximated by finite expressions containing N_c.

The main system contains the rates of capture and decay, which depend on the size of the clusters and on their thermodynamic properties. The equations are coupled by the terms $\alpha_{i+1} \cdot n_{i+1}$ which cause the main difficulty in the solution of this system, and which discern a nucleation theory from a simple theory of unidirectional growth. The solution can be given numerically or by analytical approximations. The results depend on the transport mechanism contained in the capture probability. Statistical and thermodynamic arguments for the evaluation of the rates of capture and decay will be presented in the following paragraphs.

2.2 Some principal considerations on the capture rate on a surface and on surface diffusion.

The difficulty in the treatment of atomic diffusion on surfaces is the combination of short characteristic times and very low concentrations. Therefore the question arises what meaning a diffusion mechanism could have in such a system. Another difficult question is, how the boundary conditions at a cluster edge could be formulated, if the cluster is small and if it emits and absorbs only single atoms in times which are certainly long compared to the time necessary for the diffusion from one site to the next one.

This will be illustrated in the following statements:
1. The number of adsorption sites is typically $5 \cdot 10^{14}$ cm^{-2} and will be denoted l_{ad}.
2. The impingement rate in an experiment may be $J_{im} = 10^{13}$ cm^{-2}s^{-1}; this means that the time for monolayer occupation is very long,

i.e. one minute.
3. The typical diffusion constant is $D = 10^{-5}$ cm^2 s^{-1}; the corresponding stay time in one site is $\tau_{diff} = 10^{-10}$ s.
4. The desorption rate is approximatively $\alpha = 10^7$ s^{-1}, the corresponding stay time is $\tau_{des} = 10^{-7}$ s.

The data for point 3 and 4 have been chosen for gold on rocksalt.

5. The decay rate for gold twins may be $\alpha_2 = 10^2$ s^{-1} and for clusters Au$_5$ $\alpha_5 = 1$ s^{-1}; $\tau_2 = 10^{-2}$ s, $\tau_5 = 1$ s. Whenever the binding energy to a cluster is larger than the binding energy to the surface similar orders of magnitude will apply at moderate temperatures.
6. The capture rate under the same circumstances is $J_2 = 10^3$ s^{-1}; $\tau_c = 10^{-3}$ s.

Obviously all changes at one cluster are very slow compared to the characteristic times for desorption and diffusion. So the question is, whether the few clusters and the few single atoms interact at all in the sense of a diffusion mechanism with well defined boundary conditions as has been proposed by Halpern[7] and other authors [8], or if the picture of homogeneous monomer concentration as f.i. used by Walton[9] is more reasonable.

The following arguments will show, that the diffusion equation applies to the system for all cluster sizes and for all stages of growth. Take a small surface with an orthogonal net of adsorption sites (f.i. NaCl (100)), which may be numbered by their coordinates (x_n, y_m). The surface be A, the number of monomers be $N_1(t)$, their concentration n_1. This number is increased by the impingement rate J_{imp} and by the decay rates α_i of clusters; it is decreased by the desorption rate α and by the capture rates J_{twin}, J_i. For the judgement of the changes in such a system the observation of the environment of a single cluster is useless. We rather choose a Gibbs ensemble of N_G identical surfaces, where N_G is a very great number of the order of 1_{ad}. We divide the time into intervals $\Delta t \ll \tau_{diff}$ and we observe all the systems at each interval. We are then at least in principle able to count the number of atoms located at a site (x_n, y_m) in the whole ensemble at each time interval, i.e. $l_1(x_n, y_m, t)$, and to find the probability $p(x_n, y_m, t) = l_1(x_n, y_m, t)/N_G$ at each moment. This procedure is not as artifical as it seems to be, because it is quite analogous to the real experiment, where we observe a macroscopic surface with many clusters and divide the surface into environments of similar clusters.

If the jump rate from one site in any direction is $w = 1/\tau_{diff}$, we obtain for a time $t + \Delta t$ a slightly different distribution $l_1(t + \Delta t)$ from that at time t, $l_1(t)$. From this we can again find $p(t + \Delta t)$, which is at a sufficient distance from a cluster edge

$$p(x_n,y_m,t+\Delta t) = p(x_n,y_m,t) - w.\Delta t.p(x_n,y_m,t)$$
$$+ \frac{w.\Delta t}{4}.(p(x_{n+1},y_m,t)+p(x_{n-1},y_m,t)$$
$$+ p(x_n,y_{m+1},t)+p(x_n,y_{m-1},t)$$
$$+ J_{imp}.a^2.\Delta t - \alpha.\Delta t. p(x_n,y_m,t) \qquad (9)$$

The first line represents the probability, that nothing has changed. The second line is the probability, that an atom from one of the neighbouring sites jumps into the observed site (n,m) with a probability $(w.\Delta t)/4$ for a jump in one direction. The last line accounts for the impingement and desorption, where $a^2 = 1/l_{ad}$ is the area occupied by one adsorption site and the prompt sticking coefficient is assumed unity. Equation (9) is a diffusion equation

$$\frac{\partial p}{\partial t} = J_{imp}.a^2 - \alpha.p + (w.a^2/4).(\frac{\partial^2}{\partial x^2} + \frac{\partial^2}{\partial y^2})p \qquad (10)$$

The diffusion constant D is related to the jump probability w by $D = w.a^2/4$.

Whatever the single atom will do at the surface, it has no other possibilities than those enumerated in eq.(9) each of which will be realized at a certain rate, which must be determined later. An ensemble of atoms will then behave quite orderly and the probabilities of occupation will change according to (10). To obtain the boundary conditions we observe the number of monomers l_1 in the immediate environments of clusters B_i, which may have an edge length $s(i)$, i.e. the number of adsorption sites occupied by these edge atoms is approximately $s(i)/a$. The rate of generation of an atom from one of the edge sites is $\alpha_i.a/s(i)$ if α_i is the decay rate of the whole cluster. We consider now the situation where (x_n,y_m) is $\underline{r}_b + \underline{n}a$, i.e. one step apart from the boundary position $\underline{r}_b, \underline{n}$ being the unity vector perpendicular to the boundary, and where $\underline{r}_b = (x_{n-1},y_m)$. We find in the same way as in equation (9)

$$p(x_n,y_m,t+\Delta t) = p(x_n,y_m,t)-w.\Delta t.p(x_n,y_m,t)$$
$$+J_{imp}.a^2.\Delta t-\alpha.\Delta t.p(x_n,y_m,t)$$
$$+ (w.\Delta t/4).(p(x_{n+1},y_m,t)+p(x_n,y_{m+1},t)$$
$$+p(x_n,y_{m-1},t)) \qquad (11)$$
$$+ (\alpha_i.a/s(i)).\Delta t$$

If we formally introduce a boundary probability $p_b(x_{n-1},y_m,t)$ we can obtain the diffusion equation (9,10) again from (11), provided we choose

$$w.p_b/4 = \alpha_i.a/s(i) \qquad (12)$$

The connection between the decay probability of a cluster B_i and the boundary probability, which must be introduced into the diffusion equation is thus given by

$$\alpha_i = (s(i)/a) \cdot D \cdot (p_b/a^2) \qquad (13)$$

The definition of p_b is such, that a probability of occupation $p_b(x_{n-1}, y_m, t)$ would produce the same flux as is procured by the decay rate α_i. This means also, that a cluster B_i, which is surrounded by a homogeneous concentration P_b/a^2 would statistically receive and loose an equal number of atoms. Under such conditions the processes $B_i + B_1 \to B_{i+1}$ and $B_i \to B_{i-1} + B_1$ would be equally probable. This observation will be important for the calculation of the decay probabilities.

In the frame of these considerations the capture rate is given by the probability of occupation of an adjacent site, $((s(i)+2\pi a)/a) \cdot p(\underline{r}_b + \underline{n} \cdot a, t)$ times the transition rate $(w/4) \cdot (s(i)/(s(i)+2\pi a))$

$$J_i = (w/4) \cdot (s(i)/a) \cdot p(\underline{r}_b + \underline{n} \cdot a, t) \qquad (14)$$
$$= (w/4) \cdot (s(i)/a) \cdot (p_b + a(\underline{n} \cdot \nabla p))$$

Our statistical interpretation of this capture rate is just sufficient for the validity of its use in eqs. (5 to 8). Equation (14) can be rearranged by use of the diffusion constant, $D = w \cdot a^2/4$ and of $n_1(\underline{r},t) = p(\underline{r},t)/a^2$, the local monomer concentration that must be interpreted statistically in the sense of this paragraph. If we further denote the boundary monomer concentration at a cluster B_i by $\delta_i = p_b/a^2$, we obtain

$$J_i = D \cdot (s(i)/a)(\delta_i + a \cdot (\underline{n} \cdot \nabla n_1)) \qquad (15)$$

for the capture rate, which must be calculated with the result of the solution of the diffusion equation (10), and with a determination of the boundary monomer concentration δ_i.

2.3 The general solution of the diffusion equation and evaluation of time constants.

The diffusion equation

$$\frac{\partial p}{\partial t} = J_{imp} \cdot a^2 + D \cdot \Delta p - \alpha \cdot p \qquad (10)$$

with the boundary condition $p_b = \delta_i \cdot a^2$ must be solved for possibly time dependent boundary conditions, because δ_i varies with i. The problem has cylindrical symmetry in two dimensions, which results in

$$\frac{\partial p}{\partial t} = J_{imp} \cdot a^2 - \alpha \cdot p + D \cdot \left(\frac{\partial^2 p}{\partial r^2} + \frac{1}{r} \cdot \frac{\partial p}{\partial r} \right) \tag{16}$$

Choice of $s = p - a^2 \cdot J_{imp}/\alpha$, $\quad p = s(r,t) + a^2 \cdot J_{imp}/\alpha$ results in the homogeneous equation

$$\dot{s} = -\alpha \cdot s + D \cdot (s'' + s'/r) \tag{17}$$

Separation, $s = v(t) \cdot u(r)$, gives equations for any ω

$$\dot{v} = i \cdot \omega \cdot v \tag{18}$$
$$u'' + (u'/r) - ((\alpha + i \cdot \omega)/D) \cdot u = 0$$

the solutions of which can be superposed to find a physically adequate expression. Such particular solutions are

$$v = v_0 \cdot e^{i\omega t}$$

$$u_1(r,\omega) = \int_0^\infty \exp(-\sqrt{(\alpha + i\omega)/D} \cdot r \cdot \cosh \chi) \, d\chi \tag{19}$$

$$u_2(r,\omega) = \frac{1}{\pi} \int_0^\pi \exp(+\sqrt{(\alpha + i\omega)/D} \cdot r \cdot \cos \chi) \, d\chi$$

u_1 is an extension of the modified Bessel function K_0 and u_2 of I_0. Because u_2 diverges for large r, it will not be used for the simple discussion of this paragraph although it would be necessary in the presence of more than one cluster within a range $7 \cdot \sqrt{D/\alpha}$ of the one just under consideration. Evaluation of $(v \cdot u_1 - v^* \cdot u_1^*)/2i$ results in a particular solution

$$s(r,t) = \int_0^\infty \exp\left[-\sqrt{\frac{\sqrt{\alpha^2 + \omega^2} + \alpha}{2D}} \cdot r \cdot \cosh \chi \right] \cdot \tag{20}$$
$$\cdot \sin\left[\omega t - \sqrt{\frac{\sqrt{\alpha^2 + \omega^2} - \alpha}{2D}} \cdot r \cdot \cosh \chi \right] \cdot d\chi$$

which can easily be discussed. Be the boundary conditions such that they vary with ωt or that the main Fourier component of their variation has the frequency ω. The main contributions come from small values of χ. s vanishes exponentially for large r and the range is given by

$$r_\omega = (2D/(\sqrt{\alpha^2 + \omega^2} + \alpha))^{1/2} \tag{21}$$

The influence of the variation of the boundary condition has a frequency dependent range, which is approximately equal to the range $r_o = \sqrt{D/\alpha}$ of the stationary diffusion equation if $\omega < \alpha$, and which approximates the range of purely diffusion controlled distributions, $r_d = \sqrt{2D/\omega}$, if $\omega > 2\alpha$.

The perturbations proceed in form of circular waves with a wavelength λ and a corresponding velocity of propagation $v_{diff} = \omega \cdot \lambda/2\pi$. We find for the main contributions :

$$\lambda = 2\pi \cdot \sqrt{2D/(\sqrt{\alpha^2 + \omega^2} - \alpha)} \qquad (22)$$

For fast variations of the boundary conditions ($\omega > 2\alpha$) this gives $\lambda_f = 2\pi \cdot \sqrt{2D/\omega} = 2\pi r_d$. In this case we have a real wavelike propagation of the perturbation.

If the variations are slow ($\omega < \alpha/2$) we arrive at $\lambda_s = 4\pi \cdot (\alpha/\omega) \cdot \sqrt{D/\alpha} = 4\pi \cdot (\alpha/\omega) \cdot r_o$. The range, within which the perturbation is in phase with the boundary concentration, is approximately $\lambda_s/4\pi = (\alpha/\omega) \cdot r_o \simeq (\tau_c/(2\pi\tau_{des})) \cdot r_o$. Within this range the stationary solution applies.

In all physically important cases the stationary solution is a good approximation. Whenever the monomer concentration at some distance from the nuclei is mainly determined by the desorption equilibrium, $n_1 = J_{imp}/\alpha$, we expect the second case and the validity of the stationary solution, because the coherence length $\lambda_s/4\pi$ is larger than the attenuation range r_o of the variations. In the capture regime, $\tau_{des} > \tau_c/\pi$, the first case should be expected. But in this case the boundary condition is nearly $p_b = 0$ even for twins and thus does not change with time. So ω is very small and we can again expect the validity of the stationary solution. An artificial case, which contains a maximum of short wavelength components, is the case of homogeneous monomer concentration. A comparison between the stationary solution of the diffusion equation and the homogeneous monomer concentration may be useful.

2.4 The boundary monomer concentration and the decay rate of a cluster

The discussion of the equations (12,13) has shown, that a cluster B_i would grow and decay with equal probability, if surrounded by the boundary concentration $p_b/a^2 = \delta_i$. If the clusters were a separated phase, the monomers would be in equilibrium with the condensed phase, the molecular free energies would be equal for monomers and condensed phase, the total free energy would not change from a transition of a molecule from one phase to another.

A cluster is however not a separated phase with a constant chemical potential. This has the consequence that a system, in

GROWTH PROCESSES

which a cluster B_i is in equilibrium with the surrounding monomers in the sense mentioned, is not stable. Quite contrarily such a system will start to condense, because the saturation concentration for small clusters is greater than for large clusters. Only a system, which has a very low monomer concentration, i.e. which is subsaturated, so that even in the presence of a condensed phase no condensation would occur, does not grow. Such a subsaturated system can form very few clusters by fluctuations, the relative number of which can be calculated by the mass action law. Though this contains the decay rate implicitely, their evaluation[5] can lead to difficulties with our present treatment of the transport process and we shall use another more general procedure for the determination of the boundary concentrations δ_i, which fits into the frame of the theory of diffusion transport and which leads to the decay probabilities α_i via eq. (13).

(i) Under the conditions of equal growth and decay the occupation probability $p(\underline{r}_b + \underline{n}.a, t)$ must be equal to $p(\underline{r}_b)t = \delta_i.a^2$ and so the monomer concentration around the cluster must be homogeneous.

(ii) Equal probability of growth and decay means that the change in free energy for both transitions is equal. Thus the condition for the existence of a saturation environment is, that the Helmholtz free energy of formation of a cluster B_i, $\Delta F(B_i)$, has an extremum

$$\frac{\partial \Delta F(B_i)}{\partial i} = 0 \qquad (23)$$

which is a maximum because of the positive surface energy term in the cluster binding energies. The Helmholtz free energy is chosen because cluster formation on a constant surface is considered rather than under conditions of constant monomer concentration.

The free energy of the cluster B_i is given by the partition function Z_i

$$F(B_i) = -kT.\ln Z_i \qquad (24)$$

and the free energy of N_1 noninteracting monomers on the surface A can be calculated from the partition function of such a system

$$Z_1 = (A.z_1)^{N_1}/N_1! \simeq (A.z_1.e/N_1)^{N_1} \qquad (25)$$

where z_1 is the single particle partition function for unity area. We find thus

$$F(N_1,B_1) = -N_1.kT.\ln(A.z_1.e)/N_1 \qquad (26)$$

The free energy for the formation of a cluster, $\Delta F(B_i)$, is the difference between $F(B_i) + F(N_1,B_1)$ and $F(N_1+i,B_1)$, which gives for sufficiently large N_1, i.e. $N_1 \gg i$:

$$\Delta F(B_i) = F(B_i) + F(N_1, B_1) - F(N_1 + i, B_1)$$
$$= kT \cdot (\ln Z_i - i \cdot \ln z_1 + i \cdot \ln(N_1/A)) \qquad (27)$$

The last term is the monomer concentration n_1, which may be very low provided the area A is large enough. The condition (23) permits the determination of the monomer saturation concentration δ_i

$$\ln \delta_i = \ln z_1 - \frac{\partial \ln Z_i}{\partial i} \qquad (28)$$

$$\delta_i \simeq z_1 \sqrt{Z_{i-1}/Z_{i+1}}$$

The latter expression uses $\partial \ln Z_i / \partial i \simeq (\ln Z_{i+1} - \ln Z_{i-1})/2$ and is related to a similar expression in (5), which was however derived by use of the mass action law and detailed balance of growth and decay in a sursaturated homogeneous monomer concentration. As we see here it does not depend on such arguments, which are difficult to confirm in a system with diffusion transport.

The decay rate α_i must be connected with the saturation concentration by (13)

$$\alpha_i = (s(i)/a) \cdot D \cdot z_1 \cdot \sqrt{(Z_{i-1}/Z_{i+1})} \qquad (29)$$

The actual evaluation of the quite general relations (28,29) depend on the model for the cluster and for the adsorbed state, which allow evaluation of the partition functions. Harmonic models have been used with relatively good success, though this is open to considerable improvement.

2.4.1 The model of adsorbed monomers.

The adsorption state is characterised by the number of equivalent adsorption sites l_{ad}, the desorption energy E_{des}, the mechanical desorption frequency ν_{odes}, the diffusion energy E_{diff}, the mechanical diffusion frequency ν_{odiff}, and the mass m of the adsorbed atom. Double occupation of the same site and interaction between monomers are unimportant for low concentrations n_1. For the case of fairly strong binding to the surface ($E_{diff} \geq 4$ kT) one can find[10] the unity area monomer partition function at temperature T

$$z_1 = l_{ad} \cdot (kT/h\nu_{odes}) \cdot (kT/h\nu_{odiff})^2 \cdot \exp(E_{des}/kT) \qquad (30)$$

The interpretation is simple : l_{ad} harmonic oscillator sites with frequency ν_{odes} and ν_{odiff} perpendicular and parallel to the surface plane are bound with E_{des} and constitute all possible states of an adsorbed atom. If the binding to a state become weaker ($E_{diff} < 4\,kT$), locations between the sites become important and z_1 will be more complicated. Very weak binding ($E_{diff} < kT/4$) will lead to the expression for a two dimensional gas.

For the strong binding case, the desorption rate is given by

$$\alpha = \nu_{des} \cdot \exp(-E_{des}/kT) \tag{31}$$

with

$$\nu_{des} = \nu_{odes} \cdot \nu_{odiff}^2 \cdot 2\pi m/(k.T.l_{ad}) \tag{32}$$

The diffusion constant is in this case

$$D = (\nu_{diff}/l_{ad}) \cdot \exp(-E_{diff}/kT) \tag{33}$$

with

$$\nu_{diff} = \nu_{odiff}^2 \cdot \sqrt{2\pi m/(k.T.l_{ad})} \tag{34}$$

The difference between mechanical and effective frequency constants is caused by entropy terms.

2.4.2. The cluster model.

The cluster energies which are necessary for the evaluation of the cluster partition functions are composed of kinetic and potential parts. A discussion of these energies will shed some light on the relations between the modes of growth and the different contributions to the energy of a cluster. The energy will again be given in a harmonic approximation and can be represented for a cluster B_i by

$$V_i = V_{io} + (1/2)^3 \cdot \sum_{n,m}^{(i-1)} c_{nm} \cdot u_n \cdot u_m \tag{35}$$

V_{io} is the potential energy in the equilibrium position of the cluster B_i, i.e. its binding energy. The coordinates u_n are the deviations from the equilibrium position, the second term is thus the elastic energy of the cluster. This elastic energy and the corresponding momentum terms will finally result in harmonic oscillator contributions to the partition function and to preexponential terms in the decay rate. The most important and interesting part is the binding energy.

The real binding energies for large clusters cannot be determined from first principles. Phenomenological or half experimental approximations can however give the general dependence of the bin -

ding energy on size, shape and position. One such approximation is

$$V_{io} = -i \cdot \Lambda + \Psi_i \tag{36}$$

where Λ is the energy of evaporation and Ψ_i is a correction term for the different binding of atoms at the cluster surface, which for very large clusters approaches the normal surface energy. It has been shown[11] that approximations of the form (36) agree surprisingly well with atomic calculations even for very small clusters. The term Ψ_i contains both the weaker binding of atoms at the surface to the vacuum and the influence of the binding forces of the substrate at the interface. Forces which lead to epitaxial growth like those calculated by van der Merwe and other authors[12,13,14,15,5] have to be incorporated into the surface energy in a general theory of growth. These questions are quite complicated in detail and may contain many surprises. For a general survey however some simple rules are helpful.

So it can be shown from a collection of experimental data[5], that the surface energy per surface atom, $\sigma \cdot v^{2/3}$, with v the atomic volume, is given by

$$\sigma \cdot v^{2/3} = \Lambda/6 \tag{37}$$

for most metals. For many other materials a similar relation hods

$$\sigma \cdot v^{2/3} = (z_a/2z_w) \cdot \Lambda \tag{38}$$

where z_a is the number of neighbours in an adsorption site and z_w is the number of neighbours in a growth site[16].

The experimental connection between evaporation energy and surface energy is shown in Fig. 1. On the basis of this result it can be inferred, that the energy of separation of a cluster from its substrate would be nearly equal to the desorption energy for each interfacial atom, if these atoms were not pulled out of their neighbours. This will cause a correction, which depends on the diffusion energy, on the misfit, and on the elastic constants of the cluster material. It cannot be larger than the diffusion energy. Calculations for a variety of simple models indicate, that this correction can be characterised by a parameter η with $0.4 < \eta < 0.8$. The separation energy per interfacial atom is thus

$$E_{sep} = E_{des} - \eta \cdot E_{diff} \tag{39}$$

For very small clusters or for vanishing misfit η tends to zero[14]. This relation should be applicable to both cluster growth and layer growth[5,15]. An application of (39) can give a criterium for the mode of growth and a prediction of the cluster shape[17,5]. It will

GROWTH PROCESSES

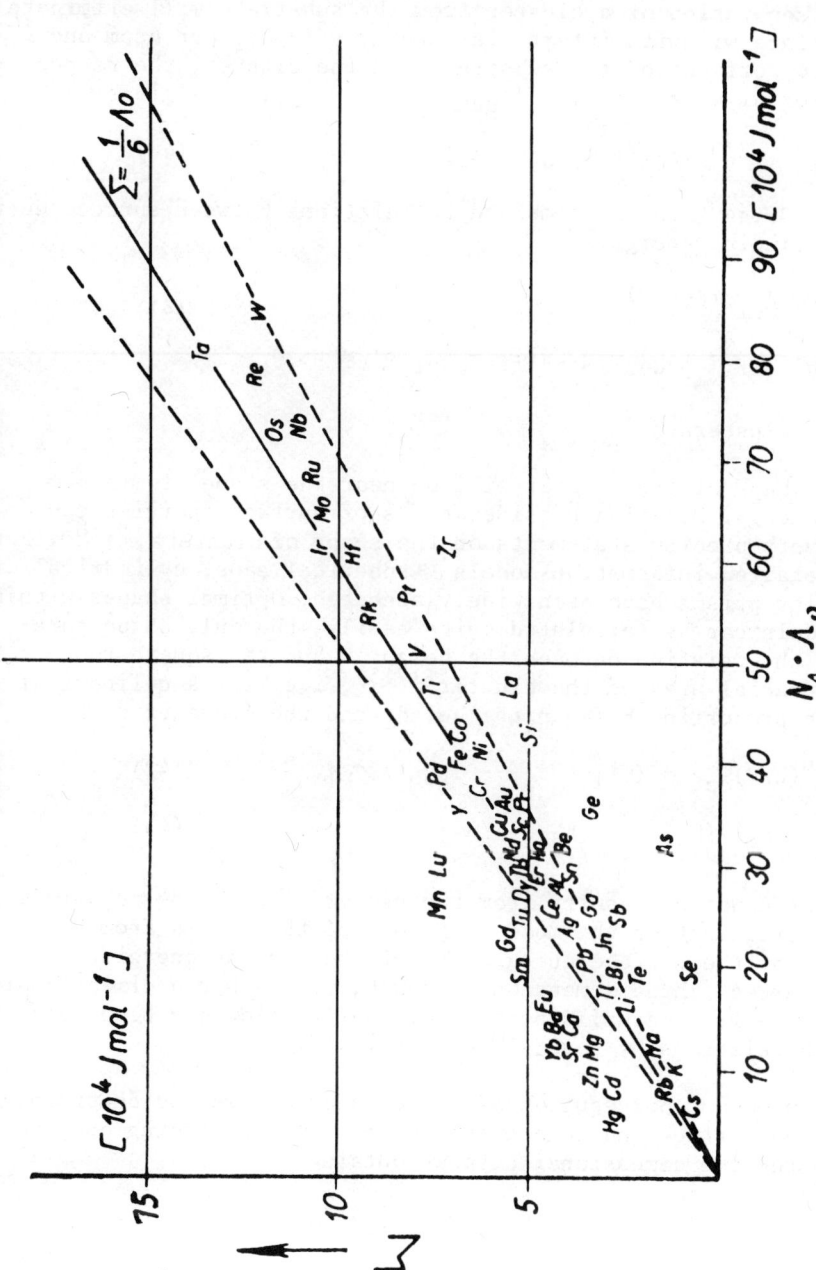

Fig. 1. Dependence of the experimentally obtained molar surface energies $\sum$ on the heats of evaporation(5).

finally enable us to calculate the surface energy term Ψ_i.

The separation of a cluster from the substrate will eliminate the interface with its interfacial energy $v^{2/3} \cdot \sigma_{if}$ per atom and create new surfaces of the substrate and the cluster, the respective energies of which are $v^{2/3} \cdot \sigma_{sub}$ and $v^{2/3} \cdot \sigma_{cl}$. This gives

$$E_{sep} = v^{2/3} \cdot (\sigma_{cl} + \sigma_{sub} - \sigma_{if}) \tag{40}$$

Eqs. (37, 39, 40) can be combined to relations between surface energies and binding energies

$$\sigma_{cl} = \Lambda_{cl}/(6v^{2/3}) \tag{41}$$

$$\sigma_{sub} - \sigma_{if} = (E_{des} - \eta \cdot E_{diff} - \Lambda_{cl}/6)/v^{2/3}$$

for metal clusters

The shape of the cluster will be near the shape of minimum surface energy. Detailed considerations of surface energies can lead to very precise statements on the shape of clusters $(9,18)$. For a less detailed information models as spherical caps, cylindrical or rectangular prisms have been widely accepted. Optimal shapes within these models can be calculated quite easily, the only shape parameter is the relation between the height h and the square root of the interfacial area of the cluster$(17,5)$, i.e. for a cylindrical prism the proportion between the height and the diameter d

$$\zeta = (h/d)_{opt} = (\sigma_{cl} + \sigma_{if} - \sigma_{sub})/2\sigma_{cl} \tag{42}$$

$$\zeta = 1 - 3 \cdot (E_{des} - \eta \cdot E_{diff})/\Lambda_{cl} \tag{43}$$

The latter equation results from the use of (41). These relations permit a statement on the mode of growth of thin films from the knowledge of the surface energies or of the binding energies: if $\zeta > 0$ we expect three dimensional growth, if $\zeta \leq 0$ two dimensional layer growth will occur. Experimental confirmation has been found for these relations by several authors (15,19,20).

A consequent persecution of these ideas allows the determination of the surface and edge contributions to the binding energy (36). In the two dimensional case we obtain

$$V_{io2} = -(2/3) \cdot i \cdot \Lambda - i \cdot (E_{des} - E_{diff}) + (2/3) \cdot \Lambda \cdot \sqrt{i} - i \cdot E_{diff} \cdot (1-\eta) \tag{44}$$

and in the three dimensional case

GROWTH PROCESSES

$$V_{io3} = -i \cdot \Lambda + i^{2/3} \cdot \Lambda \cdot (\zeta\pi/4)^{1/3}$$

$$\simeq -i \cdot \Lambda + (i \cdot \Lambda)^{2/3} \cdot (\Lambda - 3 \cdot (E_{des} - E_{diff}))^{1/3} - i_{diff} \cdot E_{diff}(1-\eta) \quad (45)$$

i_{if} is the number of interfacial atoms, which can be determined from (43) to $(i \cdot \sqrt{\pi}/2\zeta)^{2/3}$. All interaction with the substrate is contained in the last term. The binding energy is strong, if η is small. Good fit, large diffusion energy, small elastic constants of the cluster material tend to keep the atoms in the adsorption sites and to diminish η. The good fit depends on orientation and position of the clusters and also on their size. If we denote the binding energies (44,45) by $V_{io} = V(i,\eta)$ we see from these equations, that the dependence on orientation and position is given by

$$V(i,\underline{r},\phi) = V(i,1) - i_{diff} \cdot E_{diff} \cdot (1-\eta)(\underline{r},\phi) \quad (46)$$

and it is in this way, that epitaxial growth is caused by the interaction with the substrate[5]. The dependence of the potential energy on orientation and position will not be treated further here, because the analysis of epitaxial growth is not the aim of these lectures.

The calculation of the partition function requires the knowledge of the complete potential energy, eq. (35), and of the kinetic energy parts. The kinetic energy leads together with the elastic parts of the potential energy to the phonon states, the partition functions of which approach $kT/h\nu$ for high temperatures. The frequencies ν vary in principle over the phonon spectrum of the cluster, for an approximate treatment however only a mean vibration frequency ν_{vib}, the translational frequencies of the whole cluster in the surface plane ν_s, and its rotational frequency ν_ϕ are explicitely taken into account. This leads to a partition function of the cluster B_i

$$Z_i = M \cdot (kT/h\nu_s)^2 \cdot (kT/h\nu_\phi) \cdot (kT/h\nu_{vib})^{3i-3} \cdot \exp((-V(i,\eta))/kT) \quad (47)$$

The first term M is the number of equivalent and different positions, which a cluster can assume on a periodic substrate of area A. The following terms are the oscillatory partition function and the binding energy term. Eq. (47) can evidently be specified to include the dependence on orientation and position[5].

2.4.3. Evaluation of the boundary monomer concentration and the decay rate for the described model.

From eq. (28) we need $\partial \ln Z_i / \partial i$ and $\ln z_1$ for the evaluation of

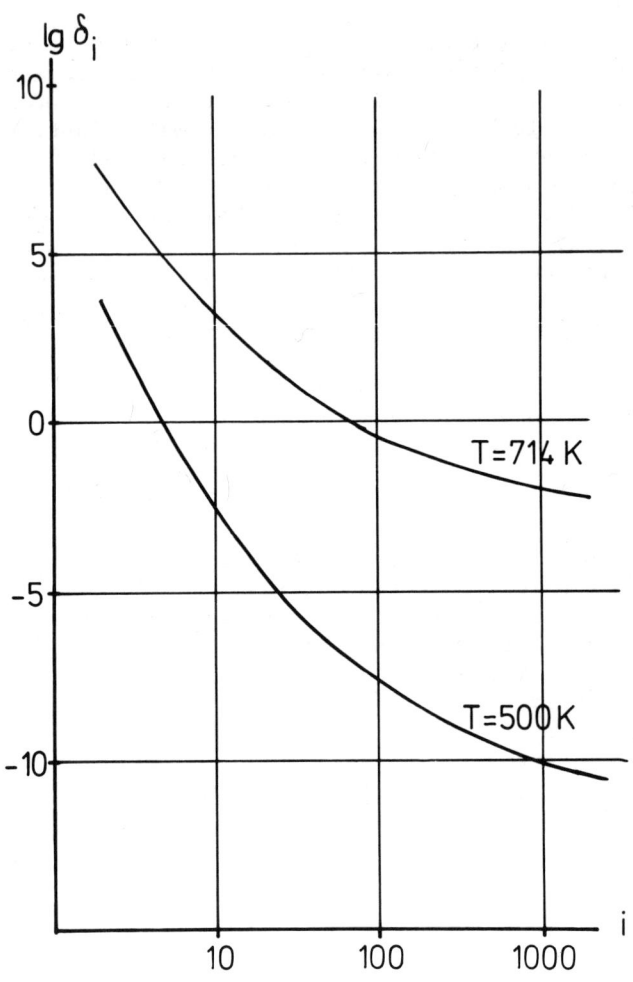

Fig.2 Dependence of saturation monomer concentration δ_i on cluster size for Au on NaCl for substrate temperature of 500°K and 714°K

the boundary monomer concentration δ_i:

$$\partial \ln Z_i/\partial i = 3 \cdot \ln(kT/h\nu_{vib}) - (1/kT) \cdot \partial V(i,\eta)/\partial i \qquad (48)$$

From (44,45) we obtain

$$\partial V_{io2}/\partial i = -((2/3)\Lambda + E_{des} - \eta E_{diff}) + (1/3) \cdot \Lambda/\sqrt{i}$$
$$\partial V_{io3}/\partial i = -\Lambda + (2/3) \cdot \Lambda \cdot (\zeta\pi/4)^{1/3} \cdot i^{-1/3} \qquad (49)$$

Together with eq. (30) this gives in the three dimensional case

$$\delta_i = \frac{1_{add} \cdot \nu_{vib}^3}{\nu_{odes} \cdot \nu_{odiff}^3} \cdot \exp\left[(E_{des} - \Lambda + \frac{2}{3}(\zeta\pi/4)^{1/3} \cdot i^{-1/3})/kT\right] \qquad (50)$$

and the decay rate can be calculated from this by

$$\alpha_i = \frac{s(i)}{a} \cdot D \cdot \delta_i \qquad (13)$$

.5. Stationary solutions of the diffusion equation and capture rates

The capture rate is statistically defined in the sense of § 2.2 and we have found an expression for J_i

$$J_i = D \cdot (s(i)/a) \cdot (\delta_i + a(\underline{n}\nabla n_1)) \qquad (15)$$

For the final evaluation of the capture rate a knowledge of the concentration gradient at the cluster boundary, $r = \rho_i$, must be obtained from a solution of the diffusion equation. In § 2.2. we have seen, that no limitation to the statistical application of the diffusion equations (9,11) exists and that these are equivalent to (16) within the limits of a linear approach. We have further seen in § 2.3. that the stationary solution with the boundary condition $n_1(\rho_i) = \delta_i$ should be adequate even for very small clusters, because the change of the cluster size is usually compared to the desorption rate.

The solutions (19) for $\omega=0$ are directly applicable to our problem, resulting in $u_1(r,0) = K_0(r/\sqrt{D/\alpha})$, $u_2(r,0) = I_0(r/\sqrt{D/\alpha})$. We thus obtain for the stationary monomer concentration

$$n_1(r) = J_{imp}/\alpha + A \cdot I_0(r/r_1) + B \cdot K_0(r/r_1) \qquad (51)$$

with the diffusion range r_1

$$r_1 = \sqrt{D/\alpha} \qquad (52)$$

The derivatives of I_0 and K_0 are I_1 and $-K_1$ respectively and we have for the capture rate

$$J_i = D(s(i)/a)(\delta_i + \frac{a}{r_1}\cdot(A.I_1(\rho_i/r_1) - B.K_1(\rho_i/r_1))) \quad (53)$$

The coefficients A and B must be determined by the boundary conditions.

From eq.(6) we have N_c clusters of all sizes on unity area. The mean area per cluster is

$$r_{Nc}^2 \pi = 1/N_c \quad (54)$$

which defines the radius r_{Nc} of the outer boundary. The cluster radius is ρ_i. For the boundary conditions the monomer concentration $n_1(\rho_i) = \delta_i$ and $n_1(r_{Nc})$ will be chosen. Another choice would be $n_1(\rho_i)$ and $(\partial n_1/\partial r)_{r_{Nc}} = 0$ which is certainly reasonable if all clusters were of the same size or if the monomer concentration vanishes at the edge of each cluster [21]. We have however a mixture of different cluster sizes, some of which are subsaturated, $J_{imp}/\alpha < \delta_i$ some are supersaturated, $J_{imp}/\alpha > \delta_i$, so the assumption of zero flux at the outer boundary is not applicable. This is physically evident from the flux which goes from the subsaturated to the supersaturated clusters.

These considerations lead to the following procedure : the outer boundary concentration is first determined by the zero flux condition for an assembly of average clusters. With this average outer boundary condition the final solutions, i.e. the coefficients A and B, are obtained.

The average cluster has a size

$$\bar{i} = N_a/N_c \quad (55)$$

which leads to a mean radius from eq. (42) with atomic volume v

$$\bar{\rho} = (\bar{i}v/2\pi\zeta)^{1/3} \quad (56)$$

and to a mean monomer saturation concentration $\bar{\delta} = \delta_{\bar{i}}$. These clusters are supercritical and uniform from the definition of N_c, N_a, so that Stowells boundary condition of zero flux[21] can be applied

$$\bar{\delta} = J_{imp}/\alpha + \bar{A}.I_0(\bar{\rho}/r_1) + \bar{B}.K_0(\bar{\rho}/r_1) \quad (57)$$

$$0 = \bar{A}.I_1(r_{Nc}/r_1) - \bar{B}.K_1(r_{Nc}/r_1) \quad (58)$$

This can be resolved to

$$\bar{A} = (\bar{\delta}-J_{imp}/\alpha) \cdot K_1(r_{Nc}/r_1)\sqrt{\overline{Det}} \qquad (59)$$

$$\bar{B} = (\bar{\delta}-J_{imp}/\alpha) \cdot I_1(r_{Nc}/r_1)\sqrt{\overline{Det}} \qquad (60)$$

with

$$\overline{Det} = K_1(r_{Nc}/r_1) \cdot I_0(\bar{\rho}/r_1) + K_0(\bar{\rho}/r_1) \cdot I_1(r_{Nc}/r_1)$$

The monomer concentration at the outer boundary is thus given by

$$n_{1B} = n_1(r_{Nc}) = J_{imp}/\alpha + (\bar{\delta}-J_{imp}/\alpha) \cdot (\bar{A} \cdot I_0(r_{Nc}/r_1) + \bar{B} \cdot K_0(r_{Nc}/r_1)) \qquad (61)$$

For large r_{Nc} ($r_{Nc} > 10 r_1$) the second term vanishes exponentially so $n_1(r_{Nc}) \to J_{imp}/\alpha$. This maximum possible monomer concentrations is reduced by diffusion for smaller distances r_{Nc}. For the monomer concentration around the average cluster this boundary condition is of course identical with Stowells. For all other clusters, subcritical or supercritical, it has however the advantage that the size of the diffusion courts depends on the cluster size and that material transport from small clusters to large ones is taken into account.

For a critical cluster we have

$$(\partial n_1/\partial r)_{\rho_i*} = 0 \qquad (62)$$

and for subcritical or supercritical clusters we have

$$(\partial n_1/\partial r)_{\rho_{sub}} < 0, \quad (\partial n_1/\partial r)_{\rho_{sup}} > 0 \qquad (63)$$

All these physical conditions at the cluster edge can be met with a fixed average monomer concentration $n_1(r_{Nc})$ calculated from (61). The different possible cases are illustrated in Fig. 3.

To complete the evaluation, we use the mean boundary concentration for determination of A and B in (51) by use of the conditions

$$n_1(\rho_i) = \delta_i, \qquad n_1(r_{Nc}) = n_{1B} \qquad (64)$$

which results in

$$A = ((\delta_i-J_{imp}/\alpha) \cdot K_0(r_{Nc}/r_1) - (n_{1B}-J_{imp}/\alpha) \cdot K_0(\rho_i/r_1))/Det$$

$$B = ((n_{1B}-J_{imp}/\alpha) \cdot I_0(\rho_i/r_1) - (\delta_i-J_{imp}/\alpha) \cdot I_0(r_{Nc}/r_1))/Det \qquad (65)$$

$$Det = I_0(\rho_i/r_1) \cdot K_0(r_{Nc}/r_1) - I_0(r_{Nc}/r_1) \cdot K_0(\rho_i/r_1)$$

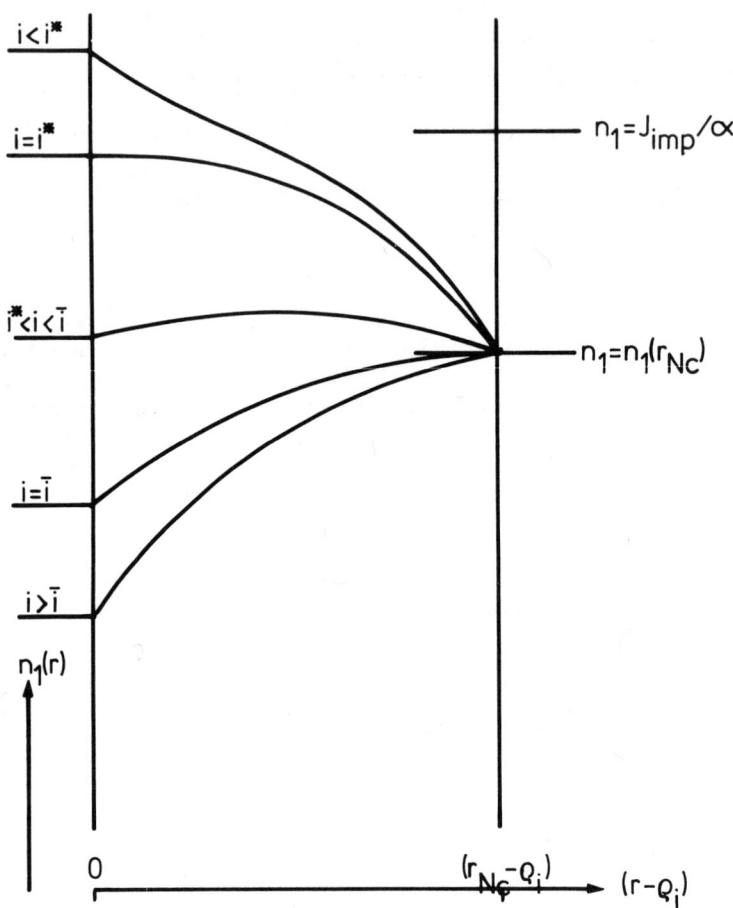

Fig. 3. Qualitative representation of monomer concentration around clusters of different sizes with a fixed boundary monomer concentration. The curvature is determined by
$\Delta n_1 = \alpha/D (n_1 - J_{imp}/\alpha)$.

GROWTH PROCESSES

For large r_{Nc} this can be simplified considerably and Halperns formula[7] is obtained

$$n_1(r) = J_{imp}/\alpha + (\delta_i - \frac{J_{imp}}{\alpha}) \cdot K_o(r/r_1)/K_o(\rho_i/r_1) \qquad (66)$$

From these formulas as well as from fig. 3 an interesting conclusion can be drawn : around a subcritical cluster the monomer concentration is larger than J_{imp}/α in the statistical sense stated above, for a critical cluster it is equal to this value and for supercritical clusters it is smaller. This means, that the capture rate for a critical cluster is equal to the value expected from a homogeneous monomer concentration, as discussed in § 2.2 and § 2.4, but that it is larger for a subcritical cluster, though less than the decay rate, and it is less for a supercritical cluster, though larger than the decay rate. Thus the diffusion mechanism diminishes the dependence of the net growth rate on the cluster size. A consequence of this is, that the nucleation rate is higher in a consequent diffusion theory, as presented here, than in a homogeneous concentration theory, and that it depends on the impingement rate and on the temperature to a minor degree than predicted by Walton[9]. It follows furthermore that the growth rate of stable nuclei will be reduced in a diffusion theory. This effect has been observed already by Halpern[7].

2.6. The rate of twin formation

The local rate of twin formation on an area dA is

$$dJ_{twin} = D(s(1)/a) \cdot n_1^2(r) \cdot dA \qquad (67)$$

The free area can be divided into N_c regions between the inner and outer boundary radii ρ, r_{Nc}, within which approximately the same monomer concentrations are found. Thus we have

$$J_{twin} = D \cdot (s(1)/a) \cdot 2\pi N_c \int_{\rho}^{r_{Nc}} n_1^2(r) \cdot r \cdot dr \qquad (68)$$

This applies to any monomer distribution and is analogous to the formula of Routledge and Stowell[21].

Because for the calculations of the solutions of the kinetic equations (4 to 8) computer work is necessary in any case, the relatively complicated expression for the rate constants given here do not add any important complication to the analysis of the system.

3. THE SYSTEM OF GROWTH EQUATIONS AND A METHOD FOR THEIR NUMERICAL SOLUTION.

The growth equations (4 to 8) are incomplete in two respects: They do not include the condensation of monomers by direct impingement on top of already grown clusters and they do not account for the coalescence of nuclei. Their solution is impeded by the infinite sums contained in eqs. (7,8), which should be approximated by finite expressions. The solution of the system is not a trivial problem, because instabilities may occur, if standard methods as Euler, Runge-Kutta, or similar are used. The content of this chapter will thus be the final formulation of the growth system and the description of the method of its numerical solution.

3.1. Condensation by direct impingement

For direct impingement on clusters a sticking probability of 1 is generally assumed. Because of the small area of the first clusters direct impingement is important only for large clusters, so that it can be fully accounted for by introduction into eqs. (7,8). We find

$$J_{i\,direct} = \rho_i^2 \pi \cdot J_{imp} \tag{69}$$

and thus the last term in eq.(7) changes to

$$\sum_{i=i^*+1}^{\infty} ((J_i - \alpha_i + \rho_i^2 \pi J_{imp}) \cdot n_i) \tag{70}$$

The additional terms $\sum \rho_i^2 \pi n_i$ is the area covered by stable nuclei, and the free area defined for eq.(8) is given by

$$A_f = 1 - \sum_{i=i^*+1}^{\infty} \rho_i^2 \pi n_i \tag{71}$$

Although these corrections are of minor importance for the observation of nucleation, they are important for the observation of condensation on surfaces.

3.2. Coalescence of nuclei

Coalescence of larger nuclei can occur if the distance between two nuclei is not greater than the sum of their radii. If the clusters migrate on the substrate, they will meet occasionally within this distance and the coalescence must be treated on similar lines as the capture of monomers[2]. A complication is, that the clusters grow and decay during their migration. If the clusters are mainly fixed at their nucleation sites, coalescence occurs by growth only.

For the calculation of the latter kind of coalescence some knowledge of the distribution of clusters on the substrate is desirable. An estimation can be given by the argument, that most clusters form as long as they are relatively far apart from each other, and that under such conditions the monomer concentration around a cluster is approximatively given by eq. (66). One result of the theory presented here is, that the nucleation rate is proportional to n_1^w, and that w is a number between 2 and 3, which depends very weakly on the size of the critical nucleus, contrarily to the case of homogeneous monomer concentration, where $w = i^*+1$. Thus we expect in the statistical sense of § 2.2, a distribution $N_C(r)$ around a selected cluster, which is given by

$$N_C(r) = N_C \cdot (n_1(r))^w / \int_0^{1/\sqrt{\pi}} n_1^w(x) \cdot 2\pi \cdot x \cdot dx \qquad (72)$$

which is normalised so that the total number of clusters on unity surface is N_C. Eq.(72) results in a certain correlation of the cluster sites.

If we consider a radius $r = 2\bar{\rho}$, it is just the radius, within which all clusters are situated that have already coalesced with the central one. The coalescence rate around one cluster is thus given by

$$-\left(\frac{d}{dt}(r^2 \pi N_C(r))\right)_{r=2\bar{\rho}} = -8\pi\bar{\rho} \cdot \left(N_C(2\bar{\rho}) + \bar{\rho}\left(\frac{\partial N_C}{\partial r}\right)_{r=2\bar{\rho}}\right)\frac{d\bar{\rho}}{dt} \qquad (73)$$

The total coalescence rate is obtained by multiplication with $N_C/2$, so that we finally arrive at

$$\dot{N}_{ccoal} = -4\pi N_C^2 \bar{\rho}^2 \frac{d\bar{\rho}}{dt} \cdot \frac{(n_1(2\bar{\rho}))^w + \bar{\rho}(n_1(2\bar{\rho}))^{w-1}(\partial n_1(r)/\partial r)_{r=2\bar{\rho}}}{\int_0^{1/\sqrt{\pi}} (n_1(r))^w \cdot 2\pi x \, dx} \qquad (74)$$

Conditions where the homogeneous monomer concentration applies are included in (74). Migration of the clusters will result in a more uniform cluster distribution than (72), and can be simulated qualitatively by reduction of the exponent w.

3.3. The equations of cluster growth

The number of eqs. (5) is infinite in principle. Those equations however, for which $\alpha_i \ll J_i$, i.e. by which the growth of very stable clusters B_i is described, are not coupled to the equation for B_{i+1} any more, because $\alpha_{i+1} n_{i+1}$ become very small. In this case the equations for all $i \geq i^* + m$ can be solved by simple recursion as it has been done successfully by Zinsmeister[21], provided m is chosen adequately. This means also, that a solution of eqs. (5) is necessary only until $i = i^* + m$, if one is interested mainly in the total number of nuclei and in condensation. Eq. (7,70) contains an infinite sum over the cluster concentration, which must be simplified if the solution of the infinite system shall be avoided. It can be shown that

$$\sum_{i=i^*+1}^{\infty} (J_i - \alpha_i + \rho_i^2 \pi J_{imp}) \cdot n_i \leq N_c \cdot (J_{\bar{i}} - \alpha_{\bar{i}} + \bar{\rho}^{-2} \pi J_{imp}) \quad (75)$$

and we find from eqs. (13,15) that

$$J_{\bar{i}} - \alpha_{\bar{i}} = D \cdot 2\pi\bar{\rho} \cdot (\partial n_1 / \partial r)_{\bar{\rho}} \quad (76)$$

The relation (75) is always true, if

$$(\partial^2 (J_i - \alpha_i + \rho_i^2 \cdot \pi \cdot J_{imp}) / \partial i^2) < 0$$

which is the case for $i > i^*$, and in most cases (75) is a rather good approximation.

With inclusion of direct impingement, coalescence, and the approximation to the infinite sum we arrive at a system of equations which can be solved by numerical methods, and the solutions of which are able to describe the growth of films from binding energies and parameters of lattice geometry.

$$\dot{n}_2 = J_{twin} - (\alpha_2 + J_2) \cdot n_2 + \alpha_3 \cdot n_3 \quad (77)$$

$$\dot{n}_i = J_{i-1} - (\alpha_i + J_i) n_i + \alpha_{i+1} n_{i+1} \quad (78)$$
$$\text{(for } 3 \leq i \leq i^* + m\text{)}$$

$$\dot{N}_c = J_{i^*} \cdot n_{i^*} - \alpha_{i^*+1} \cdot n_{i^*+1} + \dot{N}_{ccoal} \quad (79)$$

$$\dot{N}_a = (i^*+1)J_{i^*}\cdot n_{i^*} - i^*\cdot\alpha_{i^*+1}\cdot n_{i^*+1} + N_c(J_{\bar{i}} - \alpha_{\bar{i}} + \bar{\rho}^2\pi J_{imp}) \qquad (80)$$

$$\dot{N}_1 = J_{imp}(1-N_c\bar{\rho}^2\pi) - \alpha N_1 - 2J_{twin} + \alpha_2 n_2 - \sum_{i=2}^{i^*}(J_i - \alpha_i)n_i - N_c(J_{\bar{i}} - \alpha_{\bar{i}}) \qquad (81)$$

The system of equations (77 to 81) is non linear, because the capture rates (53) depend on the mean cluster distance (54) via the constants A and B calculated in (65,61) and on the mean cluster radius $\bar{\rho}$ (56), which is used directly and for the determination of $\bar{\delta}$ in eq. (61). Eqs. (79 to 81) depend furthermore explicitely on these parameters. This dependence is however not very strong, so that the system can be considered linear for short time intervals. For long times the nonlinear terms lead to (i) decrease in monomer concentration n_{1B} (61) and consequently to a decrease in all capture rates, (ii) to an increasing importance of the coalescence term in (79,74). Both effects lead to saturation and finally decrease in the number of nuclei and eventually to a linear dependence ce of condensation on time.

In special cases analytical solutions of the system are possible and have been given in (5), for a general survey of growth processes numerical calculations are however necessary and a method of easy feasability has been worked out for this purpose.

3.4. A method for the numerical solution of the growth equations

As stated above, the nonlinearity of the system of equations is not strong, so that it can be considered linear for appropriate time intervals.

For an induction period, $t_{in} < 1/\sqrt{\alpha^2 + 8Ds(1)J_{imp}/a}$, the system can be easily solved by any of the normally available methods, as Euler, Runge-Kutta or any other program for the solution of systems of differential equations. During this period the monomer concentration builds up and a fast change of the capture rates occurs. I have chosen a modified Euler method for this period. The calculations are made with a preselected accuracy and an automatic accomodation of step width. This step width does not increase essentially over the induction time t_{in} and so the further progress of the calculations with such a method is usually very slow, the order of magnitude of t_{in} beeing 10^{-7} s. The reason for this limitation of step width is the Lipschitz condition.

The variation of capture rates is on the other hand appreciable only in much larger time intervals, if the induction period has elapsed ; the same is true for the coefficients of the growth equations. It is therefore possible to solve the equations exactly for a certain time interval with constant coefficients, calculate new coefficients from the solution and compare the difference between the old and new coefficients with some prescribed limit and to adjust the step width accordingly. The coefficients calculated at the end of the correct time interval are used for the next interval. The calculation proceeds with a nearly exponential increase in step width. For the exact solution mentioned a matrix method is used. The structure of eqs. (77,78) is given by

$$\dot{n}_{i+1} = -c + \sum_k a_{ik} \cdot n_{k+1} \tag{82}$$

with

$$c_i = -J_{twin} \cdot \delta_{i1} \quad \text{(Kronecker } -\delta\text{)} \tag{83}$$

$$a_{i,i-1} = J_i \; ; \; a_{i,i} = -(J_{i+1} + \alpha_{i+1}) \; ; \; a_{i,i+1} = \alpha_{i+2} \; ; \tag{84}$$
$$\text{else } a_{i,k} = 0 \; ;$$

Because the diagonal coefficients a_{ii} are negative, saturation for each n_{i+1} will occur, so that $\dot{n}_{i+1} \to 0$ for large times. This leads to stationary solutions $x_{i,\infty} = n_{i+1}(\infty)$ and to

$$n_{i+1}(t) = x_{i\infty} + x_i(t) \tag{85}$$

where the $x_i(t)$ are negative and vanish for long times. We thus have

$$\dot{x}_i = -c_i + \sum_{k=1}^{i^*+m-1} a_{ik}(x_{k\infty} + x_k(t)) \tag{86}$$

with the asymptotic condition

$$\sum_k a_{nk} x_{k\infty} = c_n \tag{87}$$

The solution of (87), which can be performed by a matrix inversion, leads directly to the result of the stationary nucleation theory, from which the stationary Becker-Döring nucleation rate[23] can be directly obtained by eq. (6). We find

$$x_{k\infty} = n_{k+1}(\infty) = \sum_n a_{nk}^{-1} \cdot c_n = -a_{k1}^{-1} J_{twin} \tag{88}$$

GROWTH PROCESSES

The time dependent part obeys a homogeneous system of differential equations

$$\dot{x}_i(t) = \sum_k a_{ik} \cdot x_k(t) \tag{89}$$

The solution of this system is obtained by a time dependent transformation from any set of initial values

$$x_n(t) = \sum_1 TR_{nl} \cdot x_1(t_o) \tag{90}$$

where

$$TR_{nl} = (\exp(\underline{\underline{A}} \cdot (t-t_o)))_{nl}$$

$$= \delta_{nl} + a_{nl}(t-t_o) + \frac{1}{2!} \sum_\mu a_{n\mu} a_{\mu l}(t-t_o)^2 + \ldots \tag{91}$$

$$= \sum_{k=0}^{\infty} (\underline{\underline{A}}^k)_{nl}(t-t_o)^k/k!$$

For short time intervals, $((t-t_o) \cdot \max(a_{lm}) \leq 1)$, the calculations of TR can be made with small effort and great precision. For larger times, TR can be obtained from the short time solution by simple matrix multiplication, using

$$(\exp(\underline{\underline{A}} \cdot (t_1+t_2)))_{nl} = \sum_\mu ((\exp(\underline{\underline{A}} \cdot t_1))_{n\mu} \cdot (\exp(\underline{\underline{A}} \cdot t_2))_{\mu l}) \tag{92}$$

The solution of a **system** of 25 differential equations is easy and fast for large times and occurs without the appearance of spurious oscillations. The method is well known in principle, but does not seem to be used widely in the numerical solution of differential equations.

The remaining equations (79,80,81) have the general form

$$\dot{N}_c = g(t) - h \cdot N_c^2 \tag{93}$$

$$\dot{N}_a = f(t) + c \cdot N_c \tag{94}$$

$$\dot{N}_1 = J_{imp} - \alpha \cdot N_1 - \sum_{i=2}^{i^*} i \cdot n_i - \dot{N}_a \tag{95}$$

They can be solved successively, because g(t) is known from eqs. (90,88,85,79) and $b = -\dot{N}_{ccoal}/N_c^2$ is assumed from equation (74) to be a slowly varying function of time. This allows the calculation of N_c. In (94) the function (t) is known again from (90,88,85,80) and $N_c(t)$ from (93), c is $J_{\overline{i}} - \alpha_i + \overline{\rho}^2 \pi J_{imp}$ and varies slowly again. Finally the number of monomers is stationnary for the times considered and thus a straightforward solution is possible. The results are approximated by

$$N_c = N_{co}/(1+b \cdot N_{co} \cdot (t-t_0)) + \int_{t_0}^{t} f(\tau) \, d\tau \qquad (96)$$

$$N_a = N_{ao} + \frac{c}{b} \cdot \ln(1+b \cdot N_{co} \cdot (t-t_0)) + \int_{t_0}^{t} f(\tau) d\tau + c \cdot \int_{t_0}^{t} \int_{t_0}^{\tau} g(\sigma) d\sigma \, d\tau \qquad (97)$$

The first terms including the coalescence parameter b are exact solutions, if coalescence prevails, the integrals are exact solutions if coalescence is unimportant against nucleation. The complete approximations are valid, whenver $(t-t_0)^2 |N_{ccoal}| \cdot |g(t)| << N_c^2$, which can always be met and is thus suited for numerical analysis.

The nucleation is described by the integrals, which require the calculation of

$$\int_{t_0}^{t} x_n(\tau) \, d\tau = \sum_k a_{nk}^{-1} \cdot (x_k(t) - x_k(t_0))$$

$$= \sum_{k,1} a_{nk}^{-1} \cdot (TR_{k1} - \delta_{k1}) \cdot x_1(t_0) \qquad (98)$$

This gives the values of the fundamental integrals occuring in eqs. (96,97) from (79,80)

$$I_1 = \int_{t_0}^{t} n_i^*(\tau) \, d\tau = (t-t_0) \cdot x_{i^*-1,\infty} +$$

$$+ \sum_1 a_{i^*-1,1}^{-1} (x_1(t) - x_1(t_0)) \qquad (99)$$

$$I_3 = \int_{t_0}^{t} \int_{t_0}^{\tau} n_i^*(\sigma) \, d\sigma \, d\tau = \frac{1}{2}(t-t_0)^2 x_{i^*-1,\infty} +$$

$$+ \sum_{k,1} a_{i^*-1,1} \cdot (a_{1k}^{-1} \cdot (x_k(t) - x_k(t_0)) - x_1(t_0) \cdot (t-t_0))$$

These integrals are obtained exactly from the matrices calculated earlier.

If we finally keep in mind, that the capture probabilities are approximately proportional to the number of monomers, i.e. that J_i/N_1 is nearly constant, we find

$$N_1 = \frac{J_{imp}(1-N_c \bar{\rho}^2 \pi) + i^* \cdot \alpha_{i^*+1} n_{i^*+1} + \alpha_{\bar{i}} \cdot N_c}{\alpha + N_c \cdot (J_{\bar{i}}/N_1) + (i^*+1) \cdot n_i^* \cdot (J_i^*/N_1)}$$

$$\simeq \frac{J_{imp}(1-N_c \bar{\rho}^2 \pi)}{\alpha + N_c (J_{\bar{i}}/N_1)}$$

(100)

The advantage of the method described is the use of solutions, which are exact in finite intervals and which can be calculated by algebraic methods. This avoids the difficulties of the usual extrapolation methods, which result in spurious oscillations or other instabilities of the solutions. Contrarily the progress of the computer calculations with our method is very smooth and rapid.

4. RESULTS OF COMPUTER CALCULATIONS AND COMPARISON WITH EXPERIMENT

It is possible to obtain separate analytical approximations for the initial stage, the nucleation stage, and the saturation stage(5). While these are useful for special purposes, a general survey is only possible with a complete numerical solution of the system of equations (77 to 81). This will be presented here for the case of gold on rocksalt, for which
Λ = 87.6 kcal/mol, E_{des} = 20 kcal/mol, E_{diff} = 8·5 kcal/mol
ν_{odes} = 2.4·10^{12} s^{-1}, ν_{odiff} = 7.4·10^{11} s^{-1}
are assumed for best fit with the experiments of Robins et al.(3).

Two different approaches have been used in the calculations (i) homogeneous distribution of monomers, (ii) diffusion governed distribution of monomers. Both assumptions have been carried through consequently and the differences are interesting in view of the fundamental discussions of § 2.2 and § 2.3.

4.1. Computer results

The growth of gold films is represented by the concentration of the polymers n_1, n_2, n_3, n_4, n_5, by the number of stable clusters N_c, and by the number of condensed atoms N_a. The mean cluster $\bar{\rho}$ is denoted $\bar{r}$ in the figures. At low temperatures or high impingement rates the monomer concentration is limited by twin formation and capture. This capture regime is characterised by (5)

$$\alpha^2 < 8 \cdot D \cdot s(1) \cdot J_{imp}/a \qquad (101)$$

The monomer concentration attains a maximum

$$n_{1max} \simeq \sqrt{J_{imp} \cdot a/2 \cdot D \cdot s(1)} < J_{imp}/\alpha \qquad (102)$$

The general features of growth are not very different in the two cases, as can be seen from Fig.4. The twin is already supercritical in the capture regime. The number of clusters rises steeply during an induction period $t_{in} = 1/\sqrt{\alpha^2 + 8Ds(1)J_{imp}/a}$, and increases only slowly at later stages ($\sim t^{1/3}$, $\sim t^{1/7}$). The maximum cluster concentration is higher in the diffusion case, because the cluster growth is slower and thus the mean monomer concentration is higher here. The condensation is complete in both cases ($N_a(t)$). The number of polymers develops consecutively, their stationary concentration is size independent in the diffusion case while it is weakly size dependent in the homogeneous case. The reason for this is, that the capture rates are size dependent in the homogeneous case and size independent for high supersaturation in the diffusion case. This can be seen by an evaluation of (14,15) for both cases, where (66) and $K_o(x) \simeq 0.11 - \ln x$, $K_1(x) \simeq 1/x$ can be used in the diffusion case for small $x = \rho_i/r_1$. $J_i = 2\pi D \cdot (\rho_i/a) \cdot n_1$ is then obtained in the homogeneous case, while $J_i = r_1^2 \pi J_{imp}(2/(0.11+\ln(r_1/\rho_i)))$ is the capture rate for the diffusion case.

At high temperature or low impingement rates the monomer concentration is determined by the desorption equilibrium. This desorption regime is characterised by

$$\alpha^2 > 8 \cdot D \cdot s(1) \cdot J_{imp}/a \qquad (103)$$

The maximum monomer concentration is

$$n_{1max} \simeq J_{imp}/\alpha \qquad (104)$$

and is constant for some rime. An example for this case is shown in Fig. 5. In this case of high temperature the first supercritical nucleus is a cluster of six atoms. Remarkable differences between the two monomer distribution cases can be observed. The mean monomer concentration is constant in both cases and has almost the same value. The condensation is incomplete, but it is more effective in the diffusion case. This is connected to the fact that the nucleation rate in the diffusion case is 10^4 times larger than in the homogeneous case under these circumstances ($i^* = 5$). The reason for the higher nucleation rate is the local concentration around a subcritical cluster, which is enhanced from eq. (63) and Fig. 3.

Another consequence of this local enhancement of monomer concentration around subcritical clusters is the inversion of the sequence of subcritical polymer concentration in the diffusion case, where $n_5 > n_4 > n_3 > n_2$, while in the homogeneous case $n_2 > n_3 > n_4 > n_5$.

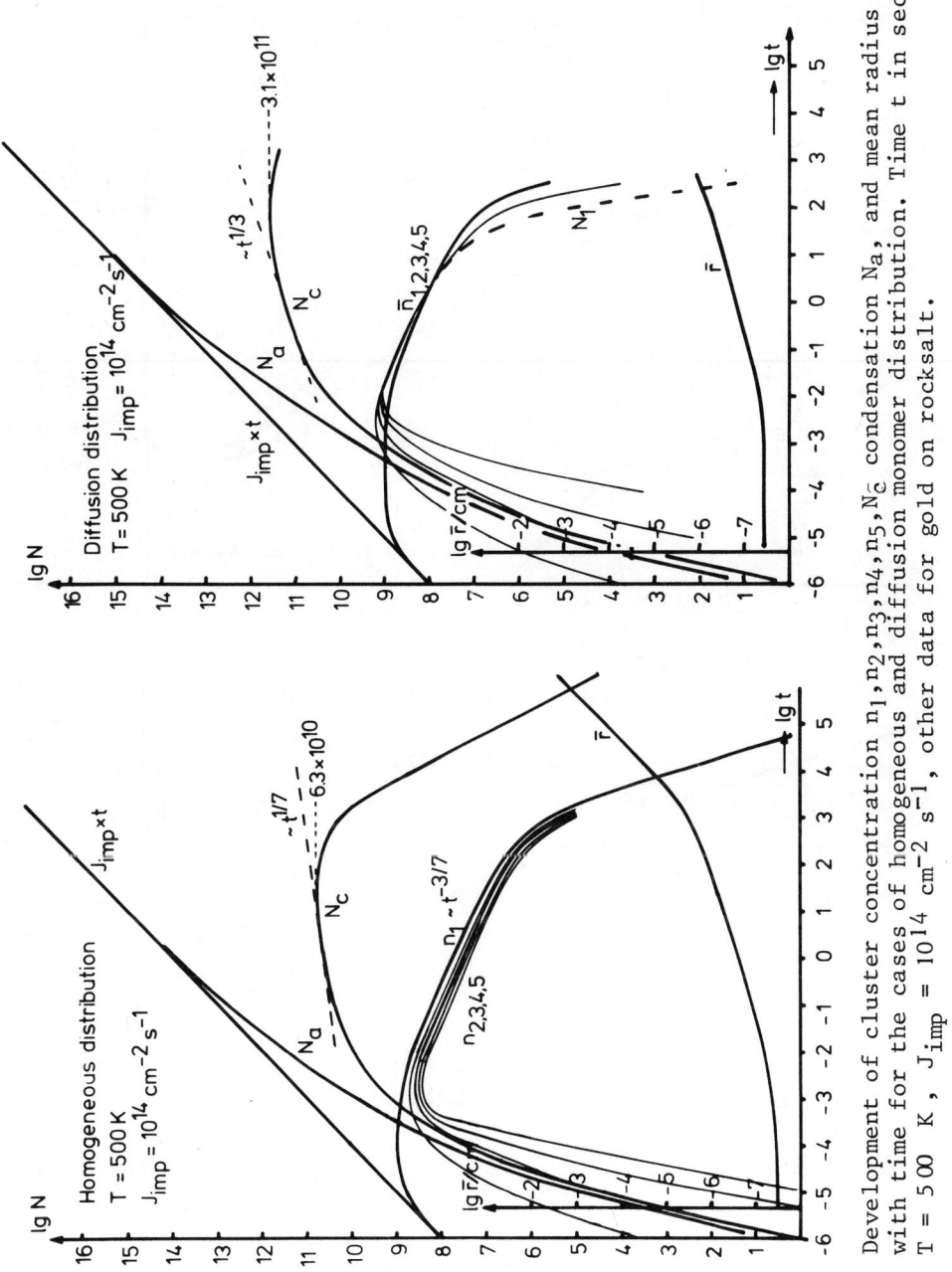

Fig.4. Development of cluster concentration n_1, n_2, n_3, n_4, n_5, N_c condensation N_a, and mean radius $\bar{r}$ with time for the cases of homogeneous and diffusion monomer distribution. Time t in seconds, $T = 500$ K, $J_{imp} = 10^{14}$ cm^{-2} s^{-1}, other data for gold on rocksalt.

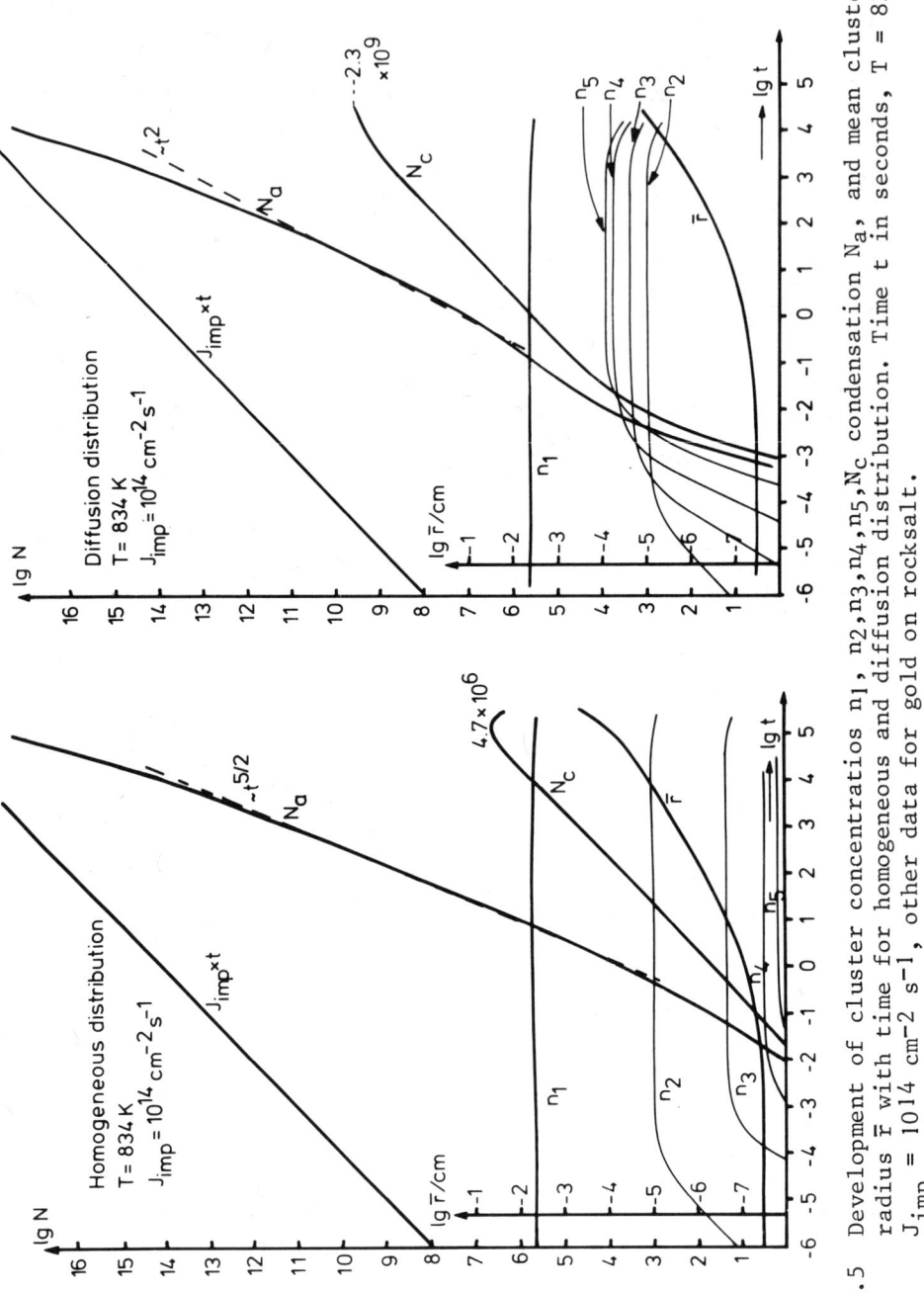

Fig. 5 Development of cluster concentratios n_1, n_2, n_3, n_4, n_5, N_c condensation N_a, and mean cluster radius $\bar{r}$ with time for homogeneous and diffusion distribution. Time t in seconds, T = 834K $J_{imp} = 10^{14}$ cm^{-2} s^{-1}, other data for gold on rocksalt.

GROWTH PROCESSES

Because the number of clusters is stationary here, an appropriate stationary nucleation theory would be able to predict the nucleation rate in both cases. The saturation effects and the maximum cluster concentration could however not be treated by such theories. The numerical result is, that the maximum cluster concentration N_{csat} would be smaller by a factor of 10^3 in the homogeneous case, the cluster radius would be slightly larger. The incomplete condensation would obey law $\sim t^{5/2}$, or $\sim t^2$ respectively, if the effect of direct impingement on stable clusters were neglected; the direct impingement obscures this difference however.

A rather complete impression on the development of the total cluster concentration N_c for one impingement rate (10^{14} cm^{-2} s^{-1}) can be gained from Fig. 6. Generally the maximum cluster concentrations N_{csat} are larger in the diffusion case. But as long as the critical nucleus is smaller than a trimer, no other remarkable differences exist between the two models. A slight, though perhaps experimentally remarkable difference is the somewhat stronger dependence of the nucleus concentration on time for the diffusion distribution at low temperatures. The differences for larger critical nuclei will be discussed later.

The development of the condensation for several temperatures is shown in Fig. 7. Both cases look very much alike, so that an experimental decision on the mechanism from condensation measurements for gold on rocksalt is very difficult. The general behaviour is in agreement with experimental results ; no condensation experiments are known for the system Au/NaCl.

From Figs. 4 to 7 it is seen, that the growth of a film may be crudely characterised by the maximum rate of nucleation $\dot{N}_{cmax}$, which is simply denoted $\dot{N}_c$ in the following figures , by the maximum cluster concentration N_{csat} and by the mean cluster radius $\bar{r}_{sat}$ at the time of maximum cluster concentration. The maximum nucleation rate is well defined theoretically, its determination might however be difficult in experiments, which are made in the capture regime. So the number of nuclei after one second might be experimentally a more useful parameter. In the desorption regime both numbers agree.

The dependence of N_{csat}, $\dot{N}_{cmax}$, and $\bar{r}_{sat}$ is plotted against the impingement rate for both monomer distributions in Fig. 8. It has been predicted for the homogeneous case[9,5] that $(\partial \ln N_c / \partial \ln J_{imp})_T = i^* + 1$, which is also obtained in the left part of Fig. 8 for 834 K. The corresponding curve in the right part does however not show this dependence. The difference is again caused by the local enhancement of the monomer concentration around a subcritical cluster. It is also observed that the maximum cluster concentration is generally larger in the diffusion case, especially so at

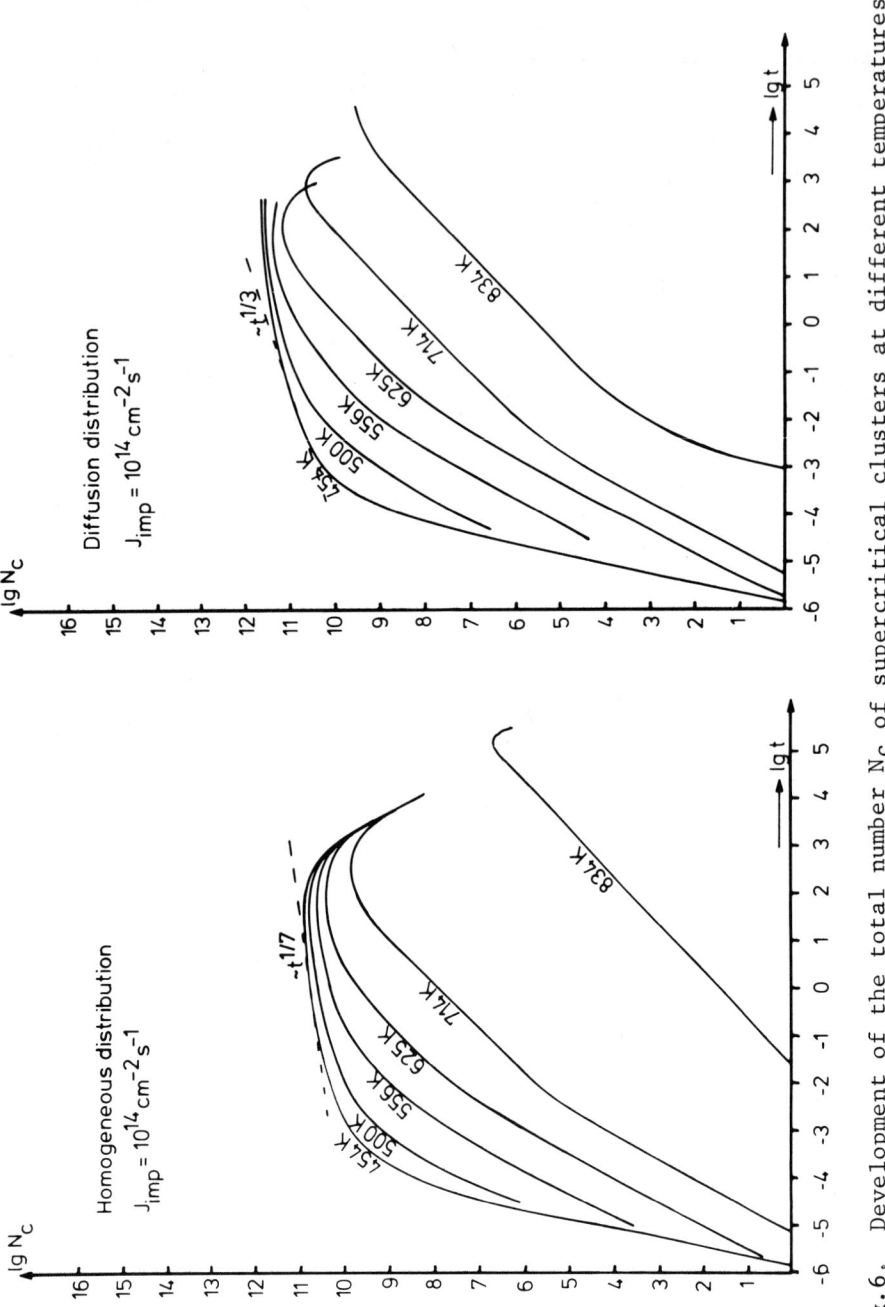

Fig.6. Development of the total number N_c of supercritical clusters at different temperatures. Time t in seconds, $J_{imp}=10^{14}$ cm^{-2} s^{-1}, other data for gold on rocksalt.

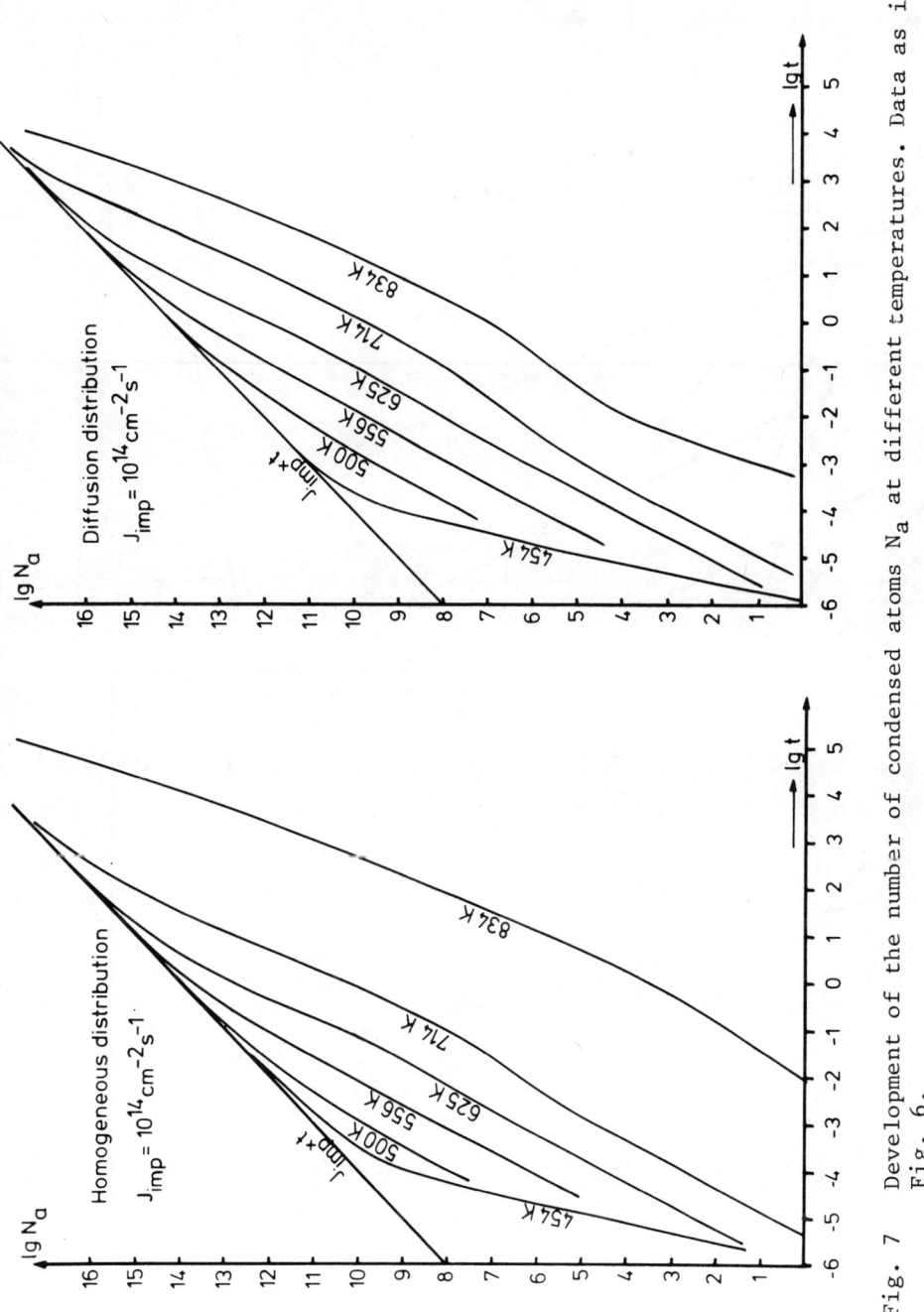

Fig. 7 Development of the number of condensed atoms N_a at different temperatures. Data as in Fig. 6.

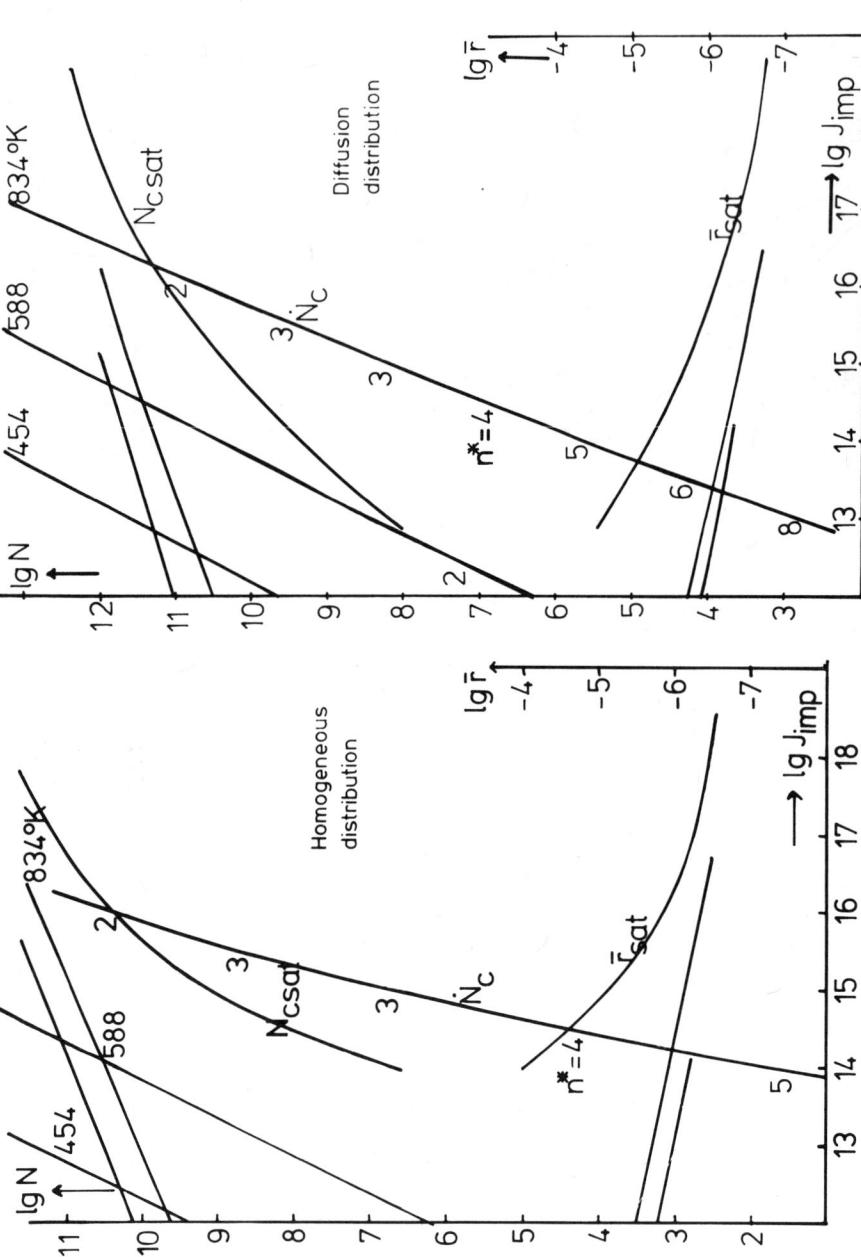

Fig. 8. Dependence of the maximum cluster concentration N_{csat}, maximum nuclearion rate $\dot{N}_c$, mean cluster radius $\bar{r}_{sat}$ on the impingement rate for three temperatures.

elevated temperatures, and that the clusters are smaller in the diffusion case. The nucleation rates are almost equal for small critical clusters in both cases, which is obviously caused by the fact, that the nucleation rate is equal to the rate of twin formation at high supersaturations.

The size of the critical cluster does not influences directly any of the measurable quantities in a consequent diffusion theory. If this theory applies, no means is visible, by which the size of the critical cluster could be determined, the Walton slope being $2 < (\partial \ln \dot{N}_c / \partial \ln J_{imp}) < 3$ for any critical cluster size. The local variations in the monomer concentration, which have been discussed in § 2.2, tend to smooth the effect of the size dependent cluster decay. This "diffusion barrier" is in many respects equivalent to a constant small critical size of $i^* < 2$, which has been postulated in previous analyses.

Corresponding observations are made in the plots of the dependence of the characteristic growth variables on $1/T$, fig. 9. The maximum cluster concentrations do not depend appreciably on temperature except at high temperatures in both cases. The dependence in the homogeneous case is stronger than in the diffusion case. The slope of $\dot{N}_c$ is for the homogeneous case given by the relation (9) $(\partial \ln \dot{N}_c / \partial (1/kT)) = -V_{i^*_o} + E_{des} - E_{diff}$; in the diffusion case no simple interpretation can be given.

The number of growth defects in a thin film will be connected to the maximum cluster concentration, so a low value of N_{csat} may be desirable. From the combination of nucleation rate and maximum cluster concentration shown in Fig. 9 it can be concluded, that this is achieved in shorter times for high temperatures and high impingement rates, i.e. the preparation of films with low defect concentration should be tried at high temperatures and appropriate impingement rates rather than by slow evaporation.

4.2. Comparision with experiments

Comparison of the considerations presented here with experiments is possible in several respects.

1. It is possible to predict the general mode of growth by application of eq.(43) to the system investigated. This has been done in some cases and relation (43) has been justified, as can be seen from Fig. 10.
The validity of this criterium is important for the qualitative understanding of film growth.

2. In the case of cluster growth a comparison with extended experiments is possible. Both the time dependence of N_a, N_c and the rela-

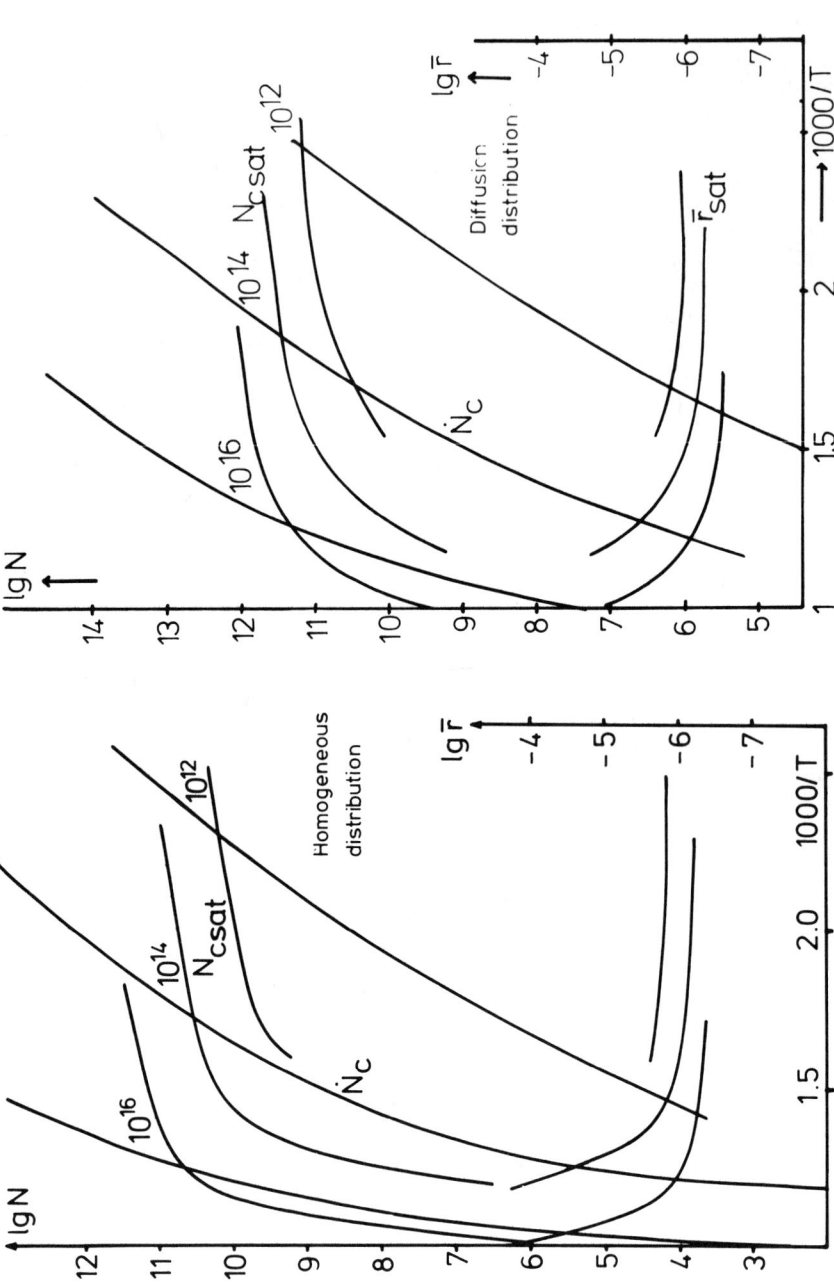

Fig. 9. Dependence of N_{csat}, $\dot{N}_c$, $\bar{r}_{sat}$ on substrate temperature T for the impingement rates 10^{12}, 10^{14}, 10^{16} cm^{-2} s^{-1}.

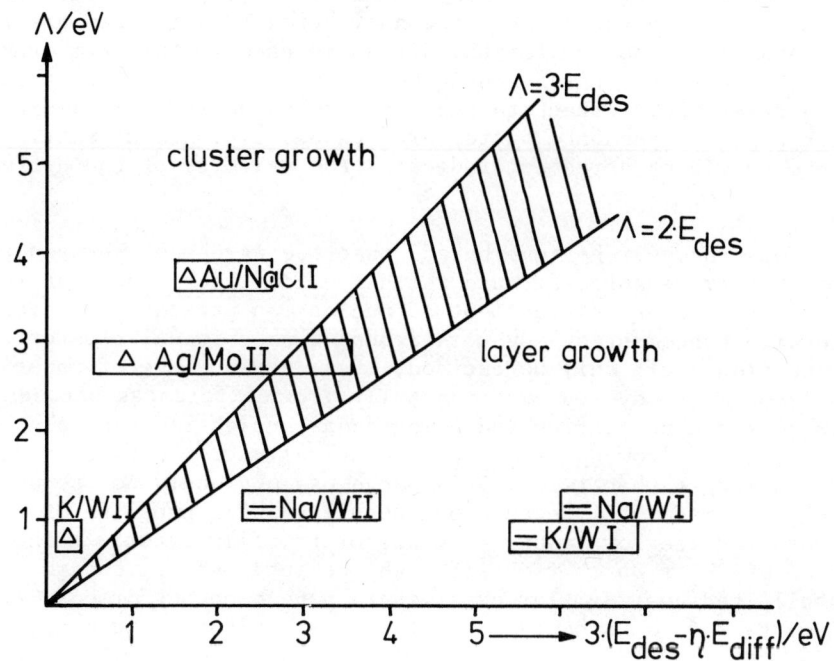

Fig. 10. Modes of growth, experimental : three dim. growth (Δ), two dim. growth (═══), uncertainties in the experimental binding energies and η are indicated by the width of the bars, border region extended because of uncertainty in η.

tions of Figs 8,9. can be used. Values of the binding energies can be chosen for best fit between theory and experiment and the overall behaviour of the experimental values can be used to check the theory. An example for this is shown in Fig. 11, where the results of Robins and coworkers are plotted together with the theoretical curves.

The experiments selected here have been made for the growth of gold on rocksalt. They are thus limited to temperature T < 650 K, because rocksalt begins to evaporate at higher temperatures. From Figs. 8,9 it can be seen, that the nucleation rate is the same for the homogeneous case and for the diffusion case in this temperature range. The maximum cluster concentration is however different for the two cases with respect to the absolute value as well as with respect to the dependence on temperature and impingement rate. Thus the excellent agreement indicates the validity of the diffusion case.

It must however be remembered, that the agreement between these and similar measurements and the diffusion theory does not really proof the diffusion treatment of nucleation presented here, i.e. the local enhancement of monomers around a subcritical cluster. This question could only be decided, if a larger range of supersaturations could be covered experimentally. For differences between diffusion governed nucleation and homogeneous nucleation can only be observed, where from E_{des}, E_{diff}, Λ critical nuclei can be expected, which are larger than twins. This range can obviously not be covered for the system Au/NaCl because of the high vapour pressure of the substrate and the low vapour pressure of the film material. The evaporation of Pb or Zn on rocksalt might be interesting, though experimentally tedious. Similar experiments have been performed[24], the amount of data is however not large enough for valid conclusions

5. FINAL REMARKS

It has been shown in these lectures, that the process of film growth can be described theroretically in some detail, starting from rather fundamental data and assumptions. The numerical results agree with the experimental findings. They are not very sensitive to the details of the nucleation theory.

The present evaluation of the size of the critical nucleus is only valid for homogeneous nucleation and in view of the interpretation of the experimental data the validity of the diffusion governed nucleation mechanism and its associated "diffusion barrier" should be investigated.

The question of epitaxial growth is closely connected to the

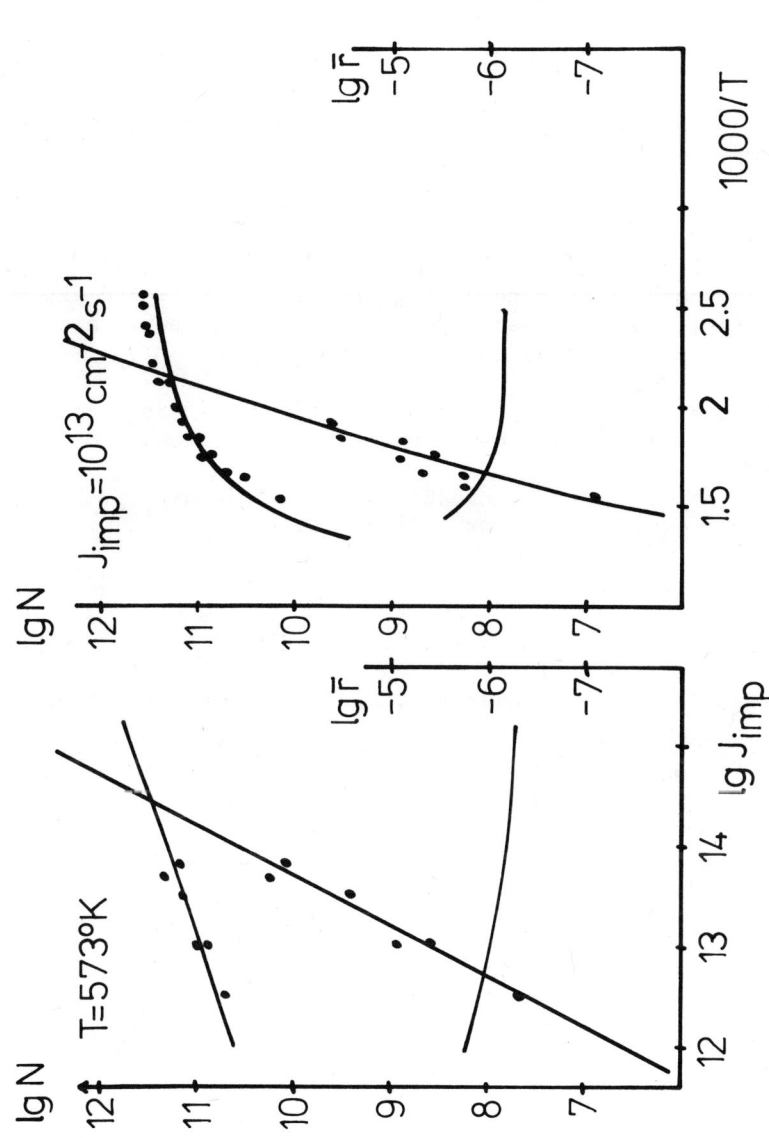

Fig. 11. Comparison between experimental and theoretical values for the nucleation rate and the maximum nucleus concentration for gold on rocksalt (3).

phenomena discussed here. Its solution depends certainly on the details of the nucleation mechanism and on the relative importance of cluster rotation.

REFERENCES

1. J.A. Venables, G.L. Price in Epitaxy, chapter 4
   J.W. Matthews ed. Academic Press (in print)

2. J.A. Venables, Phil. Mag. $\underline{27}$, 697, (1973).

3. V.N.E. Robinson, J.L. Robins, Thin Sol. Films, $\underline{20}$, 155,(1974)
   I am very much obliged to J.L. Robins for unpublished experimental data, which I used in addition to those published in[3] and in the other papers quoted there.

4. J.J. Métois, M. Gauch, A. Masson, R. Kern, Thin Sol. Films. $\underline{11}$, 205, (1972) and several other papers of this group.

5. R. Niedermayer in Advances in Epitaxy and Endotaxy, p. 21.
   H.G. Schneider, V. Ruth eds. Leipzig 1971.

6. J.I. Frenkel, Kinetische Theorie der Flüssigkeiten, Berlin 1957

7. V. Halpern, J. Appl. Phys. $\underline{40}$, 4627, (1969).

8. B. Lewis, D.S. Campbell, J. Vac. Sci. Tech., $\underline{4}$, 209, (1967).

9. D. Walton, J. Chem. Phys., $\underline{37}$, 2182, (1962)
   D. Walton, T.N. Rhodin, R.W. Rollins, J. Chem. Phys., $\underline{38}$, 2698 (1963).

10. R. Niedermayer, Habilitationsschrift Clausthal (1969).

11. B. Lewis, Thin Sol. Films $\underline{1}$, 85, (1967).

12. F.C. Frank, J.H. van der Merwe, Proc. Roy. Soc. $\underline{198}$, 205, 217 (1949).

13. J.A. Snyman, J.H. van der Merwe, Surface Sci. $\underline{42}$, 190, (1974) and Surface Sci. $\underline{45}$, 619, (1974).

14. R. Niedermayer, Thin Films $\underline{1}$, 25, (1968)

15. A. Mlynczak, R. Niedermayer, Thin Sol. Films. July 1975 (in print)

16. K.C. Wolf, Physik und Cemie der Grenzläche, Berlin (1957).

17. E. Bauer, Z. Kristallogr. 110, 372, 395, (1958).

18. M. Drechsler, N.F. Nicholas, J. Phys. Chem. Solids, 28, 2609, (1967).

19. P.W. Steinhage, H. Mayer, Thin Solid Film, July 1975 (in print).

20. K. Hartig, Thesis Bochum 1975

21. K.J. Routledge, M.J. Stowell. Thin Solid Films, 6, 407, (1970).

22. G. Zinsmeister in Basic Problems in Thin Film Physics. R. Niedermayer, H. Mayer eds., Göttingen (1966).

23. R. Becker, W. Döring, Ann.d.Phys., 719 (1935).

24. H. Poppa, J. Appl. Phys., 38, 3883, (1967).

ELECTRONIC STATES IN SEMICONDUCTORS

D. A. Greenwood

H. H. Wills Physics Laboratory, University of Bristol

Bristol BS8 1TL, UK

1. INTRODUCTION

In these lectures, we shall be discussing the nature of electronic states in crystalline and amorphous semiconductors. The Bloch states of the perfect infinite crystals are well known ; we shall be more concerned with what happens when periodicity breaks down, as it does in amorphous semiconductors, and as it does at the surface of a crystal. Finally, we shall outline some of the consequences of this electronic structure. It is intended to stress the physical principles involved, rather than details of the mathematical formalism. Indeed, for some of the problems we shall consider there is, as yet, no very adequate formalism available.

2. THE ONE-ELECTRON HAMILTONIAN

'Many-body' aspects of electronic structure will not be considered. We shall use the one-electron description for the valence electrons of a semiconductor, in which the electronic states ψ_n are solutions of a Schrödinger equation,

$$\{-\nabla^2 + V(\underline{x})\} \psi_n = \varepsilon_n \psi_n, \qquad (2.1)$$

and the potential $V(\underline{x})$ incorporates exchange and correlation effects. (We use atomic units : the unit of length is the Bohr radius $a_0 = \hbar^2/mc^2$ and the unit of energy the Rydberg , 1 Ry = $e^2/2a_0$ = 13.6 eV. In these units, $\hbar = 2m = e^2/2 = 1$).

If the potential $V(x)$ is periodic, we have to hand all the familiar methods, based on Bloch's theorem, for obtaining the energy bands. In practice, a good representation of the band structure of silicon and germanium can be obtained taking $V(x)$ is a sum of pseudopotentials centred on the ion sites X_i [1] :

$$V(x) = \sum_i u_{ps}(x - X_i) \qquad (2.2)$$

The approximation in which $V(x)$ is of 'muffin-tin' type will also be useful. In this approximation, $V(x)$ is a sum of spherically symmetrical, non-overlapping, potentials centred on the ion sites, with a constant interstitial potential :

$$V(x) = V_M + \sum_i u(|x - X_i|) \qquad (2.3)$$

A 'first principles' muffin-tin potential can be constructed by the Mattheis recipe of superposing neutral atomic charge distributions and adding an exchange potential term[2]. This method incorporates into the potential the local environment of an ion to a considerable extent, but we should not expect it to be completely satisfactory in a semiconductor without some further adjustment. The diamond type lattice is a very open structure : even if the muffin-tins touch, the interstitial, constant potential, region fills about two thirds of the volume. Nevertheless, the approximation is, both computationally and conceptually, a useful one.

In writting down a Schrödinger equation, we are adopting a physicist's approach. The energy band structure appears as a consequence of the translational symmetry of the lattice : band gaps are seen as Bragg reflections. A chemist will regard the tetrahedrally coordinated lattice of a group IV semiconductor as a large covalently bonded molecule, and interpret its electronic structure in terms of bonding and anti-bonding orbitals. They are constructed from the familiar sp^3 hybridized atomic orbitals of the theoretical chemist (Fig. 1). In this approximation, the electron states are linear combinations of orthogonal directed orbitals $|\alpha i\rangle$ at each site α, with $i = 1, 2, 3, 4$:

$$|\psi\rangle = \sum_{\alpha,i} a_{\alpha i} |\alpha i\rangle \qquad (2.4)$$

Then, in the simplest approximattion, a full valence band corresponds essentially to doubly occupied bonding orbitals. An additional electron must go into an anti-bonding orbital, which has a higher energy. Hence there is an energy gap. From this point of view, the energy gap is more a consequence of the tetrahedral environment of each ion, than the long-range order of the crystal

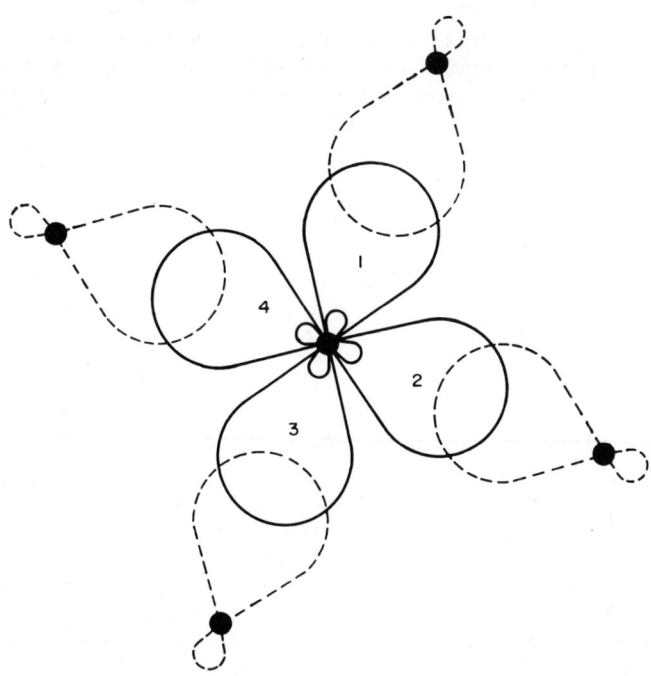

Fig. 1 Pairing of directed orbitals.

lattice. However, as we shall see later, it is not really possible to obtain a good representation of the band structure without extending the range of basis states very considerably, and the conceptual simplicity of the picture is then lost.

Retaining only the basis states of (2.4) the simple two-parameter model of Weaire[3] assumes that orbitals on different sites are also othogonal, and neglects all matrix elements of the Hamiltonian except those between orbitals on the same site,

$$< \alpha i \,|H|\, \alpha j > = V_1, \; i \neq j \quad \text{and} \quad < \alpha i \,|H|\, \alpha i > = 0,$$

and those between orbitals making a bond between neighbouring ions :

$$< \alpha i \,|H|\, \beta j > = V_2, \; \text{for } \alpha i, \beta j \text{ making a bond pair}$$

In Dirac notation, we can write

$$H = \sum_{\alpha, i \neq j} V_1 \,|\alpha i><\alpha j| \;+\; \sum_{\text{bonds}} V_2 |\alpha i><\beta j| \qquad (2.5)$$

In fitting V_1, V_2 to a real semiconductor, both parameters are negative ; the first term, for an isolated atom, gives a splitting $4V_1$, between s and triply-degenerate p orbitals, and the second term gives a splitting $2V_2$ between a bonding and an anti-bonding orbital. The Hamiltonian (2.5) formalizes a simple 'chemical' approach.

3. AMORPHOUS SEMICONDUCTORS

The band structure of crystalline semiconductors is well understood. How far can we understand electronic states in amorphous semiconductors ?

We need, first, a model for an amorphous structure. Amorphous films have been extensively investigated experimentally. The interpretation of the experiments is not universally agreed, but the continuous random network model seems favoured over alternative micro-crystallite models[4]. It is possible to generate (either by hand or by computer) networks in which each ion sits in a locally ordered, nearly perfectly tetrahedral, environment. Bond lengths and bond angles are slightly ($\lesssim 5\%$) distorted from their corresponding perfect crystal values, and the number of atoms in a ring, which in a diamond or wurtzite structure is always 6, can be either 5 or 6. Such random networks can be continued indefinitely with no long range order[5]. Real amorphous films probably differ from the ideal, completely saturated, random network by the presence of voids.

The mathematical description of the random network - other than by the enumeration of 10^{23} coordinates - raises difficulties. The pair correlation function is known experimentally, but this is far from sufficient. Third and fourth order correlation functions are necessary to describe the local tetrahedral structure, and these are not known either experimentally or theoretically. Since we are dealing with a random structure, the only quantities we can hope to calculate are ensemble averages. Thus we can set out to calculate the density of electron states,

$$g(\varepsilon) = \langle \frac{2}{\Omega} \sum_n \delta(\varepsilon - \varepsilon_n) \rangle \qquad (2.6)$$

where Ω is the volume of the system, and a factor of 2 for electron spin is included. The brackets denote an average over an ensemble of representative random networks. But we cannot calculate a particular wave function ψ_n. The density of states is the simplest physically significant quantity we can calculate ; it is related immediately to optical and transport properties. In particular, we wish to understand the presence of a gap or quasi-gap, in the density of states, which experiment clearly shows to be there[4].

ELECTRONIC STATES IN SEMICONDUCTORS

Making the approximation that the potential at all sites in the random network is the same, the Hamiltonian (2.1) and either (2.2) or (2.3) can be used, with X_i now given by the random network model. The Hamiltonian (2.5) also holds for an amorphous structure, differing from that for a crystalline structure in the enumeration of the bonds.

4. MULTIPLE SCATTERING APPROACH TO THE DENSITY OF STATES

The Weaire model Hamiltonian does indeed have a gap in the density of states for all structures, crystalline or amorphous. This can be rigorously proved, but the gap has so very much been built into the model that it is not completely satisfying an explanation. How can a gap arise if we start from (2.1), which is a more fundamental description?

One can think of the band-structure problem as a multiple scattering problem, simplified by the periodicity of the crystal. The KKR method of band-structure calculation exemplifies this, and it seems natural to adopt this approach to the amorphous case. The muffin-tin potential (2.3) is most convenient, and we set $V_M=0$ as zero of energy. In a semiconductor the energy gap region in which we are interested lies above the muffin-tin zero, so we consider only scattering at energies $\varepsilon = k^2 > 0$. Thus electrons propagate freely in the interstitial region, and are scattered by the muffin-tin potentials.

The scattering of a plane wave by a single muffin-tin potential is described by the <u>phase shifts</u> $\eta_\ell(\varepsilon)$. Given the muffin-tin potential $u(r)$, these are determined by integrating the radial Schrödinger equation out to the muffin-tin radius R_M, and matching to the appropriate free-space solution:

$$\left\{\frac{1}{R_\ell}\frac{dR_\ell}{dr}\right\}_{r=R_M} = \frac{k[\cos\eta_\ell j_\ell' - \sin\eta_\ell n_\ell']_{r=R_M}}{[\cos\eta_\ell j_\ell - \sin\eta_\ell n_\ell]_{r=R_M}} \quad (4.1)$$

Equation (4.1) we can solve for $\tan\eta$. It is possible to chose $\eta_\ell(\varepsilon) = 0$ at $\varepsilon = 0$. Fig. 2 shows phase shifts calculated for crystalline germanium. Phase shifts reflect atome structure. In a semiconductor both s- and p-phase shifts are large at the Fermi energy. The d-phase shift is small, and higher phase shifts are negligible.

Conversely, from (4.1) the logarithmic derivative of the radial wave function is specified by the phase shifts. The eigenstates of the multiple scattering problem can be constructed by matching the wave function in the interstitial region, which will be some superposition of the plane waves, to solutions inside the

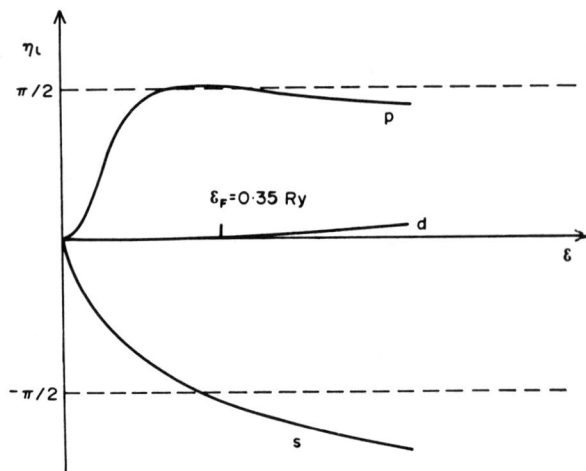

Fig. 2   Phase shifts for germanium (from unpublished calculations of P.V. Smith).

muffin-tins. Thus in principle the eigenstates of the multiple scattering problem are determined by the positions X_i, and the phase shifts $\eta_\ell(\varepsilon)$, of the muffin-tin potentials. A formalism which makes this explicit has been developed by Lloyd[6]. To understand the physics of its application to amorphous semiconductors we review a little more scattering theory.

A wave packet, scattered by a potential, is delayed in the neighbourhood of the scatterer by a time τ_W, given by

$$\tau_W = 2 \frac{d\eta}{d\varepsilon} \tag{4.2}$$

This is the <u>Wigner delay time</u>[7]. If $\eta(\varepsilon)$ is a rapidly varying function of energy, τ_W can become very large, as near a 'scattering resonance' where an electron enters a metastable state. Negative values of τ_W are possible also, but are restricted by causality : a wave packet cannot be advanced in space by a range greater than twice the range of the potential, so that

$$\tau_W > - \frac{2R_M}{u} = - \frac{R_M}{k} \tag{4.3}$$

We shall call a large negative value of τ_W an 'anti-resonance'.

ELECTRONIC STATES IN SEMICONDUCTORS

The Friedel sum rule is easily understood in terms of the Wigner Delay time. Consider a single scatterer in a volume Ω. An incoming flux of electrons of angular momentum ℓ and energy $\varepsilon = k^2$ is delayed by τ_W before being converted to outgoing flux, so that there is an accumulation (or deficit) of electron density in the neighbourhood of the scatterer, proportional to $d\eta/d\varepsilon$. At large distances from the scatterer, the electron density must remain the same. A straightforward calculation gives for the integrated density of states $N(\varepsilon)$:

$$N(\varepsilon) = \int_0^\varepsilon g(\varepsilon)\, d\varepsilon = N_0(\varepsilon) + \frac{1}{\Omega}\frac{1}{\pi}\sum_\ell (2\ell+1)\eta_\ell(\varepsilon), \qquad (4.4)$$

which is the well known Friedel sum rule. Here $N_0(\varepsilon)$ is the free electron contribution. Thus

$$g(\varepsilon) = g_0(\varepsilon) + \frac{1}{\Omega}\frac{2}{\pi}\sum_\ell (2\ell+1)\frac{d\eta_\ell}{d\varepsilon} \qquad (4.5)$$

where $g_0(\varepsilon) = \varepsilon^{1/2}/2\pi^2$.

If there are N scatterers in the volume Ω, and we neglect interference between scatterers, a zeroth order approximation to the density of states is

$$g(\varepsilon) = g_0(\varepsilon) + \frac{N}{\Omega}\frac{2}{\pi}\sum_\ell (2\ell+1)\frac{d\eta_\ell}{d\varepsilon} \qquad (4.6)$$

This is a crude approximation. Nevertheless, if a scattering resonance is important (as in a transition metal) it immediately picks out the resonant contribution to the density of states.

The basic idea of the cluster scattering method of McGill and Klima[8] can now be very simply explained : a generalized form of the Friedel sum expression (4.4) is applied, but with <u>clusters of ions</u>, rather than single ions, as the scattering elements. The clusters are chosen to represent typical configurations found in the random network model. Thus the <u>short range order</u> inside a cluster is treated exactly. The size if clusters which can be considered is limited by the resulting computational problem. Scattering between clusters is neglected, but since there is no long range order there should be some partial cancellation of multiple scattering effects.

The actual calculations use Lloyd's result[6], which replaces the evaluation of generalized phase shifts from a cluster by the evaluation of a determinant, in a mixed position-angular momentum representation. If ω_α is the volume of a cluster occuring with weight p_α

$$< N(\varepsilon) > = \frac{\varepsilon^{3/2}}{3\pi^2} - \sum_\alpha p_\alpha \frac{2}{\pi \omega_\alpha} \text{Im log}\{\det\| \delta_{LL'} \delta_{ij} + G_{LL'}(\underset{\sim}{X}_i - \underset{\sim}{X}_j) k_L^j(\varepsilon) \| \} \quad (4.7)$$

In this expression, $k_L^j(\varepsilon) = -\tan \eta_\ell^j(\varepsilon)/k$ depend on the muffin-tin phase shifts at the site $\underset{\sim}{X}_j$ and the motion of electrons between sites $\underset{\sim}{X}_i$ and $\underset{\sim}{X}_j$ is described by the 'propagator'

$$G_{LL'}(\underset{\sim}{R}) = \begin{cases} ik \sum_{L''} 4\pi i^{\ell''} h_{\ell''}^+(kR) Y_{L''}(\hat{\underset{\sim}{R}}) c_{LL'}^{L''}, & |\underset{\sim}{R}| > 0 \\ ik \delta_{LL'}, & |\underset{\sim}{R}| = 0 \end{cases} \quad (4.8)$$

where

$$c_{LL'}^{L''} = \int Y_L(\Omega) Y_{L'}^*(\Omega) Y_{L''}(\Omega) d\Omega \quad (4.9)$$

are integrals over spherical harmonics.

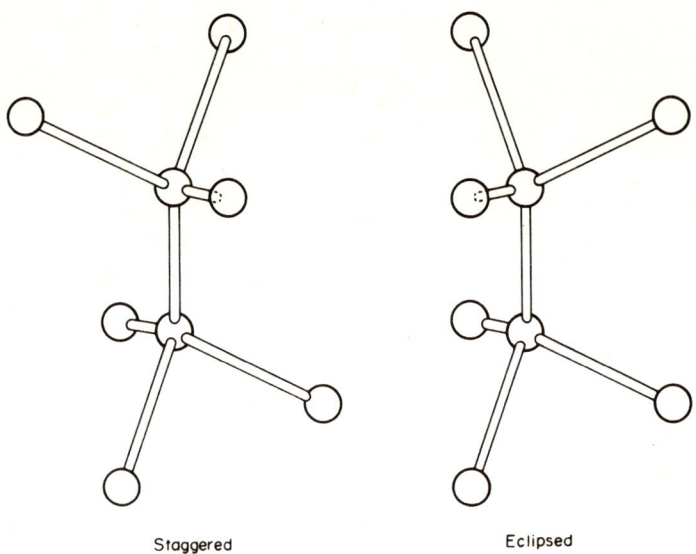

Staggered   Eclipsed

Fig. 3   Clusters of eight atoms in the staggered and eclipsed configuration.

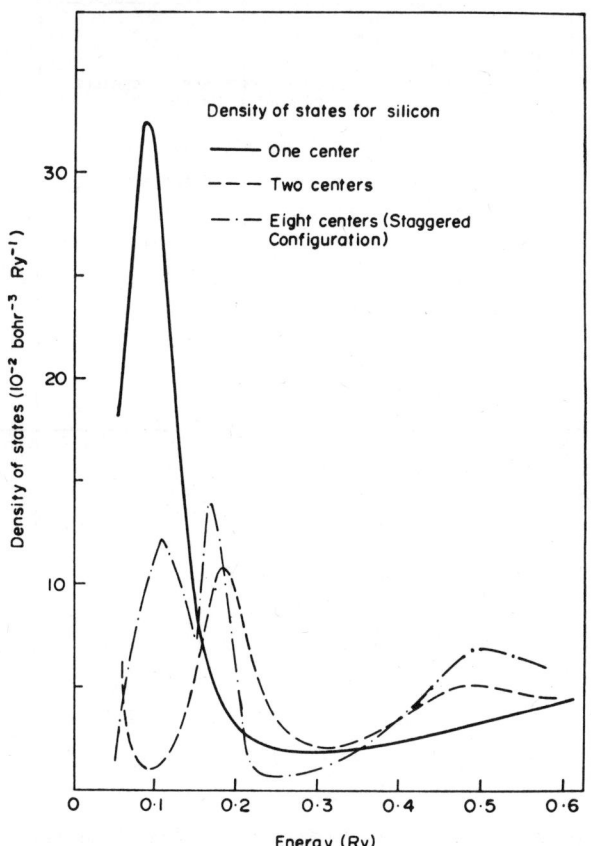

Fig. 4 Density of states for clusters of one, two and eight Si sites as a function of energy(8).

McGill and Klima took clusters of up to eight atoms in 'staggered' and 'eclipsed' configurations (Fig. 3), and performed calculations for C, Si and Ge. Results for silicon are shown in Fig. 4. Note the peak in the single atom cluster result, which corresponds to the sharp rise of the p-phase shift at this energy. The remarkable feature of these results is that, as the size of the cluster increases, a dip appears, and the density of states falls below its free-electron value. This pseudo-gap is of about the same width as the energy gap in a perfect crystal, but appears at a rather higher energy. This is shown more clearly in Fig. 5. The phase shifts used in the calculations were adjusted slightly to give a reasonable band structure for the crystal.

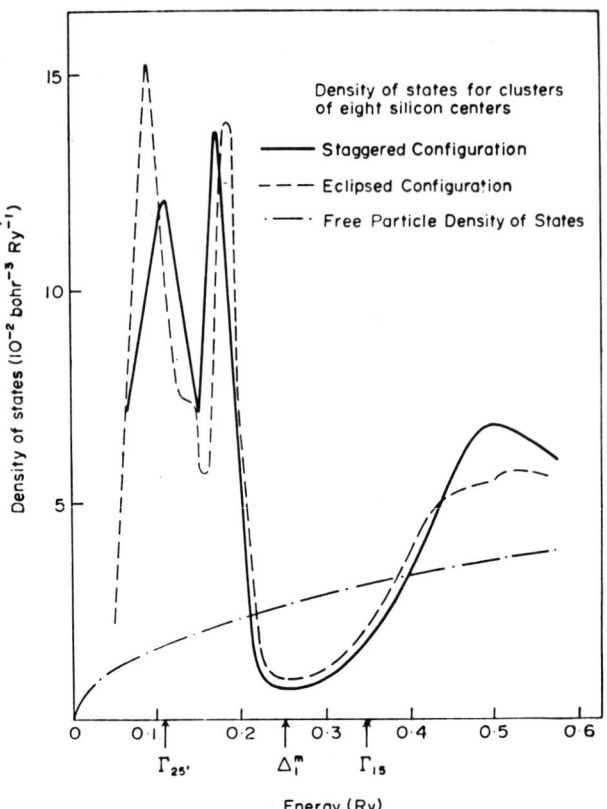

Fig. 5 Density of states for clusters of eight Si sites in the staggered configurations, and the free particle density of states, as a function of energy[8].

Calculations on larger crystals, of up to thirty carbon atoms, have been made by Keller[9]. Fig. 6 shows the behaviour of the pseudo-gap as a function of cluster size. Keller also found that the gap tended to disappear if clusters in which bond angles were highly strained were used.

These results show conclusively that short range order is sufficient alone to give a pseudo-gap. The valence and conduction bands can be thought of as resulting from resonances in the scattering by a cluster. In the gap, an 'anti-resonance' makes it extremely difficult for an electron to propagate through the crystal. It is, therefore, possible to comprehend the electronic density of states in an amorphous semiconductor from the multiple scattering point of view. The question of whether a true energy gap in the density of states should appear in an infinite random network is not one which

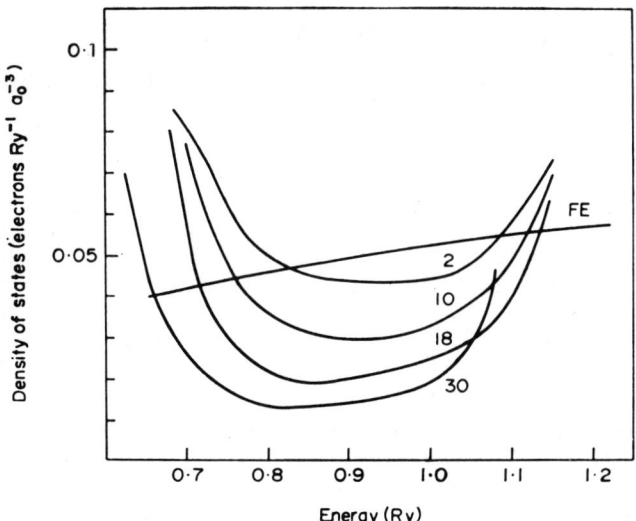

Fig. 6 Electronic density of states in the "energy gap" region for clusters of 10, 18 and 30 carbon atoms. The dashed line is the free-electron density of states[9].

computation can answer. Some way of incorporating inter-cluster scattering into the calculation is necessary so that the results become sufficiently accurate to compare with experiment. Recent analytic work by Jones[10] based on (4.7) is promising and shows how the free electron part of the density of states can be cancelled by a part from the logarithm, but the effect of remaining terms is not completely elucidated.

The Weaire Hamiltonian (2.5) shows that one can construct models in which a true gap appears between the valence band and conduction band, though a more realistic model in which the parameters V_1 and V_2 are allowed some degree of variation would lead to a tailing of states into the gap. Phillips has suggested that a carefully prepared amorphous semiconductor relaxes in such a way as to exclude states appearing in the gap, in order to minimize bond energies (11).

Theory cannot yet give an unambiguous answer to the question of the sharpness of the band edges at the gap. The interpretation of the experimental data is a matter of controversy. Paul et al[4] suggest that the Fermi level is pinned at the centre of the gap, and relative to this energy

$$g(|\Delta E|) = g(0) \exp\frac{|\Delta E|}{E_S}$$

with $E_S \sim 0.1$ eV and $g(0) \sim 10^{19}$ cm^3 eV^{-1}. Much sharper band edges - as sharp as those of a crystal- have been reported by Donovan and Spicer[12] and co-workers, but a tailing of states into the gap in an amorphous semiconductor is undoubtedly easier to accept theoretically.

5. GERMANIUM AND SILICON POLYMORPHS

The multiple scattering calculations, or calculations based on a simple model Hamiltonian, give only broad qualitative features of the density of states in amorphous semiconductors, whereas band structure calculations for perfect crystals give more detailed density of state functions. Comparison of the density of states for different tetrahedrally bonded crystalline polymorphs of germanium and silicon which differ in the nature and complexity of their unit cell, can, it is thought, give useful information on the relation between electronic structure and short range order.

This is an attractive idea since we have already established the dominant role of short range order. A drawback appears to be the technical difficulty of predicting reliable band structures in the absence of experimental data to fit the parameters of the empirical pseudopotential. It is necessary to interpolate between values of the pseudopotential form factor known for the diamond structure. (It is assumed that the pseudopotential is independent of the structure). Aymerich and Smith[13], in their study of the hypothetical germanium wurtzite structure, found that the calculated energy gap was very sensitive to the method of interpolation employed, varying between 0.11 eV and 0.60 eV for apparently equally reasonable interpolations, and presumably there are similar uncertainties in other work.

Calculations for the GeIII polymorph are most relevant to amorphous germanium. This is a metastable structure under normal conditions, but little is known about its electronic structure experimentally. In the Ge III structure there are 5-fold as well as 6-fold rings, and distorted tetrahedral bonding, so that in some respects it is similar in its local structure to amorphous germanium. The <u>detailed structure</u> in the density of states is due to long-range order. After smoothing, the forms of the valence band and conduction band are very similar to those determined experimentally[14].

ELECTRONIC STATES IN SEMICONDUCTORS

This appears at the moment to be the most promising technique for calculating densities of states which can be used in the detailed interpretation of experimental data, though of course it says nothing about the tailing of states into the gap, or the nature of the wave functions, in the amorphous material.

6. THE NATURE OF THE ELECTRON STATES

To discuss the nature of the wave functions in an amorphous semiconductor, it is useful to return to the tight-binding chemical approach in which states are constructed out of bonding and antibonding orbitals. This is a reasonable description for the valence band, but quite inadequate for the conduction band. We, therefore, concentrate on the valence band, in which the wave function can be constructed out of bonding orbitals alone to a good approximation :

$$|\psi\rangle = \sum_{\text{bonds}} b_{\alpha i} | B, \alpha i \rangle \qquad (6.1)$$

where

$$| B, \alpha i \rangle = \frac{1}{\sqrt{2}} (|\alpha i\rangle + |\beta j\rangle)$$

and $|\beta j\rangle$ is the state pairing with $|\alpha i\rangle$.

In the Weaire model, at the bottom of the valence band, bonding orbitals are all combined with the same sign, so that the interaction energy $V_1/2$ (< 0) between bonding orbitals at an atom is minimized. In a perfect crystal the Bloch functions near the bottom of the valence band are of the form

$$\psi_R(\underset{\sim}{x}) = \sum_p \exp(i\underset{\sim}{k} \cdot \underset{\sim}{X}_p) \chi_p(\underset{\sim}{x})$$

where $\underset{\sim}{X}_p$ is the centre of the pth bond orbital χ_p. Ziman has suggested (15) that in the amorphous crystal the electron states must be very similar. The wave vector $\underset{\sim}{k}$ is no longer a good quantum number, but it will be 'nearly good' and states constructed in this way may be a useful basis set. Such states are, clearly, extended throughout the crystal.

At the top of the valence band there are, in the perfect diamond structure, two bonds at each atom with coefficients of positive sign in the expansion (6.1), and two with negative sign. Thus maximum advantage is gained from 'anti-linking' matrix elements. The bands are triply degenerate at $k = 0$ (corresponding to the three ways in which four bonds can be divided into pairs). Ziman suggests that the bonds in an amorphous crystal attempt a 'best fit' to the arrangement on a perfect crystal. The triple degeneracy is probably preserved, and 'hole states' at the top of the valence band with a 'nearly good' quantum number $\underset{\sim}{k}$ will exist. Again, such states are not localized.

In the Weaire model, even if we include anti-bonding orbitals into the expansion, there is a tailing of states into the gap above the valence band only if V_1 and V_2 are allowed to vary. Such a variation would, of course, be very reasonable in an amorphous structure. We might also expect there to be 'dangling bonds' where there are deviations from the ideal random network model. Electron states associated with dangling bonds are probably localized near the structural defect, and will have energies lying in the gap.

7. SURFACE STATES

So far, we have been discussing the electronic states of bulk material, and have ignored effects which might arise from the fact that crystals, or amorphous films, have surfaces. We turn now to discuss what happens at the surface of a crystal.

It is perhaps instructive to begin with a one-dimensional model for a 'semi-infinite' crystal (Fig. 7). For x > 0 the potential

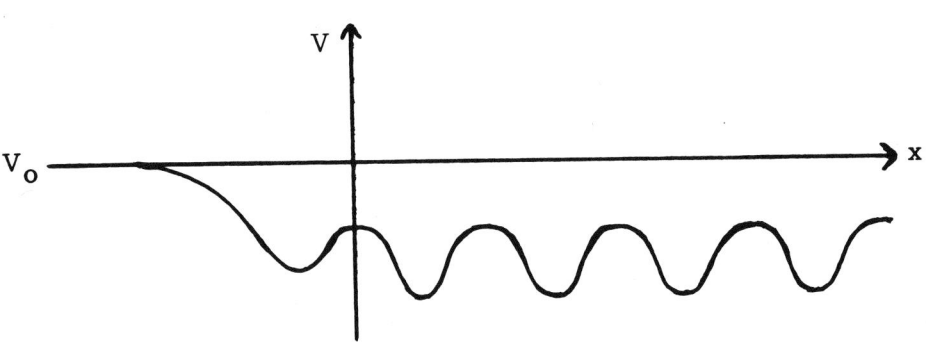

Fig. 7   One-dimensional periodic potential terminated by 'surface

ELECTRONIC STATES IN SEMICONDUCTORS

is that of a perfect crystal, with $V(x+a)=V(x)$. For $x < 0$ the potential may be terminated by some 'surface layer' before rising to the external vacuum potential V_0. In one dimension the Schrödinger equation is a second-order ordinary differential equation, and the discussion can be based on the well-known mathematical properties of such equations. At a given energy E the equation has two independent solutions. Finding appropriate eigenfunctions which satisfy the physical boundary conditions is essentially a <u>matching problem</u>. For an energy E ($<V_0$), integrating inwards from $x = -\infty$, our solution ψ_L must start like $\exp(Kx)$, where $K^2 = V_0 - E$. At $x = 0$, ψ_L must be matched to ψ_R, the solution for $x > 0$. Suppose now that E lies in an allowed energy band of the perfect crystal. Then we can write $\psi_R = \alpha\psi_k + \beta\psi_{-k}$, where ψ_k, ψ_{-k} are Bloch states of the perfect lattice satisfying

$$\psi_{\pm k}(x+a) = e^{\pm ika}\psi(x), \quad E(k) = E(-k) = E \quad (7.1)$$

Since we have two constants α, β at our disposal, we can always satisfy the matching condition, $\psi_L'/\psi_L = \psi_R'/\psi_R$, at $x = 0$. Hence, at any energy in the band there is always an eigenstate extending uniformly through the crystal and decaying away ouside the crystal, as we should expect.

Consider now an energy E lying in a bandgap. We can still find two independent solutions in the perfect lattice, $\psi_+(E)$, $\psi_-(E)$ such that

$$\psi_+(x+a) = \lambda_+\psi(x), \quad \psi_-(x+a) = \lambda_-\psi(x) \text{ and } \lambda_+\lambda_- = 1.$$

Here λ_+, λ_- are real. If $|\lambda_+| > 1$, then $|\lambda_-| < 1$. One solution decays exponentially to the right, and the other to the left. These solutions in the gap are '<u>evanescent solutions</u>'. In the infinite crystal they cannot be normalised, but in our semi-infinite crystal the solution $\psi_-(x,E)$ is physically acceptable for the region $x > 0$. It is not, in general, an eigenstate; at $x = 0$, $\psi_-'(E)/\psi_-(E)$ is determined and will not match $\psi_L'(E)/\psi_L(E)$ for an arbitrary energy in the gap. However $\psi_-'(E)/\psi_-(E)$ varies continuously with E (since the independent solutions of the Schrödinger equation from which ψ_- is constructed vary continuously with E), from its value at the bottom of the energy gap to its value at the top of the gap. For a lattice with reflection symmetry it is easy to show[16] that the state at one edge of a bandgap is odd, and at the other is even (e.g. $\cos(\pi x/a)$, $\sin(\pi x/a)$ in the nearly free electron limit). Hence $\psi_-'(E)/\psi_-(E)$ goes from zero to infinity across an energy gap. Outside the material ψ_L'/ψ_L is positive and smoothly varying with energy. One might therefore expect there to be just one energy in the gap at which the matching condition is satified (Fig. 8). Such an eigenstate lying in a bandgap is called a <u>surface state</u>, and an electron in this state is localized near the surface of the crystal.

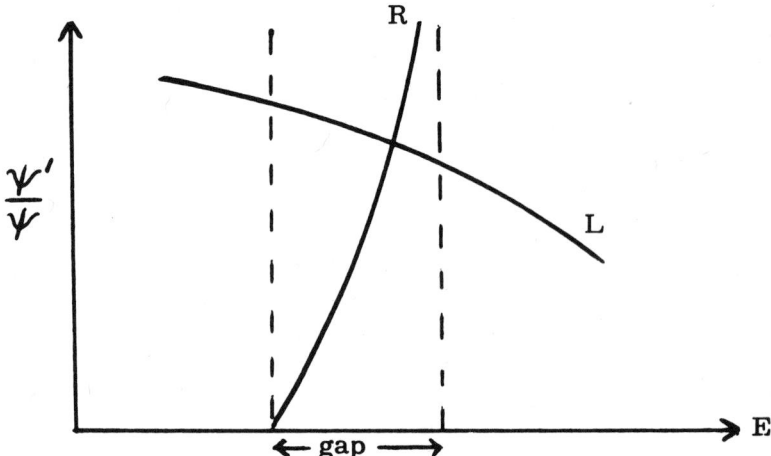

Fig. 8   Variation of logarithmic derivative with energy

Note, however, that the behaviour of ψ_L'/ψ_L can be modified by the surface layer potential so that it might, for example, become negative at $x = 0$. Thus a surface state in the gap does not necessarily exist. It depends on the nature of the surface layer. On the other hand, a strong surface potential could, clearly, bind an electron in a localized state below the bottom of the valence band. This might be a model for a surface state associated with an adsorbed impurity.

We can regard the evanescent states in the gap as Bloch states with a complex wave vector $k = k_r + ik_i$, where $k_r = 0$ or $\pm \pi/a$, for gaps at the centre of the zone or at the band edges respectively. As the energy increases across the gap, k_i increases to a maximum value and then decreases to zero (Fig. 9). This continuity follows, again, from the continuity of the solutions of the Schrödinger equation with respect to energy[17].

The algebra of various special one dimensional models is tempting and has been extensively pursued[18] ; the essential physics is, I feel, fully expressed in Fig. 8. The real problem lies in three dimensions, and is immensely more complicated. It is also an important problem, since, in a semiconductor, particularly, additional states in the gap between valence and conduction bands will have a very considerable effect on electronic properties.

ELECTRONIC STATES IN SEMICONDUCTORS

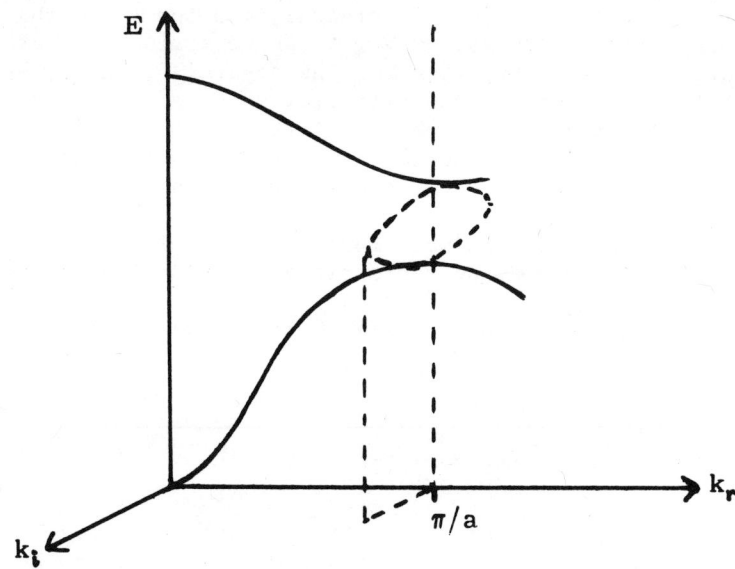

Fig. 9   Band-structure in one-dimension, showing the complex-k values for energies in the band-gap.

The features we have picked out in the one dimensional model remain. The Schrödinger equation has evanescent solutions, with $\psi(\underline{r} + \underline{a}_n) = \exp(i\,\underline{k} \cdot \underline{a}_n)\psi(\underline{r})$ where $\underline{k}$ is complex: $\underline{k} = \underline{k}_r + i\underline{k}_i$ say. Standard methods of band structure calculation can be extended to complex wave vectors, since the wave functions and energies are the analytic continuations of those with real wave vectors. To take a simple example, an effective mass approximation valid near $\underline{k} = 0$ would give for $\underline{k}$ complex

$$E(\underline{k}) = \frac{\hbar^2}{2m^*}|\underline{k}|^2 = \frac{\hbar^2}{2m^*}(\underline{k}_r^2 - \underline{k}_i^2) + \frac{\hbar^2}{m^*}\underline{k}_r \cdot \underline{k}_i$$

and $E(\underline{k})$ is real if $\underline{k}_r \cdot \underline{k}_i = 0$.

However, the topology of the energy bands in complex-k space is complicated, and an infinite set of states is needed in the construction of a surface state. In the simplest possible model of a crystal surface, the matching surface $x = 0$ is taken as a plane midway between planes of atoms in the crystal. Let us, for simplicity, take the (1,0,0) surface of a simple cubic crystal. For x>0 the potential is that of a perfect crystal, and for x < 0 the potential is taken to be a constant V_0, to be determined by reference to the experimental work function for the crystal. Then k_y, k_z remain good quantum numbers. since periodicity is maintained in the y and z directions, and we can look for surface states with fixed real k_y and k_z. At x = 0 the exterior and interior wave functions, and their gradients, have to be matched over the entire y-z plane.

Outside the crystal, the wave function is a linear combination of states of energy E and appropriate translational symmetry :

$$\psi_L(E) = \sum_{m,n} a_{mn} \exp\{\mu_{mn}x + i(k_y + \frac{2m\pi}{a})y + i(k_z + \frac{2n\pi}{a})\}$$

where

$$\mu_{mn} = \{(k_y + \frac{m\pi}{a})^2 + (k_z + \frac{n\pi}{a})^2 - (E - V_0)\}^{1/2}$$

Inside the crystal, similarly, for a given E and given k_y, k_z there is an infinite set of solutions, from which ψ_R is to be constructed :

$$\psi_R(E ; k_y, k_z) = \sum_r \lambda_r \phi_r(E ; k_y, k_z)$$

If E lies in an energy band of the constant k_y, k_z cross-section of the band structure, the set of states ϕ_r will include Bloch states ϕ_k ϕ_{-k} with real k. The matching will always be possible, though an infinite set of evanescent solutions must also contribute to the state ψ_R for the boundary conditions to be satisfied. However, if E lies in an energy gap the matching conditions lead to an eigenvalue equation for E. If this has a solution, a localized surface state exists. In this case, as k_y, k_z vary a <u>band of surface states</u> will be obtained.

Since energy gaps in different directions in k space do not all coincide, it is possible for the band of surface states to overlap with the valence and conduction bands. In a real crystal, a surface state which overlaps with the conduction or valence band may be expected to hybridize with bulk states of the same energy to modify their behaviour at the surface, and will not provide a localized state.

In a practical calculation, of course, only a finite number of states can be included, and some sort of compromise has to be made over the matching conditions. The first calculation along these lines was made by Jones [19] for the (110) surface of silicon, using a pseudo-potential scheme for the band structure, and the results are shown in Fig. 10. There are two bands of surface states, dege-

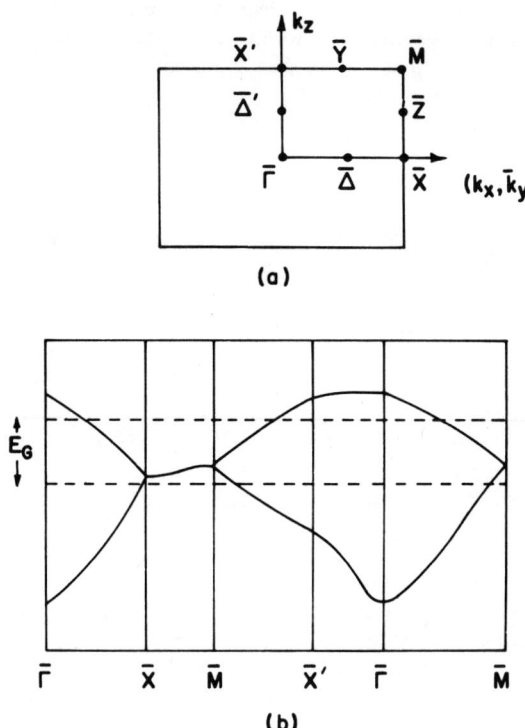

Fig. 10 Surface states on the (110) surface of silicon (a) Two-dimensional Brillouin zone (b) $E(k_x,k_z)$ along symmetry lines in the Brillouin zone. The energies at the top of the valence band and bottom of the conduction band are shown as dotted lines. (After Jones[19]).

nerate along XM, with a total width of about 3 eV, and overlapping with the valence and conduction bands. A later paper[20] extends the calculation to the (100) and (111) surfaces.

This sort of rigorous model calculation is important in that it tells us what to expect. It is, clearly, difficult to modify the formalism to study the effect of different types of surface conditions. Beeby has set up a multiple scattering formalism for a semi-infinite lattice of muffin-tin potentials[21], which takes planes of atoms parallel to the surface as scattering elements. In this approach it might be possible to consider more complex models for the surface potential. But the computational problems in applying the method are considerable, and no results have been reported. An ingenious approach by Appelbaum and Hamann[22] moves the matching plane sufficiently into the interior of the crystal that most of the evanescent solutions of the Schrödinger equation can be neglected on the interior side of the matching plane. This equation is integrated numerically through the last layers of atoms of the crystal and an optimized matching procedure is used. The potential in these layers can be varied arbitrarily (except that appropriate periodicity parallel to the surface must be maintained). The calculations can be reduced to computable size if the ionic potential can be described by a weak pseudopotential. The method was first applied to find a self-consistent surface potential for sodium, and in a more recent paper to the (111) surface of silicon[23]. We shall refer to these results below. But we need more flexible - if cruder - theoretical approaches to obtain a qualitative understanding of, say, the effect of different surface structures on surface states, or to even begin to think about surface states associated with amorphous films.

For covalently bonded semiconductors, we can use the picture of bonding and antibonding orbitals, which we have already discussed in connection with the Weaire Hamiltonian. If no rearrangement of ions takes place at the surface, then (Fig. 11) we are left with just one directed orbital which cannot be paired, associated with each surface atom. The bonding and anti-bonding orbitals construc-

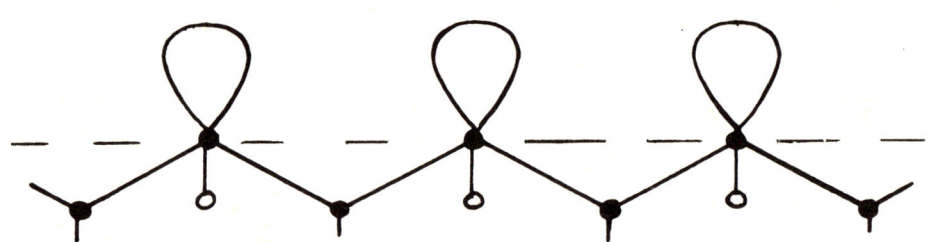

Fig. 11   'Dangling bonds' at a surface.

ted from paired directed orbitals make up the valence and conduction bands of the bulk crystal. The orbitals left over at the surface, or 'dangling bonds', interact to form a band of surface states. Since the interaction is indirect, we expect the energy of states in the surface band to lie between the valence and conduction bands. Counting spin degeneracy, we see that there are two electron states per surface atom in the surface band. In a perfect crystal, the available electrons will fill the valence band, and half-fill the band of surface states. Electrons in surface states will have a high electron density in the dangling orbitals, and a lower density in the other three orbitals of the surface atoms. However, since the surface atoms do not share the wholly tetrahedral environment of bulk atoms, we might expect the dangling bond orbitals to be somewhat modified.

Calculations of surface states in the Hall model for diamond by Koutecky and Tomášek[24] support this discussion, though, as we have stressed in an earlier section, we cannot expect numerically exact results for semiconductors from an LCAO type of calculations. The calculations of Appelbaum and Hamann[23] for the Si (111) surface are in principle more accurate, and were carried out both with surface atoms in their ideal positions, and with the last atomic plane relaxed inwards in the normal direction. Chemical bonding arguments suggest there should be such a relaxation. A smooth model potential is constructed, to be consistent with the experimental vacuum level. In both cases, a band of surface states is found which lies in the absolute band gap between the conduction and valence bands. The electron density for a state in this band is shown in Fig. 12. Rather remarkably, we see that the electron density is consistent with the 'dangling bond' picture, though the starting point of the calculation is entirely different. The relaxed structure, though not the ideal structure, is found to be 'highly self-consistent'.

In these calculations, the surface bands are found to be highly sensitive to the surface potential. The degree of self-consistency is dependent on the choice of surface structure. Further theoretical progress would seem to depend on detailed experimental knowledge of the positions of surface layers of atoms.

On the basis of this hugely simplified picture, we can speculate on the nature of surface states in an amorphous semiconductor. The number of dangling bonds per unit area of surface may be fewer, since in the formation of the structure there will be a preference for saturated bonds and a consequent lower energy. There may, then, be a tendency for states associated with dangling bonds to be localized, rather than to be extended over the surface.

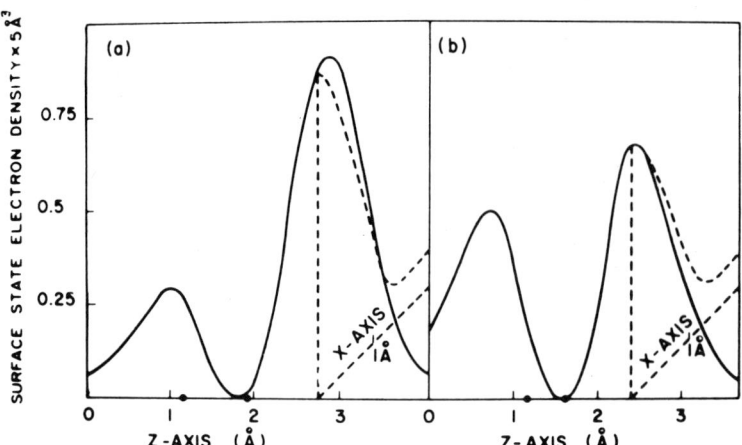

Fig. 12 Electron density in a 'dangling-bond' state plotted along a line normal to the surface and passing through a surface atom. Also shown is the density along a line in a plane parallel to the surface whose direction is between the two nearest neighbour surface atoms. (a) Unrelaxed and (b) relaxed structure. The heavy dots on the z-axis are the atom sites and the origin is the matching plane. (After Appelbaum and Hamann[23]).

8.   EFFECTS OF SURFACE STATES

In the last section, the principles which lie behind the existence of surface states were outlined, and the present state of theoretical calculation was indicated. We consider now some of the consequences of surface states.

The surfaces of real crystals will usually be more complicated than in our theoretical discussion. As well as the surface states associated with a periodic surface and extended over the surface, there may be <u>localized surface states</u> at sites where periodicity breaks down. Thus, states may be associated with randomly adsorbed impurities, monoatomic steps on the surface, or other imperfections On clean surfaces, states will vary with surface index, and with the nature of the surface. Low energy electron diffraction studies show that a surface may be 'reconstructed' from the ideal lattice, and new periodicities may appear. Unfortunately, the surface struccture cannot be found uniquely by LEED : the unit mesh is determined, but not the positions of the atoms in the mesh. One can only choose a 'reasonable' model for the surface which is consistent with the experimental data. Clean surfaces of Si and Ge offer the best hope of comparison between theory and experiment[25].

Since electrons in surface states are confined to the surface region of a crystal, the occupation of surface states can lead to a transfer of charge from the interior of a crystal to the surface. In a semiconductor, the density of charge carriers is low ; a surface charge will be screened out by a space charge layer over a distance of the order of the Debye screening length L, where

$$L = \left(\frac{KkT}{4\pi ne^2} \right)^{1/2}$$

Here K is the dielectric constant, and n is the maximum possible density of charge carriers. L is about 1 μ for intrinsic germanium (K = 16) at room temperature, and $\sim$ 100 Å for a doped sample with n $\sim$ 10^{17} cm^{-3}. Thus the space charge region in a semiconductor extends over a considerable distance (whereas in a metal the corresponding Fermi screening parameter for the degenerate Fermi gas of conduction electrons is $\sim$ 1 Å, typically).

In analogy with bulk impurity states, surface states are classified as donor-like, if they are neutral when occupied, or acceptor-like, if they are negatively charged when occupied. Bearing in mind the 'dangling-bond' picture, we can expect donor levels to lie below acceptor levels.

The occupation of surface states will be determined by Fermi-Dirac statistics. Thus the bulk doping of a semiconductor must be precisely known to calculate the distribution of electrons over levels. A specific example is illustrated in Fig. 13, which shows an n-type semiconductor with a surface band of acceptors below the Fermi-level. These are filled in thermal equilibrium so that there is a surface layer of charge. This repels conduction electrons, to leave a positive space-charge layer. In the (semi-classical) diagram, there is therefore 'band-bending' relative to the Fermi level. It is usual to assume that the position of the surface bands relative to the conduction and valence bands does not change as the occupancy of the surface states changes. It is obvious that there are difficult problems of self-consistency in constructing the atomic potential near the surface, which may make this simple picture inappropriate.

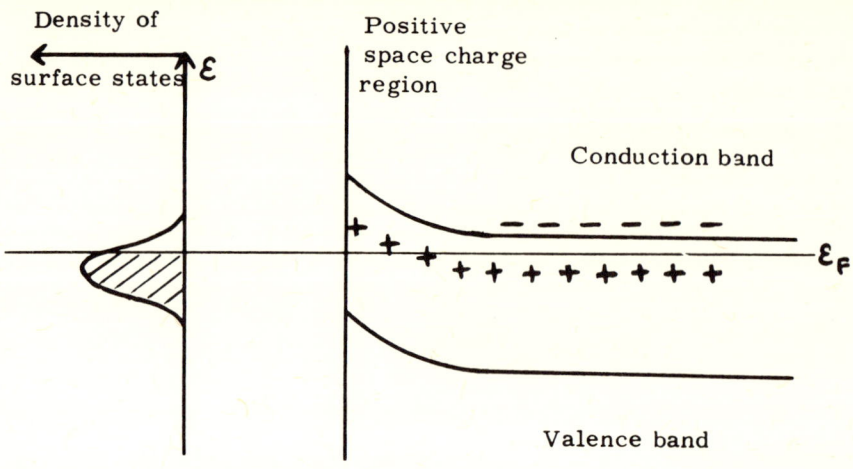

Fig. 13 Formation of a space-charge region and 'band-bending' due to the occupation of surface states.

The potential is given by the Poisson equation

$$\nabla^2 V(\underline{r}) = -4\pi\rho(\underline{r})/K$$

where

$$\rho(\underline{r}) = e\int g(\varepsilon)[f(\varepsilon + eV(r)) - f(\varepsilon)]d\varepsilon$$

Here $g(\varepsilon)$ is the bulk density of states and $f(\varepsilon)$ the Fermi function. The Fermi energy is determined by the density of surface states and the overall neutrality condition. In this equation, V is an averaged macroscopic potential, and to a good approximation will be a function only of distance from the surface. The solution is discussed in standard texts[26], though we should note that in thin films with a low density of carriers, for which L is comparable with the thickness of the film, modifications will be necessary.

Similar considerations apply in discussing the interface between a semiconductor and a metal, or a semiconductor and an insulator. The importance of surface states in the interpretation of such systems was first recognized by Bardeen[27]. An interesting discussion of a metal-semiconductor contact in terms of wave function matching has been given by Heine[28].

The extent of the band-bending, V_s, can be determined experimentally, for example by combining measurements of the work function and photoelectric threshold. However, it is clear that the relation between the theoretical calculations of the previous section, and measurements of band-bending, is not one which allows immediate detailed comparison between theory and experiment.

More directly, light can be used as a probe for surface states. Electrons from the valence band can be excited into empty surface states, or from full surface states into the conduction band. However, these effects have to be disentangled from various bulk effects. A useful technique, which can be combined with various measurements, is to vary the doping in a set of crystals, so that the Fermi energy is moved through the surface band. Field effect techniques, based on the induction of surface charge by an external electric field, can be adapted to study surface states. Experimental techniques, and data, are reviewed by Davison and Levine[18] It is agreed that surface states in Si and Ge exist, but there is no clear consensus on the distribution and nature of the states. Experiment and theory are still far apart.

In discussing amorphous films, we had difficulty enough in treating bulk states with any rigour. Elementary considerations of bonding and electron affinity suggest that surface states, perhaps localized, will exist. It may be difficult, however, to distinguish experimentally between these, and other states in the quasi-gap which are associated with dangling bonds in the bulk of the material.

REFERENCES

1. M. L. Cohen, V. Heine, Solid State Phy., 24, 38 (1970)

2. The Mattheis method is described in detail in :
   T.L. Louks, "Augmented Plane Wave Method", Benjamin, New York (1967).

3. D. Weaire, Phys Rev. Lett. 26, 1541 (1971).

4. W. Paul, G. A. N. Conell, R.J. Temkin, Adv. Phys., 22, 531 (1973).

5. R. Grigorovici, "Electronic and Structural Properties of Amorphous Semiconductors", (ed. P.G. Le Comber, J. Mort), Academic Press, London and New York (1973).

6. P. Lloyd, Proc. Phys. Soc., 90, 207, 217 (1967).

7. A. Messiah, "Quantum Mechanics", North-Holland, Amsterdam (19

8. T. C. McGill, J. Klima, Phys. Rev., B5, 1517 (1972).

9. J. Keller, J. Phys. C 4, 3143 (1971).

10. R. Jones, J. Phys. C 6, 2318 (1973).

11. J. C. Phillips, Comments on Solid St. Phys. 4, 9 (1971).

12. T. M. Donovan, W.E. Spicer, E.J. Ashley, J. M. Bennett, Phys. Rev. B2, 397 (1970).

13. F. Aymerich, P. V. Smith, J. Phys. C6, L41 (1973).

14. J. D. Joannopoulos, M. L. Cohen, Phys. Rev. B7, 2644 (1973).

15. J. M. Ziman, J. Phys. C4, 3129 (1971).

16. H. Jones, "Theory of Brillouin Zones and Electronic States in Crystals", North-Holland, Amsterdam (1960)

17. V. Heine, Proc. Phys. Soc. 81, 300 (1963).

18. S. G. Davison, J. D. Levine, Solid State Phys. 25, 1 (1970) (The theoretical discussion in this review emphasizes one-dimensional and tight-binding calculations).

19. R. O. Jones, Phys. Rev. Lett., 20, 992 (1968).

20. R. O. Jones, J. Phys. C5, 1615 (1972).

21. J. L. Beeby, J. Phys. C6 1242 (1973).

22. J. A. Appelbaum, D. R. Hamann, Phys. Rev. B6, 2166 (1972).

23. J. A. Appelbaum, D. R. Hamann, Phys. Rev. Lett. 31, 106 (1973).

24. K. Koutecký, M. Thomášek, Phys. Rev. 120, 1212 (1960).

25. M. Henzler, Surface Science, 25, 650 (1971).

26. A. Many, Y. Goldstein, N.B. Grover, "Semiconductor Surfaces", North Holland, Amsterdam (1965).

27. J. Bardeen, Phys. Rev., 71, 717 (1947).

28. V. Heine., Phys. Rev., 138, A 1689 (1965).

Other relevant review articles include :

D. A. Greenwood, "The electronic density of states in amorphous materials" in ref. 5.

V. Heine, Surface Science 2, 1 (1964).

P. Mark, Surface Science 25, 192 (1971).

Characterization

STRUCTURE DETERMINATION OF THIN FILMS

H. Raether

Universität Hamburg, Institut für Angewandte Physik

2 Hamburg 36 . Jungiusstrasse 11

1. INTRODUCTION

The concept of the structure of a thin film has different aspects. We can mean the geometrical exterior or the crystalline structure. We shall treat both these aspects in the following.

2. GEOMETRY

The ideal film has uniform thickness with a plane and smooth surface and a homogeneous density. The real film however has a varying thickness, a rough surface and it will have, especially in the case of a rather thin film (some 100 Å or less), an inhomogeneous density.

Very thin films of an average thickness of less than 100 Å, can be produced on substrates, but they are in general built up by islands more or less separated from each other. This is valid in general for metal films. If the "island-structure" films are made thicker and thicker, the islands grow in size till they touch each other and form a more or less continuous foil.

In an intermediate state, between the island structure and the continuous film, the thin film has a structure with voids (see Fig. 1) which makes these films less dense than the bulk material. This density deficit, which can reach 10% of the bulk density, disappears with increasing coalescence of the crystal units. Films of some 10^3 Å have in general the bulk density. Alkali metals show another behaviour : films of some 100 Å can be produced by vaporisation as homogeneous films only on cooled (liquid nitrogen tem-

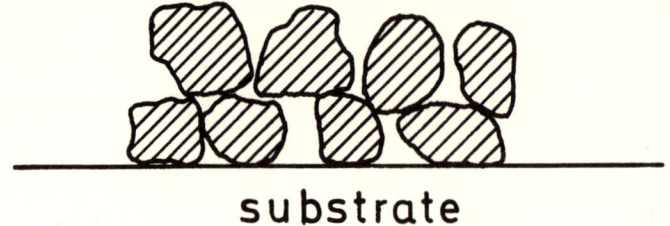

Fig. 1   Schematic Structure of a rather thin film.

perature) substrates such as quartz or sapphire which have a good heat conductivity. At higher temperatures they condense in a pronounced island structure.

In contrast to the metal films, nonmetallic material such as aluminium oxide, beryllium oxide, carbon can be produced as very thin continuous films of even some 10 Å. They are used as support for substances to be investigated in the electron microscope : metallic or other thin films, single molecules etc.

3.   CRYSTAL STRUCTURE

Having prepared a thin film we wish to know whether it consist of a monocrystal, of a polycrystalline material or of a so-called amorphous structure. In other words we are interested in the orientation of the crystalline units and their size. It is useful for the following, especially for the description of the fine crystalline state, to introduce the density $n(r)$ of atoms as a function of the distance r from an arbitrarily chosen atom. In the case of a monocrystal this is not an isotropic function ; in the directions along a lattice row $[hKl]$ it has equidistant high maxima at the sites of the atoms, indicating the high order in the crystal, (see Fig. 2). In a very fine crystal powder with units consisting of only some elementary cells however $n(\underset{\sim}{r})$ is an isotropic function of r : in the immediate neighbourhood of an atom "short range order" is still persisting whereas the long range order has disappeared (see Fig. 2b). Amorphous and liquid substances show the same density distribution as Fig. 2b. In a liquid the atoms move around and are thus in thermal equilibrium which may not be the case in an amorphous solid. There has been a lot of discusssion on the difference between the fine crystalline and the amorphous state, but we will not treat this question here.

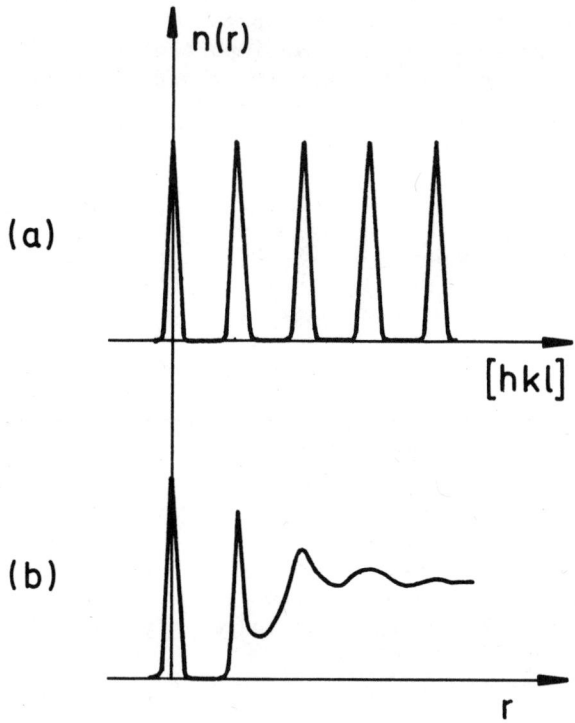

Fig. 2   The density of atoms n(r) in a perfect crystal (a) and in a highly disordered state (b)

Methods of looking at the crystal structure are

a) diffraction
b) image formation

which we discuss briefly in the following. In both cases we use either electrons or X-rays, which penetrate the volume of the film. Electrons of a velocity of at least 50 KeV are needed for transmission through film thickness of 1000 Å ; in this case the elastic scattering intensity represents still an important part of the total intensity (elastic and inelastic). Slow electrons of 10 KeV and less are not useful in transmission experimente since they lose too much energy. X-rays penetrate easily several 10^6 Å, but are less scattered by 1000 Å foils. This is due to the weaker interaction of X-rays with the single atom which consists of the scattering by the electron of the atom, whereas electrons are scattered by the nucleus, an interaction which is nearly 10^6 stronger. This high scattering probability of electrons makes them an important tool for the investigation of thin films. An further important advantage of electrons is the existence of electron lenses to construct an image, whereas we are not yet able to produce X-rays lenses.

3.1 Electron Diffraction

Some typical diffraction photographs of fast electron taken for a silver foil are reproduced in Fig. 3. They are obtained by an arrangement schematically drawn in Fig. 4 : a beam of about 50 KeV electrons passes a thin film of about 1000 Å. The diffracted electrons are photographed in a distance of about 20-30 cm from the object. The position of the intensity maxima is given by the Bragg equation $2d \sin \theta = n\lambda$ with d the lattice connstant, θ half of the scattering angle and λ the wave length of the primary electrons, which has a value of about 0,05 Å for 50 KeV electrons. The angle 2θ is therefore of the order of a few 10^{-2} rad. An other form of describing the diffraction pattern is given by the Laue equation $\gamma - \gamma_0 = 2f$, where γ and γ_0 are the directions of the incoming and the outgoing beam and f a vector of the reciprocal lattice. The construction of the Ewald sphere with radius $|\gamma|/\lambda = |\gamma_0|/\lambda = 1/\lambda$ yields immediately the diffraction pattern since the projection of the points on the photographic plate touched by the Ewald sphere represents the diffraction pattern (see Fig. 5).

One sees immediately that Fig. 3a is produced by a specimen which is nearly a monocrystal of silver. The different Laue spots (002), (020) and (200) are the projection of a plane of the reciprocal lattice on the plate. In principle there should be seen only one or very few spots corresponding to the exact adjustment of the crystal to fulfil the diffraction condition. Apparently the specimen is built up of many crystals which are tilted against each other very slightly. In Fig. 3 b-d the orientation of the different crystals becomes more and more random, so that finally powder rings are observed.

It should be mentioned that continuous rings are also observed if crystals lie on a substrate and are randomly distributed around the normal of the substrate surface, (Fig. 6). Under this condition a certain number of powder rings is lacking which correspnd to lattice planes too much inclined to the electron beam. This situation is revealed by inclining the specimen against the beam as Fig. 7b shows : those lattice planes perpendicular to the rotation axis remain in the reflection position whereas the others are out of it.

If on the other hand the specimen consists really of a good monocrystal the number of diffraction spots reduces substantially. In addition Kikuchi lines appear which are produced by the primary

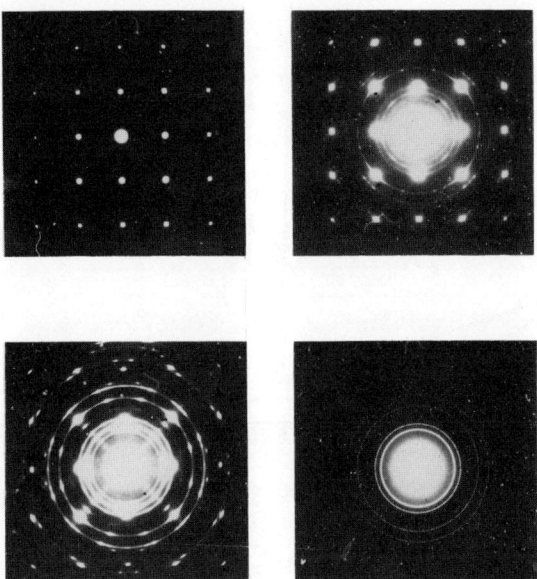

Fig. 3. Electron diffraction diagrams in transmission of thin silver films, showing the transition of a Laue diagram into a powder diagram.

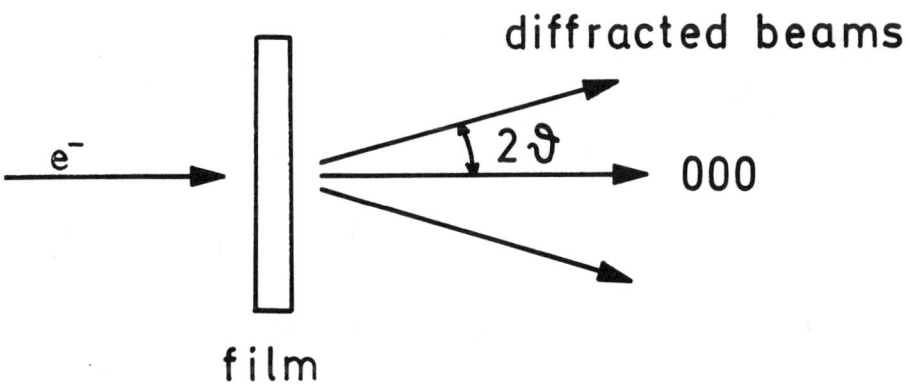

Fig. 4. Scheme of an electron diffraction apparatus.

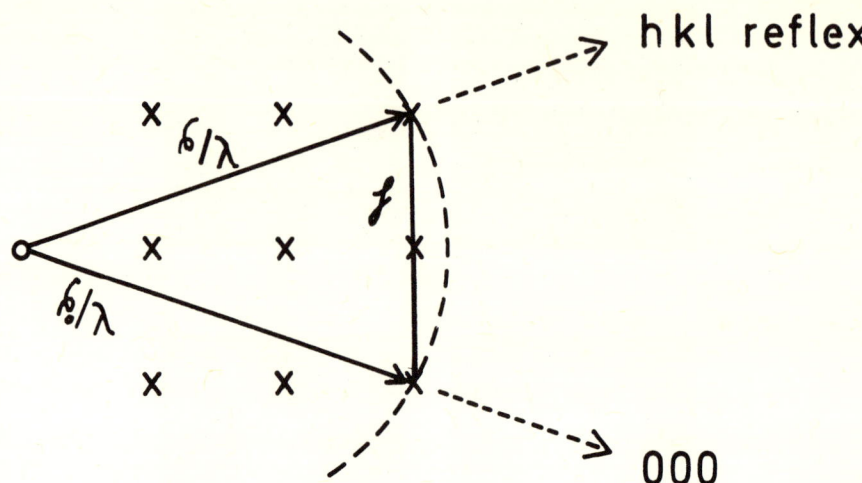

Fig. 5. Reciprocal lattice with the Ewald sphere.

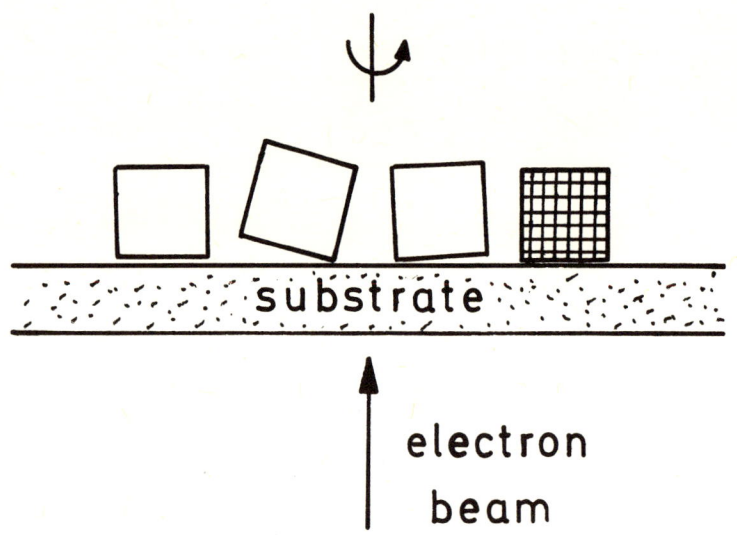

Fig. 6. The crystals are turned around the normal of the substrate but their (100) planes remain nearly parallel to the substrate surface.

STRUCTURE DETERMINATION OF THIN FILMS

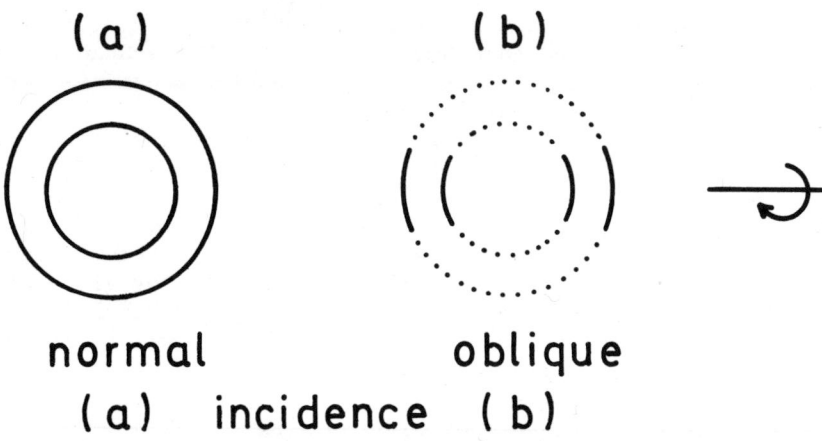

Fig. 7. (a) Diffraction diagram of the crystals of Fig. 6. The substrate is turned around an axis lying in the substrate (see at rigth) which gives a diffraction pattern as Fig. 7 (b) shows.

electrons scattered at the entrance of the specimen into a cone ; the aperture of this cone increases with the penetration depth of the electrons. Now a lattice plane which fulfils the reflection condition takes out of this cone a whole "line" of electrons and puts it at another place. One sees immediately that the Kikuchi lines become broader the more the monocrystal breaks up into slightly desoriented units each producing a Kikuchi line. With increasing desorientation of the units the lines become very broad and disappear in the background. The number of diffraction spots becomes more numerous corresponding to the desorientation as mentioned above. (1)(2).

2.2. The size of the crystals

The powder rings give information not only on the randomness of the crystal orientation but also to a certain extent on the size of the crystals. In the simple approximation of geometrical diffraction theory the width of the diffraction spots is given by the number of atoms : the angular half width can be represented by λ/Ma with the lattice constant a and M the number of atoms lying along the diameter of the crystal. This means that the powder rings become larger with decreasing M till they flow together.

The width of the diffraction spots produced by the crystal

size is independent of the value of d or of the angle θ, so every rings is enlarged by the same amount. Thus the Debye-Scherrer rings approach each other, especially at higher angles where they are close together and disappear in the background first with decreasing crystal size. Finally there remain in general two broad rings in the fine crytalline state. It may be mentioned that there exist other reasons for a broadening of diffraction spots too. One of these is a displacement of the atoms Δd varying from zero up to a maximum value which causes an enlargement of the interference maxima Δθ proportional to the reflection angle θ : by differentiating the Bragg equation one obtains Δθ = (Δd/d)tg θ which indicates the dependance on the angle θ. This has to be taken into account when interpreting the width of the diffraction maxima.

The above mentioned powder diagram with two or more broad rings can be used to derive more information than just the rough estimate of the crystal size. If we choose for example a monoatomic substance whose atomic form factor f is given and assume that the density of the atoms at distance r is n(r), we can derive rather simply the diffraction intensity as a function of the diffraction angle θ and the wave length of the irradiating beam λ as :

$$I(s) = \text{const.} \; f^2 \cdot \int \frac{\sin sr}{sr} \cdot n(r) \cdot 4\pi r^2 dr$$

with $s = 4\pi/\lambda \cdot \sin \theta$.

We can obtain the unknown density n(r) by a Fourier transform of the measured $I(s)$:

$$4\pi r^2 \cdot n(r) = \text{const.} \int I(s) \cdot s \cdot \sin sr \, ds$$

The behaviour of n(r) looks like Fig. 2 (b) : it shows (besides the maximum ar r = 0) a pronounced short range order indicating in a certain distance a shell of nearest neighbours whose number can be obtained by integrating over the first peak. If further maxima are obtained, information on further order can be derived. This procedure can be applied to liquids as well as to amorphous substances using electrons, X-rays and neutrons. The general result of these diffraction experiments is that it is not possible to distinguish between the liquid and the amorphous state as well as between the amorphous and the finest crystalline state. Other methods are necessary to distinguish between these states.

3.3. Electron Microscopic Methods

The diffraction method gives information on properties averaged over an extended region of the specimen.

More frequently we are interested in localised qualities of

STRUCTURE DETERMINATION OF THIN FILMS

the specimen such as the size and the shape of the individual crystals which wompose the film. This information can be obtained by collecting the diffracted beams whith the help of electron lenses and thus constructing the image from the scattered amplitudes, together with their phases.

Now looking at the sreen of the usual electron microscope with a polycrystalline films as object we see the boundaries of the different crystals, but we do not observe the lattice of atoms. This comes from the fact that the lenses are not good enough to gather besides the undeflected electrons the diffracted beams of at least the first order which modulate the background of the undeflected beam and produce a periodic intensity on the screen. However the quality of the lenses is sufficient to gather those beans which are diffracted by the boundaries of the crystals, whose diffraction angle is much smaller than that of the first order, and which are therefore summed up by the lenses. The diffraction angle of the boundary of a linear lattice of length Ma amounts to $\sin \theta = \lambda/Ma$ for the first minimum, see Fig. 8 and about $1,5 \lambda/Ma$ for the first maximum which is essential for image formation. This explains the ability of the electron microscope to see the crystals, but not their lattices.

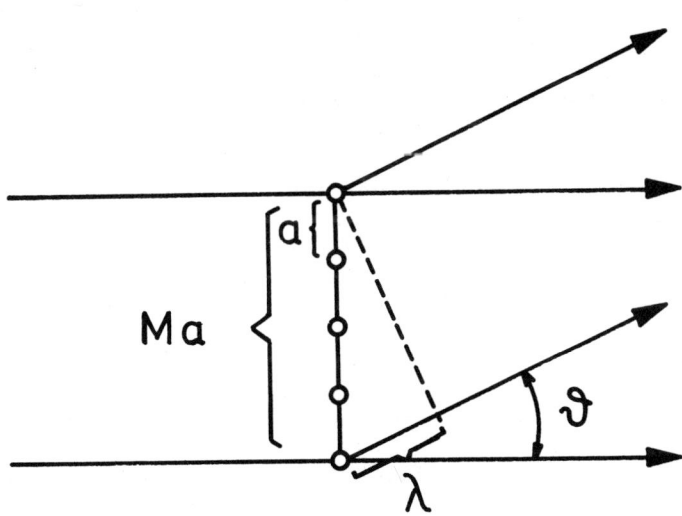

Fig. 8. Diffraction by a row of atoms. (The first subsidiary minimum is indicated.

Fig. 9. Electron Microscope image of a thin gold film. The perpendicularly crossed lines correspond to lattice planes of 2,04 Å (by courtesy of Fa. Siemens, Berlin)

The intense work on the developement of better electron lenses has successed in the last years in obtaining a spatial resolution of a few Å which is well below the spacing of some lattice planes in some crystals, so that it is possible to observe the periodic arrangement of these lattice planes as Fig. 9 shows.

We shall point out another development of electron microscopic observation ; the scanning microscope, which has become of great importance. Here the wave properties of the electrons are neglected, but the production of secondary electrons and the backscattering of electrons are of importance. The principle is rather simple an electron beam is focussed by an assembly of lenses to about 100 Å and scanned over the surface (or film) which has to be investigated The electrons, backscattered or produced by secondary emission in the focus are registered on the same side of the object point by point as Fig. 10 demonstrates ; thus the "image" is decomposed in a series of intensity values following each other in time which are gathered again on a screen storing these values ; an image of varying electron intensity is produced by this procedure and informs us about the "structure" of the surface.

This scanning electron microscope thus gives information on

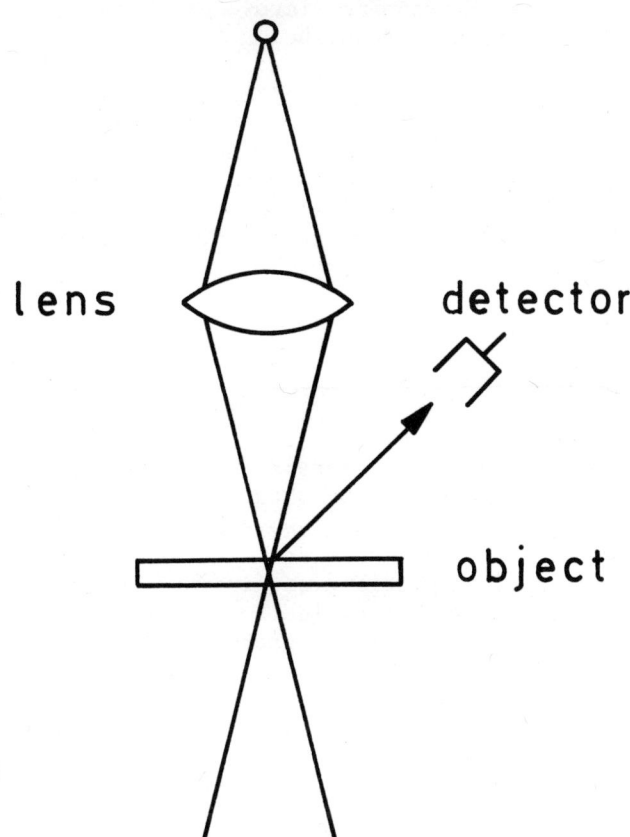

Fig. 10. Scheme of a scanning electron microscope : the primary electrons focussed on a small spot of the object produce secondary and backscattered electrons which are registered by an electron detector. By scanning the focussed spot over the whole surface one obtains an "image" of the object.

changes in the composition of the surface (the backscattering is dependent on the atomic number Z of the substance) and on the geometrical structure of the irradiated surface ; this comes from the fact that the number of secondary electrons depends strongly on the angle the incoming electrons make with the irradiated surface[3]

As already mentioned the electrons are registered on the "reflexion" side. It is also possible to look at the transmitted electrons and to scan over the film observing the varying transmittivity. However the type of scanning microscope commercially available in general is that first described.

An interesting developement of the transmission scanning microscope is the following ; if one reduces the diameter of the focussed electron beam to 10 Å or less and deposites a few molecules well separated from each other on a very thin homogeneous substrate one can register these single molecules or atomic units by scanning the electron beam over the molecules and observing them by the change in the scattered intensity. This is another way to observe single molecules[4].

4. INELASTIC SCATTERING

The diffraction method uses the elastically scattered electrons. This means that the electrons (or X-rays or neutrons) interacting with the crystal have transferred or obtained a large momentum of h.f (see the Laue equation mentioned above, h is Plank's constant) or $h/a \sqrt{\sum_i h_i^2}$ (h_i : Miller indices in the case of a cubic crystal). The transferred energy is negligibly small since the crystal has an enormous mass compared with the electrons. The inelastic processes on the contrary are in general characterised by a rather large energy transfer ($\Delta E \sim 10$ eV) and a small momentum exchange ($<$ h.f) ; they lead to an excitation of the solid as e.g. a band transition of crystal electrons, production of exitons or plasmons. To distinguish these different phenomena one has to analyse carefully the energy losses of the electrons which have passed through the film. Thus one obtains diagrams like Fig. 11, in which the intensity of those electrons which have suffered an energy loss between E and E + ΔE are registered. Peaks in this diagram indicate processes of high probability as just mentioned.

The inelastically scattered electrons thus contains rich information on the crystals composing the film ; more exactly one has to say that the information is given by the complex dielectric function $\varepsilon = \varepsilon_1 + \varepsilon_2$, since the intensity of the inelastic electrons is proportional to $\varepsilon_2 / \varepsilon_1^2 + \varepsilon_2^2$ as deduced by the dielectric theory[5].

The field of energy losses has an interesting aspect : the registration of loss diagrams of organic substances has shown that certain peaks change their intensity during the irradiation. If other work has succeeded in relating the different loss peaks to the structure of the molecule, these changes indicate where the radiation damage has occured.

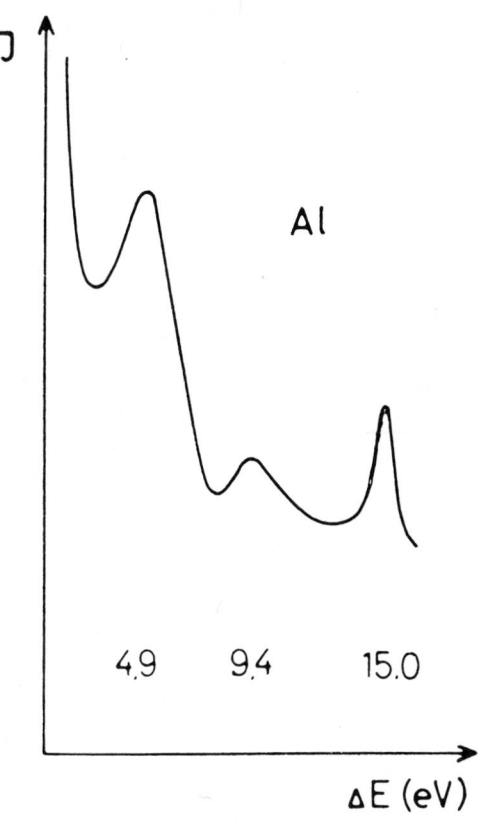

Fig. 11. Energy Loss Diagram of 50 KeV electrons of a thin Al-film : the peak at 15 eV represents the volume loss, the two peaks at 4.9 and 9.4 eV come from surface plasma oscillations.

5. THE STRUCTURE OF SURFACES

Regarding the structure of a surface several questions arise : what is its roughness, what is its composition compared with the bulk (especially in the case of alloys), are the lattice dimensions and with them the band structure different at the surface, do surface states exist ? etc. These problems are under investigation with modern means such as photoelectron emission, LEED, etc., but the results are not yet sufficiently settled to be presented here.

We shall discuss questions connected with the geometric structure, especially the roughness of the surface.

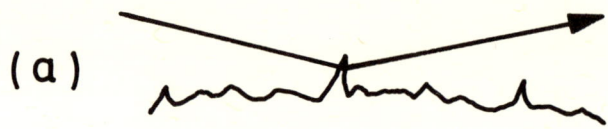

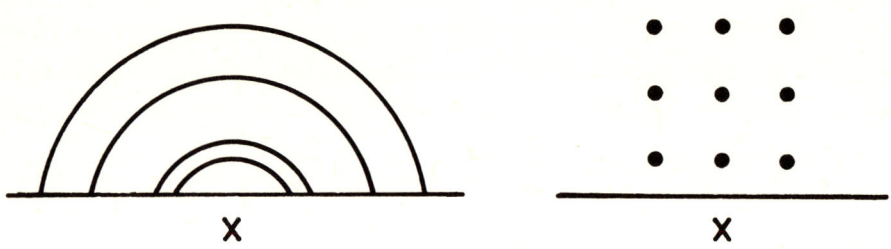

Fig. 12a. The reflexion of electrons on a rough surface :
polycrystalline (left) and monocrystalline (right).

(b)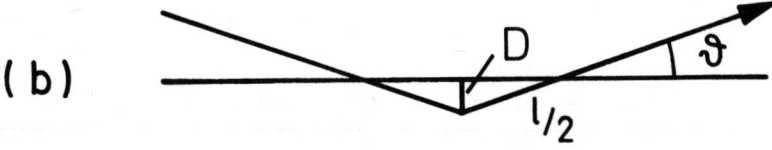

Fig. 12b. The reflexion of electrons on a smooth surface of a monocrystal.

An approximate procedure is to look at the boundary with fast electrons using the reflection method.

In the case of a rough surface the electrons hitting the surface at grazing incidence give information on the peaks of the crystalline substance powder rings are produced, in the case of a monocrystalline material (e.g. an etched boundary) a point diagram (see fig. 3) is expected ; see fig. 12. However, smooth surfaces, produced by cleaving monocrystals such as mica, graphite, NaCl etc. give a different pattern since apparently the monocrystalline surface is broken up into a mosaic by the cleaving process, in general many Bragg spots are observed as in Fig. 12 (above), but the diffraction maxima are elongated into streaks. This can be explained by the finite penetration depth of the electron waves which limits the resolution of the third Laue condition (perpendicular to the boundary) and smears out the maxima in both directions perpendicular to the surface. The penetration depth (D) is connected with the mean free path [1] of the electrons (the length after which they suffer an inelastic collision) by $D = \ell/2 \sin \theta$. ℓ is of the order of 1000 Å for 50 KeV electrons and $\theta \sim 10^{-2}$ so that $D \sim 10$ Å, a depth which changes the interference maxima nearly into streaks. (In addition the electron waves are refracted crossing the boundary, but this effect is neglected here). These streak patterns are characteristic of a smooth surface. This small value of D demonstrates the sensitivity of this method for surface observations.[2]

The same phenomenon is seen in low energy electron diffraction: here the angle θ is larger ($\sin \theta \sim 1/2$), but ℓ is strongly reduced to values of the order of 10 Å so that the same value of penetration depth results and thus similar diffraction patterns. The sensitivity of these methods makes it possible to recognise monolayers of adsorbed material on the surface if the absorbed atoms are regularly arranged on the monocrystalline substrate and become thus observable as additional streaks.

More quantitative information is obtained by the method of light scattering : we know from optics that a light beam falling on a smooth metal surface at oblique incidence splits into a reflected and a transmitted beam. We understand this phenomenon by saying that the incoming light beam excites a polarisation current at every point of the surface by the component of the electric field perpendicular to the surface. Every point radiates light like a Hertz dipole into nearly all directions. Summing up all the amplitudes radiated from the whole surface there remains only those with phase differences zero, the others being destroyed by interference ; in other words the reflected and transmitted beam are the only ones to be observed. If roughness exists on the smooth boundary this radiates in addition into nearly all directions producing diffuse light scattering by the irradiated surface.

In recent years the observation of surface plasmons has given

STRUCTURE DETERMINATION OF THIN FILMS

a new possibility of looking at surface roughness on metals : the electrons of many metals can be regarded as a quasifree electron plasma of high density (n_0). It is possible to produce density fluctuations (Δn) of this plasma which travel along the crystal as a plane wave $n = n_0 \sin(\omega t - kx)$; ω is the frequency and k the wave vector. The corresponding quanta are called plasmons. These fluctuations may occur in the volume (volume plasma oscillations) and in the surface (surface plasma oscillations). Surface waves have their wave vector parallel to the surface, along which the charge density fluctuations travel ; the electric fields belonging to these charges decay exponentially into the metal and into the vacuum. The frequency and the wave vector of these oscillations are connected by a dispersion relation $\omega(k)$ which determines also the phase and group velocity of these waves.

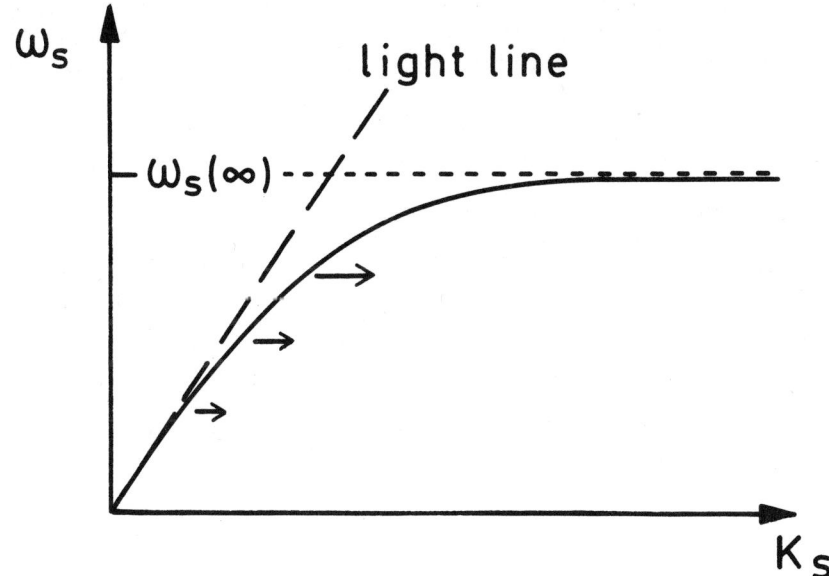

Fig. 13. The dependency of the frequency ω_s of surface plasma oscillations on the wave vector K_s (dispersion relation) in the case of a quasifree electron gas. The arrows indicate how the dispersion relation is displaced to higher K_s values on a rough surface.

Now if these waves run along a surface with a certain roughness (or any periodic perturbation), they can be scattered and transformed into light of the same frequency. By analysing the angular distribution of this light and its polarisation it is possible to deduce the mean roughness of the surface. For example silver films vaporised on a well polished quartz substrate show a mean roughness of about 10 Å. This method can be further developed to give still more information, from non-metallic surfaces too.[6]

As mentioned above the surface plasma oscillations have to satisfy a dispersion relation $\omega_s(k_s)$ which is drawn schematically in Fig. 13. Now it has been found that this behaviour is slightly but measurably changed, if the surface becomes rougher : the frequency ω_s is displaced to higher K_s values with increasing roughness. This gives a simple method of measuring the roughness of a surface, a mathod which has to be further developed [7].

6. CONCLUSION.

It was the intention of these lectures to give a short insight into various aspects of the structure of thin films and of the most important methods of its determination. The references at the end may be useful for more detailed information.

REFERENCES

1. L. Reimer, Electronenmikroskopische Untersuchungsmethoden (Springer Verlag, Berlin) (1967).

2. H. Raether, Elektroneninterferenzen, Handbuch der Physik, Vol. 32, p. 443 (Springer Verlag, Berlin) (1957).

3. L. Reimer, G. Pfefferkorn, Rasterelektronenmikroskopie, (Springer Verlag, Berlin) (1973).

4. A.V. Crewe, J. Wall, L.M. Welter, J. Appl. Phys. 39, 5861 (1965).

5. H. Raether, Solid State Excitations by Electrons, Springer Tracts Mod. Phys. 38, 85 (1965).
   J. Daniels et al., Springer Tracts Mod. Phys. 54, 78 (1970).

6. E. Kretschmann, Optics Communications, Vol. 10, n°4, p.353 (1974).

7. D. Hornauer, H. Kapitza and H. Raether, J. Phys. D. Appl. Phys. Vol 7, L 100 (1974).

PHYSICO-CHEMICAL ANALYSIS OF THIN FILMS

A. CACHARD

Université Claude Bernard /Lyon I
Département de Physique des Matériaux
43, boulevard du 11 Novembre 1918
69621 - Villeurbanne   France

1.   INTRODUCTION

In order to fully understand the behavior of thin films and to ascertain the interpretation of their physical properties, a number of basic parameters must be known and monitored.

The single most significant film parameter is its thickness and details about the measurements are given in the paper of D.S. Campbell. The problems involved with structure determination are treated in the paper of H.Raether. In this paper, we will be concerned with the analytical techniques for the determination of the atomic composition (stoichiometry), the impurities and the bondings of thin films.

A large number of analytical techniques have been developed or modified for thin film specimens which typically consist of 1 to 10 cm^2 surface area and 10 to 1000 µg material. A detailed discussion of all these techniques is beyond the scope of this paper and we will not attempt to reproduce here that which already exists in general reviews and texts[1-3]. This paper will concentrate upon rather new techniques which have been developed and increasingly used in the last few years : the nuclear microanalysis (MeV ion beams), the secondary ion mass spectrometry (keV ion beams) and X-ray or photoelectron spectroscopies.

Special attention is given on lateral or in depth resolutions, detection limits, and quantitative or qualitative possibilities.

2. COMPOSITION DETERMINATION BY MeV ENERGY RANGE ION BEAM ANALYSIS

2.1. Introduction

Basically the method consists of placing a target in a monoenergetic beam of ions of hydrogen, helium or any other light elements and of detecting the interaction products which provide information on the surface layer. When incoming ions A bombard atoms B of the target there are three main interaction possible : their trajectory can be deviated without energy loss and this is called elastic scattering ; or there can be nuclear reaction : A+B → C+D + Q, where Q is the energy dissipation, D the remaining nucleus and C the emitted species (proton, α-particles, γ rays, etc.) ; or only an electronic interaction can occur where B is excited and then desexcites via X-ray emission (the so-called ion induced X-ray emission). The activation process case where the result of the reaction is a radioactive nucleus will not be considered here since in general it needs higher energy ions.

The instrumentation required is standard in nuclear physics. A particle accelerator provides a mono-energetic ion beam, the size of which is of the order of 1 mm^2 and the intensity varies from nA (in backscattering measurements) to µA (nuclear reaction case). The interaction takes place in a target chamber in which the vacuum is of the order of 10^{-5} - 10^{-6} torr. Most of the time, the interaction products are detected using a semiconductor detector which has the advantages of simplicity, ability to be used with large solid angles, insensitivity to the charge state of the detected particles and the capability of non dispersive measurement. The electronic instruments (preamplifier and linear amplifier) required for the signal analysis deliver pulses whose heights are proportionnal to the energy of the incident particles. A multichannel pulse height analyser gives the energy spectrum of the emitted particles.

2.2. Backscattering analysis

The elastic scattering of an incoming particle of mass m by a target atom of mass M is characterized by the conservation of the momentum and of the kinetic energy. This gives a simple relationship between the energy E of the incoming particle and its energy E' after scattering in the θ direction[4-6]

$$E' = k^2 E \qquad (1)$$

where $k = \dfrac{m \cos\theta + \sqrt{M^2 - m^2 \sin^2\theta}}{m+M}$

For θ = 180° the formula is very simple:

$$k = \frac{M-m}{M+m}$$

For a given particle and scattering angle, k (and thus E') is dependent only on the atomic mass of the target atoms. Backscattering provides mass-sensitive analysis. The selectivity ($\sim dk^2/dM$) decreases with increasing M, and increases with increasing θ and m. However, the selectivity is rather poor and it allows the distinction of elements and isotopes only up to about 40. Above this weight, two elements can be distinguished only if their mass difference is larger than 10 atomic units. However, in special cases[4-7], isotopes such as ^{63}Cu and ^{65}Cu have been separated.

The sensitivity of a backscattering analysis is directly related to the differential scattering cross-section which for E < 3 MeV is given by the Rutherford formula:

$$\frac{d\sigma}{d\Omega} = 1.3 \cdot 10^{-27} \left(\frac{zZ}{E}\right)^2 \left(\frac{m+M}{M}\right)^2 \left(\sin\frac{\theta}{2}\right)^{-4} \quad (cm^2/ster) \qquad (2)$$

where z and Z are the atomic number of the incident ion and target atom, respectively. For a given ion, the sensitivity is a decreasing function of θ and E and an increasing function of Z.

Typically in backscattering analyis, one uses large angles (θ > 160°), energies in the MeV range, and heavy ions. Most of the time, $^4He^+$ ion beams are used; different analysing particles such as lithium[8], carbon[9] and oxygen[7] beams have also been used. The limit of detection is routinely in the submonolayer range and can be as low as 10^{10} atoms/cm^2 in special cases (gold on silicon using ^{12}C beams[9]).

Thick film analysis:

For film thickness larger than 300 Å, the ion beam loses energy as it penetrates the film, and the signals from greater depth will have lower energy.

In figure 1, the incoming ion beam has an energy E_0. After scattering by a surface atom the energy in the θ direction is $E'_0 = k^2 E_0$. At a depth x, the incoming particle energy is:

$$E_1 = E_0 - \left| \int_0^x \frac{dE}{dx} dx \right|$$

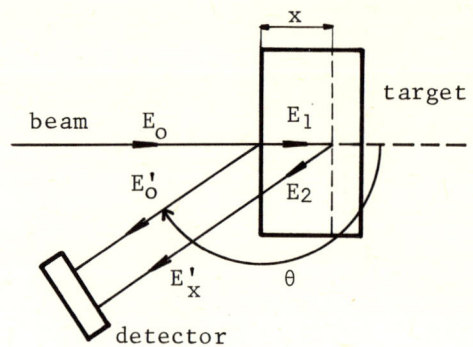

Fig.1   Particle trajectories in a thick film

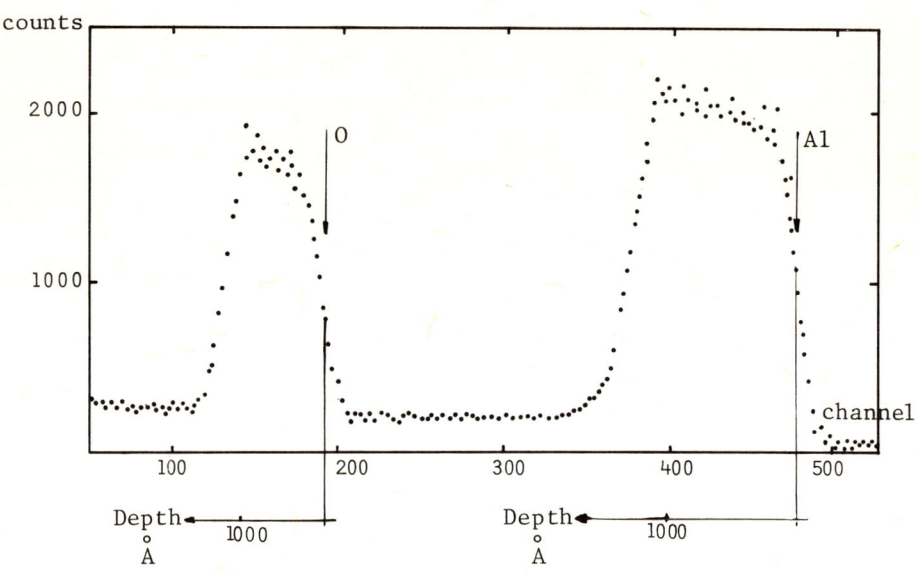

Fig.2   Energy spectrum for 1.9 MeV $^7Li^+$ ions backscattered from 1200 Å self-supporting anodic Al_2O_3

where dE/dx is the energy loss depending on the target material and on the ion. The particles scattered at the depth x have an energy $E_2 = k^2 E_1$, and emerge from the target in the θ direction with the energy :

$$E'_x = E_2 - \left| \int_0^{\frac{x}{|\cos \theta|}} \frac{dE}{dx} \, dx \right| \qquad (3)$$

E'_x is significantly lower than E'_o and the detector receives a non-monoenergetic particle beam (for a given target mass) as it does in the case for very thin films. For example, for α-particles in alumina dE/dx is of the order of 30 eV/Å which, for a 1000 Å thick film, gives an energy difference of 70 keV between surface scattered particles and back-side scattered particles. With a typical 15 keV electronic resolution, such a difference is easily measured. Figure 2 gives the energy spectrum for 1.9 MeV $^7Li^+$ ions back-scattered from 1200 Å self-supporting anodic Al_2O_3. The correspondance between the energy scale and the depth scale is obtained by the knowledge of dE/dx. The energy loss is given in semi-empirical tables[10-11] or can be measured in the case where more accurate information is needed. dE/dx depends on E, but for thicknesses lower than 0.5 µm, one can consider dE/dx to be nearly constant ; thus, the energy difference $\Delta E_x = E'_x - E'_o$ is nearly equal to :

$$\Delta E_x \simeq [S] \, x \qquad (4)$$

where [S] the energy loss parameter is given by :

$$[S] = k^2 \left| \frac{dE}{dx} \right|_{\overline{E}_1} + \frac{1}{|\cos \theta|} \left| \frac{dE}{dx} \right|_{\overline{E}_2} \qquad (5)$$

and $\overline{E}_1 \simeq E_o$ and $\overline{E}_2 \simeq E_2$

[S] depends on the mass and energy of the incident particle, on the composition and on the nucleus of the target under study. In the case of alumina [S] is different for particles scattered by oxygen and by aluminium.

Quantitative analysis :

The <u>total number of a certain atom</u> (per cm^2) can be obtained from the area in counts under the peak characteristic of that atom. For a total number Q of incident particles and a detector solid angle Ω, the area of the peak A is given by :

$$A = Q\Omega \, \frac{d\sigma}{d\Omega} \, N \qquad (6)$$

where N is the number of atoms per cm². This formula allows the determination of N as long as $d\sigma/d\Omega$ is constant; for thick films a correction should be made to account for the change of $d\sigma/d\Omega$ with energy.

The <u>composition of a compound</u> can be also obtained from the spectrum using the same formula (6). In the case shown in figure 2 the ratio of the number of oxygen to the number of aluminium is given by:

$$\frac{N_O}{N_{Al}} = \frac{A_O}{A_{Al}} \left(\frac{d\sigma}{d\Omega}\right)_{Al} \bigg/ \left(\frac{d\sigma}{d\Omega}\right)_O \qquad (7)$$

The ratio of the cross sections can be calculated from the Rutherford formula. Actual values of compositions can be obtained within 1 %.

An <u>in-depth analysis</u> (concentration profile determination) can be performed using the characteristic shape of the peak under study. The multichannel analyser gives the energy spectrum in number of counts per channel (figure 3) where each channel represents a step δE in energy. At energy E (actually between E and E + δE), the total number of counts is $\Delta n(E)$ and the "peak area" is $\Delta n(E)\delta E$. This peak area represents the number of detected particles scattered by a thin film of thickness $\delta x = \delta E/[S]$ located at a depth $x = E_0-E/[S]$. If $n(x)$ is the atomic concentration (in cm^{-3}), expression (6) can be written:

$$\Delta n(E)\delta E = Q\Omega \frac{d\sigma}{d\Omega} n(x)dx = Q\Omega \frac{d\sigma}{d\Omega} n(x)[S]\delta E \quad \text{or}$$

$$n(x) = \frac{\Delta n(E)[S]}{Qn \frac{d\sigma}{d\Omega}} \qquad (8)$$

The measurement of $\Delta n(E)$ (peak shape) and the knowledge of [S] and $d\sigma/d\Omega$ allow the concentration profile determination. Figure 4 gives the ^{4}He$^+$ backscattering spectrum from 5400 Å SiO$_2$ grown on a Si substrate and implanted with ^{64}Zn at 280 keV to a dose of 10^{15}/cm² (12). The two plateaus of silicon are indicative of the concentration of Si in SiO$_2$ and of Si in silicon. The ratio of the heights of these plateaus indicates stoechiometric SiO$_2$, and the ratio of total area of oxygen and silicon (in SiO$_2$) gives the same result. This example shows that the determination of a light element (here oxygen) in a heavier matrix (here SiO$_2$) results in a superposition of a peak on a continuous background (here due to bulk silicon) and is not the best case of the application of backscattering. The more favorable situation where a heavy impurity is

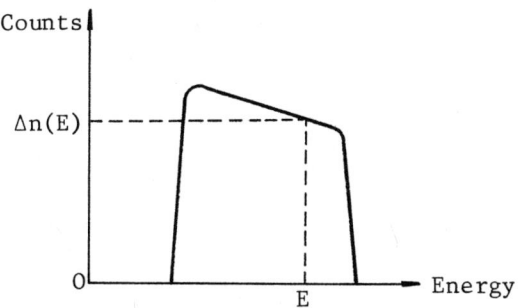

Fig.3   Energy spectrum for ions backscattered from a thick film

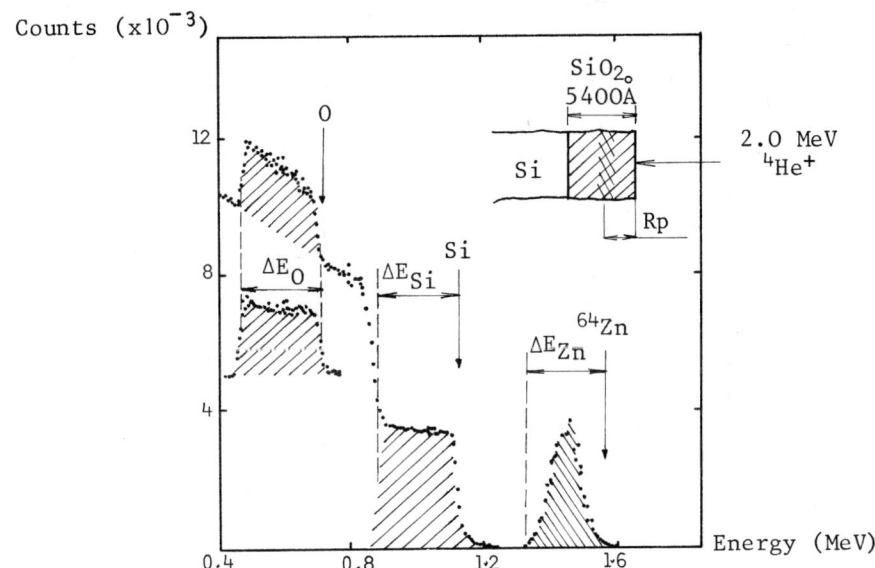

Fig.4   Energy spectrum of 2 MeV ^{4}He$^+$ ions backscattered from a sample (5400 Å of SiO_2 on Si) implanted at 280 keV with ^{64}Zn. The oxygen and silicon components of SiO_2 are the shaded areas, and ^{64}Zn is the dotted area (taken from ref. 12).

in a light element matrix, is shown by the peak of Zn located in the background free region of the spectrum. This peak is shifted to lower energy by an amount δE_{Zn} corresponding to the depth distribution of Zn below the surface.

Typically the depth resolution is of the order of 200 Å, and can be made as low as 100 Å in special cases, by proper choice of analysing ions and of element to be analysed.

In conclusion ion backscattering analysis is a method perfectly adapted to the study of films that are self-supporting or on light substrate (carbone). It is also well adapted to the study of heavy impurities in light element matrices. This method has been used to measure the composition of dielectric layers[13-16] to study the formation of anodic oxide films[17] and of silicides[18-20] and to determine the implantation profile of dopants in silicon[21,22].

2.3. Analysis by nuclear reactions

The bombardment of nuclei of mass M by MeV energy range ions of mass m can induce nuclear reactions, yielding emitted particles of mass m' and remaining nuclei of mass M'; the defect mass energy is the Q value of the reaction. Such reactions are usually written : M(m,m')M'.

The basic advantage of nuclear reaction analysis with respect to back-scattering is that medium and high Z nuclei practically do not contribute to nuclear reactions, owing to Coulomb barrier repulsion ; this allows background-free detection of light elements on heavier substrate. Due to the high positive Q values of many such reactions, the emitted particles have energies well above that of the beam, allowing the elastically scattered particles to be stopped in a suitable absorber. The second main advantage is that nuclear reactions are specific : two isotopes of the same element behave in a completely different way, and in many cases only one of the isotopes has a positive Q value, the other being ignored less regard of its concentration. This is in contrast to backscattering in which only the k factor is slightly changed.

The basic example is the first use of nuclear reactions in 1962[23] for the study of oxygen transport in anodic aluminium oxide, using $^{16}O(d,p)^{17}O$ and $^{18}O(p,\alpha)^{15}N$ reactions. The nuclei now routinely determined this way are ^2H, ^6Li, ^7Li, ^{11}B, ^{12}C, ^{14}N, ^{19}F, ^{27}Al, ^{28}Si, ^{31}P. More details may be found in references 24 to 31.

The selectivity of this method is not easily discussed due to the specificity of the nuclear reactions. The only limitations

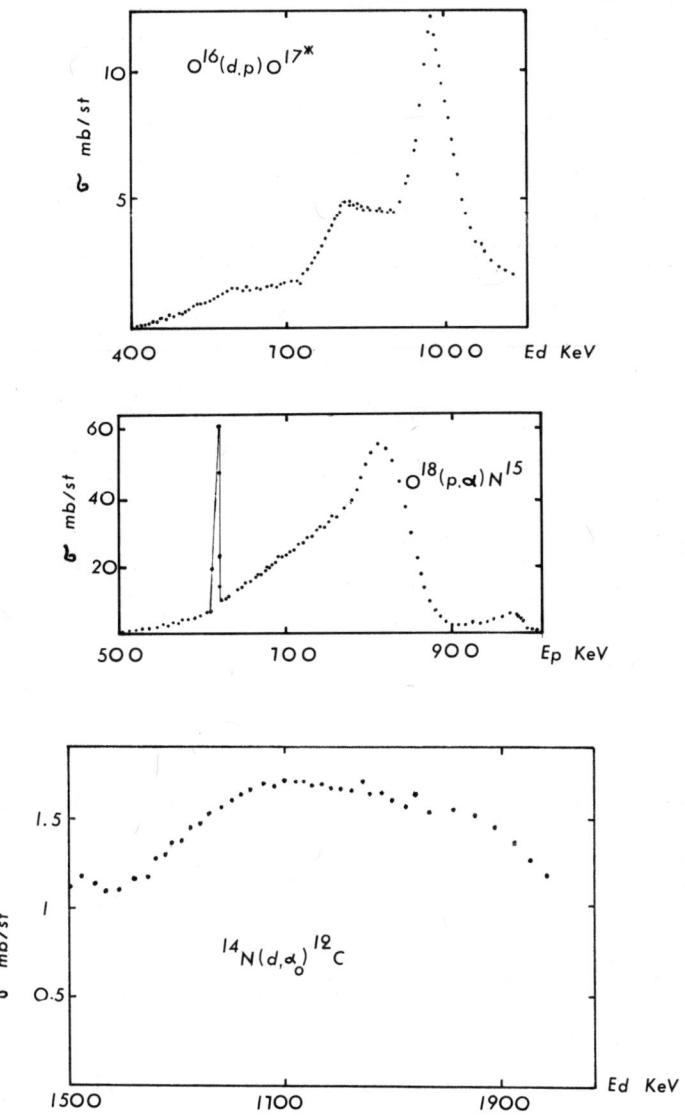

Fig.5 Differential cross-sections of the reactions used to analyse ^{16}O, ^{18}O (at 165°) and ^{14}N (at 160°). ^{16}O and ^{18}O curves are taken from ref.24, and ^{14}N curve from ref.32.

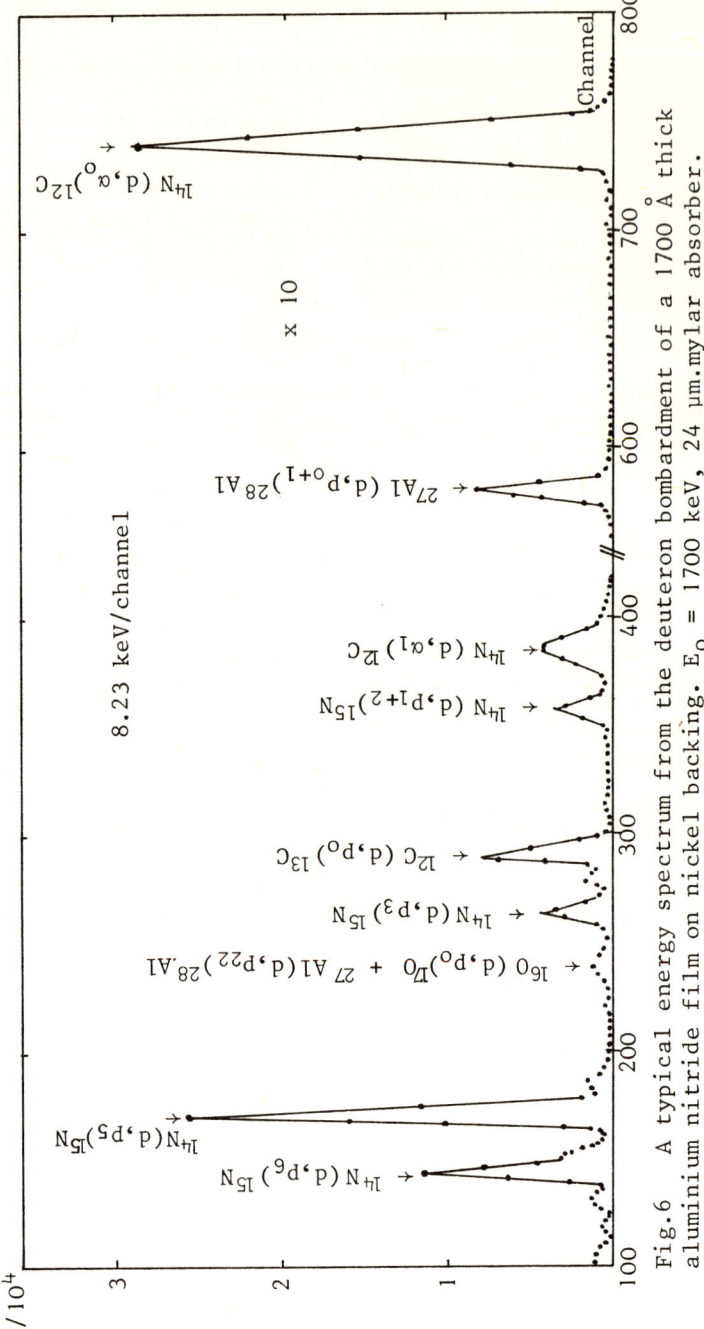

Fig.6 A typical energy spectrum from the deuteron bombardment of a 1700 Å thick aluminium nitride film on nickel backing. E_o = 1700 keV, 24 μm.mylar absorber.

are possible interferences between two peaks in the energy spectrum arising when several light nuclei are simultaneously present in the sample. Each case is to be studied, and the optimal nuclear reaction at the optimal energy must be determined.

The sensitivity, as for backscattering, depends on the differential cross-section of the reaction under study, but unlike backscattering there is no analytical expression. The cross-section curves have to be determined experimentally. Examples are shown in figure 5 for ^{16}O, ^{18}O and ^{14}N. In the case ^{16}O, there is a plateau between 800 and 900 keV for the $^{16}O(d,p_1)^{17}O^*$ reaction (this reaction is more efficient to use than that leading to the fundamental level of ^{17}O). For ^{18}O, the $^{18}O(p,\alpha)^{15}N$ reaction presents a stationary cross-section near 730 keV, and a strong narrow resonance at 629 keV. For ^{14}N, there are a variety of deuteron induced reactions, but those leading to α-particles emission are the more interesting. The α_0-particle from $^{14}N(d,\alpha_0)^{12}C$ is very energetic and less liable to interfere with other reactions ; its cross section presents a good plateau between 1650 and 1730 keV.[32]

Figure 6 gives the typical energy spectrum from the deuteron bombardment (E = 1700 keV) of a 1700 Å thick aluminum nitride film on nickel backing. The peaks of the $^{27}Al(d,p_0)^{28}Al$ and $^{14}N(d,\alpha_0)^{12}C$ reactions are located in a background free part of the spectrum and allow a precise determination of the composition of the film. Simultaneously the oxygen and carbon contaminants are visible.

Quantitative analysis :

The total number N of atoms (per cm^2) in the surface region of the sample is obtained from formula (6) as for backscattering experiments :

$$N = \frac{A}{Q\Omega \frac{d\sigma}{d\Omega}}$$

The bombarding energy is chosen near a plateau of the differential cross-section curve, so that $d\sigma/d\Omega$ appears as a constant. The absolute quantity N may be obtained by comparison to reference standards regardless of their physical or chemical states since the reaction yield depends only on nuclear cross sections.

As an example, by using $^{16}O(d,p_1)^{17}O^*$ reaction, less than a monolayer of oxygen (10^{14} atom/cm^2) can be detected within minutes with a precision of a few percent.

The composition of a compound can be obtained in the same

way. From figure 6, the ratio of the number of nitrogen to the number of aluminium can be calculated :

$$\frac{N_N}{N_{Al}} = \frac{A_N(\alpha_o)}{A_{Al}(p_o)} \cdot (\frac{d\sigma}{d\Omega})_{Al(p_o)} \cdot (\frac{d\sigma}{d\Omega})^{-1}_{N(\alpha_o)}$$

using the $^{14}N(d,\alpha_o)^{12}C$ and $^{27}Al(d,p_o)^{28}Al$ deuteron induced reactions. The ratio of the cross sections is determined from a standard of known nitrogen/aluminium ratio (in this case, aluminium oxinate[32]). The composition can be measured within a few percent precision even for films less than 100 Å thick. Such a method has been used also for SiO_x[30], Al_2O_3[31] and Si_3N_4[33,15].

The <u>concentration profile</u> may be obtained by analysing the spectrum of the detected particles. The method is the same as that developed for backscattering spectra, but for nuclear reactions the depth resolution is rather poor. This is due to the lower masses of the bombarding particles (usually protons or deuterons instead of α-particles or heavier ions common in backscattering) giving lower energy loss in the film, and to energy straggling in the absorber. Typical depth resolution is 3000 Å.

In the particular case where there exists a strong narrow resonance in the cross-section curve, the concentration profile may be obtained with better depth resolution[23,24]. Figure 7 gives the principle of concentration profile measurements using narrow resonance.

The yield of the reaction induced by protons is almost zero compared with its value at the energy E_R. If the beam energy is $E_o > E_R$ the protons, upon slowing down, reach the resonance energy E_R at a depth $x = E_o-E_R/[S]$ where [S] is the energy loss parameter for the protons in the matrix under study. For energy E_o the reaction yield comes only from a narrow slab of the film near the depth x, and the yield is proportional to the concentration C(x). By varying E_o from E_R towards higher energies, the resonance is displaced from the surface to the interior of the film. Hence the variations of the yield versus E_o give an image of the concentration profile C(x).

Such a method has been extensively used by Amsel and co-workers especially for ^{18}O tracing to study oxygen transport during anodic oxydation of metals[23,24,34]. They used the 629 keV resonance of the $^{18}O(p,\alpha)^{15}N$ reaction which allows a 200 Å depth resolution in Ta_2O_5. Figure 8 shows a typical result obtained with the $^{27}Al(p,\gamma)^{28}Si$ very narrow resonance at 992 keV. The target is a 0.8 μm thick evaporated aluminium layer on tantalum backing with a 475 Å thick anodic oxide film[24].

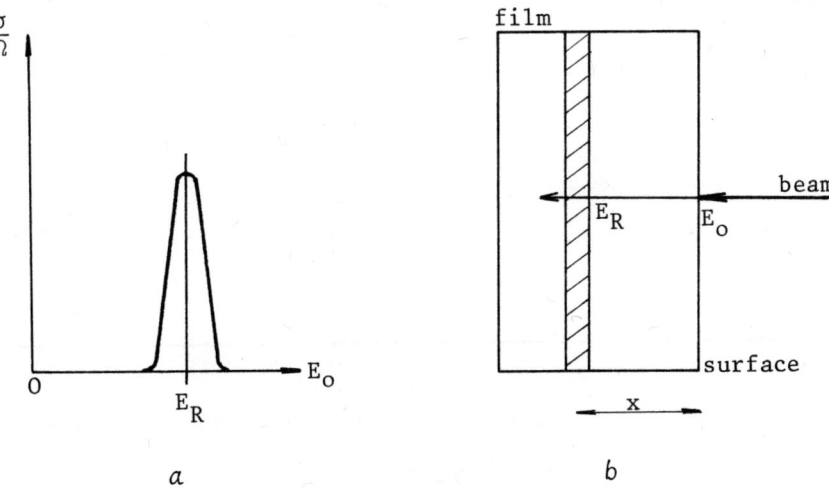

Fig.7 Principle of concentration profile measurements using narrow resonances.

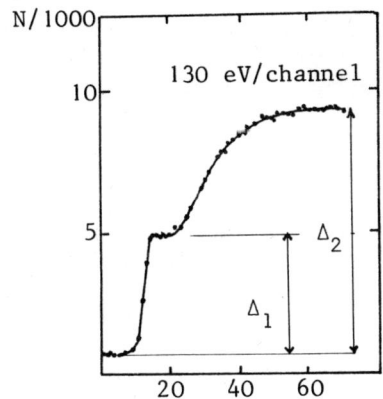

Fig.8 Typical excitation curve of the ^{27}Al $(p,\gamma)^{28}$Si resonance near 992 keV for a 475 Å thick Al_2O_3 film on 0.8 μm aluminium (taken from ref. 24).

The sharp rise and the good plateau correspond to a depth resolution of 60 Å in Al_2O_3. The ratio of the plateau heights Δ_1 in Al_2O_3 and Δ_2 in aluminium indicates an actual Al_2O_3 composition for the oxide film.

The same reaction on aluminium has been used by Dunning[29] to determine the concentration profile of aluminium implanted silicon oxide.

2.4. Ion induced X-ray emission

A target placed in a particle beam (either electrons or ions) exhibits X-ray emission due to the vacancies created by the ejection of inner orbital electrons. The subsequent reordering of the electronic shells induces X-ray emission which is characteristic of the target atom. The analysis of the energy and intensity of the emitted X-rays allows the determination of the type and the concentration of the atoms present respectively.

Ion induced X-ray emission[36,37] has two main advantages over the electron microprobe[38]. Calculations show that the magnitude of the continuous electromagnetic radiation (bremsstrahlung) which accompagnies X-rays produced by electron irradiation is reduced by a ratio of $(M_{ion}/M_e)^2$, a 10^6 factor under ideal conditions. Evenmore this bremsstrahlung varies as $(Z/M)^2$ which gives an advantage to heavier ions. In fact, however, a continuous background exists due to secondary electrons ejected along the ingoing particle path. The second advantage is that the X-ray production cross-section for heavy ions exhibits an energy cut-off above which the cross-section increases very rapidly as a function of energy. Thus, by varying the ion energy selective excitation[39] is possible.

The use of solid state Si(Li) detectors with a high efficiency and a good resolution (150 eV) results in good selectivity ; in principle all elements can be separated, except in the case where elements adjacent on the periodic table exist in the target in greatly differing concentrations. In such a case the peak of the more highly concentrated element overshadows that of the other. For example Musket and Bauer[40] have analysed vacuum annealed stainless steel by means of proton induced X-rays. They were able to separate the contributions of iron, nickel and chromium. Such selectivity cannot be achieved by backscattering analysis.

The sensitivity of this X-ray technique is very high, and Gray et al[41] claimed to have measured arsenic in silicon implanted at a level of 2.10^{13} As/cm^2.

On the other hand, the depth resolution is very poor, if any, because the result of X-ray absorption is the decrease of intensity without changes in energy. Thus, a priori we can get no informations from the peak shape. However, using the cross-section versus ion energy relationship, Musket[40] was able to obtain a concentration profile but with an in-depth resolutions higher than 0.25 μm.

In conclusion, ion induced X-ray emission appears as a promising tool for surface and thin film analyses. This is a very new technique and both the fundamental theory behind it and practical applications are presently the subject of a number of investigations.

2.5. Conclusion

Backscattering and nuclear reaction techniques have been used successfully in microanalytic studies of surfaces and thin films. To a large extent the optimum technique depends on the nature of the problem. For example, nuclear reactions are more suitable for low Z elements in heavier substrates and backscattering for heavy impurities in light matrices. For light element dielectric films such as SiO_2, Si_3N_4, Al_2O_3, etc. either of the two methods can be used by varying the substrate (heavy for nuclear reactions and carbon for backscattering). These two methods appear to be complementary.

For both methods the results are quantitative, a 1 % precision being easily reached. The sensitivity may be very high (10^{12} atoms/cm^2) and the analysis is non destructive. The results are generally independent of the physical or chemical states of the nuclei in the matrix, the only exceptions being the channeling effect in precisely oriented single crystals[35]. Depth distributions may be obtained in the first micron of a film with a resolution of the order of 100 Å, but the lateral resolution is poor (larger than 0.1 mm).

Ion induced X-ray emission, a more recently developped technique, is of great interest both as an analytical tool in itself and as a complement to backscattering analyses.

3. SECONDARY ION MASS SPECTROMETRY

When bombarding a target by a 5-30 keV beam of primary ions, the surface layers are sputtered off and a variety of secondary species is produced including neutral atoms or molecules, positive or negative ions, electrons and photons. The mass spectrometric analysis of the positive and negative sputtered secondary ions is a useful method for characterizing the target[42-44]. The primary beam size can be varied from 1 to 300 μm with scanning possibilities and the secondary ions are collected and analysed either by a classical mass spectrometer or by a direct imaging analyser[42]; both cases ensure good lateral resolution.

The ion production arises by two different processes : "kinetic" and "chemical". The former process is the transfer of kinetic energy between primary ions and target atoms, which can be ejected from the surface. Unbound electrons in the excitation volume cause the ejected ions to be neutralized before they escape into the vacuum where their state is metastable. This metastable atom becomes ionized by Auger electron emission and can be ion mass analysed. This "kinetic" ionization process predominates in inert gas bombardment of metals and semi-conductors.

The "chemical" ionization process is due to the reduction of the number of free electrons available for ion neutralization before escaping. Such reduction increases the production of secondary ions and occurs when reactive species such as superfical oxide or bulk chemical compounds (oxides, chlorides, etc.) are present.

It is assumed that the two processes exist together, and this can lead to ambiguities in the interpretation of secondary ion intensities because chemical variations in the sample can cause the intensity to change even if the actual concentration remains constant. In order to avoid such difficulties and to ensure high emission yield two techniques are currently used. In the first, a reactive species such as oxygen is introduced into the residual vacuum. In the second, the primary beam consists of oxygen ions. In both cases a surface oxide is continously formed and ensures the predominance of the chemical ionization process.

The <u>quantitative analysis</u> makes use of the erosion capabilities of the sputter ion source to obtain three dimensional analysis. Lateral analyses (x,y microanalysis) are possible with less than 1 μm lateral resolution (similar to the electron microprobe). In depth analysis is a result of the sputter removal of material by the primary ion beam. If we can assure a constant erosion rate, the ion current versus time relationship is easily converted into a concentration versus depth curve, but the correlation of time and depth requires careful attention due to the great variety of factors influencing the erosion rate. In the best cases the depth resolution is less than 100 Å.

Quantitative measurements are achieved using the comparative standard technique which requires a precharacterized sample of the element to be analysed in the matrix to be employed. Such a complicated process explains why up to now such a technique is only routinely employed in a few cases like silicon technology. Most of the time in other applications the analysis is only semi-quantitative.

However, secondary ion mass spectrometry is very useful for thin film study[42,44] mainly because of its superior lateral and depth resolutions. All elements can be analyzed with good isotopic separation and great sensitivity (10^{-15} to 10^{-19} g range).

4. X-RAY AND PHOTOELECTRON SPECTROSCOPIES

The methods described above in sections 2 and 3 give information about the atomic composition of a film but not about the chemical surroundings of the atoms. For nuclear methods this uncapability is due to the physics involved in the phenomena. However, this is not the case for ion induced X-ray analyses where the chemical shift of characteristic X-rays would be evaluated by use of a crystal detector.

Evaluation of the chemical bonding in various glass films has been performed by Pliskin[45,46] using the infrared absorption technique which deal with direct absorption of electromagnetic radiations by electric dipoles oscillating in a molecule.

Information on changes in the electronic structure of atoms due to differences in the chemical state of these atoms can be obtained using either X-ray or photoelectron spectroscopy.

4.1. X-ray spectroscopy

When an atom is irradiated by an X-ray beam transitions between electronic levels (either emission or absorption) occur which cause peaks in the subsequent energy spectrum. In the case of changes in chemical bonding, the electronic distribution around the atom changes and the observed peaks shift in energy.

For a solid only the inner shell maintains its atomic character while the outer shell exhibits a band structure. The X-ray emission results from the transition between a conduction band state (metal) or a valence band state (semi-conductor or insulator) and an inner shell energy level with rather atomic character. A broad emission band is thus observed which give a picture of the density of states of the occupied levels.

The X-ray absorption is the result of the transition between an inner orbital level and an empty state of the conduction band. Thus, an X-ray absorption spectrum gives an image of the empty level density of states.

In the case of an insulator or a semi-conductor, the energy difference between the X-ray emission and absorption band edges (with the same inner shell level) sometimes yields data about the energy gap. However, in many compounds such a difference has no real physical meaning because of the existence of intense absorption lines superimposed on the absorption discontinuity. Such peaks are indicative of transitions towards localized orbitals with atomic character.

By varying the chemical surroundings of an atom, a shift is induced either in the emission or in the absorption band edge and can give information, for instance, on the degree of oxidation of a metal in different compounds.

An exhaustive review of X-ray spectroscopy has recently been compiled by C.Bonnelle[47].

4.2. Photoelectron spectroscopy

Photoelectron spectroscopy is often referred to as ESCA (Electron Spectroscopy for Chemical Analysis). In this technique, the surface of a solid is bombarded by a monoenergetic X-ray beam and as a result photoelectrons are emitted with a kinetic energy which is roughly the difference between the primary energy of the excitation beam and the binding energy of the photoelectrons. The use of a high energy resolution electron spectrometer allows the measurement of the energy levels in the solid[48,49].

The advantages of this method from X-ray absorption are that the core levels can be seen alone giving sharp line spectra and that the absolute binding energy can be obtained. On the other hand, only occupied states are accessible, and in that sense, the two methods are complementary.

The two main applications in thin film field are the bonding determination in a compound and the band structure measurement in solids[50]. Quantitative concentration measurements can be obtained from the intensities of the emission lines. For such an elemental analysis both heavy and light elements can be detected with good sensitivity (less than 10^{-8} g). ESCA, however, is restricted to surface analysis due to the low average depth (10 Å) from where photoelectrons can be observed.

5. CONCLUSION

There is a large variety of techniques for physico-chemical evaluation of thin films, and those described in this paper have proved to be extremely useful. Of course, the choice of analytical methods always depends on the purpose of the particular investigation (atomic composition or bondings), the nature of the film and the facilities and financial support available.

It must be stressed that a single technique is seldom sufficient to satisfactorily characterize a film ; in the vast majority of cases, it is necessary to make use of several complementary methods.

REFERENCES

1. The Use of Thin Films in Physical Investigations, Edited by J.C.Anderson, Academic Press (1966)

2. Thin Film Phenomena, K.L. Chopra, Mc Graw-Hill Book Company (1969)

3. Handbook of Thin Film Technology, Edited by L.I.Maissel and R.Glang, Mc Graw-Hill Book Company (1970)

4. M.A.Nicolet, J.W.Mayer and I.V.Mitchell, Science, 177, 841 (1972)

5. M.Peisach and D.O.Poole, Anal.Chem. 38, 1345 (1966)

6. W.K.Chu, J.W.Mayer, M.A.Nicolet, T.M.Buck, G.Amsel and F.Eisen, Thin Solid Films 17, 1 (1973)

7. S.Petterson, P.A.Tove, O.Meyer, B.Sundquist and A.Johansson, Thin Solid Films 19, 157 (1973)

8. JP.Thomas, A.Cachard, M.Fallavier, J.Tardy, S.Marsaud and J.Pivot, Rev.Phys.Appl., to be published

9. F.Abel, G.Amsel, M.Bruneaux, C.Cohen, B.Maurel, S.Rigo and J.Roussel, J.Radioanal.Chem. 16, 587 (1973)

10. C.F.Williamson, J.P.Boujot and J.Picard, CEA Report R.3042 (1966)

11. L.C.Northcliffe and R.F.Schilling, Nucl.Data Tables, 7, 233 (1970)

12. W.K.Chu, B.L.Crowder, J.W.Mayer and J.F.Ziegler, Appl.Phys. Letters, 22, 490 (1973)

13. I.V.Mitchell, M.Kamoshida and J.W.Mayer, J.Appl.Phys.42, 4378 (1971)

14. O.Meyer, J.Gyulai and J.W.Mayer, Surface Sci. 22, 263 (1970)

15. M.Croset, S.Rigo and G.Amsel, Appl.Phys.Letters, 19, 22 (1971)

16. D.V.Morgan and R.P.Gittins, Phys.Status Solidi (a), 13, 517 (1972)

17. M.Kamoshida and J.W.Mayer, J.Electrochem.Soc., 119, 1084 (1972)

18. A.Hiraki, M.A.Nicolet and J.W.Mayer, Appl.Phys.Letters, 18, 178 (1971)

19. R.W.Bower and J.W.Mayer, Appl.Phys.Letters, 20, 359 (1972)

20. C.J.Kircher, J.W.Mayer, K.N.Tu and J.F.Ziegler, Appl.Phys. Letters, 22, 81 (1973)

21. J.Haskell, E.Rimini and J.W.Mayer, J.Appl.Phys., 43, 3425 (1972)

22. J.W.Mayer, L.Eriksson and J.A.Davies, Ion Implantation in Semi-conductors, Academic Press, New-York (1970)

23. G.Amsel and D.Samuel, J.Phys.Chem.Solids, 23, 1707 (1962)

24. G.Amsel, J.P.Nadai, E.D'Artemare, D.David, E.Girard and J.Moulin, Nucl.Inst.and Meth., 92, 481 (1971)

25. G.Amsel and D.David, Rev.Phys.Appl., 4, 383 (1969)

26. E.Ligeon, A.Bontemps, J.Fontenille and G.Guernet, Proc. Microelectronics Conf., Paris, 1970, Chiron, Paris, p.50 (1970)

27. B.Maurel, D.Dieumegard and G.Amsel, J.Electrochem.Soc., 119, 1715 (1972)

28. M.Croset and D.Dieumegard, J.Electrochem.Soc., 120, 526 (1973)

29. K.L.Dunning, G.K.Hubler, J.Comas, W.H.Lucke and H.L.Hughes, Thin Solid Films, 19, 145 (1973)

30. A.Cachard, J.Pivot, A.Roger, M.Talvat and J.P.Thomas, Rev. Phys.Appl., 6, 279 (1971)

31. C.Diaine, J.A.Roger, J.Pivot, A.Cachard and C.Dupuy, Congrès AVISEM, Paris (1971)

32. J.Tardy, A.Cachard, J.P.Thomas and J.Engerran, Le Vide, 173, 359 (1974)

33. A.Cachard and J.Tardy, unpublished.

34. C.Cherki, Thesis, Paris (1969)

35. D.S.Gemmel, Rev.Modern Phys., *46*, 129 (1974)

36. J.A.Cairns, A.D.Marwick and I.V.Mitchell, Thin Solid Films, *19*, 91 (1973)

37. J.D.Garcia, Phys.Rev. A4, 955 (1971)

38. E.Reuter, Surface Sci., *25*, 80 (1970)

39. J.A.Cairns, Surface Sci., *34*, 638 (1973)

40. R.G.Musket and W.Bauer, Thin Solid Films, *19*, 69 (1973)

41. T.J.Gray, R.Lear, R.J.Dexter, F.N.Schwettmann and K.C.Wiemer, Thin Solid Films, *19*, 103 (1973)

42. G.Slodzian, Rev.Phys.Appl., *3*, 360 (1968)

43. C.Evans Jr, Thin Solid Films, *19*, 11 (1973)

44. A.Socha, Surface Sci., *25*, 147 (1971)

45. W.A.Pliskin, J.A.Perri and D.R.Kerr, Physics of Thin Films, (Ed. G.Hass and R.E.Thun), Academic Press, New-York, *4*, 257 (1967)

46. W.A.Pliskin, Progress in Analytical Chemistry, Vol.2 : Physical measurement and analysis of thin films, Plenum Press, New-York, p.168-192 (1969)

47. C.Bonnelle, in Physical Methods in Advanced Inorganic Chemistry, Ed.H.A.Hill and P.Day, Interscience Publishers, p.45-73 (1968)

48. K.Siegbahm, C.Nordling, A.Fahlman et al, Nova Acta Regiae Soc. Sci. Upsaliensis Ser.IV, *20* (1967)

49. K.Siegbahm, Trans.Roy.Soc.London, Ser.A, *268*, 33 (1970)

50. T.Di Stephano and D.Eastman, Phys.Rev.Letters, *27*, 1560 (1971)

THICKNESS MEASUREMENTS

D.S. CAMPBELL

Department of Electronic and Electrical Engineering
University of Technology
Loughborough, Leicestershire, U.K.

1. INTRODUCTION

The measurement of the thickness of thin films is usually of great importance for examining the properties of the films, as it is the very thinness which often gives rise to the properties which cause the film to be different from those of bulk materials. As a result various summaries have been given previously of the techniques availables[1][2].

Methods for measuring thickness can best be divided into two, and these are :

A. Direct measurement. In this technique a direct measure of the thickness of the film is obtained after deposition. Not all direct methods of measurement are absolute as often some previous calibration against another technique has to be employed. Nevertheless, such systems are widely used.

B. Indirect methods. Indirect methods depend on the control of the rate of deposition of the film so that provided a previous calibration has been made, it is possible to prepare a layer with an exact thickness. Rate control devices by their very nature, will depend on a calibration against direct thickness measurements using either absolute or derived systems.

This paper examines the various techniques for films prepared by the various methods that are available.

2. DIRECT THICKNESS MEASUREMENTS

2.1 Interference Methods

The fundamental method of measuring films most commonly used is that using interference technique[3][4][5] and this method has provided the reference technique against which most other film thickness measurements have been calibrated. The general requirement for a satisfactory interference measurement is that the film is smooth and is supported on a smooth substrate (an optical flat). Also, a well-defined edge is required to the film. If the substrate and film are sufficiently reflective, the substrate/film combination can then be made part of an interference system as shown in Fig. 1. With the top plate of the interference system semi-silvered on the surface nearest to the film/substrate combination, interference fringes will occur, corresponding to positions of constant displacement t between the top and bottom plate. If the top plate is then tilted very slightly relative to the bottom substrate/film combination, equal path length fringes will be formed running normal to the direction of slope of the top plate. If this normal direction is arranged to be a right angles to the direction of the edge of the film on the substrate, then the interference fringes will show a distinct displacement at the step and observation of this displacement in a suitable optical microscope arrangement, will allow a direct measurement of the film thickness.

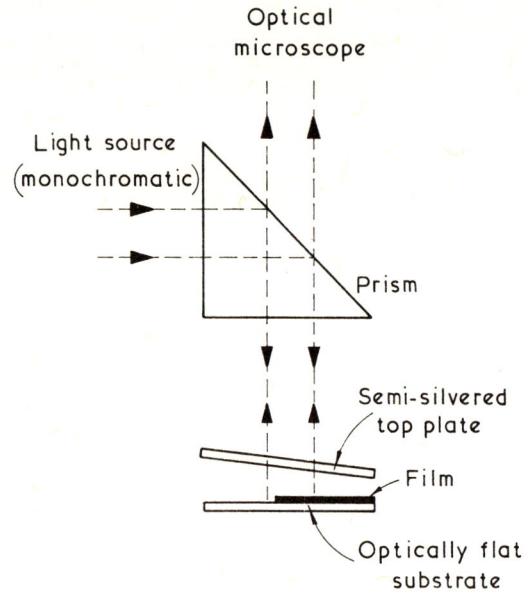

Fig. 1   Simple interferometer arrangement.

THICKNESS MEASUREMENTS

For satisfatory fringes to be obtained the surface of the substrate and film must be reflecting. Therefore it is usually necessary to evaporate a coat of silver or other reflecting metal on to the film/substrate combination, completely covering the step at the edge of the film.

The general equation relating I the relative intensity of the reflected light from a position in the system is given by :

$$I = I_{max}/(1 + F \sin^2 \pi \{\frac{2\mu t}{\lambda} \cos \Phi\}) \qquad (1)$$

where μ is the refractive index of the medium between the plates, t is the thickness between the plates, λ is the wavelength of the incident light and Φ is the angle of incidence of the incident light (see Fig. 2) F is given by

$$F = \frac{4R}{(1 - R^2)}$$

where R is the reflectivity of the film on the top plate.

It can be seen from equation (1) that fringes can be formed in a variety of ways and these are summarised in Table 1.

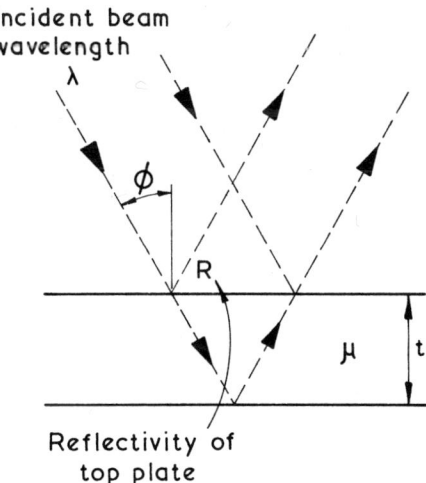

Fig. 2 Parameters of interference system.

Table 1

Summary of interference systems

Constant terms in equation (1)	Type of fringes	Name
λ, t	Equal inclination	Fabry-Perot
λ, Φ	Equal thickness	Fizeau
Φ	Equal t/λ	Fringes of equal chromatic order (FECO)
t	Equal $t \cos \Phi/\lambda$	White light Fabry-Perot

If monochromatic light is used, then it is possible to obtain very satisfactory interference effects using either Fizeau fringes or fringes of equal chromatic order. The system initially described in Fig. 1 was for the production of Fizeau fringes and these are the ones most normally used. The best sharpness in such fringes is obtained with R as high as possible, while allowing the fringe to remain visible, and the R value of 0.94 is found most satifactory. The light source used for such a measurement is usually a mercury lamp but this source is not truly monochromatic. There is a blue line associated with the mercury green which if not filtered out will cause every fourth fringe appear to have a brown edge.

Differences between successive fringes will correspond to total thickness differences of $\lambda/2$ and thus a direct measurement of the side of a step can be obtained by comparing the displacement of the fringe produced by a step with the distance between successive fringes. The accuracy of this type of measurement is limited essentially by thermal effects in the interferometer and a limiting value of $\pm$ 10 Å is usually quoted.

Fringes of equal chromatic order can also be used effectively but in this case it is necessary to employ a spectrometer to separate out the fringes. Fig. 3 shows a diagram of a typical set-up used, and fringes are obtained at the plane HK where they can be observed either with an eyepiece or photographed. A similar accu-

THICKNESS MEASUREMENTS

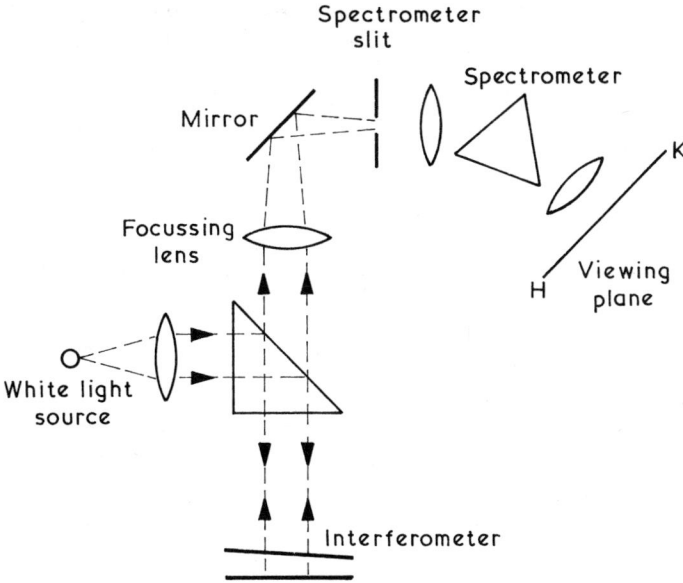

Fig. 3   Arrangement for observation of FECO interference patterns.

racy is obtained to that produced by Fizeau fringe observation.

Methods of improving the accuracy of measurement have been studied. One example is the photoelectric method of detecting interference minima that has been developed by Dyson (6,7,5). Fig. 4 shows one arrangement of his apparatus. If the beams A and B fall on identical surfaces, then the compensator can be adjusted to allow zero light through. However, if A and B travel different distances then light will be observed. The null position is best detected by applying an A/C field to the ADP and detecting this via the compensator with a photo-multiplier using a phase conscious rectifier which feeds a centre zero meter. The setting then consists of operating the compensator until the meter reads zero.

By this means it is found possible to make settings with a standard deviation corresponding to an uncertainty of film thickness of ± 0.1 Å. Such a figure is somewhat academic however because of the variation in flatness of substrates. Settings are possible to ± 1 Å with certainty and this method is the most sensitive available for determining film thickness.

Before leaving interference techniques it is worth noting that in certain cases the interference colours produced by oxide films on metal can themselves be used as a thickness guide. This

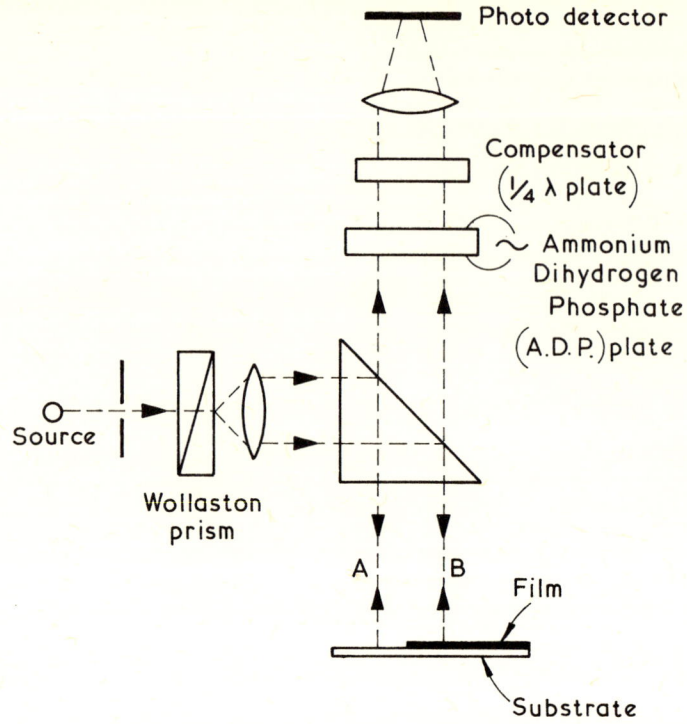

Fig. 4 Dyson's interferometer.

is particularly true for oxides such as SiO_2, Ta_2O_5 and Nb_2O_5 on the parent metal. A sensitivity of around ± 40 Å is possible around a thickness of 400 Å but the technique cannot be used for low (<150 Å) or high (>5000 Å) thicknesses.

2.2 Photometric Methods[1][29]

It is possible to determine the thickness and refractive index of a film by measurement of the transmittance and reflectance of a film. The simplest form this type of observation can take is that of a direct measurement of the intensity of a transmitted beam through the film. If the absorption of the film is known as

THICKNESS MEASUREMENTS

a function of thickness then the technique can be used for thickness determination. However, such a simple approach will only apply to a continuous film with small grain size and more generally, measurement of reflectance is required as well.

If transmittance and reflectance are known, then the mathematical equations to be solved are not complex. However, various purely numerical methods have been derived which can ease the problem.[30][31]

2.3 Polarimetric Methods

Polarimetric methods depend for their operation on the fact that a metal surface can be directly detected by measuring the ellipticity of the light after reflection from the surface. Such measurements, however, involve the knowledge of the real and imaginary parts of the refractive index of the film as well as that of the thickness under examination. A rather difficult mathematical procedure is usually necessary and although graphical and arithmetical procedures for fitting refractive index and thickness to the experimental data are available, the process is somewhat cumbersome. Details of the theory and techniques can be found in other texts.[8][9]

The method has the basic attraction that no step is required in the film, but the film itself must be transparent to light. The accuracy that can be finally obtained is considerable, figures of ± 0.5 Å have been claimed if refractive index measurements are known sufficiently accurately but ± 10 Å is more normal.

2.4 Mechanical Methods

Mechanical methods of measuring film thickness can involve either the scanning of a film surface with a probe, the weighing of a film after deposition or the rotation of a cylinder or disc. All these methods will however depend ultimately on an optical determination of thickness as a calibration.

2.4.1 Probe Systems

The most widely used probe system available for thin film measurements is that developed by the Rank-Taylor-Hobson division of the Rank Organisation, Leicester, U.K. The instrument was initially designed for the measurement of metal finishes and consists of a diamond tipped probe that is mechanically driven across the workpiece[10]. Movement of the probe relative to an average datum provided by a steel skid that also traverses the surface or an optical flat, is detected by means of a 30 kHz inductance bridge, one arm of which contains a cored coil, the core being mechanically linked to the probe. The amplified out of balance signal from the bridge is usually displayed on a recorder. Fig. 5 is a diagram of the apparatus.

Such an instrument can be applied to the measurement of film thickness provided the probe pressure can be reduced sufficiently to prevent damage to the film when the probe traverses the film and also that, as in optical measurements, a well defined edge is available[11].

Such an apparatus has now been fully developed for the specific measurement of very thin layers and is marketed under the name 'Talystep'. The pressure on the stylus has been made very light, although the diamond tip used is of very small radius. However, the load can be reduced to less than 3 milligrams. The final amplification of the system can be up to 1 million times so that it is possible to measure films down to thickness accuracies of $\pm 1 \overset{\circ}{A}$.

Other transducer systems have been used to monitor the probe movement and these have included piezoelectric devices and semiconductor displacement transduces[12]. However none of these are as yet as sensitive as the 'Talystep' inductance bridge arrangement.

2.4.2 Micro-Balances

Micro-balances have been used for measuring film thicknesses, and such systems have the advantage that they can be used for in situ measurements even in ultra-high vacuum systems.

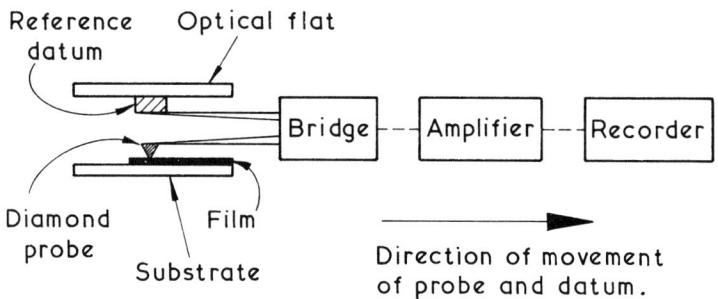

Fig. 5   Diagram of 'Talysurf' probe.

THICKNESS MEASUREMENTS

Balances can be roughly classified[13] as of the torsion type in which the deflections of the balance beam is balanced by the torsion of the wire supporting the beam or the type in which some external force, usually either magnetic or electrostatic is used to restore the balance to a zero position.

Fig. 6 illustrates the basic idea of a magnetically restored micro-balance[14]. Such a system could have a sensitivity of 10^{-8} gms per mm deflection at 3.5 metres. The apparatus can be baked to 400°C. With this sensitivity it is possible to determine the mass of a mono layer equivalent to 3 Å thick on a 1 cm square. However, the final calculation of thickness depends on a knowledge of the density of the film and the state of aggregation.

2.4.3 Rotating Systems

It is possible to use a pivotted eddy current damped rotor without restoring couple for measuring the thickness of deposition in an evaporation system[15]. If the vapour stream is only allowed to land on one side of the rotor this will cause the rotor to rotate. As the rotation generates an opposing couple proportional to the angular velocity, a measurement of the thickness deposition is given by the number of turns of the rotor. Furthermore the angular velocity is proportional to the rate of arrival of the atoms.

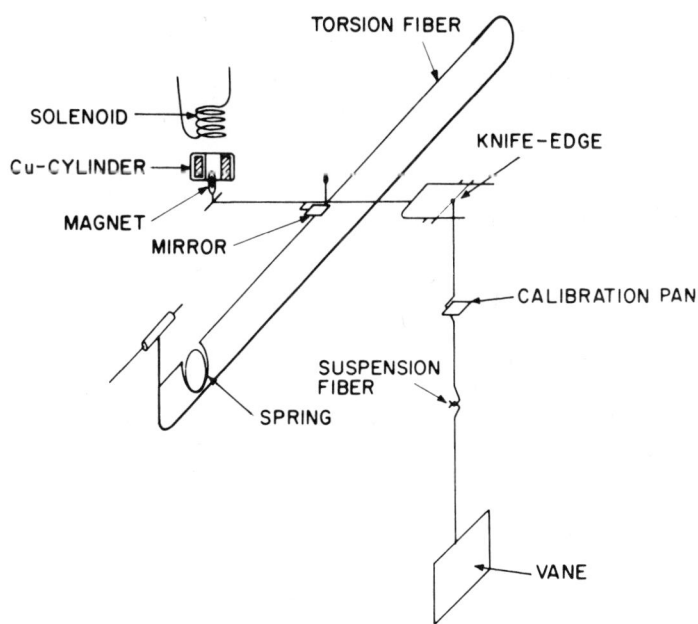

Fig. 6   Quartz microbalance (after Meyer et al. [14])

Two versions were evolved, a cylinder type for evaporation in a horizontal direction (Fig. 7) and a disc type for evaporation in a near vertical direction (Fig. 8). Fig. 9 shows a typical result using a cylinder type. A sensitivity of 10 Å/revolution was obtained with the disc type.

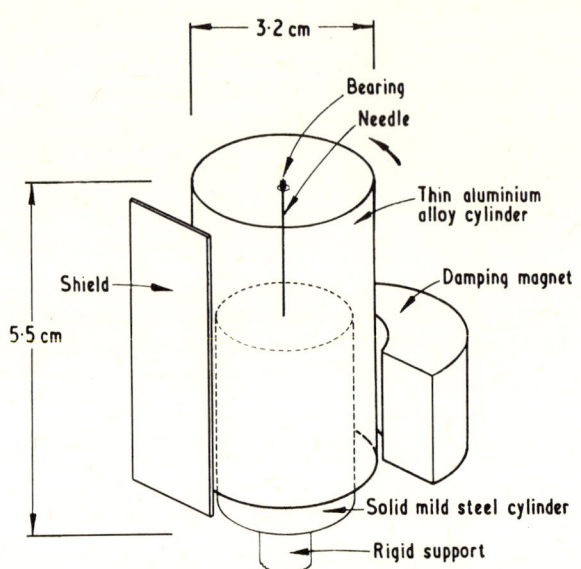

Fig. 7   Diagram of rotating cylinder thickness monitor (after Beavitt. Ref. 15)

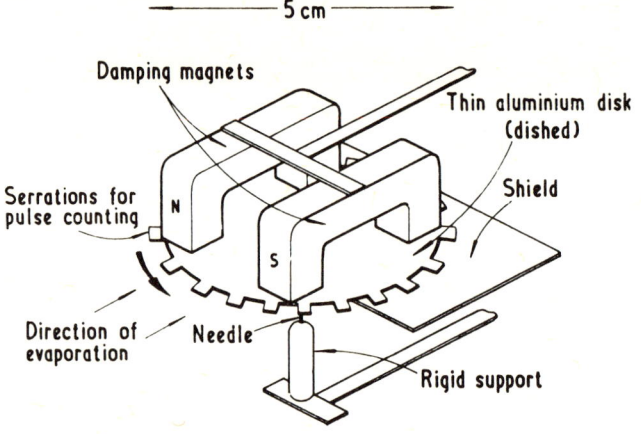

Fig. 8   Diagram of rotating disc thickness monitor (after Beavitt. Ref. 15)

THICKNESS MEASUREMENTS

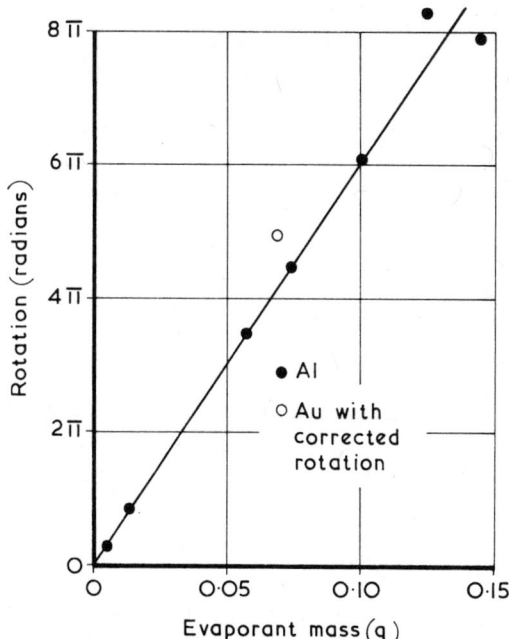

Fig. 9 Evaporant mass plotted against rotation for cylinder thickness monitor (Al evaporant ; cylinder length 2.5 cm; cylinder radius 1.6 cm ; source distance 10 cm).

2.5 Electrical Methods

2.5.1 Quartz Crystal Systems[16]

The frequency of oscillation of a quartz crystal depends on the mass of the crystal. Therefore, if a film is allowed to grow on one side of a quartz crystal plate, the frequency change, Δf, will be proportional to the mass M of the film and will be given by

$$\Delta f = - \frac{f^2}{N\rho} \frac{M}{A} \qquad (2)$$

where f is the resonant frequency of the quartz plate, ρ is the density of the quartz, A is the plate area, and N is a constant.

This equation was verified by Sauerbrey in 1959[17] and since then a large number of papers have been published on quartz crystal thickness measuring systems.

As an example of the use of equation (2), using a 6 MHz crystal it is found that an area of deposition 0.3 cm^2 in area and 1 Å thick, will give a change of 1 Hz (for film density = 5). Fig. 10

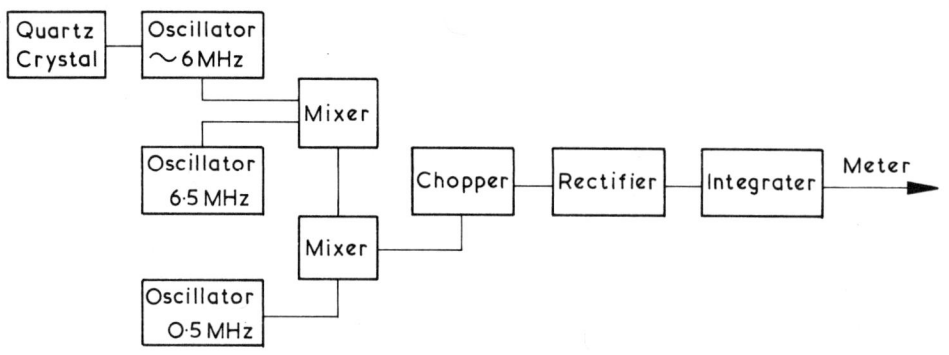

Fig. 10   Circuit diagram for use of Quartz crystal thickness measurement

shows a block circuit diagram of a typical electronic arrangement that can be used.

As with the micro-balance, a knowledge of the density of the film is required or the system needs to be calibrated against some other method, before it can be used for thickness determination and furthermore the substrate must be the quartz crystal.

2.5.2 Capacitance Measurements

It is sometimes possible to determine the thickness of a dielectric film which has initially been deposited on a conducting substrate, by depositing a second electrode on the top surface. A capacitance measurement of the resultant parallel plate structure can allow of the thickness of the film to be determined, provided the permittivity of the dielectric is known. A typical capacitance value that can be obtained is 0.03 µF per sq. cm. for a film of permittivity 6 and thickness 3000 Å.

Another arrangement of electrodes that has been used[18] is one in which a comb pattern(Fig. 11)is initially deposited in aluminium on the substrate which is usually quarz. If a dielectric film is deposited on the comb the change in capacitance can be related to the thickness of dielectric.

Fig. 12 shows a typical result for SiO deposited on a comb of line width and space 0.0075". The initial capacitance using a quartz substrate was 65 pF.

THICKNESS MEASUREMENTS

Fig. 11 Comb structure of electrodes for capacitor method of measuring thickness.

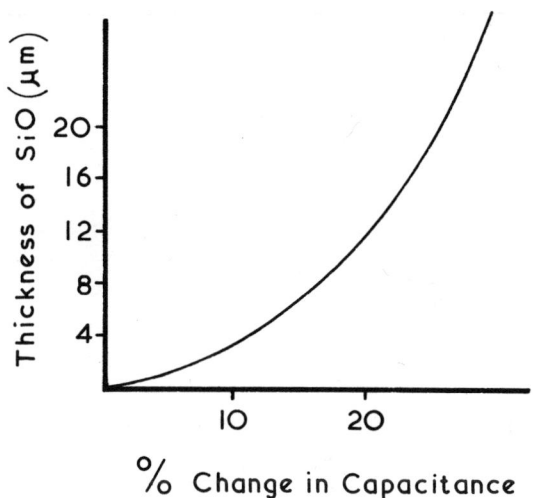

Fig. 12 % capacitance change for film thickness on comb electrodes. (SiO ; electrode width and spacing 0.0075", Quartz substrate, Initial Capacitance 65 pF).

Studies have also been made using a proximity meter approach in which the second electrode is not deposited on the film but is a plate brought into as intimate contact as possible with the top surface of the film. However, this type of system suffers from the difficulty of having, invariably, an air gap between the film and the plate.

2.5.3 Resistance measurements

The resistance of a metal conducting film is a function of film thickness and it is therefore possible to measure film thickness using a direct measurement of resistance[19]. However, this technique suffers from the disadvantage that it is not suitable for very thin non-continuous films and also that the method needs a knowledge of the resistivity of the deposited layer. Fig. 13 shows a typical resistance/thickness curve for silver on glass and it can be seen how rapidly resistance increases at a thickness of around 500 Å as the film becomes discontinuous[20].

However the system therefore is of interest in determining thickness reproducibility in metal films and has found applications in controlling the length of deposition of a metal film[21]. Control situations are possible for sheet resistance values around 100-500 Ω/sq.

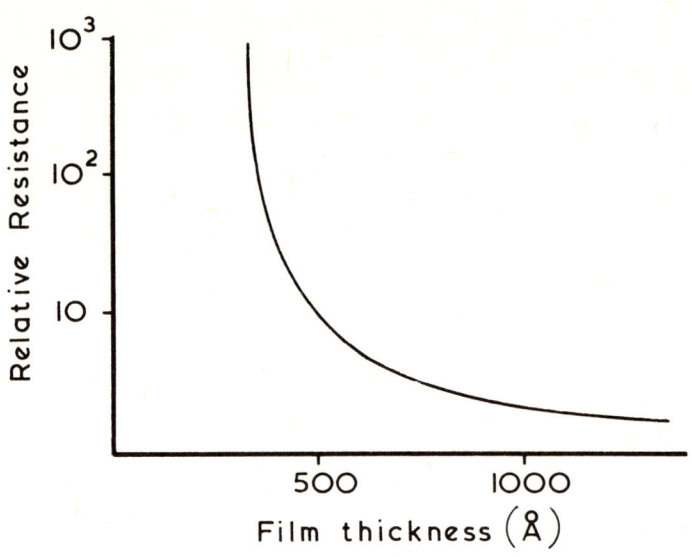

Fig. 13 Resistance against thickness for silver on glass.

THICKNESS MEASUREMENTS

2.6 Magnetic Methods

Studies have been made of the use of magnetic techniques for measuring film thickness. If the film is deposited on a magnetic material then the observation of the magnetic characteristics of a probe placed on the top surface of the film, can give data as to the film thickness. However, such techniques are generally not sensitive enough for thickness measurements on thin films.

2.7 Radioactive Techniques

It is sometimes possible to use radioactive methods for measuring film thickness [22][23][24][32]. From the various techniques available for making this type of observation, the simplest system is that of measuring the radioactivity of the depositing material. However, only a very limited number of materials have the right high saturation activity and a half life that is not too short or too long (3 hours is considered the lowest limit in order to allow time to take the irradiated film from the atomic pile, and six months is too long as the decay will cause contamination)

Gold is found to be very suitable material, as it has an activity of 8 000 millicuries per gm. and a half life of 2 1/2 days. Using detector systems that are readily available, in principle it is possible to detect 1/1000th of a monolayer by this technique, using gold.

A similar technique is to deposit the non-radioactive material on a radioactive substrate and to measure the absorption of the radioactivity due to the deposited film.

More sophisticated radioactive techniques which can be adopted for measuring thickness have been summarised by Cachard.[32] These include the use of back scattering from a 1-3 MeV ion beam, the detection of nuclear reaction products on bombardment, the use of electron micro probe, ion induced X-ray emission and secondary ion beam mass spectroscopy. All these methods, apart from the last, are non-destructive. Detection sensitivity is often to less than a monolayer and depth resolution can be to 100 Å in certain cases.

2.8 Chemical Techniques

In principle, it is possible to chemically dissolve films from their substrate and measure the changing weight of the substrate. Microchemical balance methods using such a technique on dissolvable films have been found capable of measuring 100 Å at 1 cm^2 to about 0.5%. The method is, however, destructive and depend on a knowledge of the film density.

3. RATE CONTROL SYSTEMS

3.1 Introduction

As mentioned initially, it is often possible to determine the thickness of a film by calibrating the rate of deposition. The rate control available depends on the deposition technique being used and Table 2 summarises the rate control parameters against various deposition methods.

Table 2

Rate control parameters for various deposition systems

Method	Rate control
Electroplating	Current density
Chemical reduction	Temperature, pH
Thermal growth	Temperature
Anodization	Current density
Vapour phase	Temperature of substrate
Evaporation	Temperature of source
Sputtering	Current density, Target potential

In the case of evaporation, sputtering and vapour phase deposition, direct control of rate of deposition is not as easy as with other systems and therefore other methods of measurements of rate control have been evolved. Such rate control systems will be only calibrated for a particular geometry inside the vacuum chamber and any alteration of this geometry, whether with regard to the source-substrate distance or the position of the rate measuring device in the system, must be allowed for.

3.2 Mechanical Methods

Mechanical systems have been derived for use in evaporation deposition that depend for their operation on the momentum of the arriving evaporated atoms or molecules and Fig. 14 shows a typical example.[11][25] The vane rate meter can be restored into an equilibrium position by passing a current through the meter coil, the amount of the current being dependent on the rate of arrival of the evaporated species. Such a system can be used for a feedback control to alter the boat temperature of the evaporation source, and by

THICKNESS MEASUREMENTS

this technique ± 5% control of rate of deposition is possible. The rate control with such a system can be as long as 1 Å per second though normally control is down to 10 Å/sec.

Rotating cylinders or discs can also be used for evaporation systems either with the suspension giving the restoring torque[26] or by monitoring the speed of rotation of the disc.[15] For a suspended type, Neugebauer obtained a sensitivity of 1.5 Å sec/radian, and with a disc type Beavitt obtained 0.04 Å sec/radian.

3.3 Electrical Methods

3.3.1 Quartz Crystal Systems

It is relatively straightforward to introduce a differentia - ting network into the electrical circuit associated with the ouput of a quartz crystal thickness measuring system (Fig. 10) If this is done a direct measurement of rate of deposition can be obtained. Rates can be measured to 0.2 Å/sec for Al using commercial equipment.

3.3.2 Ionisation Gauge Systems

A so-called nude ionisation gauge can be used for determining the rate of arrival of depositing atoms in a vacuum system.[27].

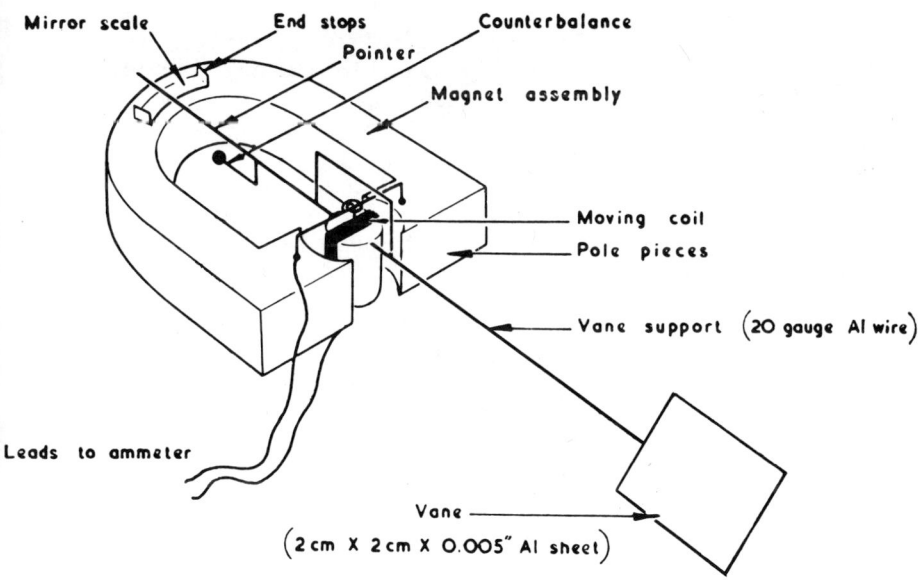

Fig. 14   Diagram of vane ratemeter (After Campbell and Blackburn Ref. 11)

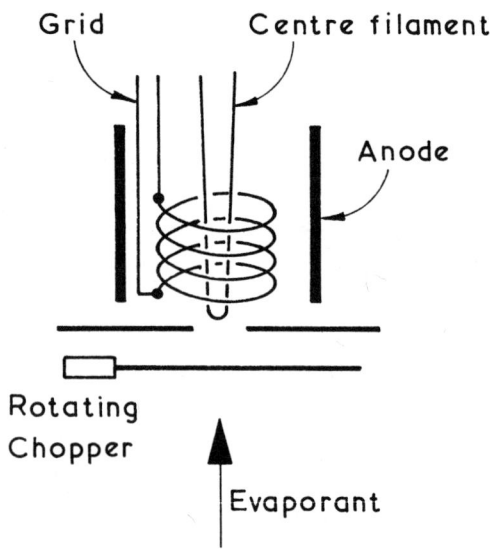

Fig. 15   Diagram of ionization gauge ratemeter.

Fig. 15 shows a diagram of the basic principle. A chopper is introduced between the source of material and the gauge with the chopper usually running between 10-20 Hz. The pulsed current obtained from the gauge is then observed and via suitable electronic circuitry can be used to provide a direct measurement of rate. Levels down to 0.2 Å/sec can be obtained.

A refinement to the nude gauge system is to use two gauges, one of which sees the vapour source and the other of which, does not. The difference between the two out-puts will then give the rate measurement directly.

It should be noted that if an electron beam source is used for the evaporation system, then sufficient of the evaporant will become ionised, and no separate filament will be required in the gauge itself.

3.3.3 Capacitance Measurements

It is possible to use capacitance measurements for rate control based on the system discussed in section 2.4.2.

It can be seen from Fig. 12 that the relationship between

TABLE 3
Summary of thickness measuring techniques discussed

(ρ = film density ; μ = refractive index of film)

	Thickness	Rate	Nature	Best Accuracy	Remarks
Interference (Fizeau)	√		F	± 10 Å	Step and reflecting coating needed
Interference (FECO)	√		F	± 2 Å	
Interference (Photoelectric)	√		F	± 0.1 Å	
Interference (Colour)	√		F	± 40 Å	Limited range and applicability
Polarimetric	√		F	± 10 Å	μ must be known. Complex
Mechanical (Probe)	√		D	± 1 Å	Step needed. Probe must not damage surface
Mechanical (Balance)	√		D	± 3 Å	ρ required
Mechanical (Cylinder or Disc)	√	√	D	± 0.5 Å	Calibrated for particular material and source temperature
Quartz Crystal	√	√	D	± 1 Å	ρ required
Capacitive	√	√	D	± 500 Å	Limited to dielectrics
Resistance	√	√	D	± 100 Å	Limited to continuous metal films
Radioactive	√		F	± 0.1 Å	Hazardous
Chemical	√		D	± 0.5 Å	Destructive. ρ required.
Ionization gauge		√	D	–	

thickness and capacitance is essentially non-linear and this means that such a device is in practice, slightly complex to use. However, using an initial capacitance of 65 pF it has been found possible to measure rates of deposition down to ± 4% over the total thickness range 1 µm to 30 µm. Rates in this thickness range could be measured up to 3 µm par minute.

4. SUMMARY

The various direct and indirect methods of measuring thickness and rate of deposition have been examined. The method actually used will depend very much on the system and requirements.

Table 3 summarizes the methods available (see also the table given by Pliskin and Zanin[28]). The table shows whether the technique can be used for thickness or rate measurements or both, and also whether the measurement is fundamental (F) or must be eventually calibrated against a fundamental system (D). The best accuracy figures given for thickness determinations will only be applicable for limited conditions of thickness.

In certain case quoted, it should be noted that accuracies of less than 1 Å are possible. It can be commented that such an accuracy is often meaningless with regard to the physical structure of the surface and these figures just indicate the reliability of results that are possible.

REFERENCES

1. K. H. Behrndt. "Physics of thin films" (Ed. G. Hass and R. E. Thun). Academic Press. N. Y. $\underline{3}$ p. 1-40. 1966.

2. W. A. Pliskin and S. J. Zanin. "Handbook of thin film technology" (Ed. L. I. Maissel and R. Glang) McGraw Hill. N.Y. p. 11.1-11.34. 1970.

3. H. E. Bennett and I. M. Bennett. "Physics of thin films" (Ed. G. Hass and R. E. Thun). Academic Press N. Y. $\underline{4}$ p. 21-41. 1967.

4. S. Tolansky "Multiple-Beam Interferometry of Surfaces and Films" O.V.P. London 1948

5. O. S. Heavens "Physics of thin films". (Ed. G. Hass and R. E. Thun). Academic Press. N. Y. $\underline{2}$, p. 229-235 1964.

6. F. Dyson. Nature, March 23rd 1963. p. 1193.

7. F. Dyson. Physica $\underline{24}$, p. 532. 1958

8. O. S. Heavens. "Physics of thin films" (Ed. G. Hass and R. E. Thun). $\underline{2}$ p. 218-227 ; 1964. W. A. Pliskin & S.J. Zanin. "Handbook of thin film technology" (Ed. L.I. Maissel and R. Glang). McGraw Hill. N.Y. p. 11.21-11.24. 1970.

9. H. E. Bennett and I. M. Bennett. "Physics of thin films" (Ed. G. Hass and R.E. Thun). Academic Press N.Y. $\underline{4}$, p.79-84 1967.

10. R. E. Reason. Symposium on the properties of metallic surfaces. Institute of Metals Monogram N° 13. p. 327. 1953.

11. D. S. Campbell and H. Blackburn. Trans. 7th National Symposium on Vacuum Technology. p. 313-318. 1960.

12. G. V. Planer and L. S. Phillips. "Thick film circuits" Butterworth. London p. 24-25. 1972.

13. K. H. Behrndt. "Physics of thin films". (Ed. G. Hass and R. E. Thun). Academic Press. N.Y. $\underline{3}$ p. 27-40. 1966

14. H. Mayer, R. Niedermayer, W. Schroen, D. Stünkel and H. Göhre. "Vacuum Microbalance Techniques" Plenum Press. N.Y. $\underline{3}$ p. 76 1963.

15. A. R. Beavitt. J. Sci. Inst. <u>43</u>, p. 182-185. 1966.

16. K. H. Behrndt. "Physics of thin films". (Ed. G. Hass and R. E. Thun). Academic Press. N.Y. <u>3</u>, p. 21-27. 1966.

17. G. Sauerbrey. Z. Physik <u>155</u>, p. 206. 1959.

18. F. Z. Keister and R. Y. Scapple. Trans. 9th Nat. Symp. on Vacuum Technology. p. 116. 1962.

19. K. H. Behrndt. "Physics of thin films".(Ed. G. Hass and R. E. Thun). Academic Press. N.Y. <u>3</u>, p. 18-19. 1966.

20. L. Maissel. "Handbook of thin film technology". (Ed. L.I. Maissel and R.Glang) McGraw Hill. N.Y. p.18.25. 1970.

21. F. W. Bishop. Rev. Sci. Inst. <u>20</u>, p. 527. 1949.

22. K. H. Behrndt. "Physics of thin films". (Ed. G. Hass and R. E. Thun). Academic Press. N.Y. <u>3</u>, p. 38-40. 1966.

23. W. A. Pliskin and S. J. "Handbook of thin film technology". (Ed. L. I. Maissel and R. Glang). McGraw Hill.N.Y. p.11.31-11.32. 1970.

24. T. P. Flanagan and J. A. Bennett. Int. J. App. Rad. & Isotopes. p. 19, Jan. 1962.

25. R. E. Hayes and A.R.V. Roberts. J. Sci. Inst. <u>39</u>, p. 428. 1962.

26. C. A. Neugebauer. J. Appl. Phys. <u>35</u>, p. 3599. 1964.

27. K. H. Behrndt. "Physics of thin films" . (Ed. Hass and R. E. Thun). Academic Press. N.Y. <u>3</u>, p. 3-5. 1966.

28. W. A. Pliskin and S. J. Zanin. "Handbook of thin film technology" (Ed. L. I. Maissel and R. Glang). McGraw Hill. N.Y. p. 11-33. 1970.

29. F. Abelès. In "Physics of thin films". (Ed. M. Francombe & R. W. Hoffman). Academic Press, New York. <u>6</u>, p. 151-204. 1971.

30. J. M. Bennett and M. J. Bostry. Appl. Opt. <u>5</u>, p. 41. 1966.

31. F. Abelès and M. L. Theye. Surface Sci. <u>5</u>, p. 325. 1966.

32. A. Cachard. In "Physics of Non-Metallic Thin Films" Proceeding of N.A.T.O. Advanced Study Institute, Corsica 19th August to 12th Septembrer, 1974. Plenum Press. London (this present volume). 1975.

Physical Properties

ELECTRONIC TRANSPORT PROPERTIES

Robert M. Hill

Chelsea College, University of London

Pulton Place, London SW6 5PR

1. INTRODUCTION

The broad classification of materials, by their ability to transport electric current, into metals, semiconductors and insulators, can be generally applied to the same materials when deposited in the form of thin films. It is not then the similarities of films to bulk that give rise to the study of transport properties but the differences between the two, and in particular, the differences that are inherent in the preparation of thin non-metallic films. As an example, if we wish to investigate high electric field phenomena, a bulk specimen 10^{-2}m thick would require a power supply capable of giving a potential of 10^6 volts to give a field of 10^8 V.m^{-1}, whereas a film 1,000 Å thick only requires 10 V for the same field. A second example is that highly disordered structures can be prepared in the form of thin films by deposition onto cold substrates. Below the deposition temperature these structures are at least metastable and their electrical properties can be investigated. Indeed, without extreme care it is difficult to prepare films which have good crystalline properties and the investigation of transport of charge in non-metallic films is essentially an investigation of the interaction of quasi-mobile charge with the non-crystalline media. Here the first example, of high field, is of interest, for if a non-linear regime of current and voltage can be induced the interaction between electric field and structure can be used to elucidate information about the electronic structure of the material. It is the extraction of this information that is of interest to us here.

We can make some general comments about conductivity in poorly crystalline materials, as in section 2, but because of the lack of a detailed theory it is necessary to set up models of behaviour and compare the predictions from the models with experimental results. This is done in section 3. Firstly, however, we shall take a general look at the properties of poorly crystalline materials in order to obtain some information to use as a basis for model structures.

1.1. Summary of Experimental Results

Four examples of the behaviour of thin non-metallic films and bulk disordered materials are shown in Fig. 1-4. These can be taken as typical of the experimental results that have been reported in the literature over the last few years. Fig. 1 shows the optical properties of the elemental semiconductors germanium, silicon and selenium, and the criterion for the choice of these particular results has only been that they, individually, show a typical feature.

The first diagram[1], (Fig. 1a) compares the reflectivity of crystalline, amorphous and re-crystallised germanium over the band of energies up to twelve electron volts. The crystalline material exhibits sharp fine structure which arises from the exact shape of the conduction and valence bands, the reflectivity being an integrated function of the product of the band states at an energy separation given by the incident radiation. In contrast, the amorphous material has a much smoother reflectivity curve, although there is still a broad peak corresponding to the maximum crystalline reflectivity, and the general form still follows the same shape. The re-crystallised material however is much more similar to the first curve with strong evidence for the presence of fine structure.

The transition from amorphous to re-crystallised is shown in more detail in the second diagram[2], (Fig. 1b). The material annealed at the low temperature is truly amorphous and shows an ill-defined absorption edge which is less than that of the crystalline material by about 0.3 eV. As the annealing temperature is raised the absorption curve moves to higher energies, for the same value of absorption coefficient, until at a temperature of about 500°C the crystalline optical gap is attained. Even here, however, at energies just above the optical gap, the absorption coefficient is still larger than that for the perfect crystalline material. Fig. 1c shows similar measurements for selenium[3], but here the range of temperatures over which the measurements have been made includes the melting point at 130°C, and shows no indication of this physical transition. However when selenium melts local order within chains is maintained although the long range order between chains is lost.

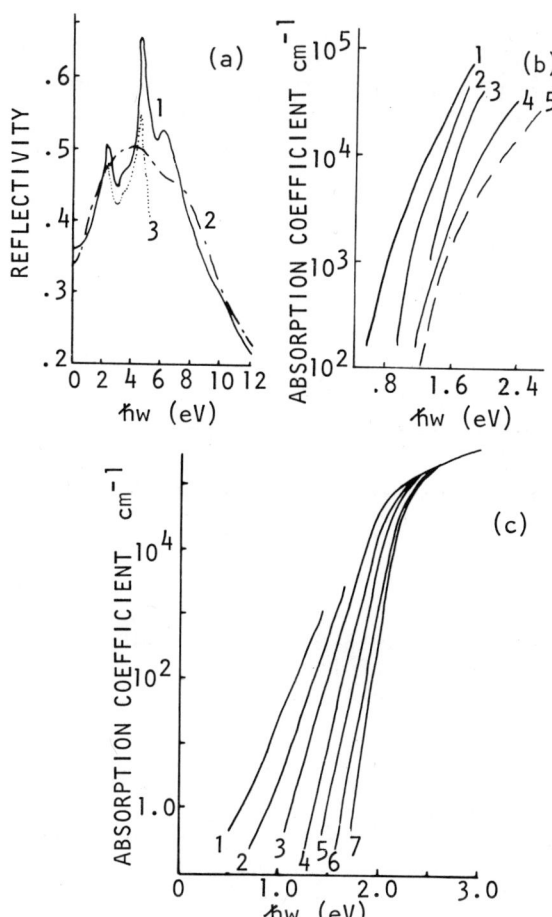

Fig. 1 (a) Reflectivity of germanium (after ref. 1). 1 - single crystal ; 2 - amorphous thin film ; " - recrystallised thin film.
(b) Absorption edge in an annealed silicon film (after ref. 2). Annealing temperatures. 1 - 20°C ; 2 - 223°C ; 3 - 500°C ; 4 - 949°C. Curve 5 is for crystalline silicon.
(c) Temperature dependence of the absorption edge in selenium (after ref. 3). Measuring temperatures 1 - 400°C ; 2 - 300°C 3 - 200°C ; 4 - 100°C ; 5 - 25°C ; 6 - -100°C ; 7 - -200°C. The melting point of selenium is 130°C and the material was non-crystalline in the solid state.

One method of examining the physical state of an amorphous material is to use diffraction measurements to construct a radial distribution function[4]. Such a curve is shown in Fig. 2 for amorphous germanium[5], the vertical lines along the radius axis indicating the positions, and relative densities, of atoms in the crystalline material. It can be seen that the first and second peaks

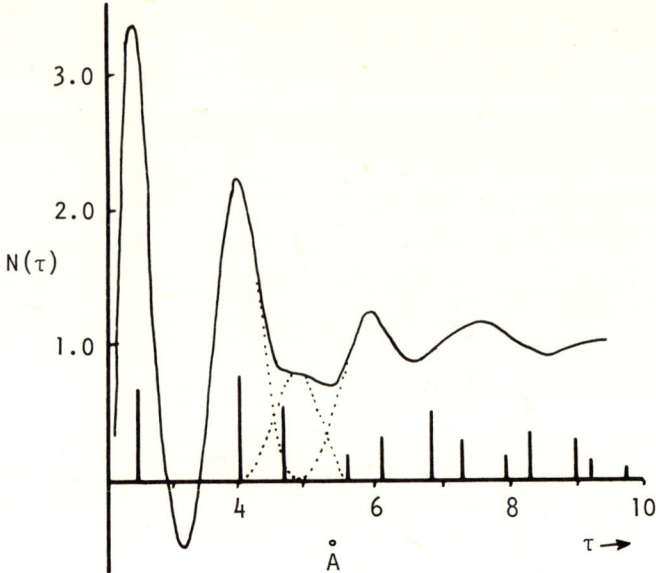

Fig. 2 Radial distribution function for amorphous germanium (after ref. 5). The positions of the atoms in the crystalline material are shown by the vertical bars.

in the RDF curve correspond reasonably well with the first and second crystalline neighbours but the resolution for succeeding neighbours is poorer and by the eighth neighbour the material looks uniform in its properties. This result is similar to that obtained in liquids and indicates a degree of short-range order but a complete lack of the normal crystalline long-range order.

For the third type of information Fig. 3 exhibits an Arrhenius plot for the alloy system Cd Tl$_x$As$_2$ as x is varied from 0.1 to 0.8,[6]. This diagram contains two pieces of information which are of interest. Firstly, there is a surprisingly small dependence of the magnitude of the conductivity on the value of x at any single temperature and, secondly, none of the plots is linear, i.e. there is no well-defined activation energy in this system. Again, both these observations are common in the field of disordered materials.

A different type of information is shown in Fig. 4. Here, the current voltage characteristics of a thin insulating film have been measured over an extensive temperature range. A casual examination of the characteristics shows that a common pattern exists

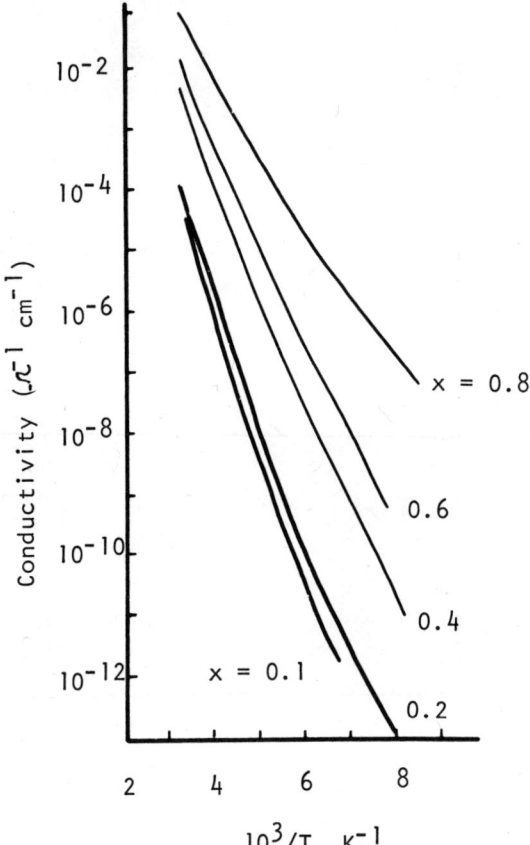

Fig. 3 Temperature variation of conductivity in vitreous $CdTl_xAs_2$ (after reference 6).

throughout the range of variables examined. At low enough voltages ohmic behaviour is observed and then there is a transition into an exponential or higher power dependence as the voltage is raised. At very low temperatures the current appears to be insensitive to the temperature but highly dependent on the voltage. This particular specimen is of silicon monoxide but very similar results have been reported on a wide range of materials prepared in thin film form.

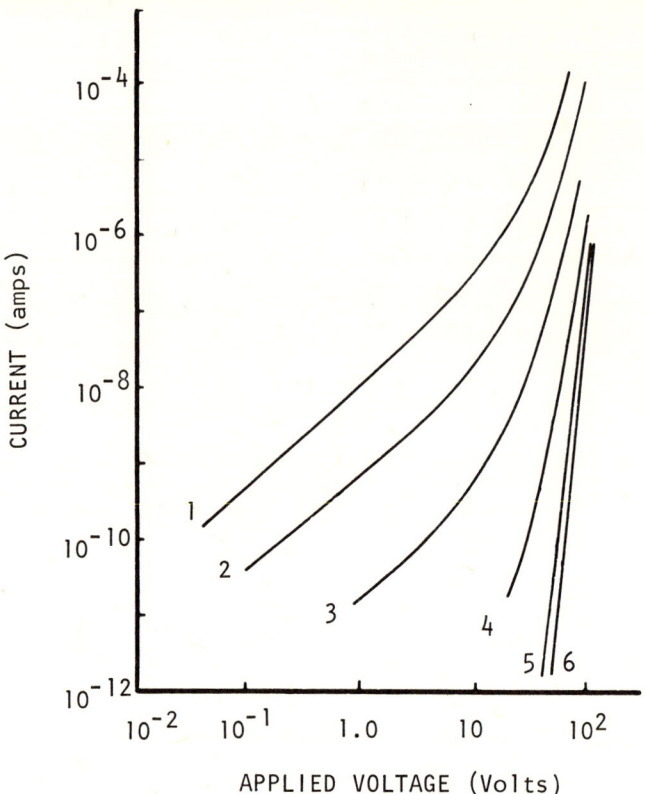

Fig. 4 Current/voltage characteristics of silicon monoxide as a function of temperature (after reference 7). 1 - 413°K ; 2 - 333°K 3 - 256°K ; 4 - 168°K ; 5 - 77°K ; 6 - 4.2°K.

We can summarise the evidence contained in Figures 1-4 and use these experimental observations as a basis for examining transport of charge. We know that the broad classification of solids into metals, semiconductors and insulators, carries over from the bulk materials into the thin film form, hence there must be some retention of energy band structure, as we have seen from the optical evidence However, there is no longer the sharp well-defined bands characteristics of a clear E-k relationship but certainly some degree of diffuseness in the region of the band edges. The results on silicon in particular show that this diffuseness can be related in some way to the degree of disorder in the material. Locally, order appears to be maintained implying that the bonding between atoms is of the same form in crystalline or disordered structures, but the

ELECTRONIC TRANSPORT PROPERTIES

loss of long-range order implies that individual groups of atoms and their nearest neighbours are rotationally distorted with respect to each other. This can only occur with physical distortion of the structure and lead to a fraction of the bonds being strained and possibly a smaller fraction actually broken, i.e. a large density of "intrinsic" defects will be present. Supporting evidence for this conclusion is given by the alloy results where the lack of sensitivity to doping, in the semiconductor sense, implies the presence of an initially high density of "impurities".

Figure 4 shows that one of the tools at our disposal is that of non-linear field effects. A simple measurement of conductivity can only yield information about the density/mobility product. Non-linear effects, if interpreted correctly, will yield information about the detail of the interaction of the carrier with the electronic structure of the material.

1.2 Examination of the concept of Localisation and Disorder

We will assume that the basic transport processes is crystalline materials are well known[8] and concentrate here on the effect of disorder. The seminal work in this area was carried out by Anderson[9] who examined a Kronig-Penney-like array of potential wells spaced a distance 'a' apart to which was added a random potential U of root mean square value U_o. In the absence of U the wells gave a band of allowed, extended, states of width J. For the perturbed system three cases were delineated,

i) $U_o/J < 0.5$. Within an electronic mean free path of length L ($\approx a J^2 U_o^{-2}$) phase coherence exists and the electron states are extended.

ii) $0.5 < U_o/J < 5$. A diffusive region for transport where no wave solution exists and the mean free path is of the order of 'a'.

iii) $U_o/J > 5$. At zero temperature all charges are localised in the potential wells and no transport occurs. However, at finite temperatures transport is possible by quantum mechanical tunneling between the wells. For this to occur, however, the carrier must capture or emit a phonon of energy equal to the difference in energy between the initial and final states. This mode of transport is termed hopping.

Intuitively we would expect a small potential perturbation in a basically crystalline medium to smear out the band edges. Within the original band regions case (i) will apply, the bands will support extended wave functions, and the carriers will still possess something like the crystalline mobility. At the edges of the bands, however, there will be relatively more perturbation yielding a diffusive region, (ii), where the physical extent of the states is finite but there is continuity in energy. For states displaced into the normally forbidden gap the local density of states will be low and the only process of transport will be hopping, either within the localised states or into the extended states within the bands.

In the extended state region the scattering of electrons is weak, and the wave number k is well defined. The uncertainty in k, Δk, is approximately proportional to the inverse of the mean free path and $\Delta k/k \ll 1$. However, as we move down from region (i) to region (ii) the mean free path decreases, $\Delta k/k \approx 1$, and wide variations from the $E^{1/2}$ density of states for the crystalline material can be expected. In the localised region k is no longer well defined, the uncertainty is of the order of k itself, and even the concept of momentum is no longer useful.

At a finite temperature Einstein's relationship can be used to define a mobility

$$\mu = -\frac{eD}{kT} = -z\frac{e\nu L^2}{kT}$$

where z is the coordination number and ν the frequency. In the extended state region this will be an electronic frequency ($\frac{2\hbar}{mL^2} \approx 3.10^{15}$ s^{-1}), assuming a mean free path L much greater than the atomic spacing. In the localised hopping region diffusion requires phonon assistance and the relevant frequency will be that of a phonon ($\sim 10^{12}$ s^{-1}) reduced by the probability of phonon capture or emission. Considering the frequency terms only the ratio extended state, to localised mobility will be given by

$$\frac{\mu_{ext.}}{\mu_{loc.}} > \frac{\nu_e}{\nu_{ph}} > 3.10^3$$

Hence a drastic change in the magnitude of the mobility is expected as we pass from extended to localised states. This situation will occur both for the bonding and the anti-bonding bands and has led to the concept of a mobility gap, equivalent to the crystalline band gap, in disordered systems, the critical energies at which the mobility changes rapidly being termed E_c and E_v for the conduction and valence band respectively.

2. GENERAL THEORY

In this section the Kubo-Grennwood formula, which is suited for a discussion of transport processes in imperfect crystalline

ELECTRONIC TRANSPORT PROPERTIES

materials, will be derived and used to determine the physical meaning of some simple experimental measurements.

2.1 Kubo-Greenwood Formalism

The general relationship for the conductivity in a medium at a frequency ω is (for example see Mott and Davies[10])

$$\sigma(\omega) = \frac{2\pi e^2 \hbar^3 \Omega}{m^2} \int \frac{D_E \cdot N_E \cdot D_{E+\hbar\omega} \cdot N_{E+\hbar\omega}}{\hbar\omega} dE \quad (1)$$

where $D(E) = \int \psi_E^* \frac{d\psi_E}{dx} d^3x$, Ω is the volume of the material per electron, N_E the density of states at energy E and ψ the electronic wave function. If we set the product $D_E \cdot D_{E+\hbar\omega}$ to be given by the square of an average value of D, $|D|_{av}^2$, where the averaging is carried out over all states E to $E+\hbar\omega$ and let the frequency go to zero, we can obtain the limiting, D.C. conductivity. Defining

$$\sigma(E) = \frac{2\pi e^2 \hbar^3 \Omega}{m^2} |D_E|_{av}^2 \{N_E\}^2$$

we can write

$$\sigma = -\int \sigma(E) \frac{\partial f}{\partial E} \, dE \quad (2)$$

which is the Kubo-Greenwood formula. The function f is the Fermi-Dirac function and $\partial f/\partial E$ ($\equiv \sinh^{-2} E/2kT$) has a bell-like shape with a maximum at the Fermi level and decays exponentially for energies greater than about 4 kT from the maximum. The physical significance of $\sigma(E)$ is that it is the conductivity of carriers at energy E and can only be taken as a really well defined at present for extended states.

A useful method of manipulating the Kubo-Greenwood integral is available if it can be shown that the product under the integral sign possesses a well-developed maximum at a particular value of energy E_0. In this case integration by the method of steepest descents[12] gives

$$\sigma = \frac{1}{4kT} \sigma(E_0) \cosh^{-2} \frac{E_0}{2kT} \int \exp\left[-\frac{(E-E_0)^2}{4(kT)^2}\right] \{1 + \tanh^2 \frac{E_0}{2kT} -$$

$$\frac{2(kT)^2}{\sigma(E_0)} \frac{\partial^2 \sigma(E)}{\partial E_0^2}\}] dE \quad (3)$$

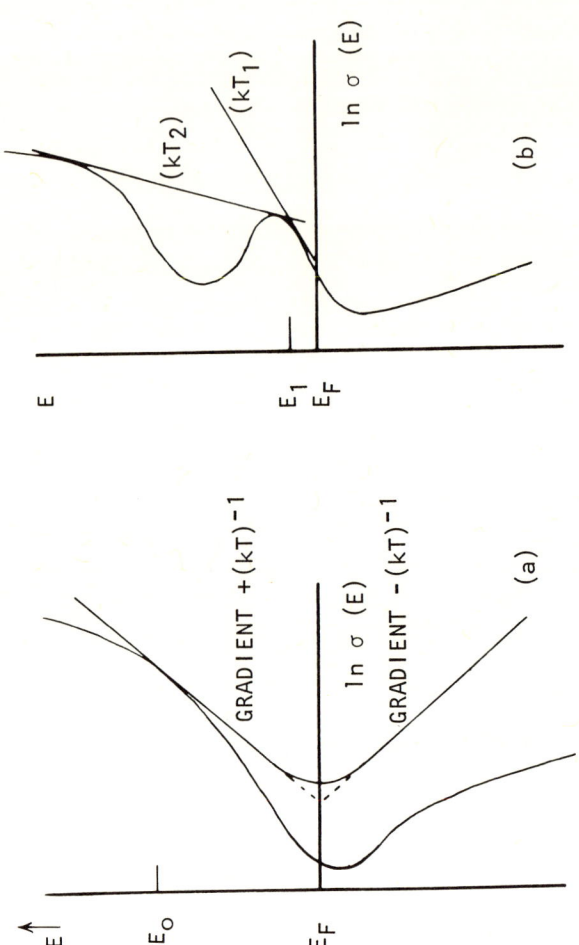

Fig. 5. Graphical solution of equation (5). The energy at which the dominant part of the current is carried, E_0, is given by the energy at which the $\ln \sigma (E)$ vs E curve has gradient kT for $E_0 - E_F > kT$. In (a) a monotonic rise of E_0 with temperature is expected but in (b) E_0 will remain in the region of E_1 for $T_1 < T < T_2$.

which for $E_o - E_F > kT$ is equal to

$$\frac{1}{4}\sqrt{\frac{\pi}{2}}\,\sigma(E_o)\,e^{-E_o/kT}(1 - \frac{(kT)^2}{\sigma(E_o)}\frac{\partial^2\sigma(E)}{\partial E_o^2})^{1/2} \qquad (4)$$

The energy at which the maximum contribution to the current is obtained, E_o, is given by

$$\{\frac{\partial\ln\sigma(E)}{\partial E}\}_{E_o} = \frac{1}{kT}\{\tanh\frac{E}{2kT}\}_{E_o} \qquad (5)$$

In equation (3) the principal temperature dependence is contained in the pre-integral terms and if $\sigma(E)$ is constant over an energy range which does not include the Fermi level and zero elsewhere, a well-defined activation energy corresponding to the minimum energy of the band will be obtained as in the crystalline case. In the more general case of $\sigma(E)$ being a continuous function of energy, equation (5) can be expressed graphically to show how the energy E_o varies with temperature. For $E_o > kT$ the right hand side of equation (5) has the value $\pm (kT)^{-1}$ and hence E_o will be given by the energy at which the gradient of a plot of $\ln\sigma(E)$ against E has this value, as in Figure 5(a). If $\sigma(E)$ monotonically increases with energy, E_o will also increase monotonically from the Fermi level as the temperature is raised from zero. However, if, as in Figure 5(b), $\sigma(E)$ exhibits negative curvature, or even peaks, E_o will be pinned over a finite temperature range and linear regions will be observed in the Arrhenius plot.

In principle, it is possible to construct $\sigma(E)$ vs E diagrams from detailed examination of Arrhenius plots but in practice it is simpler to assume simple functions for the energy dependence of $\sigma(E)$ and to look for the equivalent characteristics in the experimental activation energy, which, if the approximation yielding equation (4) applies, is identical to E_o.

2.2 Activation energy

It is common practice to regard an activation energy as only having physical significance when linear Arrhenius plots are obtained. This simple case is characteristic of crystalline solid state physics where energies are well-defined. Here, we shall show that the activation energy of transport has a more general meaning which is useful when considering materials which are not perfectly crystalline. We start by defining the activation energy to be the gradient of the Arrhenius plot, i.e.

$$\Delta E = - \frac{d \ln\sigma}{d(1/kT)} = - \frac{1}{\sigma} \frac{d\sigma}{d(1/kT)} \tag{6}$$

which in the notation of equation (2) can be written as

$$= \frac{1}{\sigma} \left\{ \int \frac{\partial f}{\partial E} \frac{\partial \sigma(E)}{\partial(1/kT)} dE + \int \sigma(E) \frac{\partial}{\partial(1/kT)} \left(\frac{\partial f}{\partial E}\right) dE \right\} \tag{7}$$

We can now consider two specific cases to simplify the physical meaning of equation (7). If the major part of the integral occurs for $|E - E_F| > kT$, then taking $\partial \sigma_E/\partial(1/kT) = 0$

$$\Delta E = \frac{1}{\sigma} \int \sigma(E) |E - E_F| \frac{\partial f}{\partial E} dE = <|E - E_F|>_\sigma \tag{8}$$

where $< >_\sigma$ denotes a conductivity averaged energy.
Hence the activation energy measured experimentally as the gradient of an Arrhenius plot, when greater than kT, represents the average energy of the carriers with respect to the Fermi energy, even when ΔE is itself a function of T. Obviously, if the carriers can only move at the bottom or top of well-defined bands, the activation energy will be constant, as observed for crystalline semiconductors and insulators. Conversely, a well-defined activation energy can be taken as evidence for band-like properties in the material under study.

The second case, $|E - E_F| < kT$, can be conveniently treated by using a Taylor's series about the Fermi energy, in which case

$$\Delta E = - \frac{k}{\sigma} \int \sigma(E) \frac{\partial f}{\partial E} T \, dE = kT \tag{9}$$

i.e., as one might expect, the average energy of the carriers is simply the thermal energy and the Fermi function is constant.

2.3 Thermopower

An alternative experimental measurement is that of thermopower. Letting dj be the current corresponding to the local conductivity $\sigma(E)$ under the action of an applied field F, we have[14]

$$dj = \sigma(E) \frac{\partial f}{\partial E} F \, dE$$

The energy carried by this incremental current is $-(E - E_F).e^{-1}.dj$ so that the total heat transport is given by

$$e^{-1} \int \frac{\partial f}{\partial E} \sigma(E)(E - E_F). \text{F. } dE = j\pi$$

where π is the Peltier coefficient. The thermopower S is defined as the ratio of the Peltier coefficient to the temperature, hence in our notation

$$S = \frac{k}{e\sigma} \int \sigma(E)(\frac{E - E_F}{kT}) \frac{\partial f}{\partial E} dE = \frac{k}{e} <\frac{E - E_F}{kT}>_\sigma \qquad (10)$$

Comparison of equations (9) and (10) shows that the activation energy and thermopower contain the same basic information but that the latter is sensitive to the sign of the carriers whereas the former is not. For either pure band-like transport or conduction at the Fermi energy it can be shown that

$$|S| eT = A k T + \Delta E \qquad (11)$$

where A is an integer of order unity, the exact value depending on the form of the density of states function[10], i.e., if $N(E) \alpha E^{x/2}, A = 1 + x$ (x = 0, 1, 2).

It would appear that measurement of both the activation energy and the thermopower should be useful in determining the detailed structure of transport as a function of energy.

2.4 Mobility

The standard method of discriminating between carrier density and mobility in crystalline materials is the use of the Hall mobility. Experimentally, measurements of the Hall constant in disordered materials, that is materials that show a small, and temperature sensitve, activation energy, have yielded little real information. The mobility values appear to be small, as Clarke observed[15] in germanium (≈ 1 cm^2 V^{-1} s^{-1}). This itself is an indication of localisation, for, if the mobility is less than about 10 cm^{2}V^{-1}s^{-1} the mean free path is of the order, or less than, a lattice spacing, hence the carrier is not moving in classical extended states. Friedman[16], starting from the Kubo-Greenwwod formula, examined mobility in a localized system by using a random phase model for transport. The effect of the Hall magnetic field was introduced by a suitable modification of the phase of the transfer integral between the localised sites,

$$J^{(H)}_{n',n} = J^{(0)}_{n',n} \exp(i\,\alpha_{n',n}) \qquad (12)$$

where n' and n are the site numbers and

$$\alpha_{n',n} = \frac{e}{\hbar c}\,\underline{H}\cdot\frac{1}{2}[n \times n'].$$

In this way Friedman was able to show that the ratio of the drift to Hall mobilities could be expressed as

$$\frac{\mu_D}{\mu_H} = \frac{J}{6\,kT}\left(\frac{z^2}{\eta \bar{Z}}\right) \qquad (13)$$

where $\eta = \overline{\cos^2\theta} = \frac{1}{3}$ and $\bar{Z}$ is the average number of sites which form a closed path.

In general $J > kT$ hence $\mu_D > \mu_H$. Friedman also showed that both are small, the Hall mobility was essentially independent of temperature, and the sign of the Hall coefficient was such that it appeared that the carriers were always n-type, as observed experimentally.

The alternative method of measuring mobility is by transit time techniques[17]. If a sheet of charge is injected into a block of material and moves under a field of magnitude F the transit time is given by

$$\tau_0 = L(\mu \cdot F)^{-1} \qquad (14)$$

where the length of the specimen is L. The magnitude of the field F has to be large enough to overcome the self field of the charge sheet and the system can be further perturbed by space charge effects. The charge is injected into the conduction band and as it drifts through the specimen interchange occurs with localised levels and the high mobility band. Once equilibrium has been established it can be shown that the effective transit time is given by

$$\tau = \tau_0\,\frac{N_t}{N_c}\,\exp + \left(\frac{E_c - E_t}{kT}\right) \qquad (15)$$

where N_t is the density of a single trapping level at an energy $E_c - E_t$ below the conduction band. For traps distributed in energy similar relationship apply, generally with the energy term appearing in the pre-exponential[10]. Obviously, in this case, the leading edge of the deformed sheet of charge will be formed from carriers which have been released from the shallowest traps and there will be a long tail to the charge distribution as carriers are released slowly from deeper traps. The simplest case of a single trapping level can be analysed with confidence but it is not such an easy matter to deconvolute the smeared out signals obtained from more complex systems.

3. SPECIFIC TRANSPORT PROCESSES

Although conceptually useful the general analysis of conductivity contained in Section II is not complete at the present time. In order to obtain more detail of the mechanisms of conduction it is necessary to postulate models of behaviour for the interaction between charges and structure, analyse these for field and temperature behaviour patterns, and critically examine experimental measurements for evidence that supports the original hypotheses. As a starting point we consider a material with diffusive band edges extending into the normally forbidden gap region. At a critical density these states will become localised[10]. If the widths of the localised tails are large the states precipitated from the conduction band may overlap equivalent states from the valence band, in which case charge transfer will occur, the conduction band like states being empty of charge and the valence band states filled. The Fermi level will then lie, and be pinned, in the region of overlap[18], (Figure 6(a)), the degree of stability of the Fermi level being directly proportional to the density of localised states in that region.

The principal characteristic of crystalline non-metallic solids is the ability to position the Fermi level by doping. Although this property does not appear to be strong in disordered solids it is still in principle, possible to dope these materials. Only if the 'intrinsic' doping is large will the effects of 'extrinsic' doping be small. Hence we are forced to conclude that the disorder gives rise to an 'intrinsic' doping from, for example, a large density of broken bonds. For the crystalline materials the ionisation energy of the dopants is given, to a good approximation, by the hydrogenic model, i.e.

$$E_i = 13.6 \, \varepsilon_r^{-2} (m^*/m) \text{ eV} \qquad (16)$$

Even in an imperfectly crystalline material we would expect equation (16) to apply as the dielectric constant is a macroscopic quantity. The effects of disorder, however, will be to broaden out the single energy level associated with non-isoelectronic impurities in the crystalline material to form a band of equivalent donor or acceptor states in the imperfect material. This arises through both

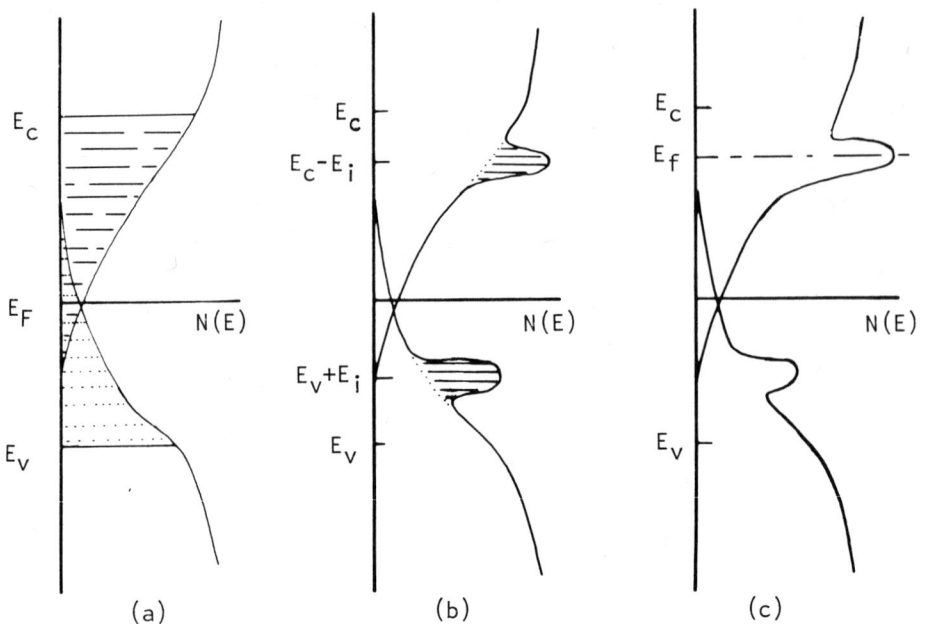

Fig. 6 Density of states diagrams for an imperfect non-metallic solid.

(a) Localised conduction and valence band states overlapping to pin the Fermi level.

(b) Donor and acceptor-like defects forming bands in the region of their ionisation energy.

(c) For a large density of donor-like defects the Fermi level will be pinned within the donor band.

interactions between the defects and because of local, spatial, variations in the energy at which extended states occur[19], these states being the datum from which the ionisation energy is measured.

In the energy band diagram we can introduce these 'intrinsic' doping states as shown in Figure 6(b). As a direct consequence of the presence of these 'impurity' states, particularly if their densities are not equal, the Fermi level will again be pinned but now within one of the band regions (Figure 6(c)).

Figures 6(a) and 6(b) represent two different types of model structures which can be examined in detail for characteristic behaviour patterns. The former is that first examined by Mott[20], whereas the latter is typical of Poole-Frenkel behaviour. These two general cases are examined in detail below.

3.2 Mott Hopping

In the absence of detailed information about the distribution of trap-like states we assume a constant density of localised states of N_1 cm^{-3} distributed randomly in space and over an energy range $2E_m$. Without loss of generality the Fermi level can be set in the middle of this energy band and we further assume that the band is wide with respect to the thermal energy. The density of states par unit energy, $N_1/(2E_m)$, we shall call N_t. Two conditions can be set for transport in this system, firstly, that of excitation of a 'free' carrier, and secondly, that such a carrier should be able to pass through the system, the continuity condition. A full examination of the random system has been made(21), but here we shall only examine the low temperature asymptotic behaviour.

a) Excitation

If one particular site is considered as datum the physically distributed sites surrounding it can be ordered in terms of their radial distance. The expected value of the radial distance of the 's' site is given by (22)

$$\varepsilon(R_s) = \left(\frac{3}{4\pi N_1}\right)^{1/3} \Gamma(s+1/3)\ \Gamma^{-1}(s) \qquad (17)$$

and for large values of $s(\gtrsim 10)$ the ratio of the gamma functions can be approximated by $s^{1/3}$. Within the radially ordered s-set the expected value for the energy difference between the datum site and the site of lowest energy is

$$\varepsilon_s(E) = 2 E_m (s + 1)^{-1} \qquad (18)$$

and again for s large the approximation $2E_m s^{-1}$ can be used.

In the low temperature regime the probability of a hop occurring can be written as (23)

$$P_h \propto \exp -(2\alpha R + E/kT) \qquad (19)$$

where R and E are the radial distance and energy associated with a hop, and α is the localisation parameter of the sites. The maximum probability of hopping occurs within the s-set where $dP_h/ds=0$, i.e. when

$$S = \left(\frac{3}{\alpha}\frac{E_m}{kT}\right)^{3/4} \left(\frac{4\pi N_1}{3}\right)^{1/4} \qquad (20)$$

and at this value of S the probability is given by

$$\exp - 3.88\left(\frac{3\alpha}{4 N_t kT}\right)^{1/4} \tag{21}$$

which differs from Mott's Law of variable range hopping only by a numerical factor. The approach used here differs from Mott's but allows some physical insight through the predicted range of hopping, equation (20). For large values of Sh, the radial parameter within which variable range hopping takes place, the temperature should be low, validating our choice of the low temperature form of the Miller and Abrahams probability function, and the density of centres in real space, N_1, has to be large. However, the density per unit energy $(2Em)^{-1}$ should be small, i.e. the states have to be of greater separation in energy than kT. If we take the conductivity as being proportional to the probability of hopping, which we would expect to be the case, the activation energy, as defined in equation (6) can be determined as

$$\Delta E_h = 0.971 \frac{(\alpha kT)^{3/4}}{(4\pi N_t)^{1/4}} \tag{22}$$

and our choice of the low temperature regime requires that this should be greater than kT.

b) Continuity

The alternative approach, initiated by Ambegaokar et al[24], is to consider that the path of a carrier will be determined by the easy hops within the disordered structure. At low temperatures particularly, only a limited range of hopping probabilities will be possible. Defining the critical probability in the same terms as equation (19), as $\exp(-h_c)$, we require that from a general site in the unformly random structure the number of sites available to which the probablity of hopping is greater than this critical value should be greater than, or at least equal to, a critical number n_c in order that a continuous path should be established, the critical number being determined from percolation theory[25].

From the definition of h_c we have that

$$h_c = 2\alpha R + E/kT$$

and as the energy of a hop goes to zero a maximum radius over which hopping can occur, R_0, is determined as $h_c/2\alpha$. The number of sites within a sphere of this radius which satisfy the probabiliy criterion is then

$$n = \sum_s \int_0^{R_0} f(R_s) \int_0^{h_c - 2\alpha R} f(E_s) \, dE \cdot dR \qquad (23)$$

where $f(R_s)$ and $f(E_s)$ are the probability distribution functions in real and energy space of the random array of sites. Taking the summation over s into the first integral sign and integrating yields

$$n = \frac{\pi}{3} N_t \, kT \left(\frac{h_c^4}{2^3 \alpha^3}\right) \qquad (24)$$

Taking the critical value of n as $1.5^{(25)}$ gives the magnitude of h_c as

$$3.46 \left(\frac{\alpha^3}{4\pi N_t \, kT}\right)^{1/4} \qquad (25)$$

regaining the $T^{-1/4}$ characteristic derived by Ambegaokar et al[24], Jones and Schaich[26] and Maschke et al[27] for this form of analysis. As we have used the same framework for the two criteria we can make a direct comparison in which case we see that the excitation condition has slightly less probability and will dominate transport by variable range hopping. Indeed, for the continuity argument to dominate the critical number n_c would have to be 2.37, which is extremely large[25].

The method used here of determining the energy at which the dominant part of the current flows is equivalent to determining the pre-integral term of equation (3) and is satisfactory as a means of investigating the first order temperature dependence.

The effect of high electric fields is to reduce the expected energy of the sth state by a amount $\varepsilon(R_s) \cdot eF$ for a state lying in the field direction. Considering only current flow in the field direction it has been shown[28] that the field introduces a correction term to the current which becomes

$$J \propto \exp - 3.88 \left[\left(\frac{\alpha^3}{4\pi N_t kT}\right)^{1/4} \left(1 - \frac{eF}{2\alpha kT}\right)^{1/4}\right] \qquad (26)$$

The usefulness of this relationship is that the localisation parameter α appears in the second term along with measurable quantities only so that it is possible, by investigating high field effects, to determine both N_t and α unambiguously.

3.2 Poole-Frenkel and Poole behaviour

The essential difference between the trap states considered in the last section and the donor/acceptor states to be considered here is in the binding force field of the centre from which emission takes place. It was implicitly assumed in the previous section that the sites through which hopping takes place have no long range potential distortion. In the limit this assumption is invalid but it is a convenient approximation which appears to be substantiated by experimental observation. When ionisation of a donor is being considered, however, it is impossible to make use of this approximation, and indeed, in equation (16) use has already been made of the coulombic force field to determine the ionisation energy. Again, two specific cases will be considered, that of a low density of ionised centres where only a single donor need be considered and the case of a large ionised density where the coulombic fields interact.

a) Low density

Field assisted ionisation of a centre which is electrically neutral in the unionised state can be described in terms of the force field by

$$F(x) = -eF + e^2(4\pi\epsilon\epsilon_r x^2)^{-1}$$

The maximum in the potential occurs at

$$x = e^{1/2}(4\pi\epsilon\epsilon_r)^{-1/2},$$

and the magnitude of the potential on this plane is $E_i - \beta_{PF}F^{1/2}$ where β_{PF} is the Poole-Frenkel constant, $e^{3/2}(\pi\epsilon\epsilon_r)^{-1/2}$.

The current can be determined by use of the Kubo-Greenwood formula,

$$J \propto eF \int_{E_i - \beta_{PF}F^{1/2}}^{\infty} \mu_c N_d \frac{\partial f}{\partial E} \, dE \tag{27}$$

where μ_c is the free band mobility and N_d the effective density of unionised donors. The energy range in which the conductivity is not zero is taken from the peak in the potential. For reasonable values of ionisation energy $E_i - \beta_{PF}F^{1/2} \gg kT$, in which case

$$J \propto \frac{eF}{kT} \mu_c N_d \exp\left(-\frac{E_i - \beta_{PF}F^{1/2}}{kT}\right) \tag{28}$$

It has been pointed out[29,30] that this Poole-Frenkel equation only applies for emission in the direction of the field. However, if consideration of the total forward hemisphere is made the correction is relatively minor. A more serious problem with the Poole-Frenkel equation is that it must be in error for such low fields that the probability of emission in the forward direction differs little from that in the reverse direction. In this case it has been suggested[31] that a hypothetical reverse peak can be used, of magnitude $E_i + \beta_{PF} F^{1/2}$ and the nett current will be the difference between that flowing forward and reverse, i.e.

$$J_{nett} \propto \frac{eF}{kT} \mu_c N_d \exp - \frac{E_i}{kT} \sinh \frac{\beta_{PF} F^{1/2}}{kT} \qquad (29)$$

which in the low field expansion gives a current proportional to $F^{3/2}$. Alternatively, it has been suggested[32] that equation (27) represents the density of carriers available in which case the reverse term should be summed to the forward term and a cosh function is obtained in place of the sinh term, with an ohmic low field expansion. In practice the high field difference between these approximations is slight, particularly when the Schubweg or mean free path of the emitted carrier, and its field dependence, are unknown and the simple sinh function with a square root field dependence in the pre-exponential term[31] appears to give as good an agreement with experimental results as the more detailed formulae.

The use of the Kubo-Greenwood approach emphasises that only carriers which are excited over the potential barrier contribute to current flow. At low temperatures the density of such carriers will be low and the only measurable current will be that which tunnels trough the distorted potential barrier. Between these two regions (Figure 7) will be a region of thermally assisted tunneling where carriers will be thermally excited within the coulombic well and tunnel through the barrier at an energy less than $E_i - \beta F^{1/2}$. Examination of the current in this regime[31] gives a $J \propto \exp-B/T^{1/3}$ characteristic whereas the pure tunneling current at low temperatures is temperature independent and essentially given by a Fowler-Nordheim[33] field dependence.

b) High density

When the density of ionised donors is large, ionisation takes place not into the extended high mobility region but into the coulombic potential associated with one of the ionised centres. Figure 8 shows the energy diagram for this situation. Because of the overlap of coulombic potentials the peak in the barrier remains in the plane midway between the centres over an appreciable field region. The distortion of the potential is then linear with the field and obeys Poole's Law[34]. It should be noted that here the zero field

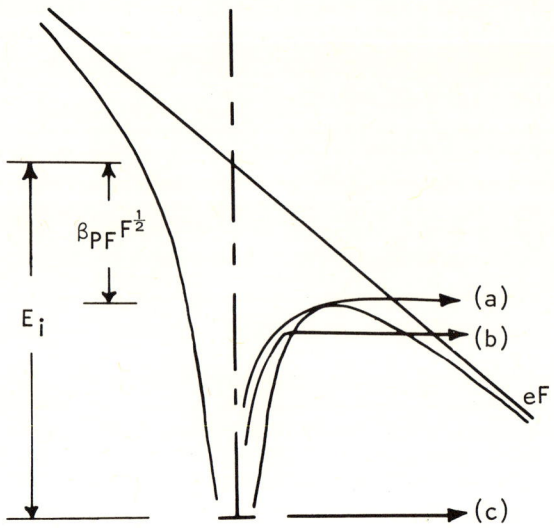

Fig. 7 The three emission processes from a donor centre. (a) Poole-Frenkel thermal emission. (b) Thermally assisted tunneling. (c) Direct tunneling at zero temperature.

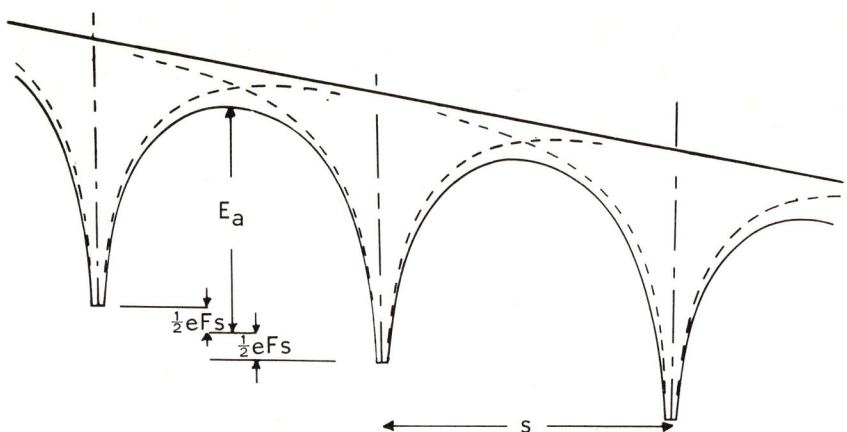

Fig. 8 Energy diagram for an array of overlapping coulombic potential wells

potential barrier, E_a, is less than the true ionisation energy of a single centre by an amount depending on the distance between the centres[31]. In this case the barrier in the reverse direction is well defined and considering emission only along and against the field direction gives a current

$$J \propto \exp\left(-\frac{E_a}{kT}\right) \sinh\frac{ef_s}{2kT} \qquad (30)$$

At high fields the barrier distorts from the symmetrical shape and a return to Poole-Frenkel behaviour is to be expected.

3.3 Injection Limited Conduction

The two specific processes of transport considered above are bulk limiting in the sense that we have implicitly assumed that the electrodes supply and extract sufficient charge not to impede current flow. For the disordered non-metallic solid this is a reasonable assumption as there will be localized states in the region of the Fermi levels of the contacts into which charge can be extracted. For more crystalline-like materials, or for materials with wide band gaps, this state of affairs is not obviously true and the current may well be contact limited. To simplify consideration of such processes we shall only look at the case of a single carrier, the electron, and injection limited currents, i.e. make use of a dielectric approach.

When the work function of the solid being investigated[35] is greater than the work function of the electrode material an enhancement layer of charge is injected into the semi-insulator; the converse situation leads to charge extraction and the formation of a depletion layer. The flat band 'ohmic' case is the limit of both situations and in general, is not to be expected. Conventionally, enhancement leads to space-charge limited flow[36] and depletion to Schottky effects[37]. However, the essential difference between these two processes is not in the form of the boundary layer but in the nature of the interaction between the injected charges. For space-charge control to be established the excess charge must be sufficiently large to interact with itself and to be definable in terms of a true local density over any small but finite elemental volume. For the Schottky emission to be the limiting process the opposite is the case, the injected charge has to be sufficiently small so that it can be treated as a set of single charges interacting only with their individual mirror image charges. The normal Schottky analysis neglects band bending due to depletion and considers only the flat band condition. This neglect is justifiable for the case of a highly insulating, very thin, film as the depletion layers will be wide and overlap extensively to give a quasi-flat band condition throughout the system.

a) Space charge Control

As electric fields are additive so are their differentials, which, from Gauss's theorem, define space charges. Hence, band bending effects can be introduced into the normal transport equations as a fictive charge which has to be added to the injected space-charge. As a boundary condition it is necessary, however, either to know how the surface charge density due to band bending, ρ_s, varies with the electrical stress or to assume that it remains constant under any experimental conditions. For convenience we assume the latter to apply. Figure 9 shows the predicted current/voltage characteristics for a range of the ratios of surface charge density to the thermally free 'intrinsic' carrier density n. For large values of this ratio the characteristic is initially ohmic, when the

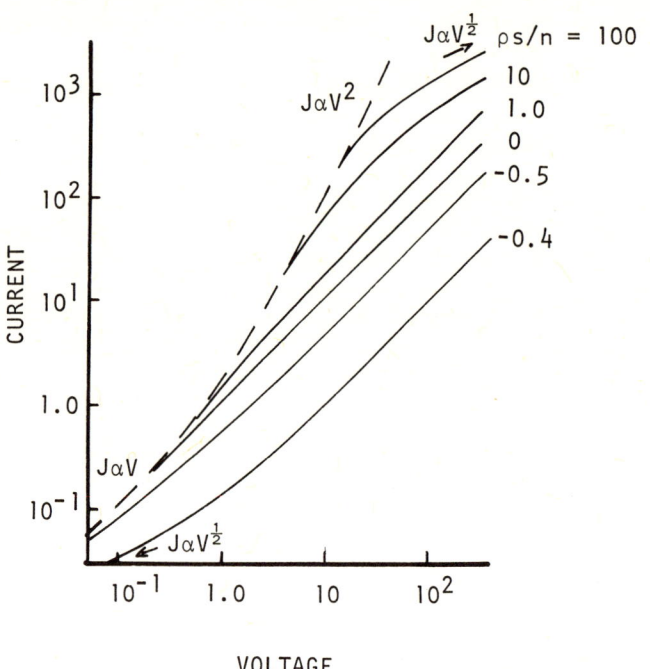

Fig. 9 Current-voltage characteristics for space-charge limited conduction. ρ_s is the density of injected charge at the injecting surface and n the thermally free carrier density. The current and voltage are in units of $J\varepsilon\varepsilon_r(n^2e^2\mu L)^{-1}$ and $V\varepsilon\varepsilon_r(neL^2)^{-1}$ respectively.

excess charge is localised at the injecting contact, but as the
space-charge cloud sweeps across the specimen and reaches the anode
the Mott-Gurney square law[39] is followed. In the limit of infinite excess charge at the injecting contact the square law is followed throughout the high field region, but for limited surface
charge a second transition to a square root voltage dependence
should be observed. This arises from saturation of the enhancement
charge pool and is a consequence of the boundary condition. However, even for moderate values of the ratio ρ_s/n the transition
to this third region occurs at sufficiently high voltages that it
may not be observed experimentally, the transition point from the
ohmic to the Mott-Gurney region occuring at the voltage $neL^2(\varepsilon\varepsilon_r)^{-1}$
which is generally within two or three orders of the breakdown voltage of most materials. In thin film specimens, however, L is small
and ultra-high field effects may be observable.

The figure also shows the situation for injection into a depletion region (ρ_s negative). In place of the ohmic region at low
potentials the square root region is observed and higher voltages
yield an ohmic region as the depletion charges is annulled.

In the absence of charge trapping a simple activation energy
is to be expected for the temperature dependence of the current.
However, if there are localised trap-like states, trap filling effects will occur[40] and anomalous power law current/voltage relationships can be expected, together with more complex temperature
dependences.

b) Single Carrier Control

The standard single carrier Schottky analysis is identical to
that already described for the Poole-Frenkel process in Section 3.2
with one important difference. In the Poole-Frenkel process the
carrier is freed from the coulombic potential generated by the immobile ionised centre. Under Schottky conditions the interaction
is with a mirror image charge (Figure 10a), obtained by summation
of the real surface charge on the injecting contact. Hence, as the
carrier moves away from the injecting surface so does the image,
by an equivalent distance, and the magnitude of the coulombic field
is halved when compared with the immobile centre case. This reduces
the Poole-Frenkel constant by a factor of two to give the Schottky
constant $\beta_s(=\frac{1}{2}\beta_{PF})$. If we consider a crystalline material the injected charge will be free to move in the conduction band, and, for
a thick specimen, the current will be given by $ne\,\mu F\,\exp(\beta_s F^{1/2}/kT)$
where μ is the conduction band mobility. For a thin film, however,
there is more than a single image charge to be considered (Figure
10b). In this case the high field characteristics are given by[38].

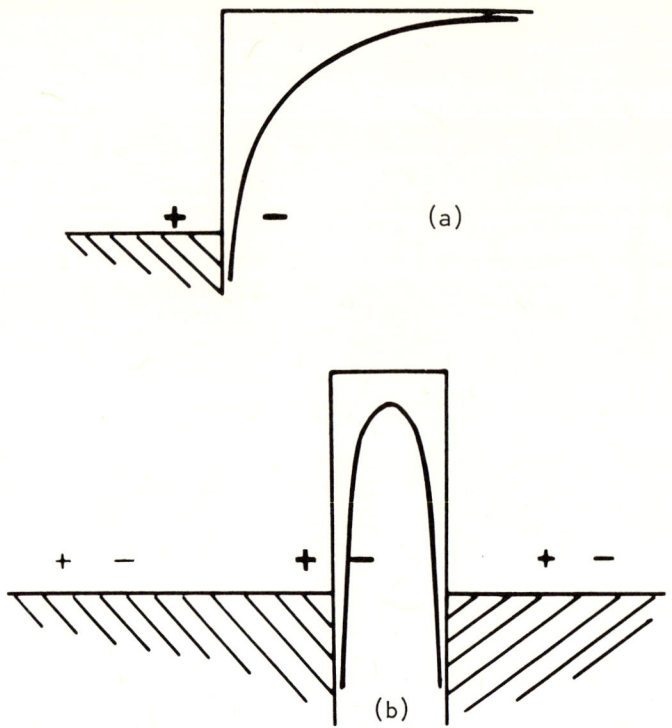

Fig. 10 Diagrammatic representation of the image charges and potential in (a) a thick and (b) a very thin semi-insulating layer

$$J = en\mu F \, (2\, \varepsilon\varepsilon_r \, FL^2)^{e(4\varepsilon\varepsilon_r L \, kT)^{-1}} \tag{31}$$

i.e., at single temperature there is a power relationship between the current and the voltage, the power being temperature dependent.

4. CONCLUSION

Critical reviews of experimental work[10,41,42] on thin films have shown that the specific transport processes discussed in Section 3. have been observed and it has also been shown that it is possible to extract useful information about the detailed band structure of thin films of non-metallic materials from transport studies.

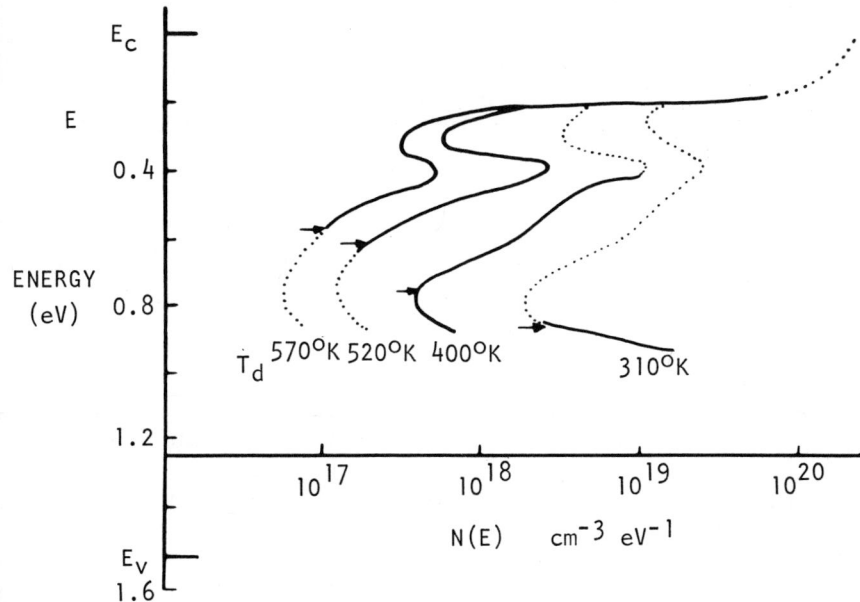

Fig. 11 Density of states for glow discharge deposited silicon films. T_d is the deposition temperature and the arrows indicate the positions of the Fermi level. The continuous lines were obtained directly from field effect experiments. (after reference 43.)

As a single example Figure 11 shows diagrammatically the form of the density of states curve obtained by Spear[43] from drift mobility measurements on glow discharge deposited silicon as a function of the temperature of deposition, T_d. In this material the position of the Fermi level was sensitive to T_d and allowed almost the full range of the normally forbiden gap to be explored under bias conditions.

The physics of crystalline materials, wheth er in the from of thin films or bulk specimens is well understood. The problem areas of poorly crystalline or disordered materials have been delineated but in order for a satisfactory theory of transport or any other physical property to be established, it is necessary for experimental work to be carefully examined and used to indicate the prime features of disorder. Thin films are ideal as specimen for this type of examination as the nature of deposition allows control over the structure, and the specimen geometry allows a wide range of experimental techniques to be applied.

REFERENCES

1. J. Tauc, L. Pajasova, R. Grigorovici and A. Vancu "Proc. Conf. Non-Cryst. Solids, Delft" North-Holland Amsterdam p. 606 (1965)

2. M. Brodsky, K. Weiser and G.D. Pettit, Phys. Rev. B1, 2632 (1970).

3. K. J. Siemsen and E.W. Fenton, Phys. Rev. 161, 632 (1967).

4. V. D. Frechette (Ed) "Non-Crystalline Solids" John Wiley, N.Y. (1960).

5. H. Richter and M. Furst, Z. Naturforschung 64, 38 (1951).

6. L. Cervinka, R. Hosemann and W. Vogel, J. Non-Cryst. Solids 3, 294 (1970).

7. A. Servini and A. K. Jonscher Thin Solid Films 3, 341 (1969).

8. e.g. J.P. McKelvey "Solid State and Semiconductor Physics" Harper & Row, N.Y. (1967).

9. P. W. Anderson, Phys. Rev. 109, 1492 (1958).

10. N. F. Mott and E. A. Davis "Electronic Processes in Non-Crystalline Materials" Clarendon Press, Oxford (1971).

11. D. A. Greenwood, Proc. Phys. Soc. 71, 585 (1958).

12. H. Jeffreys and B. S. Jeffreys "Methods of Mathematical Physics" University Press, Cambridge p. 503 2nd Ed. (1950).

13. R. M. Hill J. Phys. C 5, L267 (1972).

14. M. Cutler, Phil. Mag. 25, 173 (1972).

15. A. H. Clark, Phys. Rev. 154, 750 (1967).

16. L. Friedman J. Non-Cryst. Solids 6, 329 (1971).

17. W. E. Spear, Proc. Phys. Soc. B76, 826 (1960).

18 N. F. Mott "Electronic and Structural Properties of Amorphous Semiconductors" Le Comber & Mort, Eds. Academic Press, London (1973).

19. H. Fritzsche J. Non-Cryst. Solids 6, 49 (1971).

20. N. F. Mott Festkorperprobleme 9, 22 (1969).

21. R. M. Hill, to be published

22. P. Whittle "Probability" Penguin Books, Harmondsworth (1970).

23. A. Miller and E. Abrahams Phys. Rev. 120, 745 (1960).

24. V. Ambegaokar, B. I. Halperin and J. S. Langer Phys. Rev. B 4, 2612 (1971).

25. S. Kirkpatrick, Rev. Mod. Phys. 45, 574 (1973).

26. R. Jones and W. Schaich J. Phys. C 5, 43 (1972).

27. K. Maschke, H. Overhof and P. Thomas, Phys. Stat. Sol. (b) 62 113 (1974).

28. R. M. Hill, Phil. Mag. 24, 1307 (1971).

29. A. K. Jonscher, Thin Solid Films 1, 213 (1967).

30. J. L. Hartke J. Appl. Phys. 39, 487 (1968).

31. R. M. Hill, Phil. Mag. 23, 59 (1971).

32. G. A. N. Connel, D. L. Camphausen and W. Paul, Phil. Mag. 26, 541 (1972).

33. R. H. Fowler and L. W. Nordheim, Proc. Roy. Soc. A119, 515 (1928).

34. H. H. Poole, Phil. Mag. 32, 112 (1916) ; Phil. Mag. 34, 195 (1917).

35. D. K. Davies, J. Phys. D, 2, 1549 (1969).

36. R. H. Tregold "Space charge Conduction in Solids" Elsevier, Amsterdam (1966).

37. W. Schottky, Physik Z. 15, 872 (1914).

38. R. M. Hill, Thin Solid Films 15, 369 (1973).

39. N. F. Mott and R.W. Gurney "Electronic Processes in Ionic Crystals" Clarendon Press, Oxford (1940).

40. M. A. Lampert, Reports on Progress in Physics $\underline{27}$, 328 (1964).

41. D. Adler, C.R.C. Crit. Rev. in Solid State Science $\underline{2}$, 317 (1971).

42. A. K. Jonscher and R. M. Hill, Physics of Thin Films to be published (1975).

43. W. E. Spear, Proc. 5th Int. Conf. on Amorphous and Liquid Semiconductors, Garmisch (1973).

IONIC TRANSPORT IN THIN FILMS

Stephen J. Fonash

Department of Engineering Science and Mechanics
The Pennsylvania State University
University Park, Pennsylvania 16802

A charged defect - impurity, vacancy, or interstitial - in a thin film can drift under the influence of an electric field. As the defect moves through the material, in addition to the imposed electric field, it is subject to some sort of an oscillatory potential as a result of its interaction with the constituents of the material. Assuming that this potential is periodic allows it to be sketched as shown in Fig. 1. Here the externally applied field E has been superimposed. For the masses, or effective masses, and for the energies under consideration the wavelengths of the mobile defects are such that they are much smaller than the characteritic distance over which the potential sketched in Fig. 1 varies. Thus the motion of these defects may be treated classically.

The movement of the defect of charge q may be understood from Fig. 1 : the probability per attempt that the defect succeeds in advancing to the right one unit along this potential is $\exp[-(\phi-qaE/2)]/kT$; the corresponding quantity for going against the field is $\exp[-(\phi+qaE/2)]/kT$. If there are ν attempts per second then the net probability per second of advancing is [1]

$$2\nu e^{-\phi/kT} \sinh(qaE/2kT)$$

Here ϕ is the barrier height and a the distance between minima. Consequently this results in an electric current density J given by

$$J = 2\nu aqn\, e^{-\phi/kT} \sinh(qaE/2kT) \qquad (1)$$

for a defect density n.

From Eq. (1) it is seen that the mobility μ of the defect is

$$\mu = \frac{2\nu a}{E} e^{-\phi/kT} \sinh(qaE/2kT) \qquad (2)$$

This expression, for qaE << 2kT, simplifies to

$$\mu = q \frac{a^2 \nu}{kT} e^{-\phi/kT} \qquad (3)$$

Thus at room temperature for fields less than 10^5 V/cm, the ionic current conduction is ohmic. For higher fields Eq. (1) applies. However, for higher fields one would suspect that at some point it may become necessary to modify Eq. (1) to allow for the possibility that the field can change the number density n. Of course, for high fields Eq. (1) may be written as

$$J = \nu aqn \exp[-(\frac{\phi}{kT} - \frac{qaE}{2kT})] \qquad (4)$$

since jumps against the field are negligible.

Before examining this point of a field enhanced n some of the other problems and assumptions inherent in Eq. (1) should be discussed. First of all, since many thin film materials have high dielectric constants, it is not clear that the macroscopic electric field E is the correct field to use in Eq. (1). That is, it is obvious that a great deal of polarization is possible. Also since

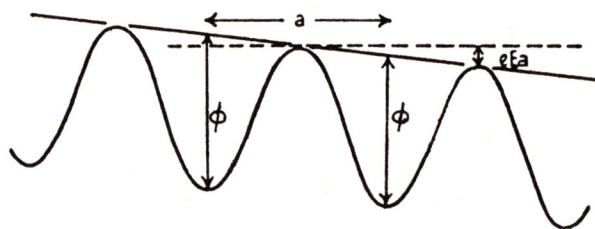

Fig. 1 Representation of total potential energy field in which the charged defect migrates.

bonding in many materials of interest is only partially ionic, it is not clear if an effective ionic charge should be used for $q^{(2)}$. However, the results of examining these effects indicate that, even at very high field strengths for wich the net activation energy (see Eq. (4)) goes from $(\phi-\frac{qaE}{2})$ to a more general power series in the electric field, the E and q to be used are the macroscopic field and the actual charge carried by the defect. This has been definitely established for the case that the potential barrier is symmetrical[1,2].

Another point of difficulty is that although a single ϕ appears in Eq. (1), for amorphous films there is a range of ϕ values. There also is a range of activation distances a, attempt frequencies ν, and local dielectric constants. Thus, to properly derive an expression for J, all possible combinations of sequential steps, parallel paths, and branches must be considered[3]. If Eq. (1) is used then for amorphous thin films, it is necessary to consider ϕ and a to be functions of the field E and the temperature T. The precise functional forms would depend on the details of the transport network assumed[3].

Further, for crystalline materials one may speak of the defect density n appearing in Eq. (1) as arising from impurities, Frenkel defects, and Schottky defects. In this case the computation of n is rather straight forward[4]. Impurities certainly can exist in amorphous thin film materials ; their migration according to Eq. (1) is easily understood. For full ionization, n would be a constant. However, if the impurity ionization were governed, for example, by electrons being liberated by the Poole-Frenkel effect then n would depend on T and E. Frenkel-like defects presumably are present in amorphous thin films also[3]. However, the computation of n is certainly not as simple as in the crystalline case due to the spread in parameter values. The same can be said for Schottky-like defects which could be attributed to regions of local nonstoichiometry[3]. In addition, for glass-like networks there may be intrinsic defects present which may be imagined as having been created by a rupture of bonds followed by an exchange of partners. The separated dangling bonds, having acquired or lost an electron, would contribute to ionic conduction[3]. Thus in thin films there are rich and varied sources of defects possible but the computation of n can be rather complex.

For the situation where the drift current is carried by an ion liberated from a fixed ion vacancy, n can also depend on the electric field, as has been demonstrated[5], by adding to the potential of Fig. 1 a binding potential between the ion and the vacancy. After a certain distance $\lambda_1+\lambda_2$ from the fixed ion vacancy, this attractive potential could be modeled as Coulombic. Thus in the direction of the field the envelope of the oscillatory potential would first increase going away from the fixed ion vacancy. Then as the external field takes over, it would decrease. The first several barrier heights (separation of adjacent relative maxima and minima) would be modified from the value ϕ ; succeeding barrier heights would be essentially ϕ.

If it is assumed that either the first (at $d_1=\lambda_1$) or second (at $d_2= \lambda_1+ \lambda_2 + \frac{a}{2}$) barrier determines the activation energy for the production of liberated ions, then, with Q_1 and Q_2 being the absolute heigths of these respective barriers, it is easily seen that Eq. (4) is valid with n given by (5)

$$n = N \exp[-(Q_i-\phi-qd_iE + \frac{qaE}{2})/2kT] \qquad (5)$$

Here N is the number of positions from which ions may be liberated per volume. Whether the first or second total barrier is higher (that is, whether i = 1 or 2) depends on the field strength. Thus an Arrhenius plot should give a discontinous change in slope, with field, at the field strength sufficient to make the second barrier lower than the first in cases where this model is applicable[5].

It has also been suggested [6] that n may be modified by considering that the activation energy necessary to liberate the migrating ion is essentially provided to overcome a Coulombic binding potential energy. Thus, with the presence of a sufficiently strong electric field the arguement proposes a Schottky barrier lowering term $2(q^3E/\varepsilon_s)^{1/2}$. This analogue to the Poole-Frenkel effect for electrons would, of course, increase the supply of migrating ions with electric field.

Although couched in classical terms, it is seen that the problems encountered in examining ionic transport in thin film are much the same as those found for the electronic transport mechanisms of hopping and percolation in thin films : difficulties in determining the number densities of carriers and their mobility. Also, just as for electronic conduction, the rate limiting process may not be characteristic of the bulk but may interfacial in nature [7,8].

Considering the experimental study of impurity ion transport in thin films, such investigations are most easily accomplished using M-I-S configurations since shifts in the C-V characteristics result with ion motion[8]. With the technique of bias temperature stressing[8-11], a typical example of which is shown in Fig. 2, information can be obtained on impurity mobilities, activation energies and number densities. The effects of mechanical stress may be examined also[11]. Due to the intense technological interest in SiO_2, ion transport in thin film structures using this material is particularly well studied with these approaches. Techniques have been developed for introducing a given ion species

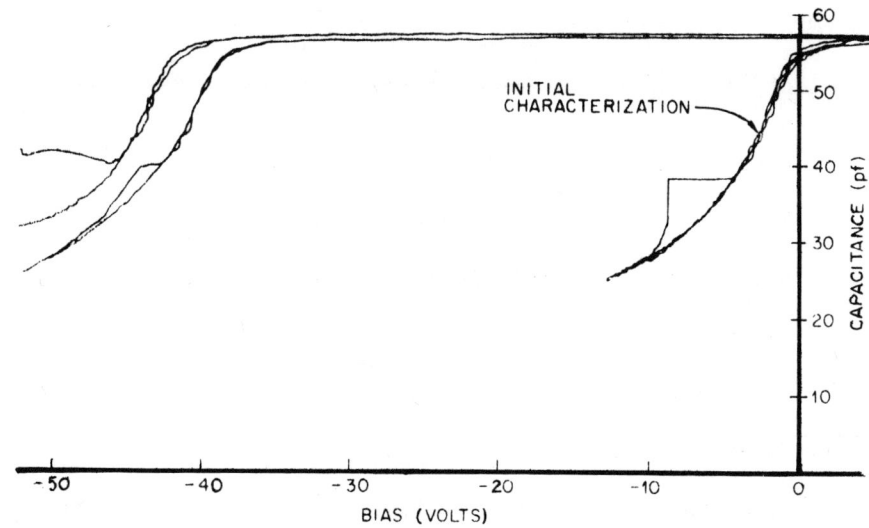

Fig. 2 High frequency C-V data for Al-SiO$_2$-Si structures in which the thin film SiO$_2$ was thermally grown. Results of bias temperature stressing are shown. The shift in C-V characterization is due to the drift of impurity ions in a dc field.

into the thin film, drifting it in a field, and collecting it again (12). Some of this work has indicated that certain species of ions are more mobile if injected into the SiO$_2$ than if present in the oxide as impurities incorporated during growth on Si. It has even been shown(13) that Na, deposited uniformly on the SiO$_2$, when drifted across the thin film, does not accumulate uniformy at the SiO$_2$-Si interface. This has been correlated with dielectric breakdown in the SiO$_2$-Si system.

REFERNCES

1. J. O'Dwyer, "The Theory of Electrical Conduction and Breakdown in Solid Dielectrics", Clarendon Press, Oxford, (1973).

2. M. Dignam, J. Phys. Chem. Solids 29, 249 (1968).

3. M. Dignam "Oxide and Oxide Films" (J. Diggle, Ed.), Dekker Inc. (1972).

4. N. Mott and R. Gurney, "Electronic Processes in Ionic Crystals"

Dover, Inc. N.Y. (1964).

5. C. Bean, J. Fischer, and D. Vermilyea, Phys. Rev. 101, 551 (1956).

6. J. Hanscomb et al., Proc. Phys. Soc. London 88, 425 (1966).

7. S. Hofstein, IEEE Trans. Electron. Devices ED-13, 222 (1966).

8. E. Snow et al., J. Appl. Phys. 36, 1664 (1965).

9. R. Castagne, C.R. Acad. Sci. (Paris) 267, 866 (1968).

10. M. Kuhn, Solid-St. Electronics 13, 873 (1970).

11. S. Fonash, J. Appl. Phys. 44, 4607 (1973).

12. M. Woods and R. Williams, J. Appl. Phys. 44, 5506 (1973).

13. T. Dunn et al., Phys. Chem. Glasses 6, 16 (1965).

14. T. Di Stefano, J. Appl. Phys. 44, 527 (1973).

DIELECTRIC PROPERTIES OF THIN FILMS : POLARIZATION AND EFFECTIVE
POLARIZATION

Stephen J. Fonash

Department of Engineering Science and Mechanics
The Pennsylvania State University
University Park, Pennsylvania 16802

1. INTRODUCTION

In thin films, with their amorphous or polycrystalline structure, a large number of dielectric polarization mechanisms are possible. These can be classified into two rather general groupings : those which occur due to the presence of interfaces and those which are characteristic of the thin film material itself. That is, this latter category does not owe its existence in any way to interfaces whereas the former exists because of interfaces. The mechanism basic to the first category is the presence of concentration modifications at electrodes and at interfaces, which may occur within the film. Although such interface effects are usually dismissed in dealing with bulk dielectric properties, they may dominate in thin film structures. Consequently their probable presence cannot be taken lightly. It should be pointed out that even in thick "bulk" samples of material it is well known that the interface effects can dominate as is the case for electrets.

In the second category are found all those mechanisms which may be considered "bulk" in origin. These include dipole and dipole-like polarization effects which result when dipolar molecules and ions are capable of occupying several closely spaced positions in a given region. Another possible polarization mechanism is that arising from imperfections ; the effective dipole resulting from a cation-vacancy pair is an example. Electronic mechanisms which can produce an effective polarization are also in this category. Of course, this includes the normal polarization arising from the distortion of the spacial configuration of tightly bound electrons; but much more importantly for many thin film materials, it includes effective polalarization which may arise from mechanisms involving itinerant electronic carriers (for example, electron transport by hopping or percolation).

In general, at given point in a thin film the flux density D, the macroscopic electric field E, and the polarization P, dictated by the response characteristics of the material, are related by[1-3]

$$\varepsilon_o E = D - P \qquad (1)$$

where ε_o is the permittivity of free space. Of course, in this equation the polarization P is due only to those mechanisms in category II ; interface effects are discussed in terms of macroscopic charge density ρ through Poisson's equation :

$$\frac{\partial D}{\partial X} = \rho \qquad (2)$$

In a static field, the static polarization P_S should be proportional to the macroscopic field E ; ie,

$$P_S = (\varepsilon_s - \varepsilon_o)E \qquad (3)$$

where $\varepsilon_s = K_s \varepsilon_o$ is the static permittivity of the material and K_s is the static dielectric constant[4]. For time varying macroscopic fields E, one would expect that P would try to follow the electric field. If first order kinetics are applicable (as then apparently ar for at last some of the phenomena in category II) this may be expressed, for the ℓ'th contributon to the polarization, as [1]

$$\tau_\ell \frac{dP_\ell}{dt} = P_{s\ell} - P_\ell \qquad (4)$$

Here τ_ℓ is a phenomenological relaxation time for this ℓ'th mechanism and $P = \Sigma_\ell P_\ell$; ie, P is the sum of all mechanisms present from category II. Thus $P_\ell = P_\ell(t)$ obeys some relationship such as Eq. (4) in a time varying field and consequently $P = P(t)$.

For a static electric field, the electric flux configuration in a thin film structure is static and only a particle current can be detected flowing through the thin film. Thus in this situation no information can be obtained on polarization effects from electrical measurements. To actually probe electrically the dielectric properties of a thin film, it is necessary to study currents passing through the film under dynamic conditions. Common configurations employed are metal-thin film-metal and metal-thin film-semiconductor structures. Considering the metal-thin film-metal configuration of Fig. 1 for definitiveness, it can be seen that by varying the difference between the electrochemical potentials (Fermi levels) of the two metal electrodes by applying a bias $V = V(t)$, changes in interface polarization can take place. Of course that polarization

DIELECTRIC PROPERTIES OF THIN FILMS

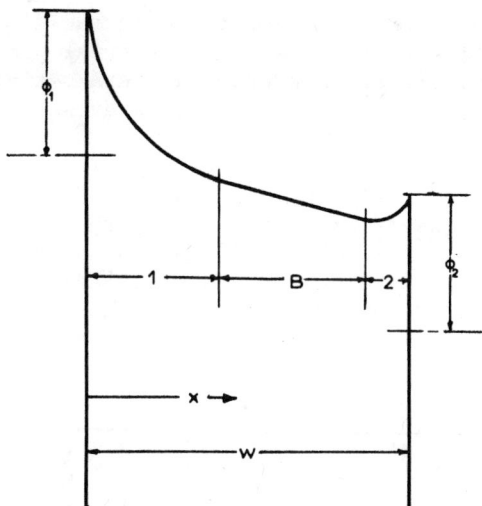

Fig. 1 Thin film in an M-I-M configuration. Regions 1, B, and 2 are shown as described in the text. A bias V is applied and positive current flows right to left.

arising from mechanisms being referred to as category II will also be varied in this manner. An additional component of current caused by these polarization changes will be detected at the electrodes in this case and information on dielectric properties can be obtained.

To put this in quantitative terms, Eq.(1) together which Eq.(2) may be used in a straightforward manner to write that the total current crossing electrode 1 of Fig. 1, for example, is given by (5)

$$I = I_p + C_G \frac{dV}{dt} - \frac{A}{W} \int_0^W \frac{\partial P}{\partial t} dx + A \int_0^W (1 - \frac{x}{W})\frac{\partial \rho}{\partial t} dx \qquad (5)$$

where I_p is the particle current crossing the plane $x = 0$, W is the width of the thin film and A is the area of the film. The geometrical capacitance C_G is given by $A\varepsilon_0/W$. In a static situation only the particle current flows. Under dynamic conditions a particle current (which may considerably different than that present in the static situation since some of this particle current may be flowing due to changes in ρ or due to changes in stored carriers) crosses $x = 0$ together with displacement currents arising from the last three therms in Eq. (5). The first of these three is the displacement current which would flow if there were no material present. The second accounts for polarization phenomena characteristic of the material itself (category II). The third term arises from the presence of interfaces (category I).

Thus by varying the field (by application of an ac signal[6], by application of a pulse[7] etc.) or even by thermally stimulating frozen-in polarization[8-10], the polarization can be made to vary and information obtained on dielectric properties as may be seen from Eq. (5). The most prevalent of these techniques is that of applying an alternating current signal δV about some dc bias V (experimentally, very often V = 0). Consequently the resulting ac quantities are related by[5]

$$\delta I = \delta I_p + C_G \frac{dV}{dt} - \frac{A}{W} \int_0^W \frac{\partial \delta P}{\partial t} dx + A \int_0^W (1 - \frac{x}{W}) \frac{\partial \delta \rho}{\partial t} dx \quad (6)$$

Further, in using ac signals to explore dielectric properties, it is freequently assumed that $\delta V < kT$ permitting linearization[5] of Eq. (6). This linearization is tacitly assumed when the current $\delta I = i e^{j\omega t}$ is attributed to two channels of flow given by[1,4,11,12]

$$i = \frac{v}{R_p} + j\omega C_p v \quad (7)$$

Here $j = (-1)^{1/2}$ and $\delta V = v e^{j\omega t}$. The quantity R_p is a differential parallel equivalent resistance and C_p is a differential parallel equivalent capaciatance evaluated at the bias V. Obviously the physical origins of these small signal equivalent circuit elements lie in terms on the right hand side of Eq. (6). On consideration of Eq. (6) it is not surprising to find experimentally that R_p and C_p are functions of the frequency ω of the ac signal.

As an alternative to Eq. (7) an effective complex conductivity $\sigma^* = \sigma' + j\sigma''$ may be defined such that the small signal current i in Eq. (7) is given by[1,4]

$$i = \frac{A\sigma^* v}{W} \quad (8)$$

Thus it follows that

$$R_p = \frac{W}{A\sigma'} \quad (9)$$

and

$$C_p = \frac{A\sigma''}{W\omega} \quad (10)$$

It is equally possible to define an effective complex permittivity $\varepsilon^* = \varepsilon' - j\varepsilon''$ (or an effective complex dielectric constant $\varepsilon^* = K^*\varepsilon_0$) (1,4) such that

$$i = j\omega \frac{A\varepsilon^*}{W} v \tag{11}$$

In this case comparison with Eq. (7) shows that

$$R_p = \frac{W}{A\varepsilon''\omega} \tag{12}$$

$$C_p = \frac{A\varepsilon'}{W} \tag{13}$$

and from Eq. (9), (10), (12) and (13) it follows that

$$\sigma' = \varepsilon'' \omega \tag{14}$$

$$\sigma'' = \varepsilon' \omega \tag{15}$$

These quantities C_p, R_p, ε^*, and σ^* are effective; they arise from and are combinations of the processes represented in Eq. (6). The task in interpreting experimental data is that of unraveling such effective, measured quantities to determine their physical origins. It is a task made quite difficult by the fact that many of the processes in category I and in category II, to the external measuring circuit, appear rather similar in behavior. However, studying these polarization phenomena-which in no way should be construed as limited to dielectrics- can provide very useful, additional information on the structure and electronic properties of non-metallic thin films. Thus it can be a very worthwhile undertaking.

2. INTERFACE EFFECTS

Interface effects, which contribute through Eq. (5)-(15) to the dielectric properties of thin film structures, arise due to modifications, near the interfaces, in the concentrations of the various charged species present in thin films. Analysis of this phenomenon is extremely complicated due to the fact that, even in the absence of any dc current (ie, V = 0), there can be space charge regions and built-in potentials V_D(diffusion potentials) at the interfaces of thin films[5, 13-15]. These are modified with the application of a dc bias and further modified if it is varied.

Considering an ac variation, it is seen from Eq. (6) that, if the term involving the bulk polarization is neglected (its effects for an ac distrubance can be represented by replacing ε_0 with ε^* as is discussed in the section on bulk effects), the quantities $\delta I_p = \delta I_p(x,t)$ and $\delta\rho = \delta\rho(x,t)$ need only to be determined to evaluate interface effects[15]. Of course the charge density ρ is composed of contributions from mobile carriers (positive ions or holes with density p_i; negative ions or electrons with density n_j) and fixed charges (a given species having the density N_k). The particle current, on the other hand, is composed of contributions from the various mobile carriers.

The basic equations governing small ac concentration modifications of the mobile carriers, in linearized form, are[15]

$$\frac{\partial \delta p_i}{\partial t} = -\mu_i \, \delta p_i \frac{\partial E}{\partial x} - \mu_i p_i \frac{\partial \delta E}{\partial x} - \mu_i E \frac{\partial \delta p_i}{\partial x} - \mu_i \, \delta E \frac{\partial p_i}{\partial x}$$

$$+ D_i \frac{\partial^2 \delta p_i}{\partial x^2} + \delta G_i - \delta R_i \qquad (16)$$

$$\frac{\partial \delta n_j}{\partial t} = +\mu_j \, \delta n_j \frac{\partial E}{\partial x} + \mu_j n_j \frac{\partial \delta E}{\partial x} + \mu_j E \frac{\partial \delta n_j}{\partial x} + \mu_j \delta E \frac{\partial n_j}{\partial x}$$

$$+ D_j \frac{\partial^2 \delta n_j}{\partial x^2} + \delta G_j - \delta R_j \qquad (17)$$

Here drift has been modeled by Ohmic conduction; a more general model for drift could be employed. The basic equations governing the concentration modifications of the various fixed species are of the form

$$\frac{\partial}{\partial t} \delta N_k = \delta G_k - \delta R_k \qquad (18)$$

Added to this system of equations is Poisson's equation

$$\frac{\partial}{\partial x} \delta E = \frac{e}{\varepsilon^*} \delta\rho \qquad (19)$$

with $\delta\rho = \sum_i Z_i \delta p_i - \sum_j Z_j \delta n_j + \sum Z_k \delta N_k$. In these equations δp, δn, δN and δE are variations of the number densities and electric field about the values p, n, N and E. The quantities G and R represent generation and recombination; the μ and D are mobilities and diffusivities, respectively. These equations are linearized forms of the continuity equation (one for each positive mobile species of valence z_i, one for each negative mobile species of valence z_j and one for each fixed species) and of Poisson's equation. Of course the quantities needed to compute the observed polarization current

DIELECTRIC PROPERTIES OF THIN FILMS

caused by these small ac variations are $\delta\rho$ given above and the particle current

$$\delta I = \sum_i Az_i e \left[\mu_i(E\delta p_i + p_i\delta E) - D_i \frac{\partial \delta p_i}{\partial x} \right] +$$

$$\sum_j Az_j e \left[\mu_i(E\delta n_j + n_j\delta E) + D_j \frac{\partial \delta n_j}{\partial x} \right] \quad (20)$$

Solving the system, Eq (16)-(19), allows these to be computed.

To solve for the δp, δn, δN and δE from the system of equations (16)-(19) initial and boundary conditions are needed. Initial conditions are simplified in part by assuming an ac variation. However, information on the initial (ie, dc) values of p_i, n_j, N_k, and the electric field is necessary as may be seen from the system of equations. Boundary conditions at the thin film-metal electrode interface are also necessary.

Thus, in general, the dc values of p_i, n_j, N_k, and E existing for some bias V must be obtained as functions of position from appropriately formulated continuity equations, Poisson's equation, and boundary conditions. This has been discussed in detail in reference[15]. These dc values are then substituted in the system Eqns. (16)-(19). Consequently, with appropriate boundary conditions, the small ac variations may be calculated.

This calculation scheme – the core of which is equations of the form (the precise form depending on the model used for drift) of Eqns. (16)-(19) – constitutes a general approach to determining the effective polarization currents induced by a small ac signal applied across a thin film. However, actual evaluation of such a system can be a horrendous task. This is true since the boundary conditions may be difficult to model (definitive information about interfaces transport is necessitated) and since the dc quantities must be calculated. The equations for the dc values are non-linear and are complicated by the fact that band bending may be present even for the case V=0 (15,20).

At least two approaches to this complex analytical problem of determining the possible effective ac polarization currents arising from interface effects may be found in the litterature. These represent two extreme models for thin films. Thus these constitute two approaches to the system Eqns. (16)-(19). One approach[13,15-18] neglects the existence of regions of large band bending (>kT) although some progress has been made to correct this deficiency of the analysis[19,20]. In this approach there is never any appearance of quantities such as ℓ_1 and ℓ_2 which are the widths, at bias V, of the regions in Fig. 1, for example, where large band bending is occuring. Actually in such analyses it is usual to treat the bands as being flat and thus to ignore completely any band bending at the interfaces. In this approach it is also usual to treat the V = 0 case only ; discussion of its extension to V ≠ 0 may be found in reference (15) and (20). This approach may be characterized by saying that the effective polarization arises from the motion of mobile charges moving back and forth under the action of the ac field, in essentially a zero diffusion potential situation.

The alternative approach neglects either completely[5,21,22] or partially[23] any modifications in concentrations which take place in region of band bending <kT. In fact this class of approximate solutions to the system Eq. (16)-(19) usually considers only modifications to charge in regions where the band bending is >kT[5,21,22] and allows for charge growth only at the edges of this space charge region. This alternative approach is essentially a depletion approximation but the charge being depleted need not necessarily be mobile carriers. It could, for example, reside in traps[22].

Using the first approach it is possible to obtain a solution to the system (16)-(19). With the initial and boundary conditions that are stipulated in this analysis, one makes the assumptions which characterize it : the possible existence of regions with large band bending is ignored (although the implications of this are fully realized[15]). That is, to make a general solution tractable, flat bands are assumed for the unperturbed situation. Further, the case where $V \equiv 0$ only is explored (i.e., E = 0)[15]. In addition the so-called Chang-Jaffé conditions[16] are imposed as the boundary conditions on the current contribution, at x = (0,w), for the i'th species ; i.e

$$\delta I_i \Big|_{x=0,w} = (2r_i\, eD_i A/W) \delta n_i \Big|_{x=0,w} \quad (21)$$

In this phenomenological expression r_i is a dimensionless discharge parameter.

From Eqs. (16)-(19), with these assumed conditions, general solutions for the effective polarization currents to be expected experimentally have been obtained for the case of one positive and one negative species present (holes, electrons, ions) with arbitrary valencies and mobilities. An immobile homogeneous charge density was also assumed present. Recombination, generation effects were neglected[15]. These solutions are such that when Eq.(6) with $\delta P=0$ is evaluated and interpreted in terms of Eq.(7), C_p and R_p are found to be extremely complicated functions. However, they simplify in a number of situations of interest for the case of $\omega = 0$. Polarization is accounted for here by ε.

DIELECTRIC PROPERTIES OF THIN FILMS

When the discharge parameters of the two mobile species are taken as equal (to say r), it is found for example, that $C_{p0} \equiv C_p(\omega=0)$ is given by Eq. (22)[15] ; here $C_G' \equiv \frac{\varepsilon}{\varepsilon_0} C_G$.

$$C_{p0} = C_G' + \frac{\frac{e}{2L_D} \cot \frac{W}{2L_D}}{[1 + r/2]^2} - \frac{C_G'}{[1 + r/2]^2} \qquad (22)$$

It is interesting to note the manner in which W appears in this expression. Obviously if the interfaces at the electrodes did not modify the concentrations adjacent to them (ie, if $r \to \infty$), there would be no interface effect.

Since L_D is the Debye length defined by

$$L_D \equiv \left(-\frac{\varepsilon kT}{e(z_n^2 n + z_p^2 p)} \right)^{1/2} \qquad (23)$$

the contribution to the effective dielectric properties of a thin film coming from the term in Eq. (22) involving L_D could be substantial – depending on the values of n and p. If the mobile charge carriers are electrons or holes (although it is difficult to conceive of a real situation where they would have the same discharge parameter except for the case of blocking contacts ; ie, r=0) the electron and hole densities would have to be something of the order of $10^{14} cm^{-3}$ or larger for this analysis to be meaningful for a thin film (W < micron). Thus in this case for wide gap thin film materials, with deep donor and acceptor levels, this origin of a contribution to the effective dielectric constant should not be important. This is especially so if there are depletion regions present. Then these have widths ℓ_1 and ℓ_2 of the order of

$$\ell_1 = \left(\frac{2\varepsilon V_D}{eN_C} \right)^{1/2} \qquad (24)$$

where N_C is the effective doping density[5] and V_D is the diffusion potential. Since N_C can be large even in wide gap materials capacitance arising from these regions would be quite important.

If the mobile carriers in this situation (Eq. 22) are ions, the large number densities required for a contribution to the effective dielectric properties of a thin film should be detectable in the material transport which would take place in a dc experiment. Some experimental effects seen in thin films have been attributed to this case with the mobile carriers being ions[25].

For the situation in which the contacts are completely blocking to both carriers it is apparent from Eq. (22) that the above comments are still valid. For the situation in which the contacts are blocking to one species of mobile carrier but not to another, it is again found that relatively large number densities of free charge are required for this phenomenon to contribute significantly to the effective dielectric properties of thin films. For example, if positive ions are blocked but the electron density at the interface is totally undisturbed by the variation about $V = 0$ (ie, $r_p=0$ $r_n \to \infty$), then, with $Z_n = Z_p = 1$, C_{po} is given by[15].

$$C_{po} = C_G' + \frac{\varepsilon}{2L_D} \frac{W}{6L_D} \left(\frac{np}{(n+p)^2} \right) + \frac{\varepsilon}{2L_D} \coth \frac{W}{2L_D} \left\{ \frac{p}{(n+p)} \right\}^2 - C_G' \left\{ \frac{p}{(n+p)} \right\}^2 \quad (25)$$

Again L_D is given by Eq. (23) and again it is intersting to note the dependence of this C_{po} on W. Obviously this resulting dependence on W predicted analytically for this case by this approach to the system Eqs (16)-(19) could be examined experimentally.

Since the quantities in parentheses are always < unity, the important difference(that is, the possible origin of larger effects) between Eq. (25) and Eq. (22) is the appearance in Eq. (25) of the factor (W/L_D). Thus, much larger effects can be found for given densities because of this factor ; however, comments made regarding Eq. (22) should also be valid in this situation.

On the basis of this discussion it would seem that experimental data, which have been obtained at room temperature for several wide gap dielectic thin film materials yet which show an apparent increase in low frequency dielectric constants which is greater than several orders of magnitude over the high frequency values, must be interpreted in terms of an effect attributable to mobile ions and defects (Fig. 2) having their concentrations perturbed at the electrode interfaces[15,20,25,26]. Similar effects have also been seen in thin films of ionic materials. The interpretation has been the same[27]. It is obvious, however, that there is range of values of C_{po} which could be determined experimentally which are extremely difficult to interpret without some independently obtained information such as that which may be extracted from dc measurements (28,29), SEM observations[30], etc. This difficulty will become clearer as the analytical predictions resulting from the second approach (that is, second model for the origins of the effective polarization are examined.

DIELECTRIC PROPERTIES OF THIN FILMS

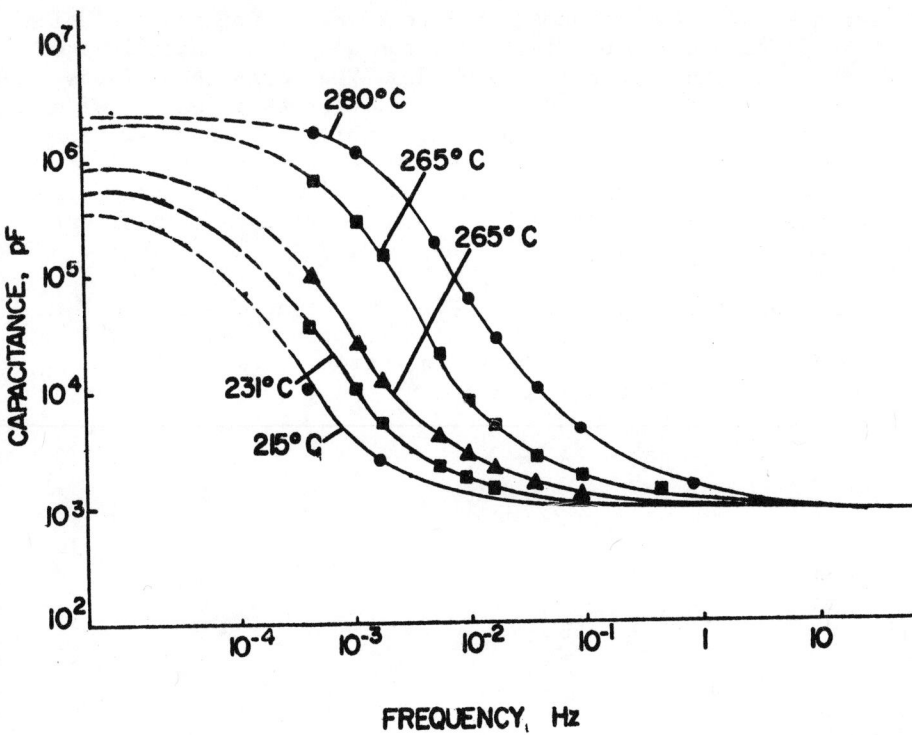

Fig. 2 Dependence of capacitance on frequency for RF sputtered SiO_2 films. Film thickness is 1400 Å. Theoretical curves and experimental data are shown. (After Meaudre and Meaudre, Ref.(25)).

For many thin film materials it is quite evident that the possibility of large band bending (and, therefore, of space charge regions) at interfaces is very good. This would occur even at zero bias[5,14,15,20,30-32]. Thus regions of band bending >kT should probably exist even for V = 0 essentially because the amorphous nature of thin films indicates that there should be states in the energy gap of these materials. These states will be able to empty or fill near the interfaces to equate Fermi levels.

An analysis of the polarization effects possible because of the existence these regions has been recently undertaken[5]. The existence of concentration modifications in regions of band bending <kT is ignored. Space charge is assumed to grow only at its edge.

For the sake of definitiveness the electronic carrier of principal interest is assumed to be electrons, the source of which is a donor level E_D below the conduction band edge. The Fermi level is assumed to lie somewhere above the donor states. With this model a wide gap dielectric thin film can be represented which has very few mobile electrons but which, through the mechanism of emptying or filling of the donor states, can develop a substantial space charge region at the interfaces. The diagram of Fig. 1 represents this situation under the general biasing conditions assumed in the analysis.

Equation (6), of course, valid as are equations of the form of Eqs (16)-(19). However, from the assumptions made it follows that δQ_1, the change in the charge in region 1, is related to E_B, the electric field in the interior, by

$$\delta I_1 + \frac{\partial \delta Q_1}{\partial t} = -\frac{\partial I_B}{\partial E_B} \delta E_B \tag{26}$$

and

$$\frac{\partial \delta Q_2}{\partial t} = \delta I_2 + \frac{\partial I_B}{\partial E_B} \delta E_B \tag{27}$$

where here I_1 is the particle current at $X = 0$ and I_2 is the corresponding quantity at $X = W$. Also since the charge in regions 1 and 2 is assumed to grow only at the edge, it is easily shown that[5]

$$\delta Q_1 = \frac{A \varepsilon}{\ell_1} \delta V_1 + \varepsilon A \delta E_B + \frac{\varepsilon A E_B}{\ell_1} \delta \ell_1 \tag{28}$$

and

$$\delta Q_2 = -\frac{A \varepsilon}{\ell_2} \delta V_2 - \varepsilon A \delta E_B - \frac{\varepsilon A E_B}{\ell_2} \delta \ell_2 \tag{29}$$

The total bias variation δV is given by[5]

$$\delta V = \delta V_1 + \delta V_B + \delta V_2 \tag{30}$$

where

$$\delta V_1 = -E_B \delta \ell_1 - \ell_1 \delta E_1 \tag{31}$$

$$\delta V_2 = -E_B \delta \ell_2 - \ell_2 \delta E_2 \tag{32}$$

and

$$\delta V_B = E_B(\delta \ell_1 + \delta \ell_2) - \ell_B \delta E_B \tag{33}$$

In these expressions E_1 and E_2 are the electric fields in regions 1 and 2, respectively. These equations are, of course equivalent to the system Eqs (16)-(19).

With $\ell_1 = \ell_1(V_1, E_B)$ and $\ell_2 = \ell_2(V_2, E_B)$ the system Eqs (26) - (33) can be solved in terms of δV by noting that the particle current crossing $X = 0$ can be written in a general way as[5,33]

$$\delta I_1 = \frac{1}{R_1} \delta V_1 + \frac{\partial I_1}{\partial E_B} \delta E_B \tag{34}$$

A similar expression follows for δI_2. Using the results from solving this system in Eq. (6) and assuming the variations are of the form $e^{j\omega t}$ allows interpretation in terms of Eq. (7). The expressions for $C_p = C_p(V,\omega)$ and $R_p = R_p(V,\omega)$ which are determined are also extremely complex[5]. However, they may be represented in terms of the equivalent circuit of Fig. 3.

For Fig. 3 the following definitions have been employed[21]

$$I_E = -\frac{C_B}{B} E_B \left(\frac{\partial \ell_1}{\partial V_1} \frac{\partial \delta V_1}{\partial t} + \frac{\partial \ell_2}{\partial V_2} \frac{\partial \delta V_2}{\partial t} \right) \tag{35}$$

$$C_L \equiv C_1 + \frac{C_B}{B} E_B \frac{\partial \ell_1}{\partial V_1} + C_1 E_B \frac{\partial \ell_1}{\partial V_1} + \frac{C_1 E_B}{B \ell_B} \frac{\partial \ell_1}{\partial E_B} E_B \frac{\partial \ell_1}{\partial V_1} \tag{36}$$

$$C_R \equiv C_2 + \frac{C_B}{B} E_B \frac{\partial \ell_2}{\partial V_2} + C_2 E_B \frac{\partial \ell_2}{\partial V_2} + \frac{C_2 E_B}{B \ell_B} \frac{\partial \ell_2}{\partial E_B} E_B \frac{\partial \ell_2}{\partial V_2} \tag{37}$$

$$R_L^{-1} \equiv \frac{1}{R_1} + \frac{\partial I_1}{\partial E_B} \frac{E_B}{\ell_B} \frac{\partial \ell_1}{\partial V_1} \frac{1}{B} \tag{38}$$

$$R_R^{-1} \equiv \frac{1}{R_2} + \frac{\partial I_2}{\partial E_B} \frac{E_B}{\ell_B} \frac{\partial \ell_2}{\partial V_2} \frac{1}{B} \tag{39}$$

$$I_L \equiv \frac{\partial I_1}{\partial E_B} \frac{E_B}{\ell_B} \frac{\partial \ell_2}{\partial V_2} \frac{1}{B} \delta V_2 - \frac{\partial I_1}{\partial E_B} \frac{1}{\ell_B} \delta V_B + \left(\frac{C_B E_B}{B} \frac{\partial \ell_2}{\partial V_2} \right.$$

$$\left. + \frac{C_1 E_B^2}{B \ell_B} \frac{\partial \ell_2}{\partial V_2} \frac{\partial \ell_1}{\partial E_B} \right) \frac{d}{dt} \delta V_2 - \frac{C_1 E_B}{B \ell_B} \frac{\partial \ell_1}{\partial E_B} \frac{d}{dt} \delta V_B \tag{40}$$

and

$$I_R \equiv \frac{\partial I_2}{\partial E_B} \frac{E_B}{\ell_B} \frac{\partial \ell_1}{\partial V_1} \frac{1}{B} \delta V_1 - \frac{\partial I_2}{\partial E_B} \frac{1}{\ell_B} \delta V_B$$

$$+ \left(\frac{C_B E_B}{B} \frac{\partial \ell_1}{\partial V_1} + \frac{C_2 E_B^2}{B \ell_B} \frac{\partial \ell_1}{\partial V_1} \frac{\partial \ell_2}{\partial E_B} \right) \frac{d}{dt} \delta V_1 - \frac{C_2 E_B}{B \ell_B} \frac{\partial \ell_2}{\partial E_B} \frac{d}{dt} \delta V_B \tag{41}$$

In addition, the small-signal circuit elements representing the interior region B, the region across which δV_B is developed, are given by

$$C_M \equiv C_B/B \tag{42}$$

$$R_M^{-1} \equiv (BR_B)^{-1} = \left(\frac{B}{\ell_B} \frac{\partial I_B}{\partial E_B} \right)^{-1} \tag{43}$$

Further, in all these definitions

$$C_1 \equiv \frac{\varepsilon}{\ell_1} \tag{44}$$

$$C_2 \equiv \frac{\varepsilon}{\ell_2} \tag{45}$$

$$C_B \equiv \frac{\varepsilon}{\ell_B} \tag{46}$$

and

$$B \equiv 1 - \frac{E_B}{\ell_B} \frac{\partial \ell_1}{\partial E_B} - \frac{E_B}{\ell_B} \frac{\partial \ell_2}{\partial E_B} \tag{47}$$

have been used, as have the definitions

$$R_1 \equiv \left(\frac{\partial I_1}{\partial V_1} \right)^{-1} \tag{48}$$

$$R_2 \equiv \left(\frac{\partial I_2}{\partial V_2} \right)^{-1} \tag{49}$$

$$R_B \equiv \left(\frac{\partial I_B}{\partial E_B} \frac{1}{\ell_B} \right)^{-1} \tag{50}$$

This involved equivalent circuit representation of $C_p(V,\omega)$ and $R_p(V,\omega)$ reduces to that of Fig. 4 if the field in the interior, region B, is much less than that in the interface regions and if changes in the field of the interior are much smaller than changes in the fields at the interfaces[21]. The extremely simple model of Fig. 4 is nothing more than a Maxwell-Wagner type of model for three layered regions.[4] However, unlike true Maxwell-

DIELECTRIC PROPERTIES OF THIN FILMS

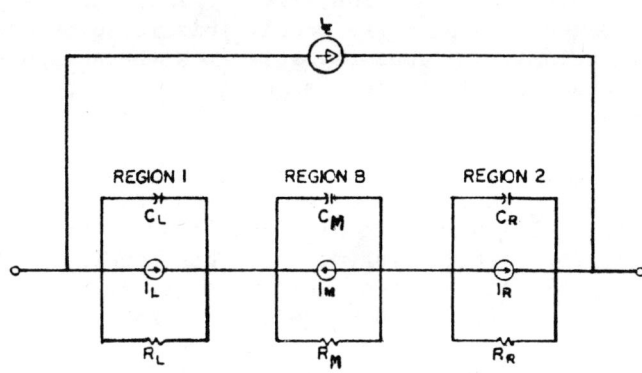

Fig. 3 AC equivalent circuit resulting from the full analysis of Ref. (5)

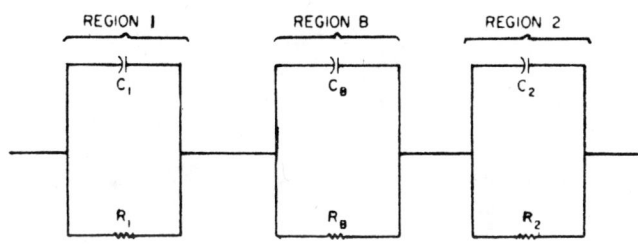

Fig. 4 Maxwell-Wagner-like model ; circuit elements are defined in the text . Fig. 3 reduces to this under two sets of conditions described in text.

Wagner models for regions of differing conductivities, these regions represented by Fig. 4 can grow and shrink with bias[5].

If Eqs. (35)-(50) are numerically evaluated for parameter values that insure $E_B \ll E_1$ and E_2 (and $\delta E_B \ll \delta E_1$ and δE_2), results such as those shown in Figs. 5 and 6 are obtained[5]. Since the circuit of Fig. 4 would apply for such a choice of parameters, the appearance of only one relaxation time for zero bias is just as expected for a Maxwell-Wagner-like model. This is so because for n layers, (n-1) relaxation times result[34]. With the application of bias, two relaxation times appear since there now are three dissimilar layers in the thin film.

For the conditions $E_B \ll E_1$ and E_2 (with $\delta E_B \ll \delta E_1$ and δE_2) the low frequency capacitance is dictated by the interfaces as may be seen from Fig. 5 and Fig. 7. Such behavior has been seen experimentally for certain systems[35]. This conclusion applies equally well to the situation of zero bias (i.e., V = 0). Equation (22) gives, however, for V = 0 a $\Delta C = C_{po} - C_{p\infty}$ which is independent of thickness. Equation (25) can give for V = 0 a ΔC which increases with film thickness. Equations (22) and (25) can explain increases in the dielective constant at low frequencies over the high frequency value which exceed several orders of magnitude. The behavior depicted in Figs. 5 and 6 is more limited ; essentially the ΔC values possible are restricted by ℓ_1. For very small values of ℓ_1 the interfaces would become invisible due to tunneling. Also ℓ_1 can be made to grow[5] if Fig. 4 is valid providing another check on this interpretation of ΔC.

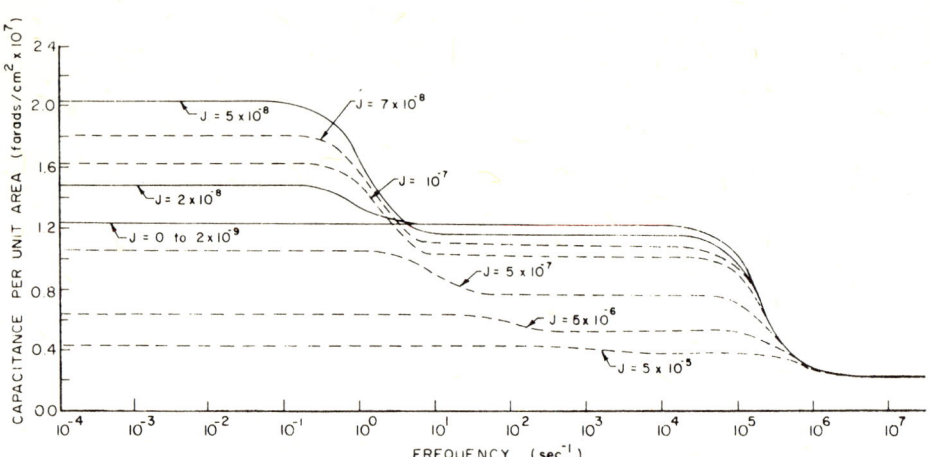

Fig. 5 Parallel equivalent capacitance ($\times 10^7 F/cm^2$) versus frequency at 300°K for a 2000 Å thick insulator. Numerical evaluation of the circuit shown in Fig. 3. is done for parameter values such that Fig. 4. is valid.

DIELECTRIC PROPERTIES OF THIN FILMS

It is interesting to note that a Maxwell-Wagner-like model may be extracted from Eqs (6), (26)-(33) without stringent conditions on the fields. This may be accomplished by writting the approximations

$$\delta I_1 \simeq \frac{\partial I_1}{\partial E_1} \delta E_1 \tag{51}$$

and

$$\delta I_2 \simeq \frac{\partial I_2}{\partial E_2} \delta E_2 \tag{52}$$

in place of expressions of the form of Eq. (34) and by defining

$$\delta V_1' \equiv \delta V_1 + E_B \delta \ell_1 = -\ell_1 \delta E_1 \tag{53}$$

$$\delta V_2' \equiv \delta V_2 + E_B \delta \ell_2 = -\ell_2 \delta E_2 \tag{54}$$

and

$$\delta V_B' \equiv \delta V_B - E_B \delta \ell_2 - E_B \delta \ell_1 = -\ell_B \delta E_B \tag{55}$$

With these statements Eqs. (26)-(33) may be rewritten as

$$-\frac{\partial I_1}{\partial E_1} \frac{\delta V_1'}{\ell_1} + \frac{\partial}{\partial t} \delta Q_1 = \frac{\partial I_B}{\partial E_B} \frac{\delta V_B'}{\ell_B} \tag{56}$$

$$\frac{\partial \delta Q_2}{\partial t} = -\frac{\partial I_2}{\partial E_2} \frac{\delta V_2'}{\ell_2} - \frac{\partial I_B}{\partial E_B} \frac{\delta V_B'}{\ell_B} \tag{57}$$

$$\delta Q_1 = \frac{A\varepsilon}{\ell_1} \delta V_1' - \frac{\varepsilon A}{\ell_B} \delta V_B' \tag{58}$$

$$\delta Q_2 = -\frac{A\varepsilon}{\ell_2} \delta V_2' + \frac{\varepsilon A}{\ell_B} \delta V_B' \tag{59}$$

and

$$\delta V_1' + \delta V_B' + \delta V_2' = \delta V \tag{60}$$

These equations are consistent with Eq. (6) and may be easily seen as resulting in an equivalent circuit like that shown in Fig. 4. However, in this case

$$R_1^{-1} \equiv - \frac{\partial I_1}{\partial E_1} \frac{1}{\ell_1} \tag{61}$$

and

$$R_2^{-1} \equiv - \frac{\partial I_2}{\partial E_2} \frac{1}{\ell_2} \tag{62}$$

define these elements in the equivalent circuit. The other elements are as defined previously. Here the three RC parallel combinations do not physically represent the three regions of the film since

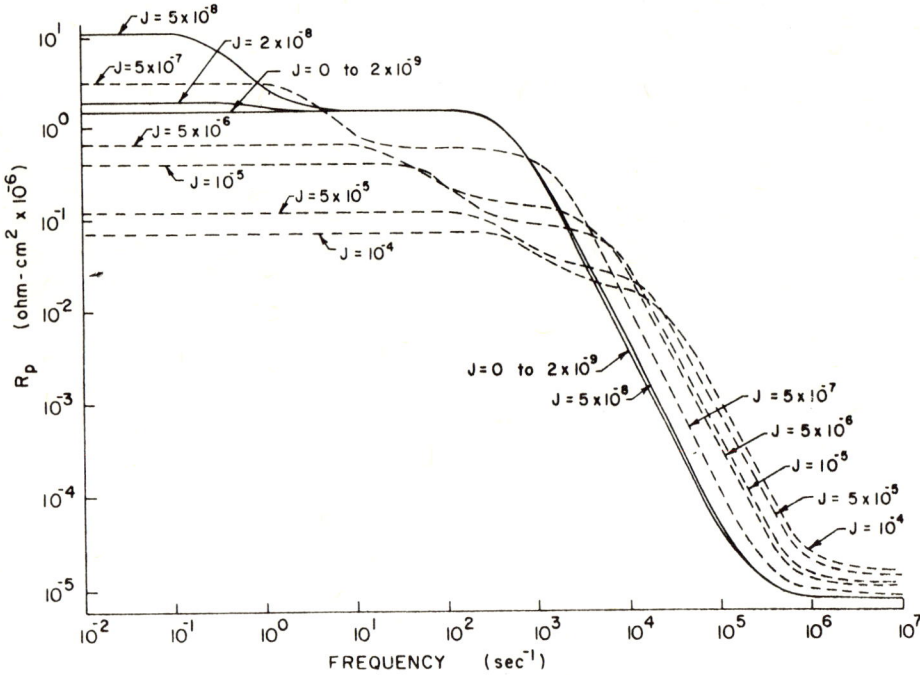

Fig. 6. Parallel equivalent resistance (x 10^{-6} cm^2) versus frequency at 300°K for a 2000 Å thick insulator. Numerical evaluation of the circuit shown in Fig. 3. is donce for parameter values such that Fig. 4. is valid.

DIELECTRIC PROPERTIES OF THIN FILMS

the transformation Eqs (53)-(55) has been employed. The circuit as a whole does represent the thin film (with its assumed interfaces) within the validity of Eqs (51) and (52). Here it is seen that C_{po} is not necessarily dominated by the interface ; i. e.,

$$C_{po} = \frac{R_1^2 C_1 + R_B^2 C_B + R_2^2 C_2}{(R_1 + R_2 + R_B)^2} \tag{63}$$

Experimental data in which a clear domination of C_{po} by the interfaces is absent but for which the existence of space charge regions has been demonstrated in a rather interesting manner[30] is shown in Fig. 8.

In general then for this approach to a system of the form of Eq (16)-(19) the results are such that Fig. 3 is valid. However, if the interior electric field (as well as changes in the interior electric fiels) is small, then Fig. 3 reduces to Fig. 4 -- a

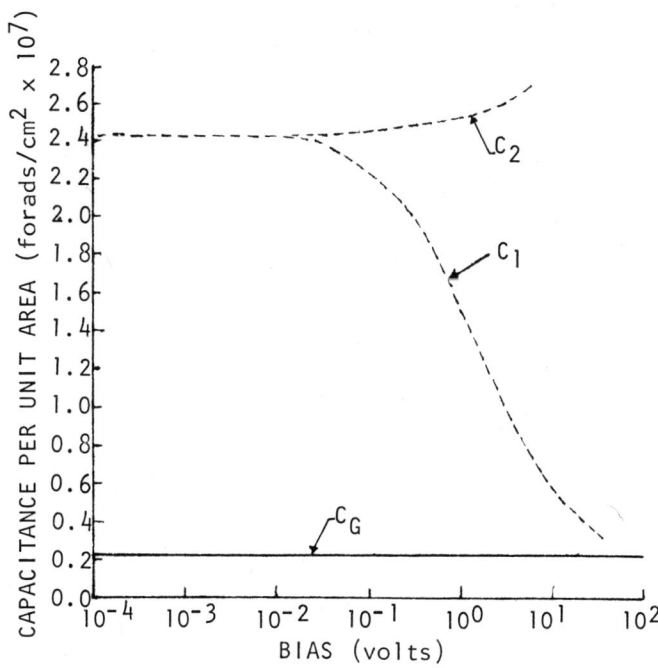

Fig. 7 The quantities C_1, C_2, and C_G ($\times 10^7 F/cm^2$) which were used in the computation of Fig. 5 and Fig. 6

Maxwell-Wagner-like model (21). Here each RC combination corresponds to a physical region in the thin film. In this situation $C_{p\infty} = C_G'$, of course, but C_{po} is dominated by the interfaces. With only the assumption of Eq. (51) and (52) the analysis again reduces to a circuit model of the type shown in Fig. 4 ; however, the RC combinations no longer necessarily represent physical regions and C_{po} is given by Eq. (63). The assumtion of Eqs. (51) and (52) is rather good in general for a reversed biased junction. For a forward biased junction it is quite good if the field in the interior is small ; however, under bias the expression for the forward biased contact should not be critical, in general, since this interface's resistance should be much smaller than other resistances in the thin film structure.

Thus the Maxwell-Wagner model can be used to explain thin film dielectric behavior in several situations and the model appears repeatedly in the literature. It is used for situations where the above

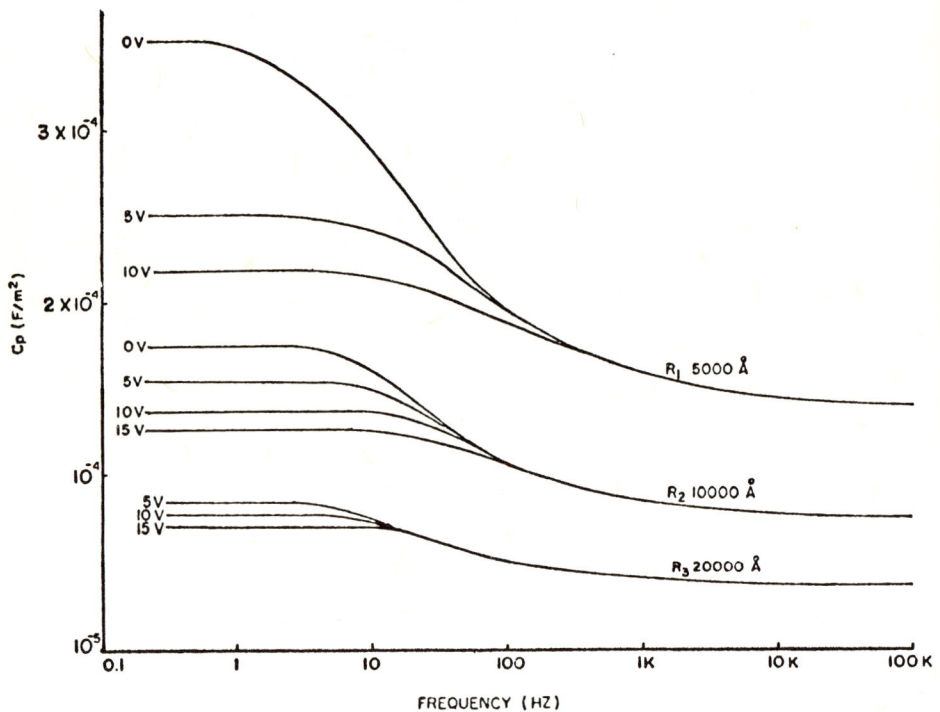

Fig. 8. Experimental data showing Cp versus frequency for a thin film with the global stoichiometry of SiO. (After ref.(30)).

analysis applies ; ie., a Maxwell-Wagner-like model. For example, Fig. 9 shows experimental data on MoO_3 in which the interfaces dominate[28,35]. It is also applied in situations where fixed regions, with differing conductivities, are believed to exist in films[36, 37]. This latter situation, of course, is that for which the model was first developed.[4]

In the discussion of the second approach presented, the finite emission time required to liberate carriers from states in the gap has been ignored. This can be a rather important omission for cases where the space charge comes from emisssion and sweeping-out of such carriers as it probably does in a wide gap dielectric thin film[5]. As may be seen from Eq. (6), if τ is a measure of this emission time, then for $\omega \gg \tau^{-1}$ the capacitance reduces to $C_G^\sim$ (again δP is being ignored here as it has throughout this section by the standard procedure of incorporating its "bulk" effects into ε).

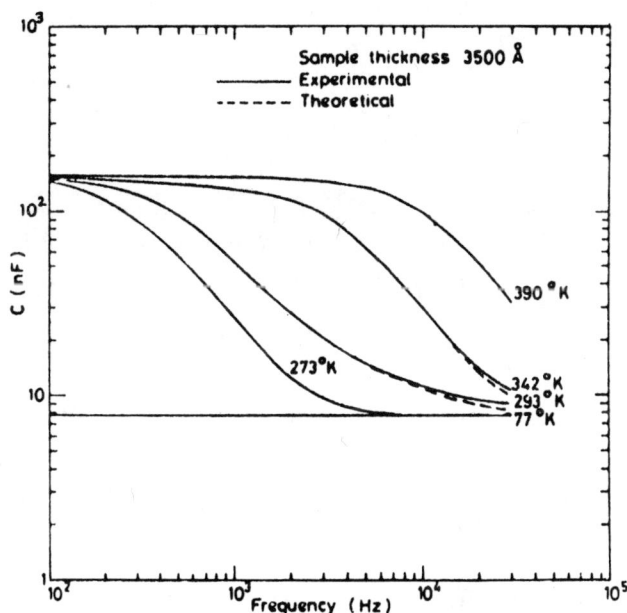

Fig. 9. Experimental data for the series equivalent capacitance versus frequency for a MoO_3 thin film sample with W = 3500 Å. Comparison is made with theoretical curves computed from Fig. with R_1 and $R_2 \to \infty$. (After G. Nadkarni and J. Simmons, Ref.(35) with permission).

3. POLARIZATION CHARACTERISTIC OF THE MATERIAL

Contributions to the polarization arising from the third term on the right in Eq (5) are characteristic of the thin film material as opposed to owing their origins to interfacial phenomena. In the case that these category II contributions are from permanent dipoles or from dipole-like phenomena for which the simple bistable configuration is applicable (can be applicable to polarization in the thin films arising from point defects-impurities, vacancies- and even to two center electron hopping[1,26]), Eq (4) is valid[1]. If an ac variation is assumed, it is seen from Eq (4) that

$$\delta P_n = \frac{\varepsilon_{sn} - \varepsilon_\infty}{1 + \omega^2 \tau_n^2} (1 - j\omega\tau_n)\delta E \qquad (64)$$

In this standard form, $P = P_\infty + \sum_n P_n$ and the definitions

$$P_{sn} \equiv (\varepsilon_{sn} - \varepsilon_\infty)E \qquad (65)$$

and

$$P_\infty = (\varepsilon_\infty - \varepsilon_0)E \qquad (66)$$

have been used; these are, of course, consistent with Eq. (3). The quantity P_∞ is the high frequency polarization which is able to follow the field[1]. If Eq. (4) is not valid (as been argued recently for many materials[38]), δP still is proportional to δE, with the proportionality factor depending on the ac frequency, if the governing equation for the temporal behavior, replacing Eq. (4) is linear.

Thus, Eq. (6) becomes (note that the sign conventions of Fig.1 require that $-\int_0^W \delta E\, dx = \delta V$)

$$\delta I = \delta I_p + \frac{A\varepsilon^*}{W}\frac{d}{dt}\delta V + A\int_0^W (1 - \frac{x}{W})\frac{\partial \delta P}{\partial t}dx \qquad (67)$$

This expression for the ac variations is of general validity; the only restriction is that the equation governing the temporal response of δP be linear.

DIELECTRIC PROPERTIES OF THIN FILMS

For the case where Eq (4) is valid the specific mathematical model for the complex permittivity ε^* appearing in Eq (67) can be written down using Eq (64) ; ie.

$$\varepsilon^* = \varepsilon_\infty + \sum_n \frac{\varepsilon_{sn} - \varepsilon_\infty}{1 + \omega^2 \tau_n^2} (1 - j\omega\tau_n) \qquad (68)$$

This expression for ε^*, resulting from the validity of Eq (4), represents the sum of n Debye-like processes.

As was mentioned in the preceding section, the discussion of category I ac behavior proceeds using Eq. (67) with ε^* replacing ε_0. If no particle current crosses $X = 0$ and if $\delta\rho$ is zero, the thin film structure still appears, in spite of no interfaces effects, to have a resistance (from Eq. (12)) as well as a capacitance (from Eq. (13)). Thus, if both category I and category II mechanisms are operative in a thin film, one is faced with their joint effects, as indicated by Eq. (67), in an ac experiment. That is, both bulk and effective polarization are present in general.

Cole and Cole[39] expressed the superposition of relaxation times indicated by Eq. (68) as

$$\varepsilon^* - \varepsilon_\infty = (\varepsilon_s - \varepsilon_\infty)/(1 + (j\omega\tau)^{1-\alpha}) \qquad (69)$$

which, although it suffers from some possible inconsistencies[40], is frequently used to interpret experimental data in terms of bulk Debye processes. In terms of Cole-Cole diagrams[39] with ε'' plotted versus ε', a semicircle is obtained for $\alpha = 0$ (one relaxation time) with its center on the ε' axis. With $\alpha > 0$, the center moves below the axis ; α being a measure of the angle between the ε axis and a radius to the closest point of intersection (to the origin) of the curve and the ε' axis[39]. For α values $\neq 0$, ε'' becomes fairly flat; therefore, fairly independent of frequency. In fact, neglecting interfacial contributions to Eq. (67) shows, using Eq. (69), that

$$C_p = \frac{A}{W} \varepsilon_\infty + \frac{A}{W} \frac{(\varepsilon_s - \varepsilon_\infty)}{\omega^{1-\alpha}} \cos(1 - \alpha)\frac{\pi}{2} \qquad (70)$$

and

$$R_p^{-1} = \frac{A}{W} (\varepsilon_s - \varepsilon_\infty)(\omega\tau)^\alpha \sin(1 - \alpha)\frac{\pi}{2} \qquad (71)$$

for the case of a distribution of relaxation times and $\omega\tau \gg 1$. Thus, depending on α, the superposition of relaxation times could lead to C_p = constant and $R_p \sim \omega^{-\alpha}$. Such behavior is commonly seen in thin films[41]. Figures 8 and 10 show similar behavior ; however here $\alpha = 2$. Consequently the explanation of Eq. (71) is not possible due to restriction of α[39]. For a single relaxation time, $\omega\tau \gg 1$ leads to both C_p and R_p being constant with frequency for the Debye model.

It is interesting to note the changes in capacitance $\Delta C = C_{p0} - C_{p\infty}$ possible from these dipole-like mechanisms : from Eq (68) it follows that

$$C/C_{p\infty} \approx \sum_n \frac{N_n M_n^2}{kT\varepsilon_o} \qquad (72)$$

Here M_n is the dipole moment for species n with the number density N_n. Thus if the N values are of the order of 10^{20} cm^{-3}, $\Delta C/C_{p\infty}$ is of the order of 20 per species using as the effective moment arm 5 Å. Consequently it is difficult to explain data such as of Fig. 2 in terms of a dipolar-like mechanism without invoking extremely large dipole densities. Data such as that of Fig. 8 could be interpreted in terms of a dipolar-like mechanism except that the activation energy ΔE (assuming $\tau = \tau_0 e^{\Delta E/kT}$ (1)) for C_p is the same as that for the dc resistance[30]. From Eqs (67) and (68) it is seen that the dc resistance for the channel of current flow due to ε'' is infinite. Also scanning electron microscope observations have indicated the presence of interfacial space charge for the data of Fig. 8[30]. There are, however, many case in the literature where dipole or dipolar-like mechanisms are used to explain observed dielectric properties in thin films[26,42,43].

In amorphous thin films, electronic conduction by hopping and percolation[41,44,45] must be considered as quite possible. The presence of these transport mechanisms would give rise to another contribution to the dielectric properties from category II type phenomena. In general electronic hopping transport, it has been argued, should lead to σ' varying as ω^n and ε' varying as ω^{n-1} with $n > 1/2$ [41]. As has been mentioned, this class of behavior is common in thin film materials for higher frequencies having been seen in SiO_x [26,30], Al_2O_3[26], and chalcogenide glasses[46] to cite a few case

For the simple case of two center hopping[26,47], this increase of σ' with ω is easily demonstrated to arise : here

$$\sigma' \sim \frac{\omega^2 \tau^2}{1 + \omega^2 \tau^2} \qquad (73)$$

and also

$$\varepsilon' \sim \frac{1}{1+\omega^2\tau^2} \qquad (74)$$

These are of the same form as Eq. (68) would yield. However, here it is possible to postulate that τ is very small; i.e., for frequencies of interest $1 \gg \omega^2\tau^2$. Consequently, σ' varies, in this case, as ω^2 and ε' is constant.

Unfortunately, interpretation of this $\sigma = \sigma(\omega)$ behavior is far from unambiguous[46]. Ionic defect hopping can also make a contribution[48]. Conductivity increasing with ω can come, as discussed, from dipolar-like mechanisms. As may be seen from the numerical evaluation presented in Fig. 6, $R_P \sim \omega^{-n}$ can come from polarization in depletion regions. It can develop from effective polarization with origins in mobile carrier concentration modification even for flat bands[13]. In fact, as pointed out in the literature, any inhomogeneous conduction mechanism — on the microscopic or macroscopic scale — can produce conductivity increasing with frequency[46].

4. CONCLUSIONS

Dielectric properties of thin films are the result of interface effects and of effects arising from the material "bulk" properties. These mechanisms add according to Eq. (5). If ac variations are used to modify the polarization, Eq. (67) is valid if the polarization obeys a linear temporal equation. The quantity ε^* appearing in Eq. (67), in cases where the Debye model is valid (Eq(4)), is given by Eq. (68) in the general situation of many Debye-like mechanisms.

Thus if, as an example, the dielectric properties of a thin film owed their existence to space charge coming from deep donors and to a "bulk" dipolar-like mechanisms described by Eq. (69), it can be seen from Eq. (67) that for low ω the resistance would be mainly due to particle current crossing the electrodes. The capacitance would be $> C_G$ since it would arise from the last two terms in Eq. (67). If the deep traps supplied the carriers, the activation energy for the dc current should be the same as that for C_P. For $\omega > \tau^{-1}$, where τ is being used here as a measure of the emission time from the deep states, the contributions from the third term on the right in Eq. (67) could become small. Thus, C_P would tend now to C_G' and $R_P \sim \omega^{-\alpha}$. The activation energy of the high frequency process would be totally unrelated to that of the low frequency process in this example.

Unfortunately many phenomena – or combinations of phenomena – can give rise to the frequency behavior described in the above example. As discussed, analysis of interface polarization arising from depletion regions indicates that it alone can produce similar behavior as can Maxwell-Wagner and Maxwell-Wagner-like models. Dipole, dipole-like, or hopping mechanisms with a constant (in ω) background conductivity can also produce this frequency behavior. However, these combinations will not produce correlations, described above, for the activation energies. It is obvious, however, that other combinations could be imagined which would.

Thus in examining thin film polarization or effective polarization response, it becomes clear that care must be exercised in searching for the physical origins of the observed behavior. The conductance as a function of frequency, bias, and thickness must be studied. A similar study must be made for C_p. Activation energies should be measured. Ac properties should be correlated with dc behavior and independent information should be obtained, if possible, from thermally stimulated currents, observations of space charge regions, etc. Thin films necessitate this kind of a continued, intensive experimental effort – together with much more theoretical work – to unravel fully the causes of their bulk and effective polarization.

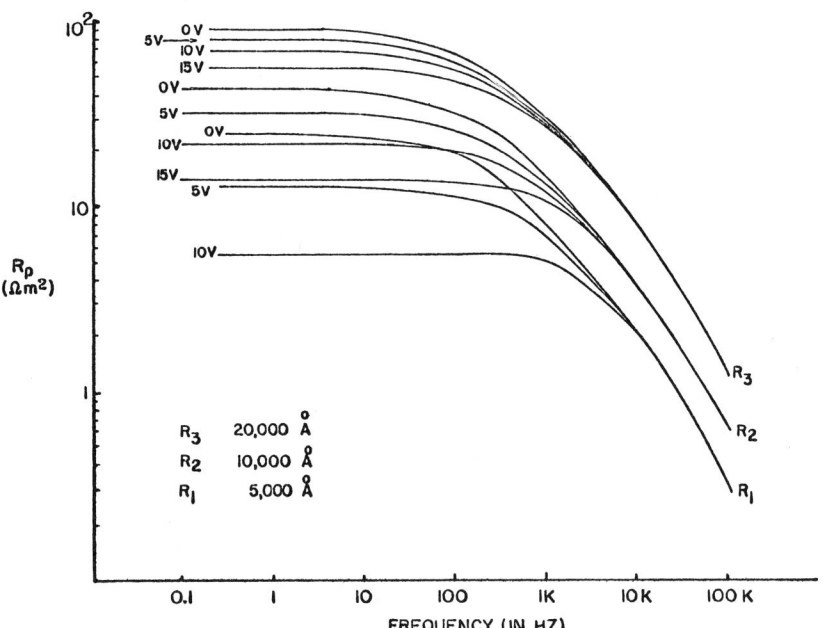

Fig. 10 $R_p = R_p(V,\omega)$ for a film with stoichiometry of SiO. Note that the high frequency behavior scales with W but is independent of bias.

REFERENCES

1. V. V. Daniel, "Dielectric Relaxation" Academic Press, N. Y. (1967).
2. H. Fröhlich, "Theory of Dielectrics", Oxford Press, Oxford (1958).
3. C. Kittel, "Intro. to Solid State Physics" Wiley, Inc., N.Y. (1968).
4. A. Von Hipple in "Handbook of Physics" (E. Condon and H. Odishaw, Eds.) McGraw-Hill, N.Y. (1967).
5. S. J. Fonash, J. A. Roger, J. Pivot, and A. Cachard, J. Appl. Phys. 45, 1223 (1974).
6. M. Perlman and S. Unger, J. Appl. Phys. 45, 2389 (1974).
7. M. Baird, Rev. of Mod. Phys. 40, 219 (1968).
8. J. G. Simmons and G. Nadkarni, Phys. Rev. 6, 4815 (1972).
9. Ai Bui, et al., Thin Solid Films 21, 313 (1974).
10. C. Bucci, R. Fieschi, and G. Guidi, Phys. Rev. 148, 816 (1966).
11. D. M. Smyth in "Oxides and Oxide Films" (J. Diggle, Ed), Dekker, Inc., N. Y. (1973).
12. P. J. Harrop and D.S. Campbell in "Handbook of Thin Films" (L. Maissel & R. Glang, Eds.) McGraw-Hill, N.Y. (1970). Contains an excellent list of references up to the late 1960's.
13. R. J. Friauf, J. Chem. Phys. 22, 1329 (1954).
14. J. G. Simmons and G. W. Taylor, Phys. Rev. 6, 4793 (1972)
15. J. R. Macdonald , J. Chem; Phys. 58, 4982 (1973).
16. H. Chang and G. Jaffé, J. Chem. Phys. 20, 1071 (1952).
17. J. R. Macdonald, J. Chem. Phys. 40, 3735 (1964).
18. F. A. Kröger, "The Cehmistry of Imperfect Crystals" Wiley, Inc. N.Y. (1964). (Contains a review of the literature on this subject up to 1963 ; see sec. 22.11).
19. E. Fatuzzo and S. Cuppo, J. Appl. Phys. 43, 1457 (1972).
20. J. R. Macdonald, J. Appl. Phys. 45, 73 (1974).
21. S. J. Fonash, J. A. Roger, and C. H. S. Dupuy, J. Appl. Phys. 45, 2907 (1974).
22. J. G. Simmons, G. S. Nadkarni, and M. C. Lancaster, J. Appl. Phys. 41, 538 (1970).
23. J. Maserjian, J. Vacuum Sci. Technol. 6, 843 (1969).

24. G. Jaffé, Phys. Rev. 85, 354 (1952).
25. M. Meaudre and R. Meaudre, Solid-St. Electron. 16, 1205 (1973).
26. F. Argall and A. K. Jonscher, Thin Solid Films, 2, 185 (1968).
27. J. C. Macfarlane and C. Weaver, Phil. Pag. 13, 671 (1966).
28. G. Nadkarni and J. Simmons, J. Appl. Phys. 43, 3741 (1972).
29. J. Simmons, G. Nadkarni, and M. Lancaster, J. Appl.Phys. 41, 545 (1970).
30. J. A. Roger, C. H. S. Dupuy, and S. J. Fonash. J. Appl. Phys. 46 (1975) ; J. Solid State Chem. 12, 238, (1975).
31. J. G. Simmons, J. Phys. Chem. Solids 32, 1987 (1971).
32. J. G. Simmons, J. Phys. Chem. Solids 32, 2581 (1971).
33. S. J. Fonash, Solid-State Electron. 15, 783 (1972).
34. J. Volger in "Progress in Semiconductors" Vol 4 (A.F. Gibson Ed.) Heywood and Co., London (1960).
35. G. Nadkarni and J. Simmons, J. Appl. Phys. 41, 545 (1970).
36. I. S. Goldstein, J. Appl. Phys. 45, 2447 (1974).
37. M. C. Lancaster, Thin Solid Films 8, 213 (1971).
38. A. K. Jonscher, Private communication.
39. K. Cole and R. Cole, J. Chem. Phys. 9, 341 (1941).
40. J. Macdonald, J. Chem. Phys. 36, 345 (1962).
41. A. K. Jonscher, J. Non-Crystalline Solids 8-10. 293 (1972).
42. G. Pfister and M. Abkowitz, J. Appl. Phys. 45, 1001 (1974).
43. A. Willoughby and F. Yuen, Thin Solid Films 13, 199 (1972).
44. J. Ziman, J.Phys. C (Proc. Phys. Soc.) 1, 1532 (1968).
45. D. Adler, L. Flora, and S. Senturia, Solid. St. Communications 12, 9 (1973).
46. A. Owen and J. Robertson, J. Non-Cryst. Solids 2, 40 (1970).
47. M. Pollak and T. Geballe, Phys. Rev. 122, 1742 (1961).
48. E. Snow and P. Gibbs, J. Appl. Phys. 35, 2368 (1964).

THRESHOLD SWITCHING:

A DISCUSSION OF THERMAL AND ELECTRONIC ISSUES

H. K. Henisch and C. Popescu

Materials Research Laboratory
The Pennsylvania State University, University Park
Pennsylvania 16802, U.S.A.

1. INTRODUCTION

Papers of joint authorship ordinarly result from shared research, but the present case is an exception. Shared principles are its basis, of course, but it represents above all a search for common ground, approached from opposite sides of the interpretational spectrum. The opposite sides are represented by thermal and non-thermal (electronic) models of threshold switching, an issue which has been widely (and, at times, hotly) debated by protagonists at international conferences and in the literature. In the course of such debates, misunderstandings abound, and some of the sweetest victories are scored against "opposing" models not actually advanced by the other party. In the present didactic context, an attempt at clarification therefore seems particularly appropriate. Once found, the common ground must serve at the starting point for more discriminating experimentation. Inasmuch as our agreement falls short of total unanimity, we each take our cue from the clergyman who ended a bitter dispute with a colleague, saying : "It is clear that we cannot settle this issue, and we must therefore agree to disagree ; you will have to continue to serve the Lord in your way, and I shall continue to serve Him in His."

Threshold switching, in the form here discussed, is a phenomenon specifically associated with thin films. It can take place in an astonishing variety of materials, but few of these appear to be suitable for commercial applications ; indeed, only the multicomponent glasses (typical composition $Te_{40}As_{35}Ge_7Si_{18}$) have come close to marketable devices, following their invention by S.R. Ovshinsky[1] (All the detailed results here discussed refer to them).

Nevertheless, the hope and prospect of commercially viable devices has in large part helped to stimulate the wide-spread interest in high resistivity materials in general and chalcogenide alloy glasses in particular. The view is often expressed that this interest can and will be sustained only if it can be shown beyond doubt that the phenomena involved are primarly electronic, and, conversely, that it would cease if it were shown that they are primarly thermal. This dichotomy has its simple attraction, but we cannot support it, because it seeks to draw excessively far-reaching references from the right answer to the wrong question. Although there are such connections in limiting cases, it is wrong to use the terms "electronic" and "thermal" a priori as if they were synonymous with "stable" and "unstable" respectively. Device design is always a matter of compromise in which the conflicting demands of performance and durability must be reconciled, and the fact that thermal aspects are important in electronic devices of many different kinds has not automatically precludes such a compromise. Moreover, whereas there is no problem about envisaging a "purely thermal" process, a "purely electronic" process is clearly impossible. Once and electronic process is demonstrated, we are immediately concerned with "thermal overtones" in varying degree, from the functionally trivia to the functionally essential. In the case of threshold, which involves power dissipation within a volume of a few cubic microns, the functionally trivial limit is unlikely, and that is of course, one reason why thermal models of various kinds have been so popular

It is, of course, possible to envisage models in which switching arises from the simultaneous action of thermal and electronic effects, neither mechanism being by itself sufficient to secure the existence of a discontinuity. Such a model would have to be called "electrothermal" rather than thermal. Accordingly, the choice is between a thermal model (albeit modified by electronic processes e.g. by the field dependent conductivity), and electronic model (albeit modified by self-heating), and an electrothermal model (implying a cooperative process). Though the most discussed in qualitative terms.

It was certainly hoped during the initial phase of this research that the relative importance of thermal and non-thermal aspects was a matter which could be quickly and unambigously resolved by experiment, but this hope has not yet been fulfilled to everyone's satisfaction. To be sure "proofs" of one sor or another fill the literature, but they carry conviction mostly so kindred spirits, rarely to outright opponents. Why this should be so is not hard to see. Because some heating is inevitable, the demonstration of heating effects as such proves nothing. One must also realize that a switch is a two-thermal network and, in the last analysis, the only electrical measurement we can make on it is a measurement of resistance. Resistance as a function of voltage, current, time,

THRESHOLD SWITCHING

temperature, rate-of-address, pulse duration represent some of the possible variations, but the fact is that the consequences of electronic disequilibrium are expected to be rather similar as far as their effect on overall resistance is concerned. Experimental results therefore leave a good deal of room for interpretation, and, in the past, features which have seemed essential, characteristic and conclusive to one party have often seemed trivial to another. However, the field of threshold switching cannot continue to grow in this way, and attempts to define decisive issues are urgently needed.

2. PHENOMENOLOGICAL CHARACTERISTICS

For practical device purposes, thershold switches must be made entirely by thin film techniques, and unless alternative stable electrodes can be found, this involves the preparation and manipulation of graphite films. That this can be done has been demonstrated by Ovshinsky and co-workers[2]. However, most of the results in the literature (indeed, an overwhelming proportion of them) have been obtained on other types of structures, usually as shown by the insert on Fig. 1. Assembled and encapsulated units of this kind have been known to be highly stable, necessarily, within an appropriate range of operating conditions, and on them is based the (very reasonable) hope that stable all-thin-film devices can also be produced.

The V-I characteristics of threshold switches have often been described and only a brief summary is needed here, e.g. see Shanks[3]. A schematic representation of a switch is shown on Fig. 1. It corresponds to the picture which can be traced out point by point by applying pulses of varying amplitude and using series resistances over a wide range of values. Under AC conditions, the oscilloscope display gives a similar picture, but some controversy is attached to the exact manner in which the ON-state terminates near the minimum holding current I_{MN}. Under pulse conditions, Henisch and Pryor[4] found it to be as shown by the full line in Fig. 1. However, under alternating signals, the slope is often reported as negative (broken line), just before the return of the OFF-state. There is a general belief that the exact conditions of measurements (including circuit capacitance) and the rate of external voltage withdrawal influence the behavior, but precise relationships have never been ascertained, as far as we know.

The V-I relationship is also known as the "primary characteristic". In addition, switches have a variety of secondary characteristics, some of which are illustrated by Fig.2. Thus, switching at minimum voltage (the threshold point) is statistical[5] as far as the switching delay is concerned (Fig. 2a). Voltages higher than threshold can be applied, with corresponding reductions of the

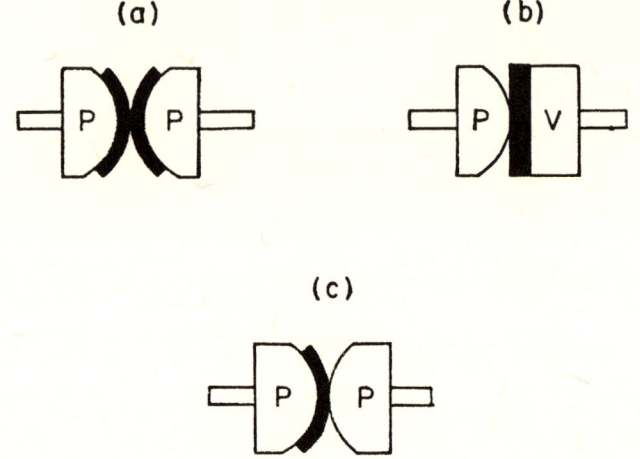

Fig. 1 (a-c). Threshold Switching; Primary Characteristics. Switch structures with solid electrodes.

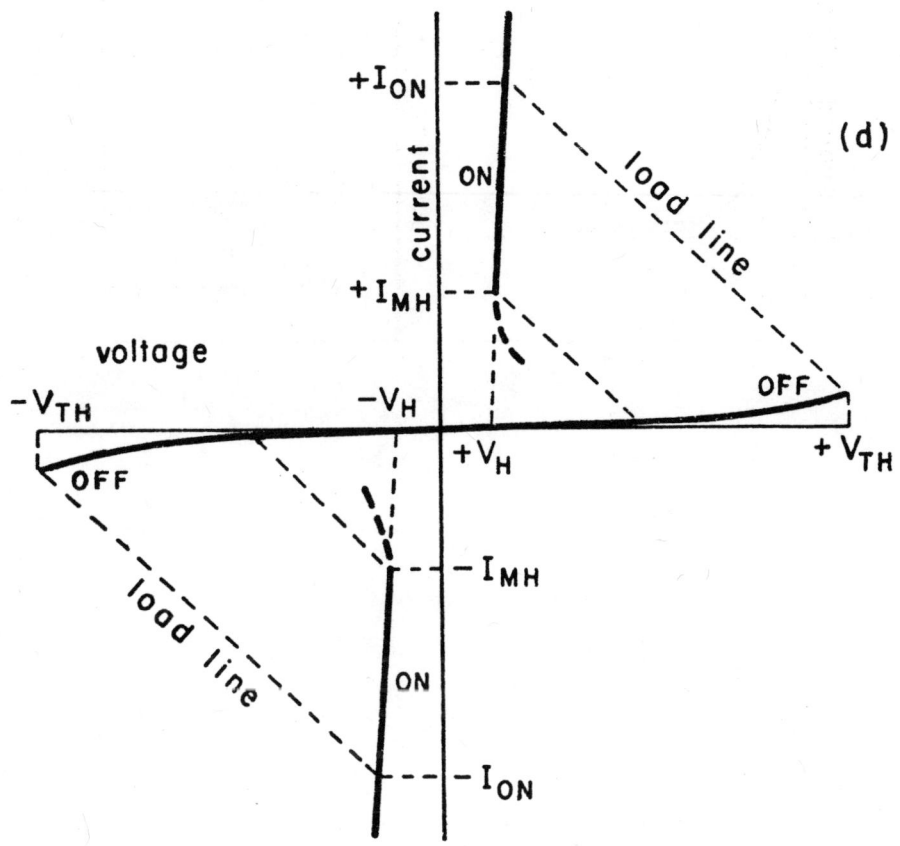

Fig. 1 (d) Threshold Switching; Primary Characteristics. Schematic voltage-current characteristic of threshold switches made with multicomponent chalcogenide alloys.

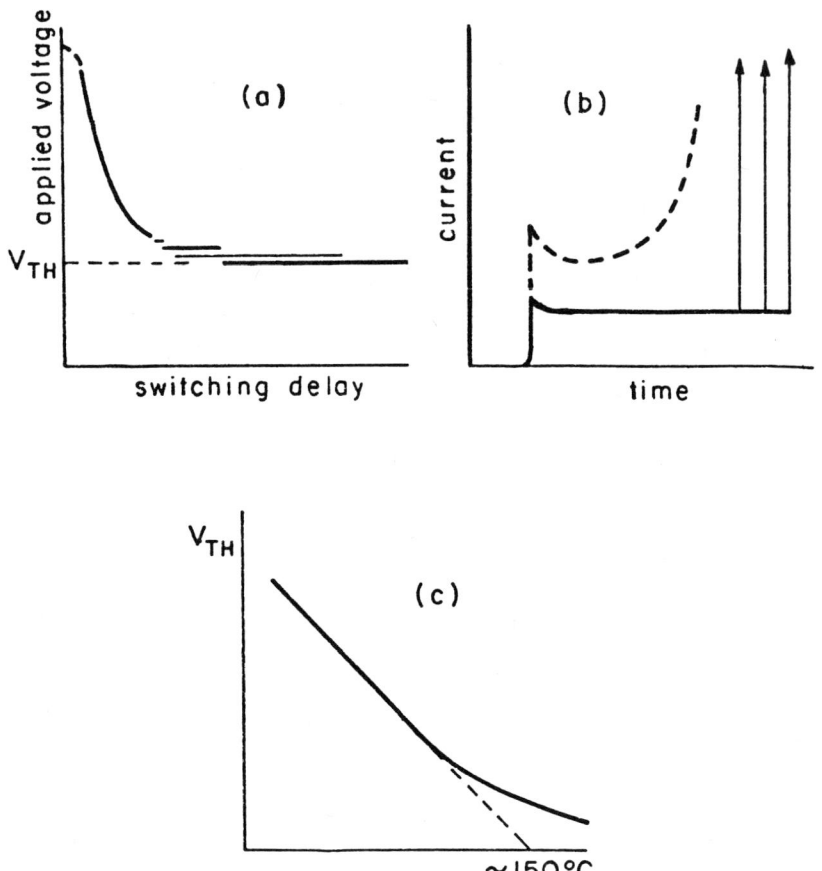

Fig. 2. (a), (b), (c) Threshold Switching; Secondary Characteristics (multicomponent chalcogenide alloys).

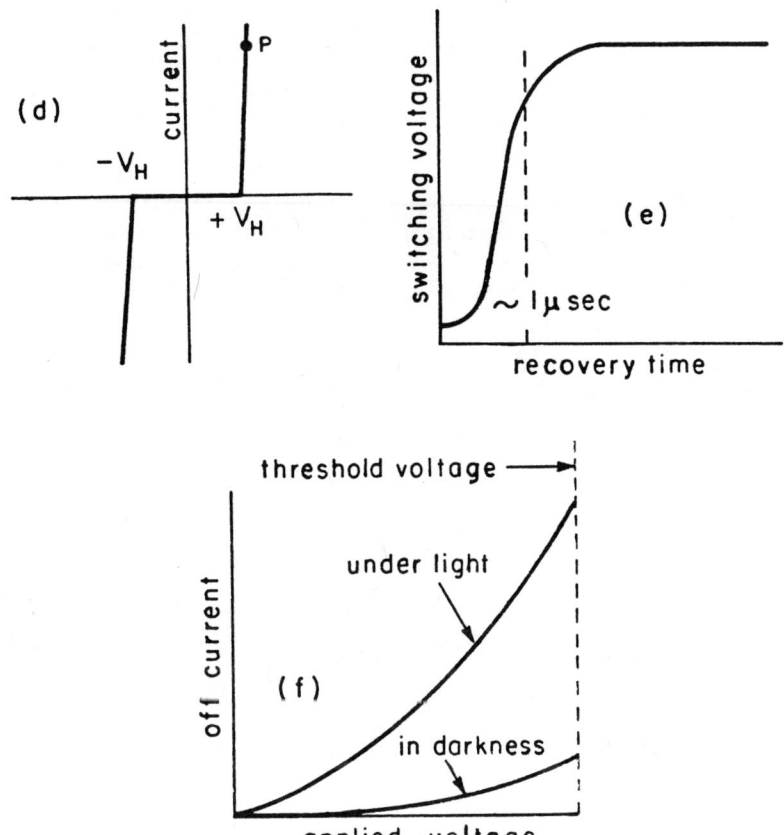

Fig. 2 (d), (e), (f). Threshold Switching; Secondary Characteristics (multicomponent chalcogenide alloys).

switching delay. With increasing overvoltage, the switching delay
ceases to be statistical and soon becomes well defined. During the
switching delay within the statistical regime, the current is constant (2b), which means that there is no general increase of temperature. A highly localized increase, responsible for only a small
fraction of the total current, is not ruled out by this observation.
The threshold voltage V_{TH} (measured with AC) diminishes with increasing ambient temperature in a manner which is virtually linear
over a 100°C range from -50°C upward, a non-linearity then sets in,
as shown on Fig. 2c. Extrapolation of the linear region sometimes
yields a temperature close to the glass transition temperature, a
fact which has given rise to a good deal of speculation[6][7].
However, it is not known whether this coincidence is fortuitous or
significant. Such measurements are most conveniently made on encapsulated units, but this procedure introduces additional uncertainties
arising from the thermal expansion of the capsule. Moreover, the
relationship between threshold voltage and ambient temperature is
more complicated than is generally realized. On un-encapsulated
units, one of us (CP) and also Stubb and co-workers[8] have reported
finding a certain low-temperature range over which the threshold
voltage remains constant while the threshold power strongly decrease

When the ON-state is displaced by rapid transient, a V-I
relationship is obtained as shown on Fig. 2d. Any point of this
relationship is accessible from any other point within 10 nanoseconds or less. The high resistance portion has been called the
"blocked ON-state"[4][9]. Starting with any operating point, (say),
P_1, the blocked ON-state can be reached by means of a suitable voltage pulse and allowed to prevail for periods up to 0.3 to 0.4 μsecs
after which point P_1 can be restored with a minimum of delay (e.g.
10 nanoseconds or less), and without switching. Switching is necessary only after longer periods of current interruption, as shown
in Fig. 2e. This curve is independent of the energy dissipated
during the preceding switching pulse, and the position of the knee
is independent of temperature.

There are other secondary characteristics, but those listed
are probably the most important. In addition, there are some basic
(room temperature) observations upon which all interpretations
ultimately depend :

    (a) The OFF-current, at any rate before the first (ever) switching event (see below), is proportional to the electrode
area ;
    (b) The OFF-state resistance is roughly proportional to the
film thickness, as is the threshold voltage V_{TH}, before
as well as after the first swtiching event (forming) ;
    (c) The V-I characteristic in the ON-state is insensitive to
changes of film thickness ;
    (d) The ON-state characteristic in the ON-state is independent
of electrode area.

THRESHOLD SWITCHING

Though experiments as a function of film thickness are highly informative in principle, they are not easily performed, because there is no simple way of ensuring that thick films have precisely the same bulk and surface structure as thin films. Published results must be viewed with this caution in mind. Point (c) implies that most of the film thickness is free of field, and the last observation implies filament formation. On these point all models are agreed. The term "filament", though widely used (including here) is actually misleading, inasmuch as the conducting volume often has a length of 1 micron and a cross-section of 1 micron^2. Observation (a) means that the current distribution is uniform. Observation (b) means that what matters for the threshold point is a critical field at a given temperature, but whether this is the only condition is uncertain and remains to be ascertained. Item (b) has some further implications, namely that there is no appreciable contact resistance in the OFF-state and that the field distribution is uniform. Accordingly, any observed behavior in the OFF-state can be regarded as bulk behavior[10],[12]. This concerns particularly the field dependence of the conductivity which, at high fields, can be approximated as

$$\sigma = \sigma_o \exp(F/F_o) \tag{1}$$

Beyond threshold, on the unstable part of the characteristic which may be explored with short overvoltage pulses, the field dependence is even steeper. It is believed[13] to arise from a mobility effect (rather than from an increase of carrier concentration), but its precise origin is not yet understood. This constitutes one of the most challenging problems in solid state physics.

3. THERMAL THEORY FOR UNIFORM STRUCTURES

Thermal theory begins with the heat balance equation

$$C \frac{\delta T}{\delta t} = j^2/\sigma + \nabla(K\nabla T) \tag{2}$$

where C = specific heat, j = current density, and K = thermal conductivity. In the steady state, it says no more than that whatever energy goes in must come out. Its integration leads in principle to a three dimensional temperature distribution contour for any current, and ultimately to a complete voltage current relationship. In that sense, it is always desirable to examine how far agreement with experiment can be obtained on this basis alone. However, the integration needs boundary conditions, and when these are formulated in the most general case, eqn. (1) cannot be explicitly solved. The procedures are therefore open : (i) the introduction of approximations which will permit explicit solutions, and (ii) numerical solutions by computer, of which the latter is by far the more satisfactory. At least two sets of computed solutions are in fact available, one by Kroll and Cohen[14], and one by Popescu[15]. They differ somewhat in the assumptions made concerning the nature of the bulk conductivity and the degree of thermal coupling. Thus Kroll and Cohen assume a bulk conductivity in the form

$$\sigma = \sigma_o \exp (F/F_o) \mathcal{J}(T) \qquad (3)$$

where $\mathcal{J}(T)$ is taken from experimental results and could be interpreted as arising from an activation energy which varies with the temperature. Popescu's conductivity has a constant activation energy but the ratio of his thermal coupling constants (lateral to longitudinal) can be varied at will. Lateral coupling depends, of course, on the thermal conductivity of the material and on the film thickness. A priori, its importance must tend to zero as the film thickness tends to zero. Longitudinal coupling depends on the thermal connection between the electrode surfaces and the surrounding ambient. Perfect lateral coupling would imply a uniform temperature and current density, as shown by curve (b) on Fig. 3. Reduced lateral coupling leads to filament formation and the appearance of a distinct branching point, which may be as low as the turnover point. There are, of course, longitudinal thermal gradients, but it can be shown that these gradients are relatively unimportant in determining the V-I relationships.

In accordance with these considerations, thick specimens are expected to exhibit voltage turnover, followed by a (stabilizable) negative resistance region. Thin specimens are expected to exhibit a threshold voltage, followed by an experimentally inaccessible region and, in due course by a stable ON-state. The branching point marks the onset of filament formation. On a temporary basis filament formation can be avoided. Experiments with high overvoltage pulses[1] of very short duration should therefore reveal the characteristic which corresponds to the case of a system completely isothermal with the ambient (see Fig. 3). Practically all the current is carried by the filament. Between the threshold point and point V_{min}, the filament temperature increases, the diameter decreases. After that, the current can increase only because the filament widens, its temperature coefficient being by then close to zero.

The curves on Fig. 3 which correspond to a small lateral-to-longitudinal coupling ratio are evidently similar to the primary switching characteristics on Fig. 1. (They predict an ON-state which terminates in accordance with the broken, rather than the full line. A field dependent conductivity modifies the shape of the characteristics, but does not otherwise contribute critically to the switching process. In the ON-state the average field is low but $\sigma(F)$ may still have a modifying role, inasmuch as a relatively high field

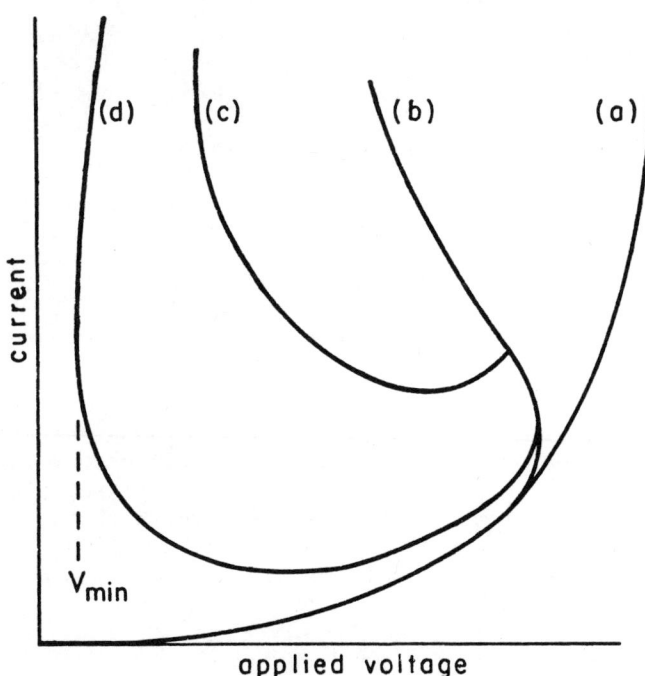

Fig. 3. Schematic Voltage-current Characteristic Calculated on the Basis of Thermal Theory (after C. Popescu). (a) Limiting of system without heating; (b) System with uniform temperature and current distribution; (c-d) Filamentary systems for diminishing ratio of lateral to longitudinal thermal coupling.

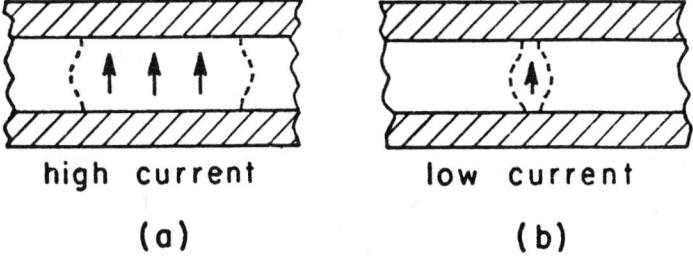

Fig. 4. Schematic Representation of Current Distribution for High (a) and Low (b) ON-currents.

may exist across the cooler portions of the glass near the electrode as suggested by Kroll.

As far as agreement with the primary characteristic is concerned, the thermal model must be regarded as highly successful. However, granted that thermal switching is possible in principle, the question remains whether such a model can account convincingly for the known secondary characteristics. This will be further discussed below.

4. ELECTRONIC THEORY FOR LIFETIME SEMICONDUCTORS OF UNIFORM STRUCTURES

Thermal models aims at a single relationship which describe the entire V-I characteristic in terms of bulk properties. Of course if it were considered necessary at any time, contact and interface effects could be introduced as additional refinements. In contrast, electronic models concern themselves separately with three issues:
  (i) the shape of the OFF-state characteristic,
  (ii) the nature of the instability at the threshold point, and
  (iii) the nature of the ON-state.
Of these issues, (i) need not be considered in detail here, however interesting it may be (as it certainly is) in principle. It can be set aside because it is known that threshold switching can occur in systems, i.e. those consisting of polymer films[18], which show only linear pre-threshold characteristics. Pre-threshold non-linearity is therefore not a pre-condition for switching.

Most progress has been made with (iii), the most detailed model being that of Mott[19], following suggestions by Henisch[20,21] and Lee[22]. It concerns itself with the manner in which the ON-state could be maintained, once established by some mechanism (which remains to be discussed). In accordance with clause (c), Section 2, there is almost no electric field in the inclose to the electrodes, possibly across space charge regions. At one time, these were seen as regions of trapped charge, but such a system would be expected to exhibit a strong temperature dependence. In fact, the results on Fig. 2(e) do not, which suggests that trapping centers are not involved. The alternative is a region of free charge carriers, in excess of the concentration demanded by neutrality. Mott[19] has shown on a semi-quantitative basis that such layers can be stably maintained in the presence of double injection and current flow, assuming a near-intrinsic material. The appearance of space charges is, of course, immediately likely whenever current passes through a boundary between media of different conductive properties. Exactly how dense and how spatially extensive these charges are can be established only through computer-aided calculations, which have not yet been done with complete generality.

The notion is that carriers from the electrodes tunnel through these space charges regions and thereby maintain an electron-hole plasma of high density in the interior of the film. As has been pointed out[21], this implies that most electron traps are full of electrons and most hole traps full of holes, the oucome being a population inversion. The electron-hole plasma will tend to lose carriers through recombination (some radiative, as discussed below), and will be compensated by injection. When the two rates balance, the system is stable. However, unless a certain minimum injection rate is maintained, the carrier loss through recombination will "win", which means that the system will revert to the OFF-state. This is the electronic explanation of the minimum holding current. The critical quantity is, of course, the ratio of transit time to carrier lifetime. The minimum holding current should correspond to equality to the two times, the transit time being, of course, field- (and therefore current-) dependent. It will diminish sharply as the ON-current diminishes, partly because of the lower average field bu also for geometrical reasons (Fig. 4), as suggested by Mott.

The V-I relationship in the ON-state would be governed by the tunneling characteristic of the barriers and by the field distribution as between barriers and bulk. If the voltage drop across the bulk is neglected (for this purpose only) one can calculate plausible V-I relationships which can be fitted to experimental results without difficulty. Characteristics which take account of barriers and bulk have not yet been computed and, in the circumstances, comparisons between calculated and observed results are not as realistic as they should be.

The absence of field in the interior implies an absence of space charges and that is <u>a priori</u> an astonishing claim considering that these high resistivity materials contain traps. However, according to the Cohen, Fritzsche and Ovshinsky[23] model, at any rate in its simplest form, the concentrations of electron traps and hole traps in a multicomponent glass are <u>automatically</u> equal (This is believed to make it possible for both types of traps to be full or almost full without generating the major space charges which one would otherwise expect). The model would thus help to explain the special position which the multicomponent chalcogenide glasses appear to occupy in threshold switching.

It remains necessary to discuss electronic "options" for (ii) above, the nature of the threshold point. The statistical character of the switching delay makes impact ionization an attractive possibility in principle. The suggestion is that the threshold field is the critical field required for impact ionization. Once the critical field is available, there will be a random waiting time (switching delay) for an event. At higher fields (overvoltages) that time should be shorter, because carriers in less-than-ideal location should be able to initiate the avalanche. If this were correct as it stands, then illumination should diminish the delay, in fact it does not, but it has been shown[13] that it leaves the threshold field unchanged, as one would expect on the present basis. To account for the independence of switching delay on illumination, one would have to argue that the initiating carriers arise from localities which are always occupied (e.g. from deep traps) or else from a locality in which the carrier lifetime is very low (e.g. the electrode interfaces). The matter has not yet been satisfactorily resolved. However, despite the remaining gaps, there is no doubt that an electronic model is feasible, based in part on conventional semiconductor ideas and in part on the available information on the band structure of multicomponent glasses. The same processes should occur in single crystal materials, provided they are accurately intrinsic, and it has been shown that this is indeed so[24]. A prediction is therefore fulfilled, but lest this be interpreted as a decisive triumph of electronic theory, it must be pointed out that the intrinsic material is also a material of high resistivity and high temperature coefficient, factors which, in the absence of more detailed evidence, permit other explanations.

Of course, thermal considerations would eventually have to be superimposed upon any electronic model, but this has not yet been attempted.

5. ELECTRONIC THEORY FOR RELAXATION SEMICONDUCTORS OF UNIFORM STRUCTURE

The concept of relaxation semiconductors was introduced by Van Roosbroeck and co-workers[25-28]. It concerns materials in which the dielectric relaxation time τ_d exceeds the carrier lifetime τ_ℓ. Such materials are expected to behave very differently from normal ("lifetime") semiconductors, particularly in their response to injected minority carriers. Van Roosbroeck has, for instance, predicted a majority carrier depletion region as the outcome of minority carrier injection. On the basis of such considerations, he has also proposed a theory of threshold switching or, at any rate, a theory of the ON-state[29]. The new concepts are believed to be of greatest interest for our understanding of highly resistive semiconductors, particularly when measured and explored in thin film form. However, the analyses so far available do not give a realistic picture of the consequences, inasmuch as they are based on drastic simplifications, some of which are now known to be impermissible. In particular, the role of diffusion processes must be considered, as must the contribution of minority carriers to the current. Moreover, the presence of a minority carrier injecting electrode on one side of the film implies minority carrier extraction on the other side, a process which yields less easily to analysis. Once again, the complete equations cannot be explicitly solved ; a computer analysis is in progress[30] and until it is complete,

THRESHOLD SWITCHING

the significance of relaxation concepts for switching cannot be properly assessed. The matter will therefore not be further pursued for the moment, beyond pointing out

(a) that there are electronic mechanisms potentially active in switching other than those described above, and
(b) that an analysis of thin film systems consisting of semi-insulating (relaxation)materials, may not ultimately be possible without consideration of the relaxation effects. The expectations arising may turn out to be just a compelling as those arising from Joule heating.

6. CRITIQUE OF EXISTING MODELS

All the existing models have shortcomings, and though this fact is not denied by anyone, it has proved difficult to distinguish between shortcomings which are intrinsically associated with a model (and thus cannot be remedied by any conceivable elaboration) and those arising only from the available approximations and simplifications. No dilemma if this kind is more vexing than that arising from the observations of radiative emission. Kolomiets and co-workers [31], and more recently Vezzoli and co-workers [32] have reported observing radiative emission from the ON-state. In the former case the wavelength peak corresponded to half the mobility gap, in the latter to the entire gap. For the Kolomiets radiation to arise from thermal causes, the source would have had to be at 1500°K, i.e. far above the glass transition temperature. Moreover, in both cases, the output was proportional to current, not to the power dissipated. Similar observations have been made by Preudenziati and co-workers [33] in the course of experiments on switching in boron. The emitting volume is, of course, very small (e.g. 10^{-12} cm^3) in all these cases and the optical signal correspondingly weak, but some such emission is clearly expected on the basis of electronic models. In particular, it is expected from the population inversion in the ON-state discussed above. On the face of it, this evidence is conclusive, and advocates of electronic models are entitled to find it in the ultimate vindication of their view. However, we do not in fact know whether it is intrinsically linked with switching process. More evidence is needed, under a greater variety of conditions. If it were found that radiation is always or generally associated with the ON-state, one would have to conclude that an electronic desequilibrium is involved. Further research of this kind constitues the best chance of resolving the basic issue.

Thermal theory is at this stage more mathematically polished and permits more detailed predictions. Among these are the prevailing filament temperatures, corresponding to various ON-state currents. They are high, e.g. over 500°C, and even for quite reasonable operating currents far above the accepted value of the glass transition temperature ($\sim$ 150°C). This remains true, even though the temperature estimates can be lowered by making additional assumptions, e.g. by assuming that the device structure is non-uniform. Such a non-uniformity can also explain the independence of V_{TH} upon illumination and its exsistence is a priori likely, either because it is built-in during initial preparation or else because it is the result of a "first-switch" forming process. Thermal theory can thus deal with a part of the problem, but the high filament temperature is in conflict with the high degree of stability which threshold switches of careful manufacture are known to exhibit. Additional assumptions concerning possible mechanisms which inhibit phase transformation and crystallization processes have to be introduced to maintain the thermal position.

The detailed nature of the changes introduced during "forming" and, indeed, the extent to which forming processes are active during the first (ever) operation of different switching systems remain uncertain. The most drastic manifestations[34] involve a decisive "first switch" event, after which the system is permanently modified. Günthersdorfer[35] has shown that this modification involves (or, at any rate, can involve) a certain amount of melting. In contrast, other workers[36] have reported no drastic "first switch" effects of this kind, only a gradual modification of the system over the first few hundred switching operations, and sometimes no forming at all[37]. The matter deserves further investigation.

Figure 2e shows the recovery of V_{TH} after a previous switching event. For switches of 1 micron thickness, the recovery time (knee of the curve) is about a microsecond. In terms of thermal theory, the origin of the recovery curve is clear : it is a consequence of cooling. It should, then, depend on the filament temperature attained during the previous switching event, but does not[38]. Whatever the after-effects of that event may be, they are not linked with the current passed or the energy dissipated during that event. Thermal theory, in its present form, leaves the matter open.

At low temperatures, slight polarity effects have been noted, depending on the direction of the preceding switching event, but at high temperatures (room-temperature and above) the after-effects appear to be non-polar. Electronic theory ascribes them to residual space charge, left over from the ON-state pattern of the previous event, decaying slowly at low temperatures (and hence observable), fast at high temperatures (and hence unobservable). Thermal theory has at this stage no competing explanation. However, electronic theory is not free from corresponding embarrassments. Thus, it is hard to see how a field configuration corresponding to the ON-state could establish itself in a time smaller than the carrier transit time across the film, which is believed to be about 10^{-7} seconds. In fact, switching itself takes place within times which are orders of magnitude shorter, e.g. 10^{-9} seconds or less. Neither theory offers a conclusive explanation of switching with 1-10 nanosecond

THRESHOLD SWITCHING

pulses. Under such conditions, the switching voltage tends to become constant, as indicated on Fig. 2a. This is within the expectations of electronic theory, inasmuch as it implies a critical field under conditions in which heating was believed to have been avoided. However, as far as they go, the available results can also be interpreted thermally. The constant voltage is then ascribed to the field dependent conductivity. Switching voltage is therefore not a unique and sufficient definition of the switching condition ; what is needed is a measurement of total energy dissipated during the short pulse.

The two theories have, of course, quite different interpretations of the minimum holding current. This point of the characteristic has been investigated much less than the threshold point but is of great significance. On the basis of electronic theory, the minimum holding current should certainly increase with increasing film thickness, because thicker films are associated with greater transit times. Going in the opposite direction, the minimum holding current should tend to zero as the film thickness tends to zero, and this is actually observed. Thus very thin films exhibits no switching process, but a static V-I characteristic, very similar to the transient charcteristic shown on Fig. 2d, and highly suggestive of tunnelling. Thermal theory must call upon additional ad-hoc assumptions in order to explain the transient ON-state (Fig. 2d), not because it is a priori at fault but because it does not deal with anything but bulk properties, governed by thermal inertia. Any experiment which does not relate to such properties is necessarily outside the compass of thermal theory. On the basis of electronic theory the transient ON-characteristic is the direct consequence of the contact barriers. A similar dichotomy arises in connection with phenomena observed at the interfaces between amorphous and crystalline semiconductors[40,41], phenomena which strongly suggest some form of electronic interaction.

It is clear enough that the resolution of these problems calls for further experimentation, but the design of experiments is no longer a simple matter. The obvious measurements have already been done, and since they proved indecisive, more sophisticated procedures will have to follow. Among the most urgent needs are:

(a) further work on the light emission associated with switching;detailed spectral characteristics for various contacts and switching materials; temperature dependence, etc.;

(b) further work on the nature of the ON-state close to the minimum holding current, including dependence on circuit conditions and on rate of voltage withdrawal;

(c) further work on contact effects, in darkness and under illumination;

(d) further work on the nature of the high-field conductivity, over a wide range of pulsed fields and temperatures, to determine its association (if any) with the switching process;

(e) further work on switching short pulses of high overvoltage to determine the dependence of the switching response on energy input;

(f) further work on switching characteristics as a function of the material constants;

(g) further work on the thickness dependence of all the parameters involved (co-planar and sandwich structur

ACKNOWLEDGMENT

We would like to express our grateful thanks to each other, for many useful discussions and for gracious forbearance in the face of intolerable provocation.

REFERENCES

1. S.R. Ovshinsky, Phys. Rev. Letters 21, 1450 (1968).

2. S.R. Ovshinsky, personal communication.

3. R.R. Shanks, J. Non-cryst. Solids 2, 504 (1970).

4. R.W. Pryor and H.K. Henisch, J. Non-cryst. Solids 7, 181 (1972).

5. S.H. Lee and H.K. Henisch, J. Non-cryst. Solids 11, 192 (1972)

6. C.B. Thomas, A.F. Fray and J. Bosnell, Phil. Mag. 26, 617 (1972).

7. P.J. Walsh, et al., J. Non-cryst. Solids 2, 107 (1970).

8. T. Stubb, T. Suntola and O.J.A. Tiainen, Solid State Electronics 15, 611 (1972).

9. H.K. Henisch, R.W. Pryor and G.J. Vendura, J. Non-cryst. Solids 8-10, 415 (1972).

10. J. Marshall and A.E. Owen, phys. stat. solidi (a) 12, 181 (1972).

11. P.J. Walsh, R. Vogel and E.J. Evans, Phys. Rev. 178, 1274 (1969).

12. M. Telnic, L. Vescan, N. Croitoru and C. Popescu, phys. stat. solidi (b) 59, 699 (1973).

13. H.K. Henisch, W. Smith and M. Wihl, Fifth Int. Conf. on Amorphous and Liquid Semiconductors, Garmisch (1973). Proceedings in print.

14. D.M. Kroll and M.H. Cohen, J. Non-cryst. Solids 8-10, 544 (1972).

15. C. Popescu, personal communication.

16. T. Kaplan and D. Adler, J. Non-cryst. Solids 8-10, 538 (1972).

17. D. Buckley and S.H. Holmberg, personal communication.

18. W.R. Smith and H.K. Henisch, phys. stat. solidi (a) 17, K81 (1973).

19. N.F. Mott, Phil. Mag. 24, 911 (1971).

20. H.K. Henisch, E.A. Fagen and S.R. Ovshinsky, J. Non-cryst. Solids 4, 583 (1970).

21. H.K. Henisch, Scientific American 221, 30 (1969).

22. S.H. Lee, Appl. Phys. Letters 21, 544 (1972).

23. M.H. Cohen, H. Fritzsche and S.R Ovshinsky, Phys. Rev. Letters 22, 1065 (1969).

24. R.W. Knepper and A.G. Jordan, Solid State Electronics 15, 45 (1972).

25. W. Van Roosbroeck and H.C. Casey Jr., Phys. Rev. B5, 2154 (1972).

26. W. Van Roosbroeck and H.C. Casey Jr., Proc. Tenth Int. Conf. on Semiconductors, Cambridge, Mass. (U.S. Atomic Energy Commission), 832 (1970).

27. W. Van Roosbroeck, Phys. Rev. Letters 8, 1120 (1972).

28. H.J. Queisser, H.C. Casey Jr., and W. Van Roosbroeck, Phys. Rev. Letters 26, 551 (1971).

29. W. Van Roosbroeck, J. Non-cryst. Solids 12, 232 (1973).

30. C. Popescu and H.K. Henisch, Phys. Rev. B11, 1563 (1975).

31. B.T. Kolomiets, E.A. Lebedev, N.A. Rogachev and V. Kh. Shpunt, Soviet Physics-Semiconductors 6, 167 (1972).

32. G.C. Vezzoli, P.J. Walsh, P.J. Kisatsky and L.W. Doremus, in preparation.

33. M. Prudenziati, G. Majni and Alberigi Quaranta, Solid State Communications 13, 1927 (1973).

34. L.A. Coward, J. Non-cryst. Solids 6, 107 (1971).

35. M. Günthersdorfer, J. Appl. Phys. 42, 2566 (1971).

36. J.R. Bosnell and C.B. Thomas, Solid State Electronics 15, 1261 (1972).

37. R. Pryor, Ph.D. Thesis (Physics), The Pennsylvania State University (December 1972).

38. H.K. Henisch and R.W. Pryor, Solid State Electronics 14, 765 (1971).

39. S.R. Ovshinsky, personal communication.

40. G.J. Vendura and H.K. Henisch, J. Non-cryst. Solids 11, 105 (1972).

41. K.E. Petersen, D. Adler and M.P. Shaw, in preparation.

MECHANICAL PROPERTIES OF NON-METALLIC THIN FILMS

R. W. Hoffman

Professor Department of Physics

Case Western Reserve University - Cleveland, Ohio 44106

1. INTRODUCTION

The importance of mechanical properties of thin films has been established over the past several decades. In the first place, many properties are substantially modified in a condensed flm. The literature contains many examples [1] illustrating shifts in band-gap in semiconductors, transition temperature for super-conducting films, or expected magnetic anisotropy. Any property which itself is strain sensitive, may well be modified in a deposited film. Even a property which is originally isotropic may have lower symmetry when the effects of the strain are considered. The control of the stresses by changing deposition conditions is one of the goals now partially reached in the search to develop new properties.

More recently, mechanical constraints introduced by the substrates have been used to produce or stabilize new film structures that are unknown in the bulk.[2]. In this way properties or devices may be constructed that fit special needs and are unvailable by conventional techniques. This design, through the mechanical constraints in the system, is one of the more exciting new areas.

One of the earliest evidences of the importance of the mechanical properties in the various modes of failure. These failures may be the more obvious ones of actual fracture of the film or buckling as a result of the loss of adhesion to the substrate. Dislocations or cracks may be introduced in the substrate at high stress levels. Somewhat more subtle effects such as strain-enhanced diffusion or electromigration may also occur.[3]

A complete review would define the stress and strain system, consider the techniques for measurement of the internal stress, as well as present the data and consider the various models for the origin of the stress. The tensile properties should also be treated, including the elastic modulus, the tensile strength, plastic deformation, and the fracture mode. Indeed, the earlier general reviews by Hoffman[4,5], Cambell[6,7], Buckel[8], Kinosita[9], Scheuerman[10], and books[11], have proceeded along these lines. We will not attempt to reproduce that which already exists in these review papers. We shall update them, however, and call attention to the more specialized reviews, especially in those areas where the information is pertinent to non-metallic films.

We also consider our task to discuss the stress and strain distributions found in a film so that a given property may, in turn, be calculated. New experimental techniques and data will, of course, be included.

Although at this time we begin to understand, at least in principle, the mechanical properties of metallic films, our state of knowledge for non-metallic films is much more rudimentary. In the first place, the properties seem to be not very reproduceable from laboratory to laboratory and, indeed, seem to depend sensitively on the conditions of deposition. In addition, instabilities or pronounced changes in the internal stress may be seen when compound films are exposed to the atmosphere or other surroundings after removal from the system. All of these make for a much more difficult problem in understanding the origin of the internal stress because the structure of non-metallic films is not so well characterized. We shall see that stoichiometry and rapid diffusion of ambient gasses are partially responible. In spite of these difficulties, a qualitative understanding is emmerging.

2. THE ELASTIC PROBLEM

    Formulation of the Elastic problem

    Since, as we will see, the mechanical properties of a film on a substrate are primarly a result of constraints in introduced by the substrate, we wish to understand both qualitatively and quantitatively the nature in which the forces are distributed in the film and substrates and then transmitted across their interface. Even though many problems of practical importance can be solved in terms of isotropic elasticity, we wish to set up the problem generally enough that we can consider cases of anisotropic strains. We must also take account of single crystal films and substrates in order to include epitaxial cases.

Only strains may be experimentally measured. Hence the stresses we talk about are calculated on the basis of measuring deformation, knowing the boundary conditions and assuming that the sample is in equilibrium. Following Nye[12] and Smith[13] we use a mutually orthogonal coordinate system with the x_1' and x_2' axes in the plane of the film and the x_3' axis normal to the film plane.

The coordinate system is defined in Fig. 1. The first subscript refers to the direction of the force on a face perpendicular to the second index. Thus, σ_{11}', σ_{22}' are the normal stresses in the plane and σ_{33}' is the normal stress perpendicular to the film plane. The shear stresses σ_{12}', σ_{23}' and together with rotational equilibrium, represent force systems attempting to decrease the angle between the axes indicated. The corresponding normal strains are ε_{11}' ε_{22}' and ε_{33}' and shear stress ε_{12}, ε_{23}, ε_{31}. We point out that the tensor shear strains (ε's) are half the engineering shear strains (γ's) since the γ's represent the decrease in angle between two original orthogonal directions. The primes are used since the orthogonal primed axes are sample axes and are not the crystal (unprimed) axes.

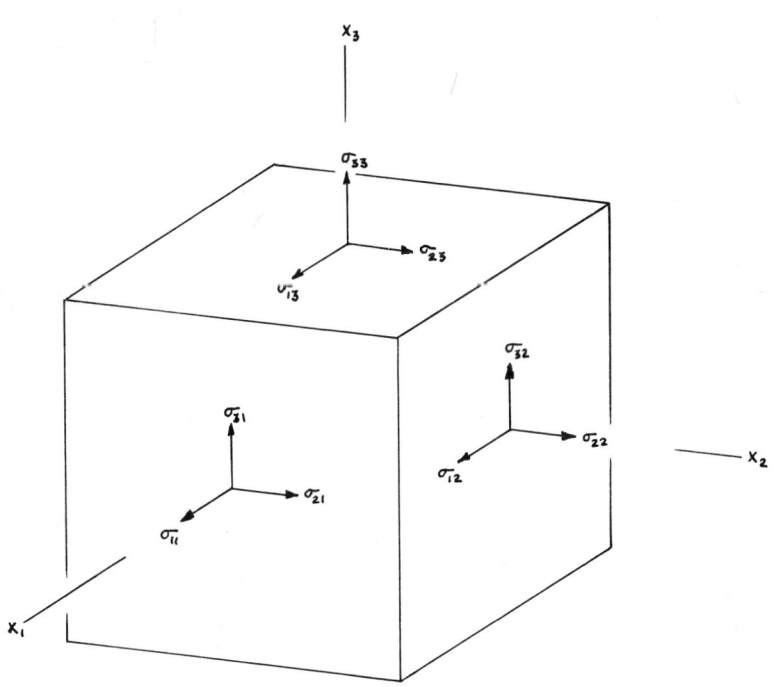

Fig. 1   Definition of stress system

Since elasticity is a fourth-rank tensor property,

$$\varepsilon'_{ij} = s'_{ijk\ell}\,\sigma'_{k\ell}$$

and (1)

$$\sigma'_{ij} = c'_{ijk\ell}\,\varepsilon'_{k\ell}$$

where the s's are the elastic compliances and the c's the stiffnesses for the particular sample orientation. Fourth rank tensor properties can be quite complex in the case of low symmetry crystals. The 81 possible components may be reduced by well-known equilibrium and symmetry arguments.

Even in the cubic system three constants are independent and two are needed for an isotropic specimen. The elastic equations are often reduced to the two index notation[12,13]

$$\varepsilon'_i = s'_{kj}\,\sigma'_j$$

(2)

$$\sigma'_i = c'_{ij}\,\varepsilon'_j$$

where the subscripts 1, 2 and 3 stand for normal components and 4, 5 and 6 refer to the shear components for both stresses and strains. The reader is reminded that the arrays s_{ij} and c_{ij} are not second rank tensors in this notation.

For the (001) orientation in the cubic system, we may drop the primes and find the independent constants s_{11}, s_{12} and s_{44} or c_{11}, c_{12} and c_{44} listed in the handbooks. For a general orientation the primed constant must be calculated by transforming from the crystallographic axes where the elastic constants are tabulated to the primed sample axes. Nye, among others, gives this prescription.

Young's modulus defined as the ratio of the longitudinal stress to the longitudinal strain, i.e., $1/s'_{11}$, is often desired. For the cubic system[13].

$$\frac{1}{E_{\ell_i}} = s_{11} - 2(s_{11} - s_{12} - \tfrac{1}{2} s_{44})(\ell_1^2\ell_2^2 + \ell_2^2\ell_3^2 + \ell_3^2\ell_1^2)$$

(3)

where $\ell_1\ell_2\ell_3$ are the direction cosines of the arbitrary direction ℓ_i referred to the crystallographic directions. As the quantity

($s_{11}-s_{12} -1/2s_{44}$) is positive for all cubic metals except mobybdenum, Young's modulus has a maximum in <111> and a minimum in <100> directions. Young's modulus is independent of direction in the plane for (111) orientation. Nye lists expression for Young's modulus for crystals of lower symmetry.

Vook and Witt[14] give expressions for Young's modulus in several simple directions. Turley and Sines[15], following Thomas[16], separate the given stiffness in a constant and orientation-dependent part and then suggest several rotation methods to calculate the transformation. Polar plots are given for several shear constants for Cu, Mo, and Si.

For convenience in our discussion and following equations, we will drop the primes from the notation. In order to obtain a better feeling for the problem, let us consider the simplifications of an isotropic elastic substrate. Firlmy attached to this substrate is a film which is also elastically isotropic. We shall regard the dimensions in the plane as being semi-infinite. If we now consider a homogeneous stress in the film, and invoke the usual boundary conditions of no forces at the free surfaces or edges, we have the situation of Fig. 2, where we depict the forces acting on an interior section of the film. As we shall see later this corresponds in reality to the situation found in most systems except near the edges of the film itself, or perhaps at the boundaries of crystallites within the films. In this figure we see that the film is in a state of tensile stress indicated by the total force per unit width of the film F.

We note that this tension is a biaxial one. As one traverses the interface in a direction normal to the plane of the film, there is a discontinuity in the stress but the strain is continuous. If the film is in tension we find that the substrate is in compression just below the interface. In fact, if the film is quite thin compared to the substrate, the neutral plane is a third of the substrate thickness from its free surface.

Thus our oversimplified picture of a film under the uniform tensile strain would give us the situation indicated in Fig. 3. An interior volume element would see the normal stresses, σ_{11} and σ_{22} but no shear stresses. If we imagine cutting through the film in the center we see a uniform tensile stress, assuming the substrate were rigid. If, on the other hand, the origin of stress were concentrated at the interface, we would see decreasing tensile forces as indicated schematically in the figure when we moved away from the interface. The free edges of the film can support no forces. Hence, as seen in the right hand side of the figure, we would expect shears to exist only near the edges of the film.

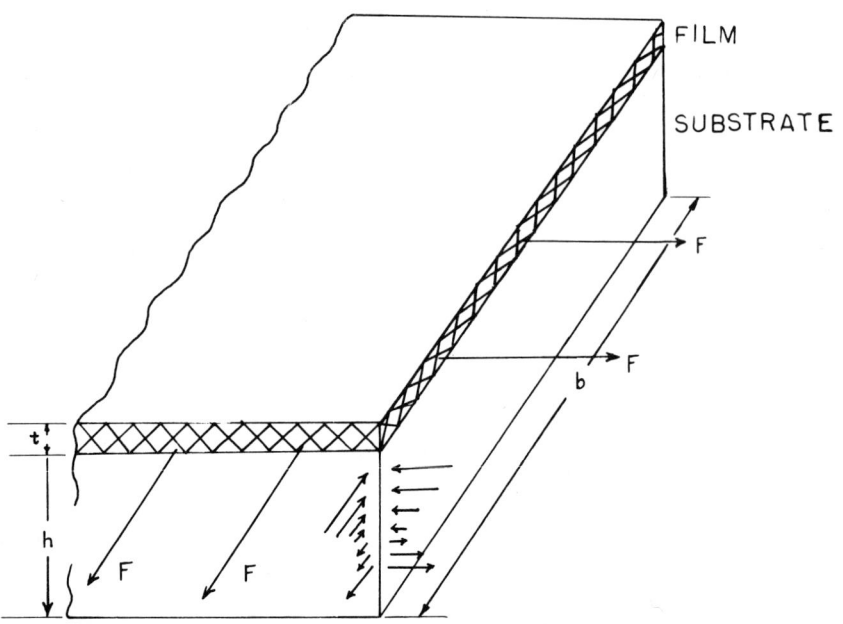

Fig. 2 Forces acting on an interior section of the film-substrate composite

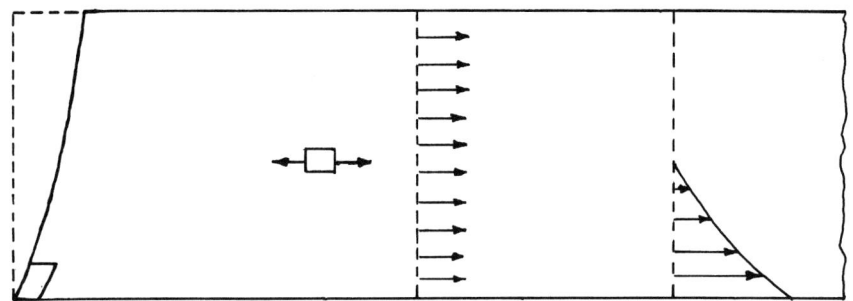

Fig. 3 Idealized stress distribution within a constrained film

MECHANICAL PROPERTIES OF NON-METALLIC THIN FILMS 279

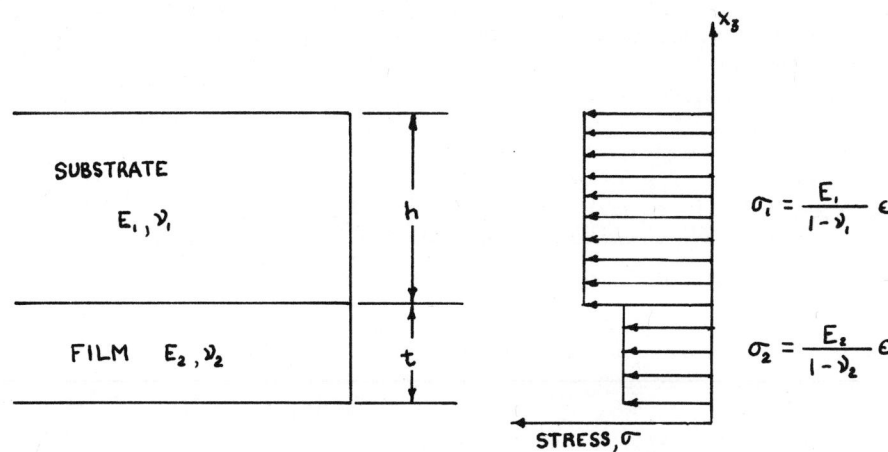

Fig. 4 Stress distribution for longitudinal applied force.

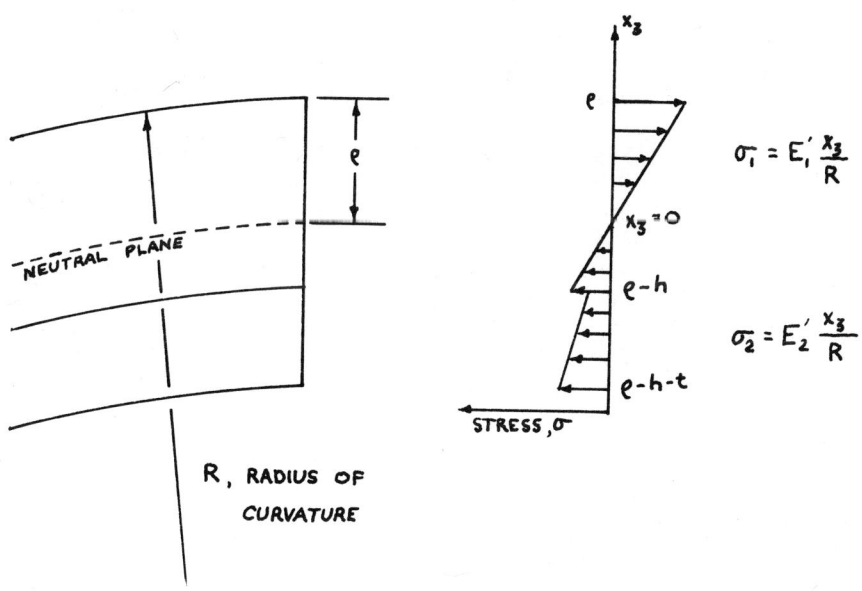

Fig. 5 Stress distribution resulting from bending.

Stress Distribution

Let us consider some of these points in more detail. There are two problems that need to be attacked. One is the stress distribution within the interior of the substrate and film, and the second, the effects near the edges. We will consider the first of these. As pointed out by Brenner and Senderoff,[17] and treated in the review by Hoffman, using the case where the elastic constants of film and substrates are the same, and later extended by Doljack,[18] the problem is one of longitudinal forces in the film superposed on bending. It is instructive to look at each of these separately.

If we apply an external force to the composite plate longitudinally, so that all of the fibers strain equally, then the stresses across the cross section will be those indicated in Fig. 4 for the case of a biaxial stress system. A discontinuity in the stress exist at the interface but the stress in each of the substrate and the film is uniform.

In the case of pure bending, when only an external moment is applied to each of the elemental pieces and no external force is applied, the elastic response is only a bending strain. This produces the stress distribution as shown in Fig. 5. It is seen that the bending gives rise to non-uniform stress in the direction perpendicular to the film, arising from the fact that the stretching in the longitudinal fibers depends on the distance from the plane of zero bending strain. If the plate were not a composite one this plane of zero bending stress would just be in the middle.

We now relate these longitudinal and bending relaxations to the actual case of depositing a film on the substrate. Consider first a substrate clamped against bending and contraction during the deposition of the film. In this case the stress distribution in the film after deposition is the intrinsic stress. This distribution is indicated schematically in Fig. 6, and of course, the distribution of the stress need not be constant throughout the thickness of the film. The total force exerted by the film is be integral of the stress over the thickness of the film ; and equal to the product of the average stress and the film thickness.

$$F = St = \int_0^t \sigma(t) \, dt \qquad (4)$$

where (t) represents the stress distribution as a function of the thickness t.

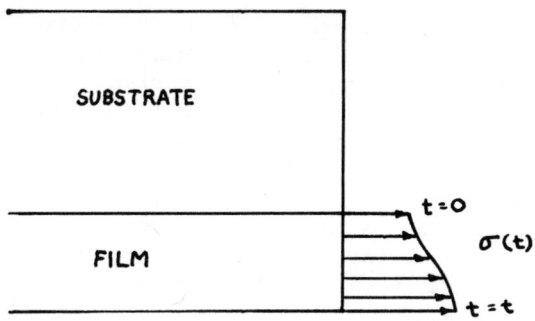

Fig. 6 Intrinsic stress distribution for deposition with substrate clamped against bending and contraction.

If we now release the clamping preventing contraction, but still do not allow anay bending to take place, the stress distribution throughout the plate is modified as in Fig. 7. The uniform stress distributions σ_1 and σ_2 are produced by the net contraction of the tensile stress distribution, $\sigma(t)$ in the film. The relaxed stress distribution in the film $\sigma'(t)$ is just the original distribution reduced by constant stress σ_2. Equilibrium is achieved when the total force on the cross section depicted in Fig 7, is zero.

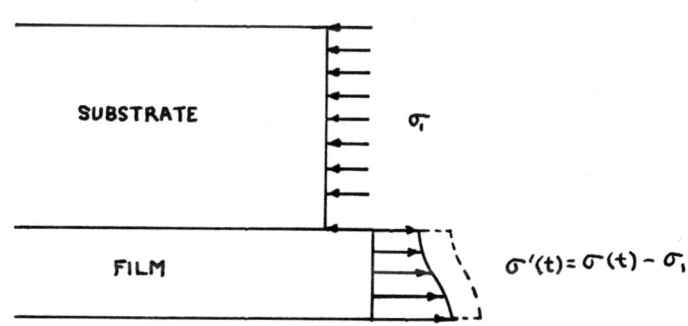

Fig. 7 Relaxed stress distribution upon releasing longitudinal clamps.

If the remaining clamps are now removed and the plate is allowed to bend, it reaches an equilibrium radius of curvature. This bending of the film substrate plate gives rise to a relaxation of the stress as was indicated in our discussion of pure bending. In essence, the additional contraction which results from the necessity to achieve bending equilibrium decreases the actual stress in the film even more. And, furthermore, makes it non-uniform as a function of thickness. The radius of curvature may be determined by equating the moment of the film to the moment produced by the stresses resulting from the bending. The plane of zero elastic strain, or neutral plane, is now shifted, and may be determined from the condition that there is no net external force exerted across any cross-section and thus the sum of the bending forces must be zero.

Equations for the radius of curvature and the relieved stress distribution are given by Brenner and Senderoff[17] for the case of identical elastic constants of the film and substrate and Doljack[18] and Klokholm[19] for the general case. The stress distribution will be modified by a bending distribution as depicted in Fig. 4. If $\sigma(t)$ is constant, Klokholm calculates the relaxation correction M by which the simple Stoney[20] average stress S should be multiplied to obtain the true average stress.

$$M = \frac{\beta^3(\eta+1)}{(\beta^3+\eta)(\beta+\eta) + 3\eta\beta(\beta+1)^2} \quad (5)$$

where β is the substrate to film thickness ratio h/t and η is the ratio of elastic constants $E_f/1-\nu_f$ and $E_s/1-\nu_s$. E is Young's modulus and ν Poisson's ratio. The subscripts f and s refer to film and substrate. For $\beta > 1000$ and any η or $\beta > 10$ and $\eta < 0.2$ N is sufficiently close 1 to be experimentally negligible.

In practice it is difficult to clamp a substrate sufficiently well to meet both the conditions of no bending and no expansion during deposition. A much more common case is that with the substr. completely free during deposition. During deposition an initially flat plate begins to bend slightly relieving the stress in the film already deposited. The continuous bending proceeds, giving rise to a stress relief in the film in addition to the contraction and bending just discussed. Under these conditions the relief in each layer is different when the deposition is completed. If the mechanism giving rise to the stress in the first place were originally a process uniform with thickness then the result of such a stress relief would clearly give rise to larger stresses near the outside growing edge of the film as indicated in Fig. 8.

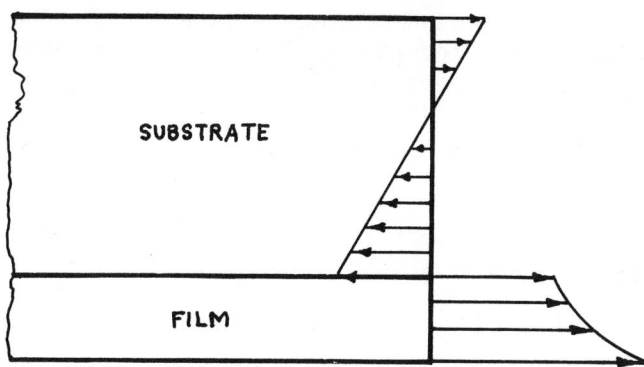

Fig. 8 Relaxed stress distribution for free cantilevered substrate during deposition. After Brenner and Senderhoff[17].

Brenner and Senderhoff, in fact, suggested that this would be an explanation for the fact that many films curl when detached from the substrate in a way to suggest that the outer layer has the higher stress. Fortunately, these stress relief terms are not large in practical cases since the film is generally much thinner that the substrate. On the other hand, we must keep in mind that they will give rise to stress gradients within the film. The stress distributions have been calculated by the authors previously referenced, and the errors are less than the previous case for the same geometry.

Chaudhari[21] gives an approximate treatment of the relaxation following the thermal stress analysis of Timoshenko. The stress distribution in the film may be wtritten as :

$$\sigma_{11} = \sigma_{22} = \frac{\varepsilon_{11} E}{1-\nu} \left\{ 1 - \frac{C}{D} \left[1 + \frac{3(C-D)(x_3-D)}{D^2} \right] \right\} \qquad (6)$$

For this treatment ε_{11} is the assumed uniform strain arising from thermal contraction or other mechanism, E and ν are the usual elastic constants assumed the same for film and substrate, $C = t/2$, the film half-thickness, and $D = h+t/2$, the substrate plus film half thickness. x_3, the coordinate normal to the film, is 0 on the free surface of the film.

The first term in the equation represents the stress as one would calculate it considering a rigid substrate. The second term represents the contraction or compressional relaxation while the third term represents the bending relaxation. The stress distribution in the substrate is given by

$$\sigma_{11} = \sigma_{22} = \frac{\varepsilon_{11}E}{1-\nu} \frac{C}{D} [1 + \frac{3(C-D)(x_3-D)}{D^2}] \qquad (7)$$

The results are presented in Fig. 9 for several values of C/D. As the stress in the substrate is reduced by the factor C/D, it is obvious that in most cases of deposition the film is much, much thinner than the substrate so that the stress in the substrate is only a very small fraction of the stress in the film. If we consider a film thickness of 1 μm and a substrate thickness of 1 mm, than the maximum value of the stress in the substrate is less than 1/2 % of the stress in the film.

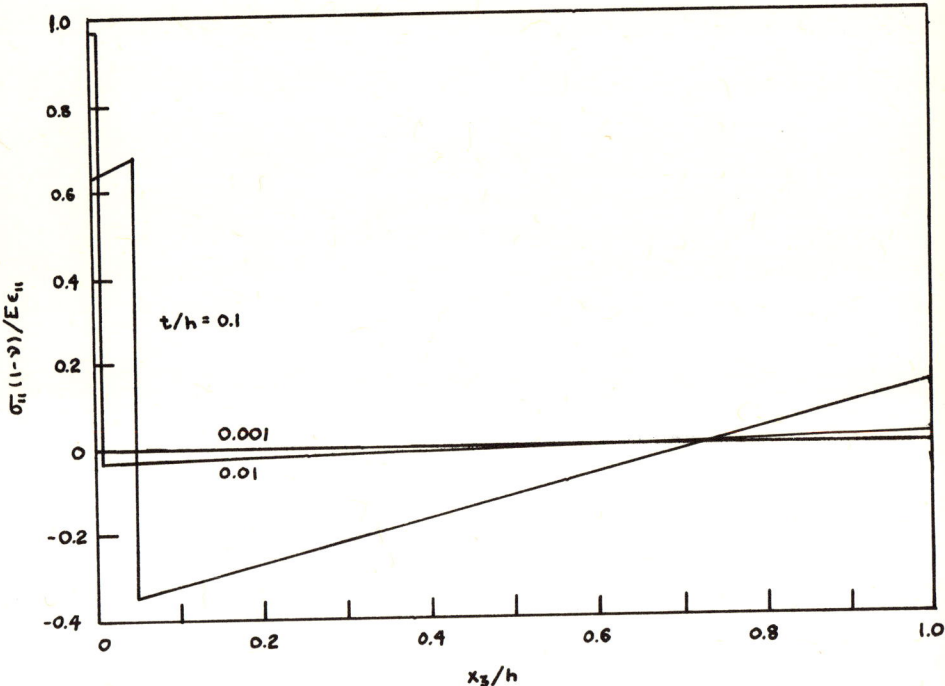

Fig. 9   Approximate stress distribution for several values of film-substrate thickness ratio. After Chaudhari[21].

As Chaudhari has pointed out the critical feature of whether plastic flow will take place in the substrate is the value of critical resolved shear stress. If the stress in the film is 1% of the shear modulus and we consider the case where the ratio of the film thickness to the substrate thickness is 10^{-4}, the maximum value of stress in the substrate would be 10^{-6} μ. For comparison, most bulk single crystals will begin to undergo a plastic deformation when the shear stress is of the order of 10^{-5} or 10^{-4} times the shear modulus μ. Thus, when the film thickness is very thin, compared to the substrate thickness, we can neglect plastic flow by a dislocation mechanism in the substrate. When the ratio of the film thickness to the substrate thickness is larger than 10^{-4} we have to consider the particular substrate in question to determine whether or not plastic flow may take place.

There are two cases of practical importance where one has to be concerned about plastic deformation in the substrate. One of these, of course, is the obvious one where the material in the substrate is such that it easily undergoes plastic deformation, alkali halides for instance. Stress concentrations at propagating crack tips may also introduce dislocations in single crystal substrates. Dislocations and other local defects may also be introduced in the substrate during growth. The second case takes place when the substrate is comparable to the film in thickness. As one can see by examination of Fig. 9, the relaxation in the film and substrate is sufficient to cause a non-uniform stress distribution within the film as well.

Similar distributions have been calculated by Oel and Frechett (22) for both large and small area planar interfaces for the case of comparable film and substrate thickness. Model experiments with thermal strains induced in birefiringent glasses gave good agreement with the calculations except when diffusion took place during sealing at high temperatures. In this case, far from merely smoothing the stress gradient, diffusion-caused stresses had large effects in magnitude and even sign of the stresses.

No detailed treatment concerning the sharpness of the interface seems to exist. The large stress gradient in this region may play a role in interfacial dislocation generation or mass flow and other effects, and represents an area of needed future study.

Edge Effects

We now consider the second part of the question, namely, the spatial details of the stress distribution. Of course, if the mechanism giving rise to the stress in the first place is not uniform, then the stress distribution will be non-uniform as well. Such distributions will be discussed later under origins of the

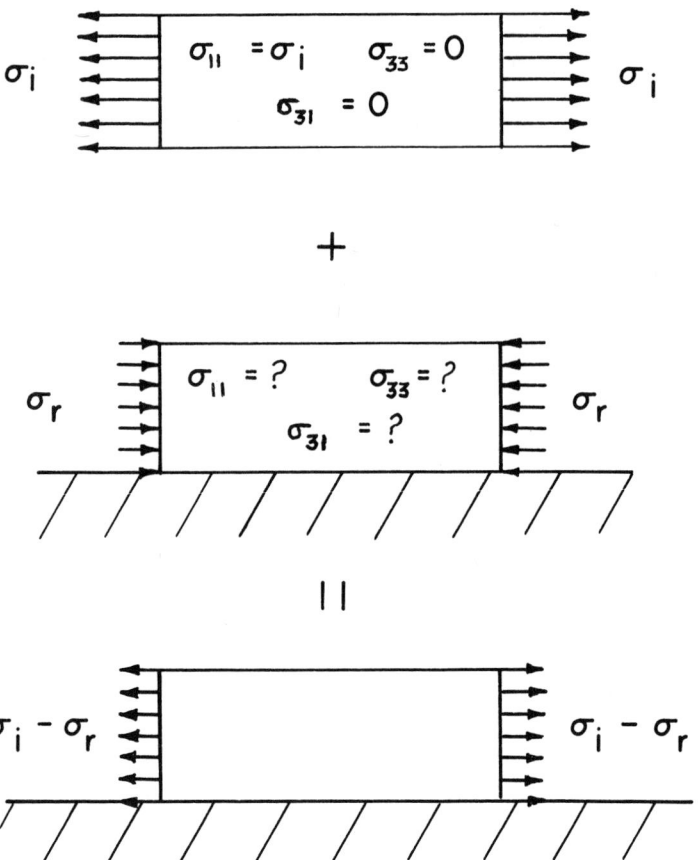

Fig. 10. Boundary traction method applied to a thin film under uniform intrinsic stress constrained to a rigid substrate. After Aleck[23] modified by Doljack[18].

intrinsic stress. However, we shall consider here the case where we initially have a large stress $\sigma_{11} = \sigma_{22}$ that lies in the plane of the film is isotropic, and nearly uniform throughout the thickness, and is uniform over most of the area of the film. If we look at the cross section of the film-substrate composite, we see that the film appears to be a rectangular plate attached along its long edge to what we shall assume to be a semi-infinite rigid substrate. As a result, a slab of thin film attached to the substrate is under a state of stress nearly identical to that of a rectangular plate clamped along an edge and having undergone thermal expansion.

This latter problem has been treated approximately by Aleck[23]. We shall consider the question of thermal expansion later, but at the moment, we shall use the results of this treatment to ask the question as to the stress distribution near the edges of the film. According to Aleck, the problem is first converted to one of boundary tractions. The film is detached from the substrate and allowed to relax. To bring it back to a uniform state of stress, one imagines applying a uniform stress of magnitude σ_i normal to the ends of the film as shown in the top of Fig. 10. Next, the film is attached to the substrate and a uniform compressive stress σ_r is applied to its ends. σ_r in this case is equal to σ_i because at the edge if the film there are no net applied normal boundaty tractions. The solution to the problem of the specified surface traction thus replaces the original stress problem. Doljack has solved this

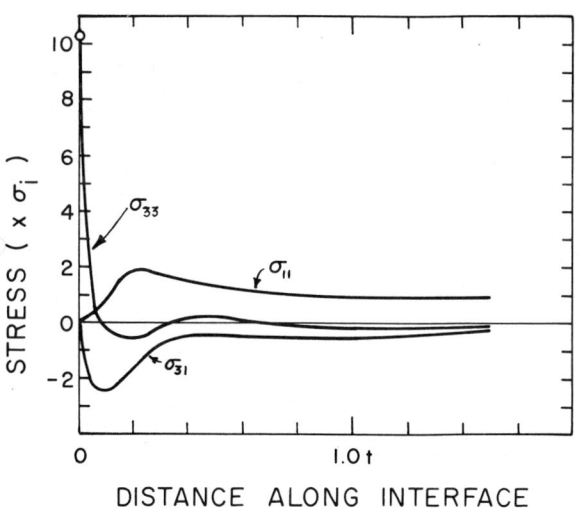

Fig. 11 Interfacial stresses near the free edge of a semi-infinite film. After Doljack.[18]

problem based on the approximate solution of Aleck for the case of thickness of the film being much thinner than its length. Fig. 11 shows the results of these calculations. The pertinent pictures are these : σ_{11} is just equal to the intrinsic stress until one gets within a distance of the order of 10 thicknesses to the edge. A large shear exists very close to the edge. In addition, a concentrated stress normal to the interface lies at the attached corner. We point out that there exist no non-zero stresses or stresses normal to the interface once one gets to a distance several times the thickness away from the edge. The presence of an extremely large normal force near the edge points out its relationship to failure at scratches or film edges. We will return to this phenomenon later.

Haruta and Spencer[24] have observed the spatial stress distribution by X-ray topography which is sensitive to the strain gradients, and qualitatively confirmed the rapid change near the film edge. A substantial decrease in strain was found after annealing.

To summarize then, we conclude by detailed calculations that as long as the substrate remains essentially rigid, and as long as we are not concerned with the stress distribution very close to the edges of the film, our simplified point of view which says only the components σ_{11} and σ_{22} are presents, is indeed the correct one.

3. TECHNIQUES FOR STRESS MEASUREMENTS

Introduction

Although the literature primarily quotes values for the stresses in films, it is actually strains that are measured. Two direct ways exist. First of all, a measure of the deformation of the substrate upon which the film has been deposited, or secondly, by diffraction techniques, a measurement of the elastic strain within the film, or in some cases, substrate. These two ways need not yield precisely the same values because of the distribution of forces across grain boundaries and whithin the grains of the fine grain deposit affect each technique differently.

A summary of the various methods used prior 1970 for measuring the deformations has been well documented in the reviews by Scheuerman[10] Cambell[7], and earlier by Hoffman[4,5,25,26]. We see no point for reproducing that information here. Many different techniques and their sensitivities were compared. In the last several years progress in measurement technique has been in several areas. First, the production of automated systems for measuring the stress on the routine basis on production samples[26]. The search for a technique of measurement of the stress on a localized scale on real structures has resulted in increased use of x-ray techniques. Holographic methods are not used commonly, perhaps because of some of the stringent vibrational requirements associated with that technique.

There has been recent progress in the refinement of the calculations used to relate the observed deflections of the stresses in the films. As the substrate temperature during deposition is perhaps the single most important parameter affecting the stress, its control or the correction for the various temperature factors in the measurement has also been of extreme importance, and progress has been made along these lines. We consider each of these areas in more detail.

Cantilever Plate Method

Perhaps the most common method for measuring the stress is the cantilever beam technique. Many have contributed to this type of measurement ; it exists in a wide variety of detection schemes. Because good data is needed for further progress in understanding the origin of the internal stresses, and also because there is still a lack of approciation for some of the subtleties in the experiment, we describe the cantilever method in detail ; the information may be extrapolated to other geometries.

In the cantilever method the deflection of the freee end of a thin plate is measured; or alternatively a force applied to the free end to restore the beam position to some fixed point. The other end of the beam is imagined to be clamped rigidly. This technique is often used because the deflection can be monitored continuously during deposition of the film and thus obtain information about the stress distribtuion. If the stress in the film is tension, as is commonly the case, then the film applies both a compression to the substrate and a bending moment. The stress is called tensile if the surface traction compresses the plate. This definition comes about because the plate applies a tensile stress to the thin film to prevent it from elastically relaxing. The resultant bending will be such that the film in tension finds itself on the concave side. The commonly used equation relating the deflection at the end of this cantilever beam to the force/unit width of the film is :

$$\delta = 3 \frac{1-\nu_s}{E_s} \frac{\ell^2}{h^2} F \qquad (8)$$

where ℓ is the length of the cantilever leam, E_s Young's modulus and ν_s Poisson's ratio for the substrate, and δ the deflection of the free end. It should be pointed out, although this is the presently accepted cantilever beam relationship, occasionaly the $1-\nu_s$

term, arising from the biaxial film stress, seems to be still neglected.

The concern, then, is how well this relationship actually describes the free end of the beam. The problem is one of properly describing the curvature in the transverse direction of the beam when it undergoes a large longitudinal curvature. This problem has been in the literature for a long time having been discussed as early as 1632 by Gallileo. St Venant in 1864 presented the first mathematical formulation. The more recent treatments by Ashwell and Greenwood[27] and later by Bellow, et al.,[28] have given experimental justification for the treatment put forth by Lamb in 1891.

The results may be summarized as follows : consider a plate of width b, length ℓ, and thickness h, bent to a longitudinal curvature 1/R by uniform moments applied to the ends. If the deformed shape is to be truly cylindrical, that is, the transverse strips are to remain straight, then moments equal to the moment in the longitudinal direction times Poisson's ratio are required along the long edges. Since such moment are not present the plate tends to assume a curvature ν/R in the transverse direction. This case is known as "anticlastic bending", and is illustrated in Fig. 12 . The transverse distorsion results in longitudinal membrane forces. As the curvature becomes larger the effect to these forces must be considered. It results in a stretching in the neutral plane and the Poisson effect eventually becomes cancelled leading to the case of

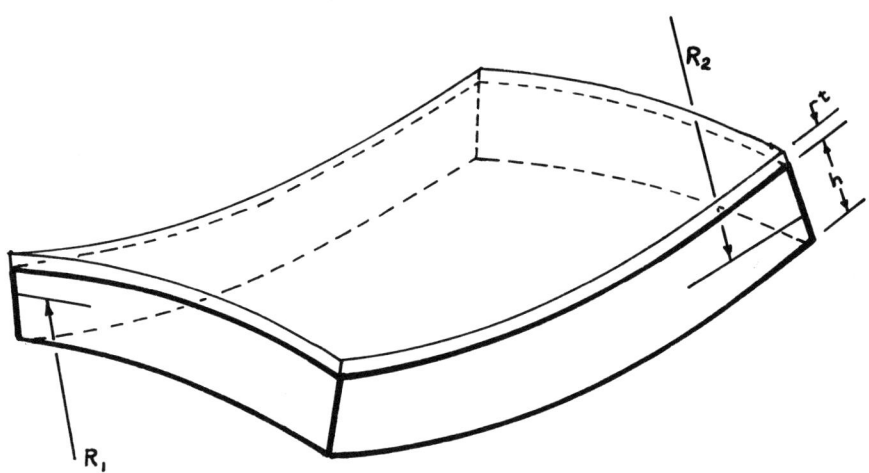

Fig. 12   Anticlastic bending of a flat plate . From Bellow.[28]

cylindrical bending. The pertinent parameter in this problem is the ratio b^2/Rh. For values of this ratio from 0 to 1.6, the anticlastic case occurs. If the ratio is larger than 1000, then the surface can be considered cylindrical.[27] Under cylindrical conditions, the deflection of the end of the plate becomes smaller by the factor of $1/(1-\nu^2)$. As the transition from anticlastic to synclastic bending is a gradual one, the appropriate relationship must be known for the beam of our particular experiment. At present numerical calculations have compared favorably with experimental results for ratios up to 50.[28] By connecting the radius of curvature to the dimensions of the beam approximately through the rigidity relationship, one can write the ratio b^2/Rh in the form $2(b/\ell)^2\delta/h$. It is seen that for a beam whose length to width ratio is a reasonable 10, one may apply the anticlastic bending formula to the situation where the deflection is about 100 times the thickness of the beam itself. It appears that for practical purposes, unless one is dealing with a beam of rather square geometry, that the anticlastic formula will meet most practical pusposes, and equation (1) or its equivalent for other geometry will suffice.

For the geometry of the cantilever beam the boundary conditions require that three edges be free while the fourth is clamped. Thus the deflection and slope of the plate as is emerges from the clamp are rigdly fixed. In practice, this seeems to be a much more difficult condition to meet experimentally than is usually realized. The problem is often encountered when one uses a concentrated load at the end in order to determine Yong's modulus for the substrate. Differences of as high as 20% are encountered between the apparent modulus values for the same beam measured by a cantilevered loading a a free beam loading. Rottmayer and Hoffman[29] pointed out that it is extremely important that the beam be gripped in the clamps such that no sliding would take place. Indeed, a rigid glueing is sometimes necessary. Resonance techniques are sometimes used to obtain values for E.[7,9].

Springer[30] has calculated the cantilever experimental arrangement on the basis of the linear bending theory of plates. The basic assumptions are that $\varepsilon_{33} = 0$, that is, there is no strain in the direction perpendicular to the plane. Secondly, plane sections remain plane upon bending which implies that no serious non-uniform stretching occurs and, thirdly, normal sections remain normal. This means that the transverse shear deformations are small. With these assumptions one proceeds as follows. The function for the shape of the plate is assumed, the stress distribution is given, and the total work done in forming the plate to a given shape is calculated and set equal to the strain energy stored in the plate for this configuration. Numerical calculations were necessary, and the result shows that the deflection calculated by the usual anticlastic relationship was 7% larger than the numerical calculations. This difference is probably due to the fact that the boundary conditions at the clamp is such that the beam may not have any cross curvature and the transition from this behaviour to the anticlastic situation probably does not become apparent until about one beam width away from the clamp. The net result is to reduce the force necessary to bend the beam at the clamp and thus cause a greater deflection to appear at the end.

In practice it is probably not worthwhile to go through the effort of the numerical calculations and one should use the anticlastic relationship. It should be pointed out, however, that in careful experiments the error of about 7% in the bending equation may be comparable to other experimental errors. In order to compare numerical values for stress obtained amongst various laboratories one should make certain that the elastic equations used are described in the publication.

Detailed calculations of the deflection for another geometry have also been carried out. A free circular plate was first considered by Finegan and Hoffman[31] for the case of determining whether the stresses were anisotropic. As will become more obvious in the next section, temperature control demands that any plate be attached to a thermal sink. Doljack and Hoffman[32] have solved the deflection equation for a circular plate clamped along an inner circumference. One solves the problem by superposition of the deflection of the free plate and the deflection corresponding to the forces applied by the central clamp, under the condition that the intrinsic stress may be anisotropic. The plate itself wants to bend to the familiar elliptic paraboloid. The clamp at the inner radius applies those moments and shears to the plate that are necessary to change the deflection and give zero slope at the inner radius. The solutions can get rather complex because one must satisfy a fourth order differential equation, and enforcing the boundary conditions results in solving two sets of four simultaneous equations for four unknowns in each set. The resulting solutions are shown in Fig. 13, wher it is seen that the centrally clamped circular substrate deforms in a manner very similar to the free circular substrate, with the zero deflection point moved out to the clamp.

It is also worthwhile noting the solution to the problem where the plate is clamped around its outer circumference. It is easy to show that under these conditions the clamp applies opposite bending moments that precisely cancel those applied by the surface traction. Thus, no deflection takes place. If a concentric hole is cut in the center of the plate, the plate will deflect and the solution will be the same as the problem previously solved with the roles of inner and outer radius interchanged. As one would expect, the overall bending produced by the same surface traction is less than obtained with the central clamp configuration.

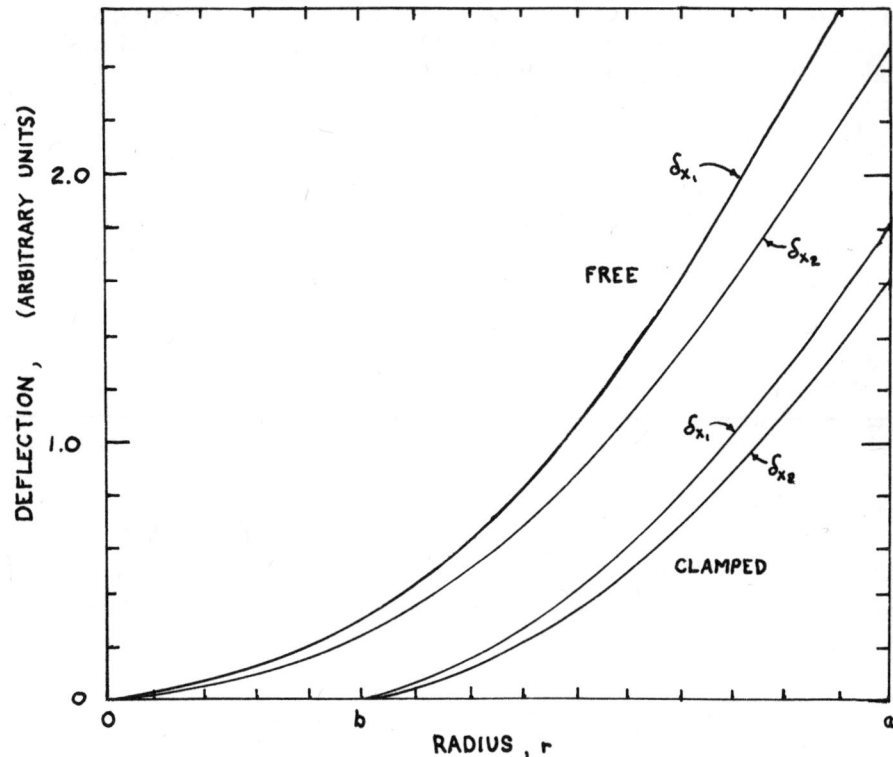

Fig. 13 Anisotropic bending of a free and centrally clamped circular substrate. Doljack and Hoffman.[32]

The last configuration that is worth mentioning is the case of a free circular plate with a thin film deposited over the central circular area of the plate. The solution to this problem shows that as the outer diameter of the plate becomes very large, the bending of the plate in the region of the film does not decrease significantly. although in practical problems the films deposited are seldom circular, the extrapolation of this statement is important for the case of many discreet devices on a single substrate.

The second area where a marked improvement has come about in the last few years is in the temperature stabilization of the substrate during deposition. Since the early work of Campbell and his co-workers in measuring the stresses in the early stages of deposition, it has been realized that both thermal and momentum effects are important. More recently, Kinosita[9] and his coworkers have shown how to make corrections to the deflections obtained during deposition. In their case for depositions of low stress metals on extremely thin mica substrates, the corrections were substantially larger than the observed deflections and in some cases of the opposite sign.

To illustrate the thermal contributions, let us take the cantilever beam of length ℓ, width b, and thickness h, and condense the film of density ρ or mass per atom m, at a rate R.

The momentum effect can be calculated by knowing the temperature T_m of the source, because that determines the velocity of the Maxwellian distribution of atoms as they leave. Assuming the arriving metal atoms do not rebound from the substrate, on can integrate the effect along the length of the beam and find the resultant moment. The result for the implied force and deflection from momentum consideration is equal to:[34]

$$\delta_m = 2.38 \frac{\ell^4 \rho}{E_s h^3} \left(\frac{kT_m}{m}\right)^{1/2} R \qquad F = 0.79 \frac{\ell^2 \rho}{(1-\nu_s)h} \left(\frac{kT_m}{m}\right)^{1/2} R \quad (9)$$

From the point of view of the experiment in which the deflection is measured as a function of the thickness of the film, this contribution would be a constant compression from the time the film starts to deposit until deposition is completed, assuming the rate of deposition is held constant, and the source temperature does not change.

Before we can say anything about the thermal effects, we have to establish the temperature history of the substrate. Maki and Kinosita[35], Yoda[36,37] and Namba[38] have measured the temperature rise of 12 μm thick mica substrates during the deposition of silver films, by measuring the temperature with an iron-nickel thermocouple on the back side of the substrate. As indicated in Fig. 14, the shape of the curve is sensitively dependent upon the rate of deposition. Indeed, in films such as silver, the increase in the reflectivity of the thermal radiation may result in a decrease in the film temperature as the film becomes thicker. Note that the temperature rise may be quite large, perhaps 50° and saturate at thicknesses corresponding to the order of 1000 Å. Similar experiments in our laboratory with extremely fine wire thermocouples bonded to the back side, show an increase in the temperature of the free end which is almost linear in the deposition thickness for deposition rates of the order of 25 Å per sec.

Normally, the cantilever beam is clamped at one end with a

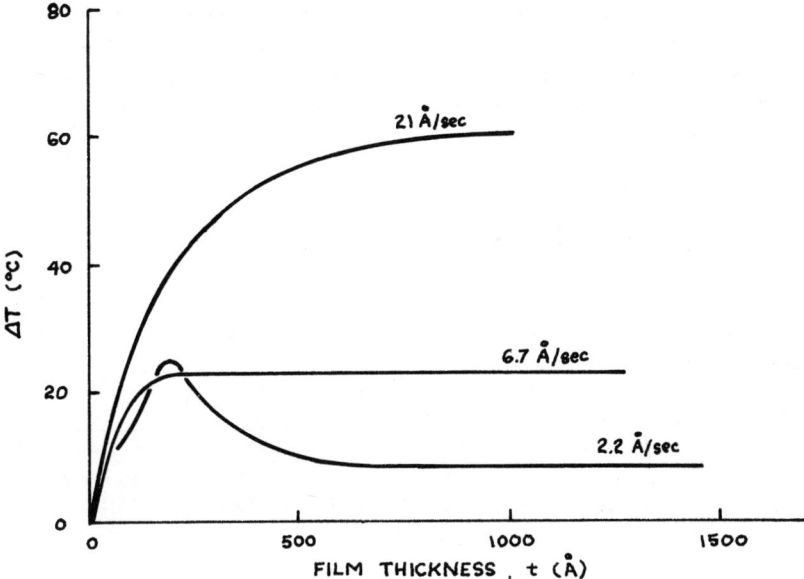

Fig. 14 Temperature history during deposition of Ag on very thin, poorly conducting substrate. After Maki and Kinosita.(35)

large thermal mass acting as an heat sink. At room temperature we assume the incident power must be conducted away from the substrate which produces the temperature gradient along the length. Assuming radiation negligible one may integrate the heat flow equations resulting in the temperature difference along the length

$$\Delta T_2 = \frac{P \ell^2}{2K h} \tag{10}$$

where P is the power per unit area incident on the substrate and K the thermal conductivity of the substrate. P includes the condensation energy as well as the energy radiated from the source and surroundings. For 'hot' sources the condensation energy is usually negligible, but the kinetic energy may be considerable in the case of high bias sputtering. For the evaporation geometry used by Springer,(30) P is 20 milliwatts per cm^2 resulting in a calculated temperature rise of 33°C. This estimate is perhaps 30% high compared to experimental evidence presumably resulting from the fact that steady state conditions are not reached and radiation losses are neglected.

Elastically, there can be several effects resulting from thermal expansion. First, the effect of differential thermal expansion between the film and substrate if the temperature of the specimen is uniform, but changed from T_{so}, the temperature of the substrate at the begining of film deposition. Secondly, in a cantilever experiment, the temperature gradient along the length of the beam given rise to a non-uniform bending, and thirdly, there is a temperature gradient normal to the film plane. All of these conctribute, but the first two are interrelated.

Rottmayer,[39] Kinosita and Doljack have treated the first effect, under the model that the intrinsic stress is temperature independent and no stress relief takes place as the temperature is changed upon further deposition. A given layer deposits thermal stress-free, so there is a thermal stress gradient in the film from maximum at the interface to zero at the free film surface. The thermal strain may be calculated from the expansion coefficients α_f and α_s of the film and substrate. The result is, for a rigid substrate, equivalent stress or end deflection

$$S_1 = \frac{E_f}{1-\nu_f}(\alpha_f-\alpha_s)(T_s-<T_s>)$$

$$\delta_1 = \frac{3\ell^2 E_f(1-\nu_s)}{E_s(1-\nu_f)} \frac{t}{h^2}(\alpha_f-\alpha_s)(T_s-<T_s>) \tag{11}$$

where T_s is the final substrate temperature, assumed equal to the film temperature and $<T_s>$ is the thickness averaged substrate temperature.

If the substrate temperature is linearly related to the thickness, then this net differential thermal contribution vanishes when cooling to the average substrate temperature $(T_{so}+T_s)/2$ after the deposition is completed.

We consider now the case of a thermal gradient along the length of the plate. Elastically, this problem is not the same as the differential thermal expansion contribution that arises by changing the temperature of the substrate and the film by an amount ΔT_1. The reason for this is, of course, that the temperature gradient is not constant, giving rise to a non-uniform bending even though the average strain is still the same. The solution to this bending problem has been approximated by Alexander[34] giving rise to the smaller stress

$$S_2 = \frac{1}{6}\frac{E_f}{(1-\nu_f)}(\alpha_f-\alpha_s)\Delta T_2 \tag{12}$$

if the temperature difference ΔT_2 between the free end of the cantilever beam and the clamped end is established before the film is

TABLE I

Comparison of Extraneous Effects in Cantilever Technique*

Effect	Temperature Change °C	Deflection cm	Force/ Unitwidth dynes/cm	Percent of 1000 Å Film
1000 Å Film	---	2.5×10^{-2}	10^5	
Substrate gravitational	---	4×10^{-2}	1.6×10^5	160
Film gravitational	---	4×10^{-5}	1.6×10^2	0.2
Momentum R=20 Å/sec	---	5×10^{-5}	2×10^2	0.2
Uniform temperature increase, Case I	10	3.7×10^{-4}	1.5×10^3	1.5
Gradient along length - Case 2	30	1.8×10^{-4}	7.5×10^2	0.7
Gradient through thickness, Case 3	$5\ 10^{-4}$	5×10^{-6}	20	0.02

*Calculated for a representative metallic substrate, 0.01 cm thick by 5 cm long and $\alpha_f-\alpha_s = 5\times10^{-6}\ °C^{-1}$

Stress in 1000 Å reference film is assumed to be 10^{10} dynes/cm^2

deposited. We see that this effect, too, may be eliminated entirely if the film and substrate have identical expansion coefficients.

These effects may have either an apparent tensile or compressive contribution but for the case where the α_f is larger than the α_s, the situation realized most often in practice, one finds that the contribution is an apparent compressive one. When related to the initial temperature T_{so} the third thermal effect arises from the temperature gradient through the thickness of the substrate. Although this contribution is extremely difficult to measure a worse case estimate of this gradient can be computed assuming that the incident power must be conducted from one side to the other. This gives rise to the equation $\Delta T_3 = P\ h/K$. Assuming the thermal expansion strains very linearly through the substrate the following expression is obtained for the deflection and equivalent force.

$$F_3 = \frac{1}{6} \frac{E_s}{(1-\nu_s)} h \alpha_s \Delta T_3 \qquad \delta_3 = \frac{1}{2} \frac{\ell^2}{h} \alpha_s \Delta T_3 \qquad (13)$$

For our experimental conditions the temperature gradient is 0.5 millidegree C. Table I summarizes these effects and compares them to the force/unit width for a high stress material, such as nickel at a thickness of 1000 Å. Unless the substrates are exceptionally flexible to obtain high sensitivity, we see that the momentum effect can be neglected for reasonable rates of deposition. Since the intrinsic stress is itself a function of the temperature of substrate during deposition, it is then important to maintain the temperature as constant as possible. In terms of good substrate design one would then pick a substrate of high thermal conductivity and make the cantilever beam as short and thick as possible. The limitation to such an approach is usually the sensitivity of the deflection measurement.

Experimental Verification

Data for a cantilevered beam of the approximate dimensions used in our previous illustrations have been obtained by Springer and Hoffman.[40] For the case of a nickel substrate initially maintened at 32° C, the free end rose by 23° C. This temperature rise decreased as the temperature increased owing to the contribution of radiation losses. At approximatively 150° C the temperature increase vanished and at temperature above that, additional power had to be supplied in order that the substrate temperature did not fall during the deposition. A reproduction of the actual data for the case of a nickel film on a nickel substrate was shown.[40] A small compressional effect upon opening the shutter did not have a corresponding feature after the close of the shutter. Thus we conclude from the lack of a final transient that the compressional effect at the beginning of the film formation is a real phenomenon.

For this experiment $\alpha_f - \alpha_s = 0$ and only a temperature gradient normal to the plane would produce an effect.

Even tighter temperature stabilization was achieved in the experiments of Doljack and Hoffman[32] with 2.4 cm dia., 0.015 cm thick (111) Si substrates. In part this was achieved by a shorter distance to the heat sink plus water cooling of the heat sink. With this geometry the temperature difference between the inner and outer radius was no more than 2°C over the temperature range from 1°C to 200°C. The worst case increase in the average temperature was 10°C over this same temperature range. The average temperature increased linearly with film thickness up to about 100 Å and then remained constant for a deposition rate of 20 Å/sec. In these experiments, $\alpha_f > \alpha_s$ and differential thermal expansion gives a bending moment that decreases intially to some compressive value before coming constant when the substrate temperature reached its equilibrium value. Upon terminating the deposition, the force curve would rise toward a tensile value as the substrate temperature cooled to its original temperature. For the circumstance where the temperature increase took place up to roughly half the final thickness of the film the apparent tensile transcient at the end of the deposition would be about 4 times the amount of the apparent compressive value. Maki and Kinosita[35] analyzed the more difficult experimental case of silver films on extremely thin mica substrates. Large transcients with the thermal time constant took place after the close of the deposition. As was pointed out by the authors, the negative values of deflection are inexplicable. Furthermore, the decreasing deflection during the course of the deposition indicated that relaxation in the film was taking place. The silver films used in these experiments were probably responible for the aging effects at extremely low temperatures.

In addition to the techniques treated in some detail in this section, we now list some of the techniques that have been used recently for deflection measurement. Many of them have been contructed for special measurements and it is suggested that the references be consulted for details. For <u>in situ</u> measurements during deposition, the force restoration techniques[40,41] and optical interferometers, in some cases using laser sources,[32,42,43,44] are the most widely used.

Holographic techniques, as proposed by Haines and Hildebrand[45] for macroscopic objects, were developed by Magill and Young.[46] To avoid the requirements of flatness and high reflectivity, Glang et al.[47] have surface profiled Si waters in orthogonal directions using a light-section microscope. The profile has also been determined by inversion of local radius of curvature data found by an optical lever method proposed by Axelrod and Levinstein.[48]. Moire techniques have been useful in determining the principal stress direction in anisotropic films.[49]

Stress determinations using the bulge method were developed by Beams[50] in his studies of mechanical properties of metal films. Jaccodine and Schlegel[51] used this technique on an unsupported SiO_2 film by etching away the Si substrate, or by removing the oxide from one side and observing the resulting deflection. These concepts have been extended to a more local basis first, by Lane[52] using a moat geometry and later by Lin and Pugacz-Muraszkiewicz[53] who have developed a technique to make deflection measurements on a microscale. Their technique uses standard photoresist methods to etch out a small bar-shaped sample and then free it from the substrate by etching with a second selective solution. If the film were initially under compression, upon being freed from the substrate is expands and deflects giving rise to considerable stress relief. The authors have analyzed the elastic problem in detail and have developed an expression relating the maximum displacement of the free oxide film, observed optically as indicated in Fig. 15, to the strain in terms of elliptic integrals of second kind. They have found that approximately 10% of the total stress remains in the deflected sample. With much less sensitivity this technique can also be used for films under initial tension by measuring the change in length of an undercut beam, after it is released from the substrate. The technique has been applied to films under compression down to thicknesses of 250 Å for the case of SiO_2 on silicon and the results will be reported later.

A similar free-film technique is also reported by Wilmsen et al.[54] who considered the stability of the thermal SiO_2 film in relation to the thickness and width.

We summarize this section on the most commonly used technique for the quantitative measurement of stress in thin films by remarking that with sufficient care, the thermal difficulties may be overcome, the boundary conditions used in the calculations can be realized experimentally, and that interpretative data may be taken continuously during the deposition of a film in order to give information as to the stress distribution. The bending plate techniques have thus been very useful in studies dealing with the origin of the internal stresses and have even been used in automated systems on production samples through the use of fiber optics.[55] Micro-techniques have led to localized stress distributions.

Thus we conclude with proper substrate design and choice of high thermal conductivity substrates, temperature effects may be reduced to the point where they are negligible for high stress

MECHANICAL PROPERTIES OF NON-METALLIC THIN FILMS 301

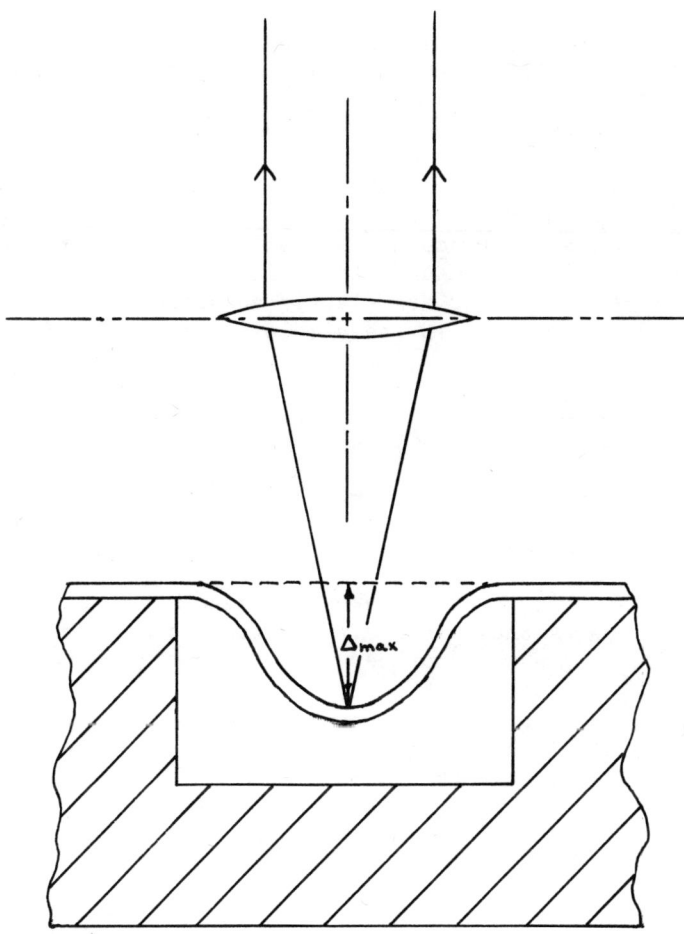

Fig. 15 Moat technique for local measurement. Reference[53].

deposits, even if thermal expansion coefficients are not matched. It is still true, however, that for measuring the stress in the early stages of growth or for materials which have a very low intrinsic stress, these problems have not been completely eliminated

X-ray Methods.

In order to follow the change in stresses through various processing steps, or in order to determine local changes in a stress as a function of position in the sample, x-ray techniques have been used. The more familiar one is to measure the lattice constant, usually with a diffractometer arrangement which determines the spacing between planes lying parallel to the plane surface. The strain in the film is most commonly measured but in a few cases the strain in the substrate has actually been used to infer the stress in the film. As usual, the stresses are calculated from the strain in the appropriate elastic constant, in this case the stress in the film plane in the simplest formulation is given by

$$\sigma_{11} = \frac{E_f}{2\nu_f} \left(\frac{a_o - a}{a_o}\right)_{33} \qquad (14)$$

The elastic strain normal to the film is determined from the difference between the measured latticed constant of the film and the bulk lattice parameter.

Sufficient care must be taken in these measurements that changes in line shapes and shifts arising from stacking faults or heterogeneous strains are not confused with the homogeneous component.

Recent progress in diffractometer methods follows. Zosi[56] measures the strain as the function of the angle from the normal to the film. Using only the position of the line centers, he finds a linear relation between the strain $\varepsilon_{\phi\psi}$ and $\sin 2\psi$. The stress in the ϕ direction is then determined from

$$\frac{d(\varepsilon_{\phi\psi})}{d(\sin 2\psi)} = \left(\frac{\nu_f + 1}{E_f}\right)\sigma_\phi \qquad (15)$$

for a biaxial stress. In this expression ψ is the polar angle and ϕ the azimuthal angle.

Bush and Read[57] simplify the expressions derived by Taylor[58] and also relate the film stress to normal and inclined diffraction angles upon specimen rotation. These treatments assumed no stress gradient normal to the film.

In order to directly measure the elastic strain, and thus eliminate the difficulty of using tabulated values for a_0 the film may be detached from the substrate thus relieving the elastic strain and the lattice constant measured a second time. Kamins and Meieran[59] have successively used this technique for the stress in silicon films on sapphire substrates, through various processing treatments. The Nelson-Riley extrapolation function has been used to determine the lattice constant. Microbeam techniques coupled with scanning motion should allow the determination of the local elastic strain with a reasonable resolution. No such experiments seem to have been done.

Cullity[60] formulates the more general problem to consider the importance of line shapes. Borie[61,62] considers the effect of a linear variation of strain through the thickness. McDowell and Pilkington[63] derive a five-parameter nonlinear distribution from the line shape which should be able to consider differences in structure at the free and substrate surface. In their results with gold films, immense elastic strains of 2.5% tension exponentially decaying within 100 Å of the glass substrate interface, slowly shift to compression in the body of the film and a large 2.5% compression near the free surface. The antisymmetric strain distributions in the thickness are a result of the symmetric diffraction peaks. but the authors offer no explanation for large strains and the lack of agreement with deflection experiments.

Within recent years an understanding of the topological x-ray contrast of imperfections in crystals has emerged through a sudy of both kinematical and dynamical effects.[64] As applied to stresses, contrast may be observed when a film deposited on a single crystal substrate provides a sufficient gradient to locally distort the crystal and give rise to intensity fluctuations. As previously decribed, Haruta and Spencer[24] used the technique to examine the stress distribution at the edge of a deposited film. Meieran and Blech[65,66] used Borrmann contrast to explain the diffraction contrast along interfaces. Schuttke and Howard[67] regard the model as oversimplified and analyze the transmission geometry using the scanning oscillator technique following the treatment of Penning and Polder.[68] Dynamical effects and the lattice curvature resulting near the edges of an etched diffusion window give rise to the observed contrast. The strain gradient is a maximum at the window edge. Reference to Fig. 16 indicates that when the radius of curvature and the diffraction vector $\vec{g}$ are pointing in the same direction enhanced blackening is found on the film. If the substrate is found to be in tension when the film must be in compression, as illustrated in the figure. This technique provides good spatial resolution for the sign of the stress but quantitative information as to the magnitude is lacking. The authors also give illustrations of kinematical contrast at the boundaries where the

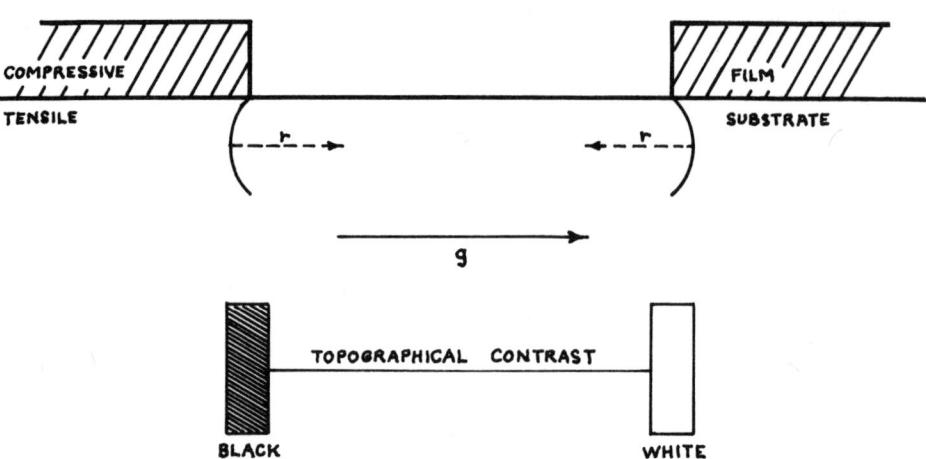

Fig. 16 X-ray topological contrast according to Schuttke and Howard.(67)

adhesion is lost. Dynamical images obtained by either adjusting the substrate thickness or choosing the appropriate radiation give rise to unambiguous interpretation of the regions that have failed at the interface. It is worthwhile pointing out that localized adhesion losses were determined by this technique when the optical appearance of the film was mirror like and without any observable flaws. The authors also pointed out that these techniques make it possible to map the stress buildup in planar semiconductor devices. It is then possible to obtain information about the relative stress magnitude in the stress direction after each processing step, and the stresses associated with the diffusion processes may be determined.

A x-ray double crystal arrangement in which the resultant topographs show the non-uniform variations of lattice orientation near near the surface has been developed by Zeyfang(69) for thin silicon crystals bonded to glass. The origin of the strain may be inferred from the distinction between a uniformly distorted surface or one which has non-uniform distortions.

A modified scanning x-ray topographic arrangement has been developed by Rozgonyi and Ciesielka(70) which uses a feedback system to maintain a wafer orientation while the beam traverses the specimen. The radius of curvature may be determined with a spatial resolution of mm with simultaneous x-ray topographs. Stresses in both very thin films ($\sim$100 Å) and substrate are calculated as well as effects of diffusion and ion implantation.

MECHANICAL PROPERTIES OF NON-METALLIC THIN FILMS

Because of the availability of x-ray topographical techniques for semiconductor defect studies, their increased application to stress systems caused by deposited films, diffusion, or implantation is expected. Both lattice parameter and topological methods are non-destructive, have good spatial resolution and sensitivity, but are not convenient nor rapid enough to be used during deposition.

Other Techniques

A new technique by EerNisse[71,72] has been developed for simultaneous measurement of stress and mass change. This double quartz crystal resonator technique uses AT cut and BT cut crystals such that the sum of the frequency shift is proportional to the mass change and the difference of the frequency shift is proportional to

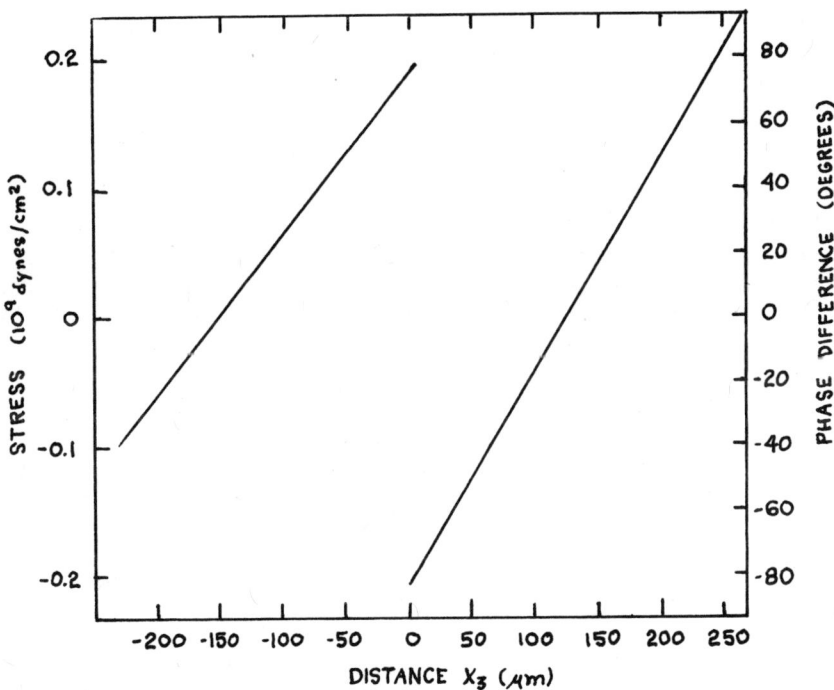

Fig. 17   Stress distribution from optical birefringence measurements of Reinhart and Logan[73] in GaAs.

the change in the thin film stress. The sensitivity is quoted at 125 dyne/cm for a 10 Hz frequency shift. The technique is suited best for stress studies when the mass change is quite small, and has been used in connection with implantation studies on silicon samples.

Any physical property may be used to determine the stress, and examples are beginning to appear. Note that this is the inverse of the usual care in which the 'unusual' properties of a film are being explained in terms of the stress, and represents are more sophisticated point of view. Oel and Frechette[22] have done this for macroscopic glass discs using birefringence to measure the strain. Reinhart and Logan[73] calculated the phase difference and elastic constants from the crystal piezo-optical tensor coefficients, and compare with data to determine the interface stress of thick epitaxial layer structures. Fig. 17 gives the stress distribution near the interface for $Al_{0.1}Ga_{0.9}As$ on (111) Ga As.

An increased use of such 'indirect' measures of the strain is anticipated.

4. THERMAL STRESSES IN THIN FILMS

Introduction

The techniques for measurements that we have discussed in the preceeding sections have allowed the determination of a strain and the calculation of an associated stress. We now turn to the questions of the origin and the various mechanisms that can contribute to these strains. As a matter of definition the total stress in the film is regarded as being made up of two possible contributions. These are defined as the differential thermal expansion stress whose origin is understood, and whose value is generally amienable to calculation, and, the intrinsic stress which is the manifestation of all other contributions. We may write $\sigma_T = \sigma_{DTE} + \sigma_i$, where the intrinsic stress, σ_i, may be composed of several terms. Clearly, σ_T is measured and a property responds to the total stress. The contributions from either term may be tension or compression, with the resultant total stress in the film being determined by the relative magnitudes.

In Fig. 18, we consider in an idealized way the case of a film deposited on a substrate at temperature T_s and cooled to the final temperature T_m. If the thermal expansion coefficient α_f of the film is larger than the expansion coefficient of the substrate α_s then the contribution to the stress from differential thermal expansion increases. On the other hand, the intrinsic stress is often a decreasing function of the temperature, so that the total stress in

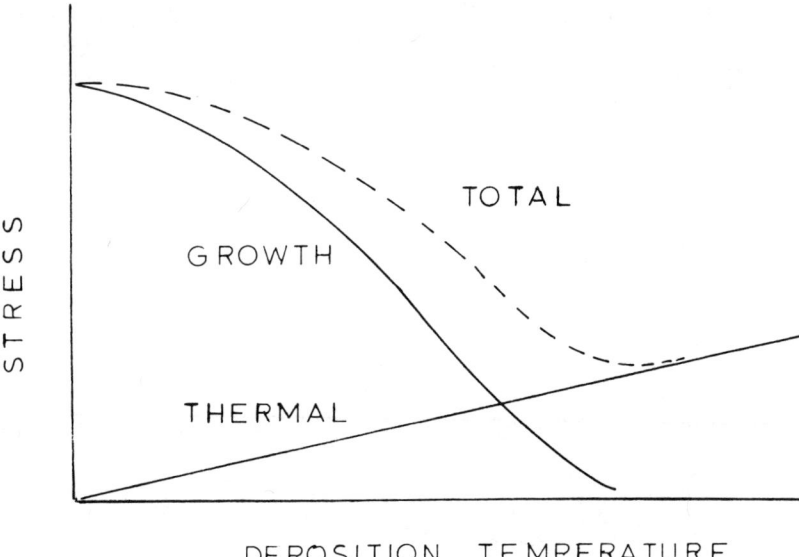

Fig. 18 Thermal and intrinsic stress contributions.

the film may, in fact, have a minimum at some intermediate temperature.

Even today the literature far too commonly considers differential expansion as the only contribution to the stress. For the higher melting point metals, for example, intrinsic stresses are quite large up to substrate deposition temperatures of several hundred degrees C. In other systems the diffusion of impurities into or out of the film at elevated temperatures give rise to substantial stress effects. Even in the case of thermally grown SiO_2 on Si at high temperatures, the intrinsic stresses have an important role.

On the basis of our earlier sections we see that the fundamental origin of all of the strains in films lies in the constraint that the film is being tightly bonded to the substrate and after this bonding has taken place there is a subsequent volume change in the film (or strictly a planar distortion). Thus, in addition to the same assumptions made in solving the elastic problem, we must evaluate the strain itself. Normally this is simplified by assuming the substrate is rigid and that neither contraction nor bending takes place in the film substrate. Under this assumption the change in lengths of the substrate resulting from the stresses on it is negligible, but if the film and substrate were of comparable thickness, then both of these relaxation effects must be taken into account.

Thermal Expansion Stress

Normally, the thermal expansion strain is considered to be isotropic. In general, the coefficient of thermal expansion relates the strain tensor to a small uniform temperature change. Thus, the thermal expansion tensor is symmetrical since the strain tensor is. (12) It is worth noting that since the thermal expansion of a crystal must possess the symmetry of the crystal, it cannot destroy any symmetry elements and this is why the class of a crystal does not depend on the temperature. The thermal expansion tensor may be referred to principal axes and thus reduce the number of coefficients to 3. For most substances the principal thermal expansion coefficients are all positive. But in a few crystals, for instance calcite, or silver iodide, some coefficients are negative.

We may write the nine thermal strain components in terms of the thermal expansion tensor.

$$\varepsilon_{ij} = \alpha_{ij}\Delta T \tag{16}$$

where the identical subscripts refer to the normal components and the mixed indices to the shears. The thermal expansion tensor is often reduced to a column matric to be consistently used with the six single index strains ε_i.

$$\varepsilon_i = \alpha_i \Delta T \tag{17}$$

If we consider the restrictions of crystal symmetry,[13] only in the triclinic and monoclinic systems are the shear coefficients other than zero. The orthorhombic system has three independent normal expansion coefficients, the tetragonal, trigonal, and hexagonal systems have two, and the cubic system has an isotropic thermal expansion. We must point out however, that even though the thermal expansion may be isotropic, because the elastic constants are fourth rank tensor properties, anisotropic stresses will be obtained even in a cubic crystal. Only if the sample may be considered to be elastically isotropic, that is in practice a polycrystalline sample with small grain size, will the resultant thermal expansion stresses be isotropic. Once the thermal strains are determined, the stresses are calculated from the appropriate elastic constants. The equation

$$\sigma_{DTE} = \frac{E_f}{1-\nu_f}\varepsilon_{11} = \frac{E_f}{1-\nu_f}(\alpha_f - \alpha_s)(T_s - T_m) \quad (18)$$

is appropriate for the rigid isotropic case and takes account of the biaxial stress generated by the linear strain $\varepsilon_{11} = \varepsilon_{22}$.

The sign is such that the tensile stress is positive. Obvious extensions are needed when the temperature range considered is so wide that the α's are no longer constant. Vook and Witt[14,74] carried out calculations of the strain normal to the plane of the film for arbitrary orientations of cubic crystals. In these calculations which were also extended to include the strain energy it was assumed that the shear strains in the film were zero. For copper and silver, the metals considered, the strain energies were found to be in the same relative ranking as Young's modulus. Vook and Witt point out that the observed line shifts in an x-ray diffractometer experiment when the temperature is changed correspond directly to the calculated thermal strain, even though other contribution to the line positions as a result of faulting may be present.

To this point we have considered the specimens to be semi-infinite in size and edge effects have not been important. Aleck[23] has converted the thermal stress problem in a rectangular plate clamped along an adge to one of boundary stress conditions. His two dimensional solutions indicate there are large shear stresses present in the film within about two film thicknesses of the edge, and a stress normal to the plane of the film concentrated at the intersection of the edge of the film with the film substrate interface. The distributions along the interface (clamped edge) are qualitatively similar to the data in Fig. 11 if normalized to the one-dimensional thermal stress $E \Delta\alpha \Delta T$. The large variation of the normal stress σ_{33} as a function of the distance from the free surface at the edge of the film is given in Fig. 19.

Zeyfang[75] has extended Aleck's approximate solution to plates of arbitrary lengths, rather than the semi-infinite plate used by Aleck. He also gives detailed stress, strain and displacement distributions for plates with length to thickness ratios of 2, 5, and 15. Compared to the semi-infinite case where there are edge effects only, when the length of the film becomes short enough, nonuniform stress distributions are found throughout the volume of the film. The maximum stress in the plane of the film decreases as the sample length decreases and actually reverses sign on the free surface of the film for samples that are only twice as long as their thickness. The maximum shear stress decrease with increasing plate lengths and the stress normal to the plane in the center of the interface increases with decreasing plate length. Zeyfang

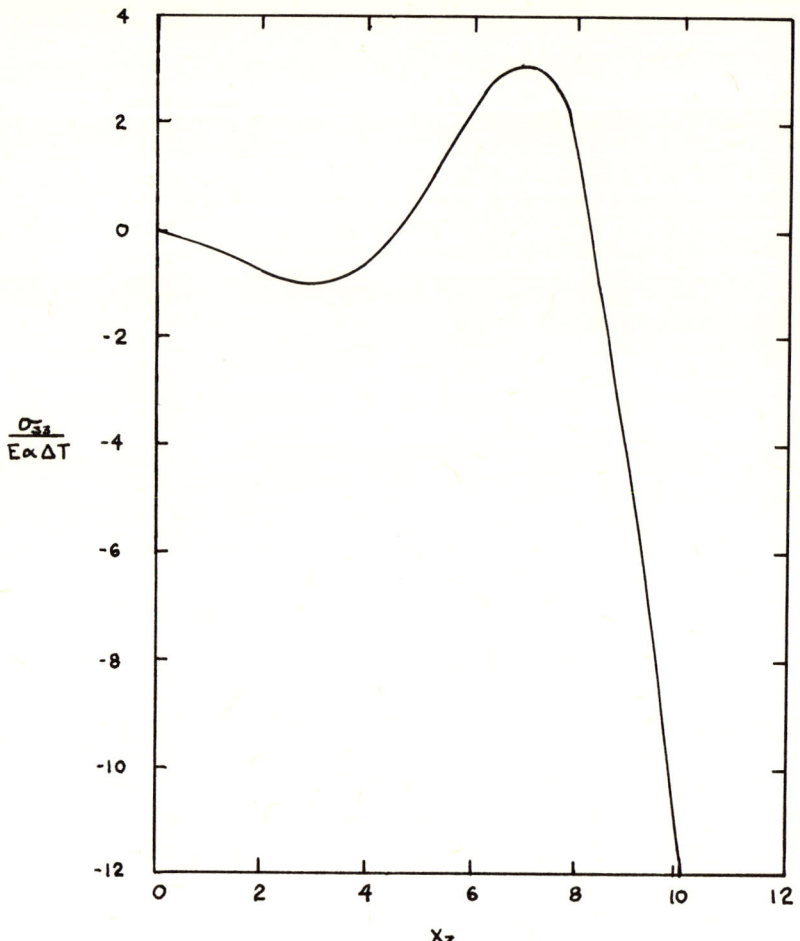

Fig. 19 Thermally generated normal stress at a free edge as a function of distance from the free surface. After Aleck.(23) Coordinate $X_3 = 10$ corresponds to the interface.

points out that for plates whose length to thickness ratio is greater than 15, it is possible to think of the distributions as being composed of a uniform part for the central part of the sample and edge terms, but such a treatment is not valid for shorter samples. The shape of a half film under the exaggerated thermal expansion condition is shown in Fig. 20. Because of the singularity of σ_{33} in linear elasticity theory at the interface edge, a cylindrical cavity was cut out and the displacements approximated by surface displacements. For such short samples it follows that the upper suface no longer remains plane under a thermal expansion strain.

In carrying out the calculations of thermal expansion contributions the questions arises as to the values of the thermal expansion coefficient to be used. There seems to be no evidence for abnormal expansion coefficients in deposited films. There are cases where the total stress effects do not seem to coincide with calculations of the thermal contributions, but these are attribu-

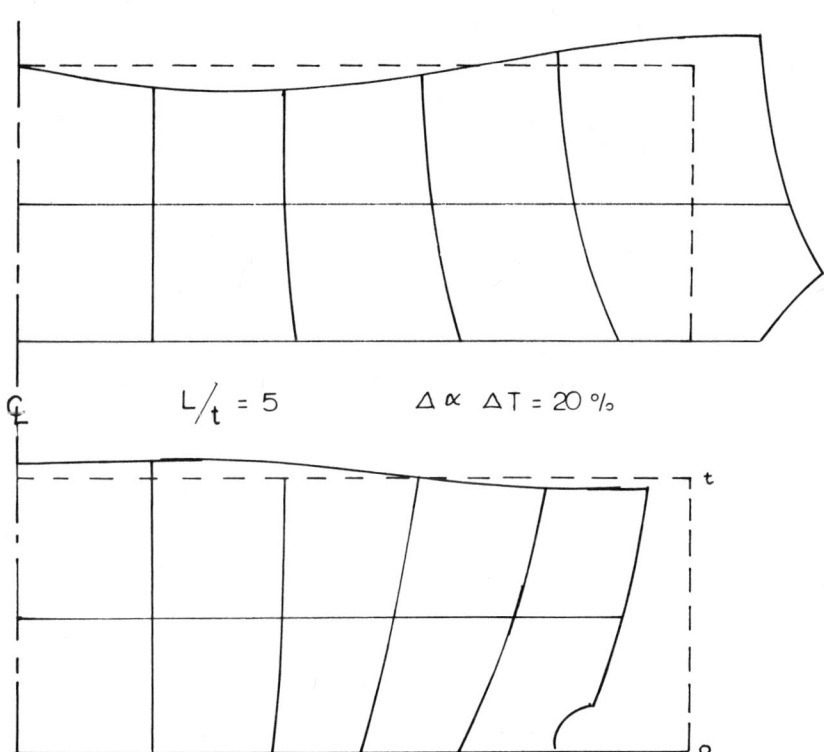

Fig. 20 Thermal expansion distortions of short films according to Zeyfang.[75]

ted to irreversible intrinsic stress changes in the film upon heating. Hence, tabulated values of the expansion coefficient seem quite suitable in practice. Feder and Light[76] have reported an apparatus for directly measuring the differential thermal expansion between a film deposited on the substrate by observing the optical fringe shifts corresponding to bending the sample. For the case Ge/Ga As film couple the difference between the thermal expansion coefficient is demonstrated to be less than 1 part in 10^8.

One of the goals of the calculation of thermal stresses is the prediction or explanation of properties from data on bulk, strain-free samples. Schlötterer[77] determined the resistivity change in epitaxial Si on spinel in terms of the piezo-resistance coefficients for isotropic thermal stress. A second example is the anisotropic Hall mobility found in silicon on saphire films by Hughes and Thorsen[78,79]. After calculating the thermal expansion stresses arising from the anisotropic thermal contraction of Al_2O_3 in cooling from the deposition temperature, they also use the piezoresistance effect to relate the Hall mobility. The approximate 10% anisotropy of the Hall mobility in two orthogonal directions seems adequately described in terms of thermal strains. Hughes[80] carried out detailed calculations of the effect of stress on resistivity for general orientations in several crystal systems.

Thus it appears as though both isotropic and anisotropic thermal expansion strains, and their effect on transport properties are amenable to calculation.

4. ORIGINS OF THE INTRINSIC STRESSES

The review articles by Hoffman[4,5,6] Campbell[6,7] Buckel,[8] and Kinosita,[10] all have the goal of trying to assimilate the massive experimental literature on measurements of the stress and attempting to interpret them in a few simple concepts. In addition to the thermal stress, Buckel lists six other processes which may produce stress in the following way.

1. Incorporation of atoms, for example, residual gases or chemical reactions.

2. Differences of the lattice spacing of monocrystal and substrates and the film during epitaxial growth.

3. Variation of the interatomic spacing with crystal size

4. Recrystallization processes

5. Microscopic voids and special arrangements of dislocations.

6. Phase transformations.

In addition, all of these processes may occur both during condensation and growth of the film and under annealing conditions afterwards. A search of the literature also indicates that most of these models have been developed to explain the stress in metallic films, pure metals or some simple alloys. Especially with nonmetallic films, where the stoichiometry and the structure is still far from being well established, the understanding of the origin of the stress is now unsatisfactory. For these reasons we review the models for the intrinsic stress very briefly, calling attention to those cases where there is some confirmation in the literature for non-metallic films. More importantly, the discussion may give a physical basis for the application of these concepts to non-metallic systems. Many of the models find it relatively easy to explain shrinkages and hence tensile stresses in the film, but can explain compressive behavior only with difficulty.

Several different kinds of processes may operate in the first classification. Nakajama and Kinosita[81] have measured the stress in silver films deposited at 10^{-3} and 10^{-6} Pascal and find the following behavior. The dependence upon the thickness is not greatly influenced by the deposition parameters before the film becomes continuous. The residual gas pressure during the deposition has no appreciable influence on the stress behavior. However, after the deposition the stress will remain at its constant tensile value if the sample is kept in a vacuum of 10^{-6} Pascal but if kept in a 10^{-3} Pascal surroundings, a compressive relaxation to a smaller value of tension in the film is found. Furthermore, this compressive contribution will recover with a time constant of the order of hours if the pressure in the system is reduced to 10^{-6} Pascal. The authors suggest the reason is the absorption of residual gas in the film, and the stress effects correlate with changes in the electrical resistivity.

Actually, the presence of residual gas, oxygen and water vapor in particular, has long been suspected in the case of those metallic films showing compression when deposited in poor vacuum, or when annealed in air after deposition. A more direct quantitative confirmation of the effect of gas incorporated in the film during the deposition process has been forthcoming in the work with bias sputtering which we shall treat in some detail in a later section. Ion implantation which clearly gives rise to dilation also falls in this category.

Because of their technological importance, the growth stresses in thermal and anodic oxide films must be included. A recent review by Stringer[82] considers the stress generation and relief in grown oxide films on metals. The sign of the stress in the oxide layer may be predicted by the ratio of the volume per metal ion in the oxide to the volume of metal atom in the metal or the Pilling-Bedworth ratio. Most grown oxide layers are compressive but MgO is correctly predicted to be tensile although actual stresses are much less than calculated by this ratio. These differences may arise from ion motion, particular anisotropic growths, or, perhaps more likely, plastic flow taking place during the oxidation process. A second mechanism includes the oxygen migration down grain boundaries in the metal oxide in order to react with metal ions diffusing through the bulk. This mechanism, illustrated in Fig. 21, also leads to a compressive growth stress in the oxide according to Rhines and Wolf.[83] The similar impurity migration down grain boundaries has been suggested as the origin of the compressive stress in metallic films.[32]. A similar situation

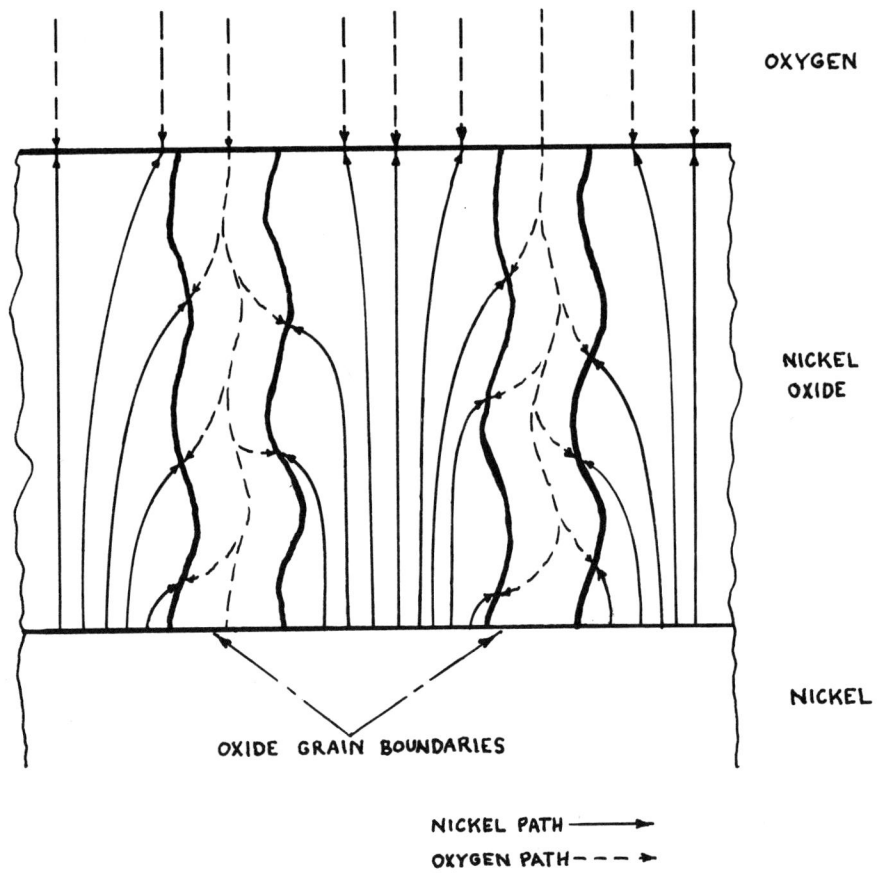

Fig. 21 Schematic diagram of diffusion paths. Following Rhines and Wolf[83] and Stringer.[82]

is found in the growth of anodic oxide films, which have recently been reviewed by Dell'Oca et al.,[84] following earlier articles by Young[85] Vermilyea.[86] The growth of inter-metallic phases following diffusion has just been confirmed as a stress mechanism. [137]

One of the best documented origins of internal stress occurs during epitaxial growth as a result of the efforts of Van der Merwe[87,88,89] Matthews,[90,91] Jesser and Kuhlmann-Wisdorf,[92,93] Honjo[94] and others. An interfacial energy,[92,93] is calculated which depends on the differences of the lattice spacing of the substrate and film, as well as the elastic constants of each material. For small mismatches or thin films, the total energy in the system will be minimum if the film develops a uniform elastic strain. However, if the film becomes thicker or if the mismatch is more severe, the energy of the system may be lowered by the creation of an array of interfacial dislocations which may relieve a large fraction of the strain in the film. A total or partial elimination of misfit dislocations may take place as schematically indicated in Fig. 22, where the energy gain is ΔE and the critical misfit is seen. It is worth noting that forces normal to the interface contribute only about 10% of the interfacial energy. Recently the misfit dislocation energy is calculated[88] from a periodic interfacial potential which is a generalization of a Peirls-Nabarro model rather than the older pair-wise or elastic models.

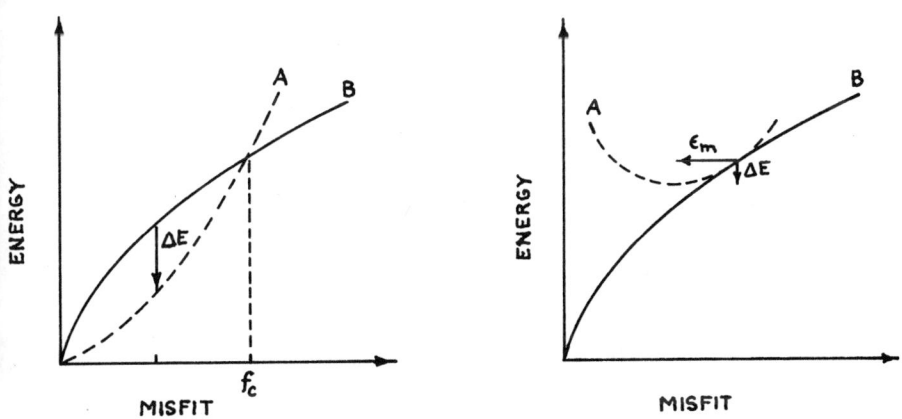

Fig. 22 Energy gain E arising from total or partial elimination of interfacial dislocations. Curve A is the homogeneous elastic strain energy, and B the misfit interfacial energy. After Van der Merwe.[87]

Jesser and Kuhlmann-Wilsdorf[92,93] have shown that a similar formulation holds for films in the isolated island stage of growth. The interfacial dislocation mechanism for the origin of intrinsic stress is attractive because it may be formulated quantitatively and compared with experiment. As a general mechanism, however, it must be criticized on two grounds. First, the obvious one that it is not easy to see how such a mechanism would operate for growth of polycrystalline films and, secondly, in this model the strain is localized and decays exponentially from the interface and thus does not project the constant stress often experimentally found. Nevertheless data for 50 Å polycrystalline alkali halide films on glass seen to follow this model.[95]

A stress model for heteroepitaxial films combining thermal stress and assuming the film lattice constant is constrained to match the substrate has been developed by Besser et al.[96] They have examined many CVD magnetic oxides and developed a physical model in which the film stress is entirely determined by uniform misfit at room temperature independent of the deposition temperature up to a critical value and entirely by thermal expansion for larger misfits. This idealized model works well when the thermal expansion

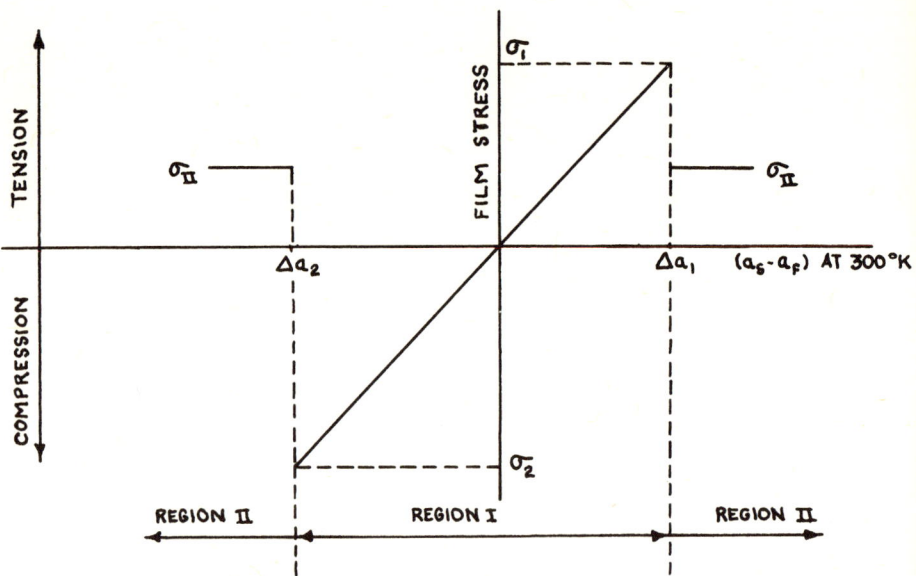

Fig.23. Epitaxial misfit-thermal strain model of Besser et Al.[96]

coefficients of film and substrate are nearly the same. It takes account of the thermal expansion in region I by using the room temperature lattice constants but assumes the complete relaxation of the mismatch stress in region II.

Fig. 23 indicates these two regions, where

$$\sigma_I = \frac{E_f}{1-\nu_f} \frac{a_s - a_f}{d_f} \quad \text{and} \quad \sigma_{II} = \frac{E_f}{1-\nu_f} (\alpha_f - \alpha_s) \Delta T$$

Carruthers[97] proposed a similar model including the effects of plastic deformation occuring at strains larger than 0.005. He fits the thermally-activated glide relation derived by Matthews et al.[91] in which the misfit strain is the driving force to the data of Besser et al. Carruthers' expression for the planar strain is

$$\varepsilon_{11} = \left(\frac{a_s^{T_m} - a_f^{T_m}}{a_f^{T_m}} \right)(1-\xi) + (\alpha_f - \alpha_s)(T_s - T_m) \tag{12}$$

where

$$\xi = \left| \frac{a_f' - a_s}{a_f - a_s} \right|_{T=T_s} \qquad \xi \text{ is}$$

the fractional strain recovery, one finds a number of experiments relating the lattice constant a_f' and the lattice parameter of the film material a_f both evaluated at the deposition temperature T_s, Braginski et al.,[98] gives additional data for the magnetic oxides.

In the third category, one finds a number of experiments relating the lattice constants with the crystallite size and interpreting the rsults in terms of a surface energy. Cabrera[99] formulated the thermodynamics following the classical paper by Herring.[100] Lindford and Mitchell [101] introduce the concept of interplanar potentials which relate the surface work to macroscopic parameters of elastic constants and thermal expansion coefficients. Wasserman and Vermaak[102] give recent data and references for the surface stress as observed by electron diffraction. The surface energy provides a recognized contribution generally giving a compressive stress in the isolated island stage of growth, [99] however, unless the growth of a deposit can be formulated in terms of a continual nucleation and growth framework it is difficult to see how this category can contribute to the stresses in fhick films. It remains then a difficult experimental area because of the vacuum environment, but in need of future work.

The surface stresses in Ge, InSb and GaSb have been measured using the bending of thin sample by Taloni and Haneman.(43). The effects of surface stress as correlated with bending plate measurements have been discussed by Hoffman.

Classification 4, recrystallization processes, contains a number of different models that differ in their details. Klokholm and Berry,(105) suggest that the stresses are generated "by the annealing and constrained stinkage of disoredered material buried behind the advancing surface of the growing film". The magnitude of the stress is determined in this model by the amount of disorder at the time the atom become constrained and the subsequent annealing during condensation process. The temperature dependence of the intrinsic stress is determined by the kinetics and it is held that no stress would be found in films deposited at extremely low temperature substrates, and at substrate temperatures during deposition higher than about 1/4 of the melting temperature. This form is approximately followed for metals. In this model, films deposited on cold substrates would have no stress since the atomic rearrangement is no longer possible.

The helium temperature experiments of Buckel [8] indicate that simple metals growing in the normal crystalline phase, even though they have very small crystals and have a high degree of disorder, to develop large stresses when deposited at these very low temperatures. It is often stated that amorphous films have low stress. Gallium and bismuth may be frozen in amorphous phases if the substrate temperature is below about 20 K. These data as reproduced in Fig. 24, show that for low substrate temperatures a small stress is developed when the islands join at an average thickness of about 60 Å but no additional stress is produced on further growth. For higher substrate temperatures the usual tensile stress is found. The amorphous Bi films are quite unstable, and with increasing film thickness a spontaneous crystallization often takes place which yields large compressive stresses. These experiments, as well as calculations, convincingly exclude the presumption that the surface temperature of the film is substantiantly higher than the substrate temperature. Because of the present importance of amorphous films, and the general behaviour that even alloys seen to grow without intrinsic stress,(106) we report Buckel's result that Sb films condensed on low temperature substrates exhibits rather high tensile stresses. Because the Sb vapor from which the film is built up consists almost completely of Sb_4 molecules, Buckel has reformulated the general hypothesis about amorphous films to read "amorphous phases with a short range order similar to that of a liquid (frozen liquids) grow without internal stress", and interprets the statement on a microscopic void model.

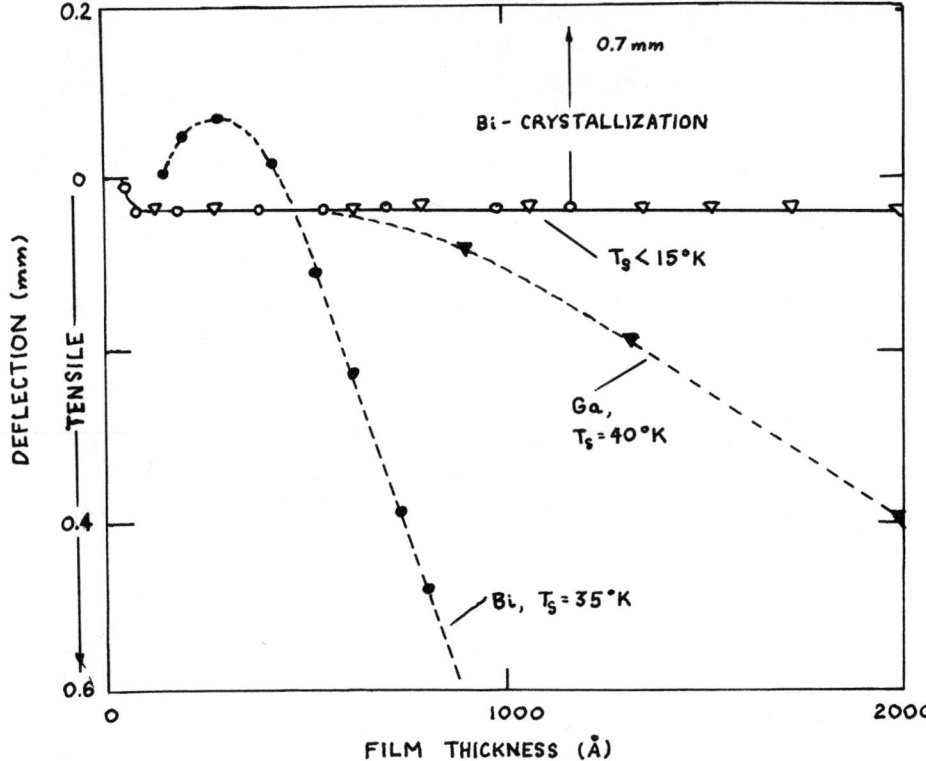

Fig. 24. Stresses in low-temperature depositions as measured by Buckel[8].

Chaudhari[107] has considered the effect of grain growth on the stresses in films when densification occurs by the elimination of a grain boundary. If the initial grain size is very small a tensile stress is generated and the final grain size is determined by the minimization of the sum of the strain energy and surface energy. When the initial grain size is above this critical size, perhaps 20 Å, no energy minimum is found and grain growth is not restricted. In this model an initial elastic compressive strain aids recrystallisation, whereas the experimentally found initial tensile strain is not favorable for grain growth.

Wilcock et al.,(108) have determined the stress in the early growth stage for Ag and Au. They also consider the stress to arise from the elimination of grain boundaries and suggest only 10% of its boundary need be eliminated to account for the observed stress.

In a series of papers Hoffman(30,32,40) and his coworkers have developed a different concept of a grain boundary model. In this model this strain is generated as the adjacent surfaces of two grains come into contact during growth. The problem may be formulated in terms of a grain boundary potential but this is formally equivalent to relating the strain energy to the difference between the surface energy of the two crystallites and the energy of the resultant grain boundary. In addition to values of elastic constants and energies this model needs only a value of the final grain size of the film for a quantitative calculation.

The grain boundary potential is demonstrated in Fig. 25. The depth of the potential is given by $2\gamma_{sv}-\gamma_{gb}$, where γ_{sv} is the

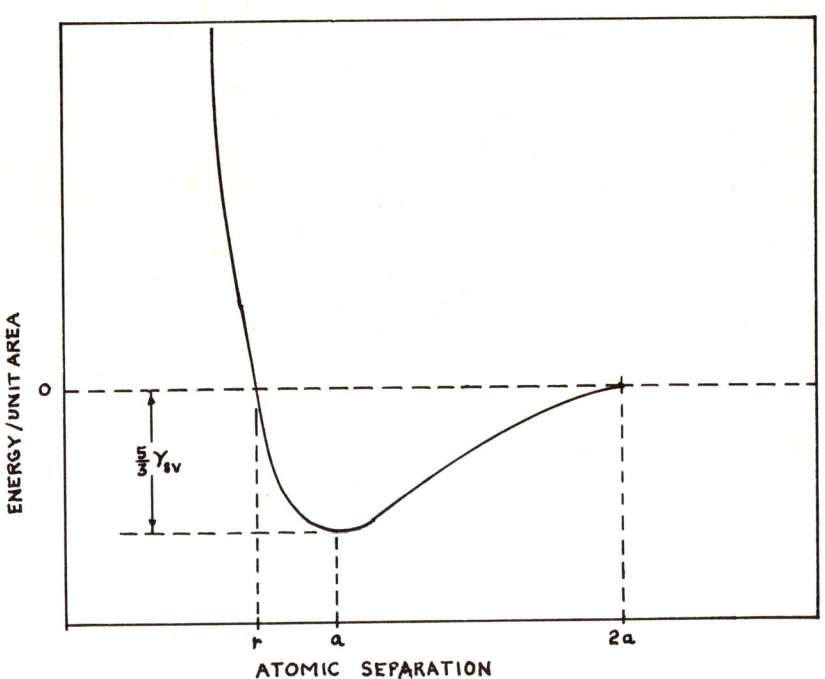

Fig. 25 Grain boundary potential. After Doljack, Springer, and Hoffman.

surface free energy and γ_{gb} the grain boundary energy. As an average grain boundary has an energy $\sim \gamma_{sv}/3$, the depth of the potential is $5/3\gamma_{sv}$ at the nearest neighbor distance a. An arriving atom may populate any position where the potential is negative if it lies between r and a then the atom relaxes outwards ; if it arrives at a position between a and 2a, then a contraction will take place at the boundary. In both cases a strain energy is produced, and the minimum produced between the strain energy and potential energy defines the net strain. As the potential is asymmetric a tension in the film is produced of

$$<\sigma> = \frac{E_f}{1-\nu_f} \frac{<\Delta>}{d}$$

where the average displacement $<\Delta>$ is obtained by integrating over the range of the potential. Good agreement has been obtained in both the temperature dependence and the magnitude for the case of nickel.[40]

Classification 5, voids and special dislocation arrays, becomes a convenient catch-all. Voids in films have been postulated to explain magnetic anisotropy and low values of density,[109,110] even though the x-ray density is typical of bulk material. With dielectric films evaporated at angles of incidence, voids are popular in order to explain the so-called porosity of the film. Nevertheless, detailed experiments to show the size and shape of such voids have usually failed. The same kind of general remark may be made about dislocations because there is a well-defined stress field arond a given dislocation. We may postulate a special dislocation array in order to give rise to any stress distribution that we desire. However, with the exception of the interfacial dislocations mentioned earlier, the dislocation lines found in films by electron diffraction are generally normal to the plane of the film. Thus, generalized statements relative to voids and dislocations, are generally of little use in detailed analysis of the stress problem. Exceptions to this may be found in the case of stacking faults, and other such documented defects in a particular film in question. Abrahams et al.,[111] have proposed an oriented array of inclined dislocations to explain the stress gradient that curls a film upon removal from the substrate. Saito et al.,[112] have postulated a periodic distribution of screw and edge dislocations obliquely piercing the film plus added terms to have a traction-free surface. The maximum stress does not occur at the boundary of the dislocation.

Category 6, contains the expected first-order phase transformation with its atteded volume change that must take place after the film is formed for a stress to be generated. Buckel[8] using the Oswald step rule of crystal growth, suggests the film may pass through metastable liquid phase during condensation, and the internal stresses thus arise from a different in density from the liquid itself. Thus, one would expect that a frozen in liquid as the first phase formed exhibits no internal stress, consistent with many results already presented for amorphous metals. As most metals melt with a decreased density, normally a tensile stress will be observed, and this is also experimentally found. Amorphous alloy films also show a very low stress.[106] This model has had some success in predicting the sign of the stress but as yet has not been formulated in a quantitative fashion.

Without specifically stating, our discussion so far, has led one to believe that the intrinsic stresses would be isotropic in the plane of the film for randomly oriented polycrystalline samples where the elastic constants are isotropic. The experience with metals has indicated that this is true when the films are deposited at normal incidence. Stress differences of perhaps 20% are found for metals when the evaporated beam is incident at 45° with respect to the substrate normal.[113] These early experiments with magnetic metals were concerned with the anisotropies resulting from the anisotropic stress. Smith et al.,[110] suggested that the anisotropy arises during the film growth from a random nucleation in combination with some shadowing effects. This mechanism very likely operates for specimens deposited near grazing incidence but Hoffman and coworkers suggested that anisotropic nucleation and growth would allow the grain boundary mechanism to operate at smaller angles of incidence. Dielectric films have much larger anisotropy effects as we shall see in a following section.

V.   SUMMARY OF THE DATA

Introduction

In view of the fact that the importance of the stresses has been known for such a long time, it is surprising that so few papers giving data exist in the literature. Even tody the pioneering work of Turner and Truby,[114] Heavens,[115] and Smith, Blackburn, Campbell,[116] Ennos,[42] Carpenter and Campbell,[95] Kinosita et al.,[117] and Scheuerman[10] contain perhaps 3/4 of the films studied if we exclude the epitaxial situation. In examining even the recent papers it is still true that improper expressions have been used to relate the stress to the observed parameters, and sufficient information is seldom given to separate the thermal stress from the intrinsic stress. As it is the total stress which affects a given property, this practice may be excused if one is interested in a particular processing treatment, but certainly an understanding of the origin of the intrinsic stresses is not aided by such experiment

In order to focus our attention let us classify materials as optical films, alkali halides, epitaxial films, and films which perform a passivation, isolation or dielectric function. It will be realized, of course, this is not an all inclusive way of classifying non-metallic films, and furthermore, a number of redunancies will occur when we look at an actual material. The purpose in this classification is to see if there are generalizations we can make within each category which prove useful in trying to interpret the data. Table II contains an updated list following the earlier tabulations of Hoffman,[4] Campbell, and Scheuerman.[10] When only one author made measurements for a particular material, it was simple to quote a value for the stress. Now that several authors have reported for the same material, but often carrying out the deposition at different temperatures or different techniques, there may be considerable disagreement as to the magnitude of the stress and, in some cases, even the sign. Thus, the individual papers must be consulted to see if the conditions of the reader's interest match with those of the literature. We wish to again caution that the intrinsic stress is indeed a structure sensitive property, and seldom is the complete characterization of the film carried out with sufficient detail that such a cause and effect relatioship can be found. It is well known that there are significant effects from the ambient atmosphere when a film is taken from vacuum surroundings and that there are longer time effects associated with diffusion of impurities along grain boundaries or through the lattice. Although a few recrystallization studies have been reported, in general stress data is not available for post-deposition treatments.

We have also eliminated consideration in Table II stresses induced in materials as the result of ion implantation as these are not inherently a thin film effect, although measured by the same techniques. These is no question that the incorporation of impurity atoms during sputtering or ion beam deposition can lead to the same kind of behavior, and some of these data are included and will be discussed. We have also eliminated from consideration the growth stresses in what are commonly called "oxide scales on metals". These predominatly thermally-grown oxides, a subfield in themselves, have been reviewed in considerable detail by Stringer. (82) Likewise we have excluded anodic oxides.

After this rather lengthy introduction we turn now to the data for optical films. A general review of optical films by Ritter[103] will appear soon and laser window coatings have been reviewed by Young.[104] The average stress is generally displayed as a function of thickness, in some cases the force per unit width or its derivative is plotted. The average stress usually increases rapidly, reaching a maximum value at a thickness of a few hundred angstroms, and then becomes a slowly decreasing function of the thick-

TABLE II

Index to Stress Literature

Film	Substrate	Reference
AgCl	Glass	114
AgF	Glass	114
AgI	Glass	114
Alph*	Glass	116
AlF$_3$	Glass	114
Al$_2$O$_3$	Al	159
	Si	54, 163, 164
	Silica	10
B$_2$O$_3$	Glass	116
BaF$_2$	Glass	114
BaO	Glass	116
C	Glass	131
CaF$_2$	Glass	114, 116
	Mica	115
	Silica	42
CdS	Glass	116
CdTe	Mica	165
	Silica	10, 42
CeF$_3$	Glass	116
	Silica	42
Ce$_2$O$_3$	Glass	114
	Silica	42
Chiolite	Glass	114
	Silica	42
Cryolite	Glass	114, 115, 132
	Silica	42
CuI	Glass	114, 157
Fe$_3$Si	Fe	162
Ge	Silica	10, 42
	Glass	56
	Mica	166
KBr	Glass	95

*Ph = phtalocyanine

TABLE II (con't)

Film	Substrate	Reference
KCl	Glass	95
KF	Glass	95
KI	Glass	95
LiF	Carbon on glass	134
	Cellulose	135
	Glass	95, 114, 116, 131, 132, 134
	Mica	115
MgF_2	Glass	114, 116, 132, 136, 143
	Mica	115, 117, 137
	Silica	42, 151
MgO	Silica	10
MgPh*	Glass	116
MoO_3	Glass	114
NaBr	Glass	95
NaCl	Glass	95
NaF	Glass	95, 132
$PbCl_2$	Glass	114, 116
	Silica	42
PbF_2	Glass	116
PbTe	Glass	138
	Mica	138
RbI	Glass	95
Sb_2O_3	Glass	114
Sb_2S_3	Glass	114
Si	Sapphire	59, 78, 123, 124
	Quartz	71
	Si	59
	Spinel	33, 77
Si_3N_4	Si	48, 128, 129, 158
SiO	Glass	116, 122, 134, 139, 140, 141, 142, 144, 145
	Nickel	146
	Silica	10, 42
	Al	161
Glass	W	133

TABLE II (con't)

Film	Substrate	Reference
SiO_2(CVD)	Si	67, 148, 160
SiO_2(Sputtered)	Si	55, 67, 148, 149, 164
SiO_2(Reactive Evap)	Si	153
	Silica	10, 152
SiO_2(Thermal Decomposition)	Si	149
	Ge	149
SiO_2(Thermal Oxide)	Si	10, 51, 52, 53, 54, 67, 127, 148, 149
SiO_xN_y	Glass	153
	Si	130, 158
SnO_2	Glass	114
$SrSO_4$	Glass	114
TaO_x	Ta	154
Ta_2O_5	Ta	155, 156
	Glass	156
Te	Silica	10
TiO_2	Glass	153
TlCe	Silica	42
TlI	Silica	42
ThF_4	Silica	10, 42, 152
($ThOF_2$)	Silica	147
ThO_2	Silica	10
ZnS	Glass	114, 116
	Mica	115, 117
	Silica	10, 42, 147, 152
ZrO_2	Silica	10

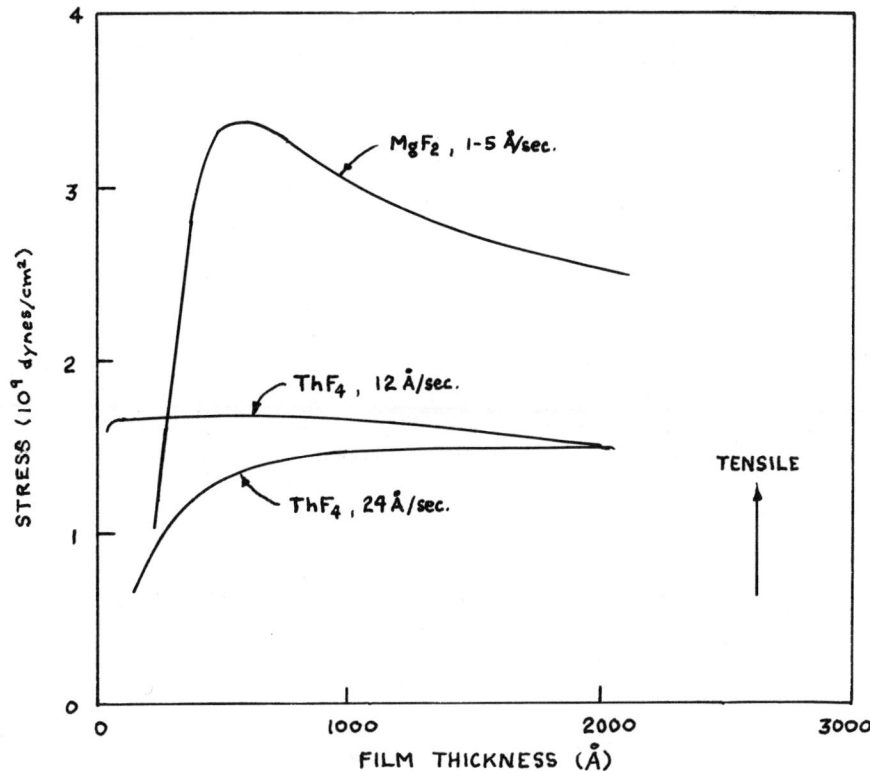

Fig. 26 Representative data for the stress in optical films. From Ennos.(42)

ness as indicated in Fig. 26 for several materials. For films deposited in a liquid-nitrogen trapped diffusion pump system at pressures of about 10^{-4} Pascal, Ennos(42) finds a significant increase in tensile stress with a time constant of about 100 seconds after the deposition is stopped. This transient very likely results from thermal relaxation of the substrate although the author experimentally found no initial transient. Significant changes are also found when the film is exposed to air over a period of perhaps 30 minutes.

Of the common optical materials only ZnS is deposited under compression, although Kinosita et al.,(117) indicate that a tensile stress may be found for low deposition rates. Fig. 27 summarizes some of these results for ZnS and it is interesting to point out that for the same deposition rate, data from different laboratories

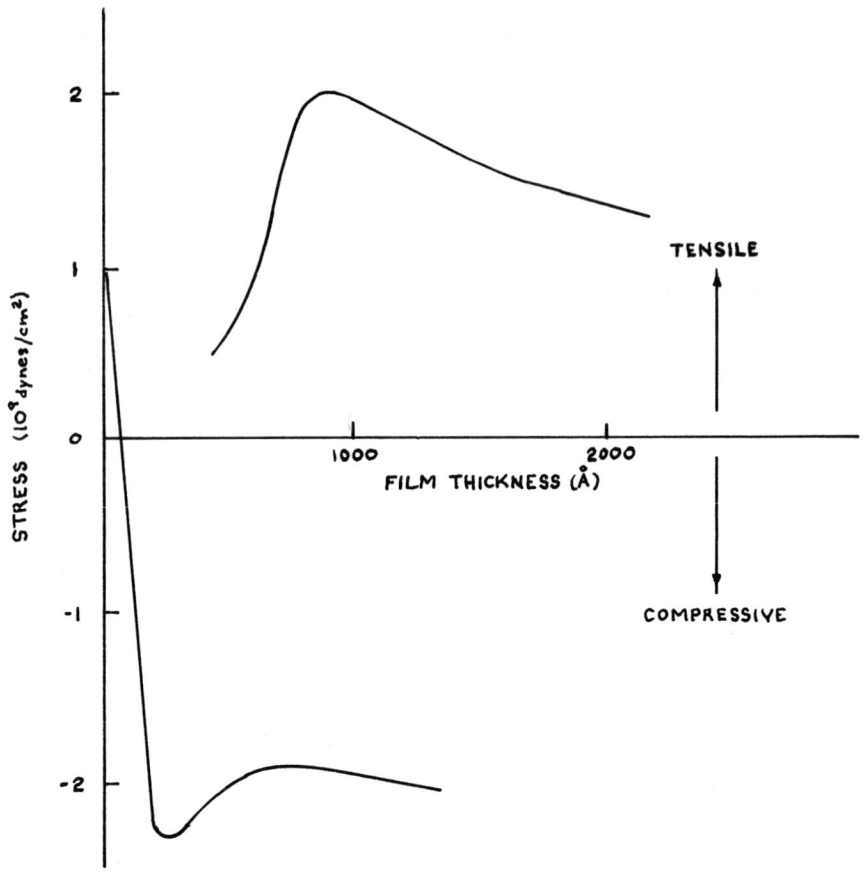

Fig. 27 Stress in ZnS films. Tensile data from Kinosita et al. (117) Compression from Ennos.(42)

can be obtained which are most a negative of each other. Blackburn and Campbell[46] have found a linear increase in the compressive stress in ZnS films with the increasing rate of deposition in the range of 5 to 30 Å/sec. The intrinsic stress in ZnS is also a rapidly decreasing function as a substrate temperature is raised from 80 to 150°C.

These differences point out the subtleties in the dependence of the stress on the deposition conditions. It is known[118] that the evaporation geometry, especially as related to the self-gettering is important for determining the morphology. ED, SIMS, index of refraction, and density measurements have recently been made for ZnS by Preisinger and Pulker.(119) The columnar crystallites have a diameter which slowly increases with thickness above 500 Å. The non-stoichiometric excess of S was found to be due to S_2^- bound to the surfaces of the crystallite columns.

Pulker and Jung(120) have suggested a cylindrical column model for the structure of ZnS and other optical films based on electron microscopy and water vapor absorption data. The observations suggest that an explanation for the stress might be sought in the impurity-grain boundary model.

Hill and G. R. Hoffman(122) have made an extensive evaluation of silicon monoxide. In the first place, they find a linear force thickness curve for constant evaporation conditions, suggesting that some of the behavior for thin films less than 1000 Å thick may result in the difficulty in establishing steady-state conditions.

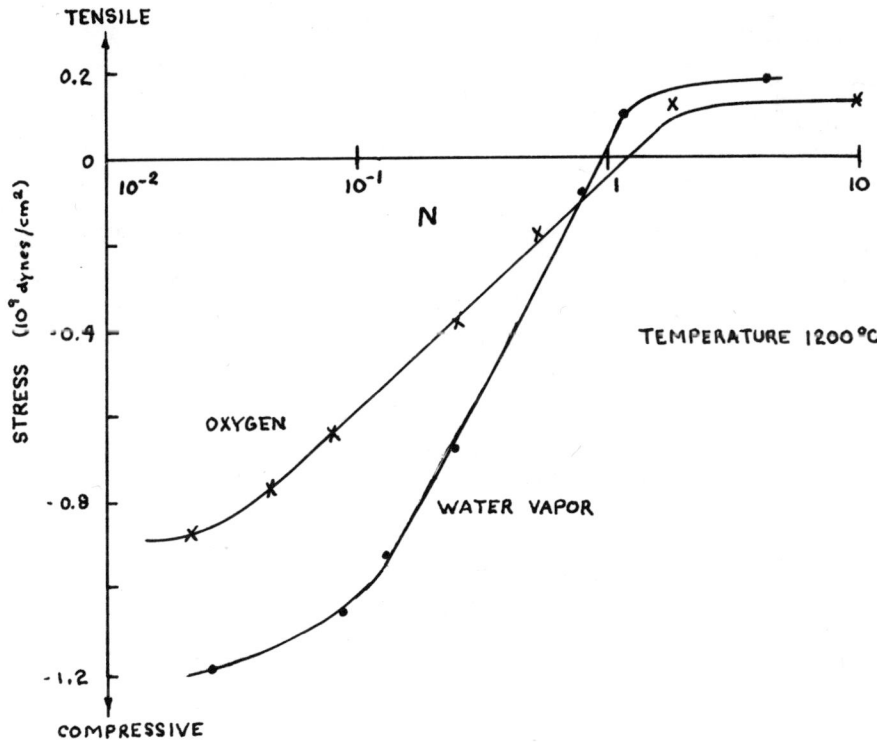

Fig. 28 Stress in SiO films as a function of SiO to O arrival Rate. After Hill and Hoffman.(122)

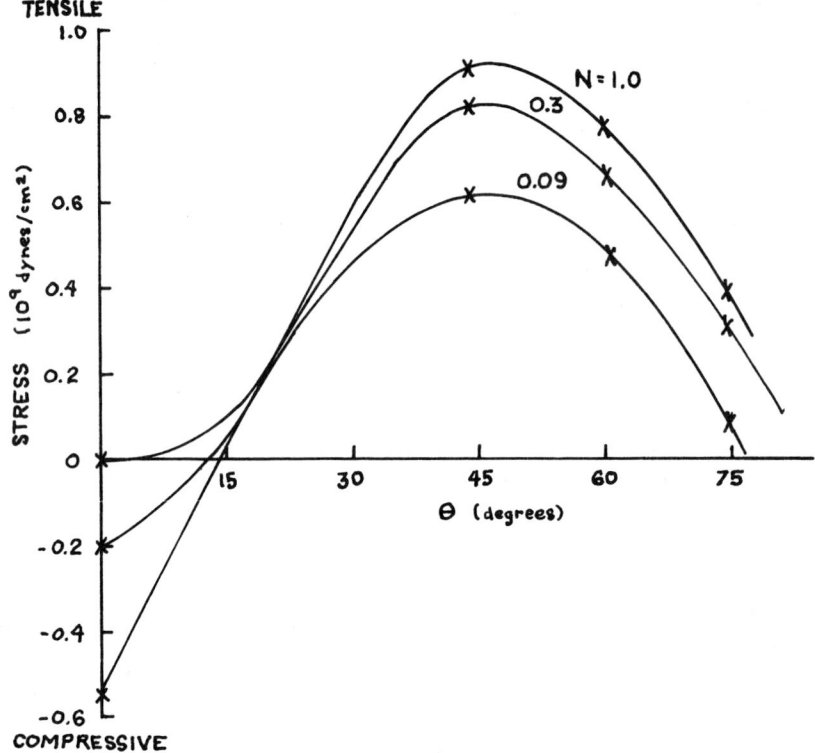

Fig. 29 Angle of incidence effects in SiO. After Hill and Hoffman(122)

Furthermore, they establish that the stress is a sensitive function of the residual gasses when the arriving silicon monoxide flus is comparable in magnitude. That the stress was indeed a function of the atmosphere during deposition, was strikingly demonstrated by changing the atmosphere during deposition with the resultant change in the sign of the stress in the material being deposited. Their data is summarized in Fig. 28 where the stress is plotted against the logarithm of the ratio, N, of the arrival rate of silicon monoxide to residual gas molecules at the substrate. Dielectric constant measurements suggest that the tensile stress occuring when the ratio N > 1 may be thought of as the intrinsic stress of silicon monoxide, whereas increasing the pressure of oxygen results in a silicon dioxide deposit. The earlier source temperature dependence found by Novice and Priest and his co-workers may now be understood in terms of the ratio N. Furthermore, these results confirm the earlier rate and residual gas pressure measurements of Blackburn and Campbell.(116)

Hill and Hoffman also report that the film stress remains stable as long as the speciment is under vacuum, but on the readmission of atmosphere the stress becomes more compressive, independent of its initial value. The increase in compression upon exposure to ambient may be several times the initial compressive stress, and to explain this fact they suggest that such films have a low density which makes them more susceptible to oxidation.

Hill and Hoffman have investigated angle of incidence effects where there is a dramatic increase in the tendency to form films in tension near 45° incidence even under oxidizing conditions (Fig.29). At the same time and enhanced compressive change results upon exposure to the atmosphere. These results are in general agreement with the earlier ones of Priest et al.[121] in which the stress anisotropy was measured using two orthogonal substrates, and again emphasize the sizeable morphological changes at modest angles of incidence.

Hodgkinson and Walker[210] have irradiated SiO films with ultraviolet and found that prolonged irradiation in vacuum will change the initial compression to tension, but the stress reverts back to compression when air is admitted. Films irradiated in air also had a substantial reduction in the compressive stress, and related changes in retractive index and UV optical absorption. The stress changes under irradiation are associated with the rearrangement of oxygen already gettered during the deposition with a resultant tensile contribution which partly counteracts the existing compression.

Although the data for other optical film is less complete, we suggest that the purer the film the greater the tendency for tension, and oxygen or other impurities tend to give compressive contributions.

The intrinsic stress in alkali halide films has been investigated by Carpenter and Campbell.[95] Data in the form of force and average stress as a function of thickness are presented for many of the alkali halides. Lithium salts show tension and potassium salts compression throughout the thickness range covered. Sodium salts have a large compression below a thickness of 1000 Å and a tension above. The common features exhibited by all the curves are a stress maximum which is usually compressive and occurs at a thickness of about 250 Å and a constant stress region for thicknesses greater than about 1000 Å, with a relative insensitivity to deposition rate. LiF has been studied by a number of investigators as summarized in Table II. Carpenter and Campbell point out that there appear to be two different stress producing mechanisms and suggest a misfit model for the first stage, even though the substrates were soda-lime glass. The average stress as a function of lattice constant as plotted in Fig. 30 goes through zero, at the oxygen nearest-neighbor distance in the glass. Although surface tension contributions may contribute to the stress in this isolated island stage of growth it is not possible to explain the systematic behavior shown in the figure on this basis. No satisfactory explanation has been made for the intrinsic stress in thicker films.

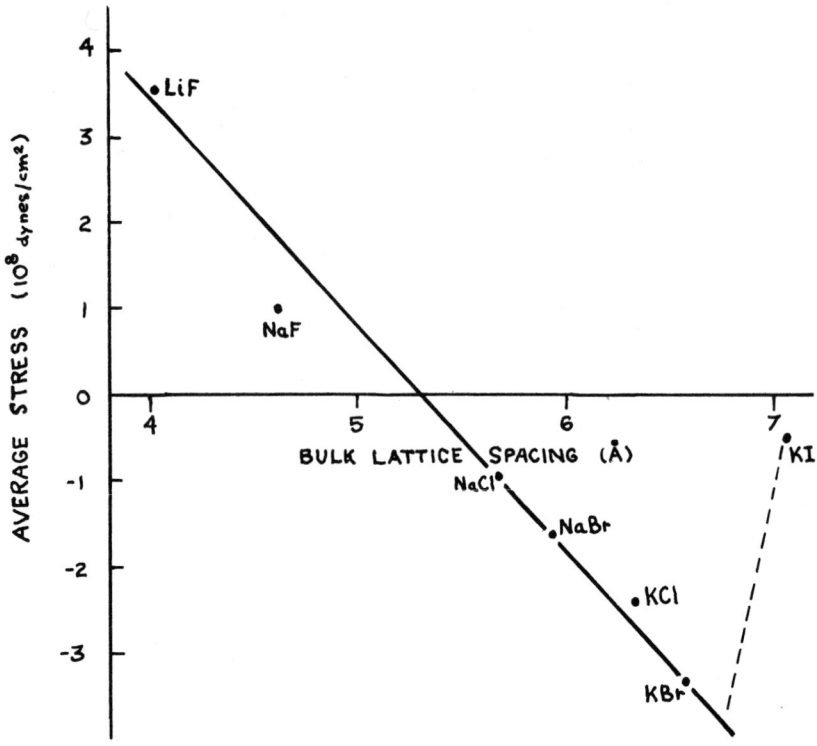

Fig. 30 Average stress in very thin alkali halide films correlates with epitaxial effect on glass substrates. After Carpenter and Campbell.(95)

Rather than to try to review the entire literature relative to epitaxial films, we consider a few illustrative examples. When the film is very thin or consists of small nuclei, a uniform elastic strain is predicted and has been observed for metals. The introduction of misfit dislocations to reduce the total energy localizes the strain at the interface. This problem has been treated in various formulations by Van der Merwe and his colleagues and discussed earlier. However, the fact that many epitaxial films are deposited at high temperatures and then observed at room temperature means that the thermal stresses are also often very large even though care is usually taken to minimize the difference in expansion coefficients. Thus, although the stress gradient near the interface may be especially important in determining properties which are sen-

sitive to that region it is often not the major contribution to
the total stress in a thicker film. One should really ask the question, "are there any data for the stress in epitaxially grown films
that require the presence of a volume intrinsic stress"? This is
a difficult experimental question to answer. First of all, the intrinsic stress is generally a decreasing function of the substrate
temperature and is likely to be very small at elevated temperatures.
And, secondly, relaxation processes undoubtedly operate which may
prevent an intrinsic stress, if generated, from being observed.
Dumin,[123] in the case of silicon on sapphire, was not able to
obtain the necessary precision. Other workers,[124,59,78] have
also found large compressive stresses ($\sim 10^{10}$ dynes/cm^2) for the
Si/Al$_2$O$_3$ system. Kamins and Meieran find if the epitaxial films
are electrolytically stripped from the substrates the constraint
is lifted and the strain is reduced to a level equal to the resolution limit of the diffractometer technique. In very case, thermal
expansion is sufficient to account for the compression. Dumin does
not mention any stress anisotropy, Kamins and Meieran, using a
diffractometer technique, mention but cannot measure an anisotropy,
and Hughes and Thorsen calculate the stress anisotropy resulting
from the anisotropic thermal expansion and confirm it by mobility
measurements.

In the case of silicon on spinel, larger compressive stresses
were observed. Schlötterer[77] found isotropic stresses while Robinson and Dumin[167] notice anisotropic deformations, again with
the magnitude close to that expected from a thermal origin. The
recent stress birefringence measurements of Rinehart and Logan[73]
on Al$_x$ Ga$_{(1-x)}$/GaAs structures also indicate that the thermal expansion stress dominates the room temperature stress situation.

The epitaxial garnet films also respresent a well-studied system because of their use in bubble devices. Zeyfang,[125] using a
double crystal diffractometer concluded the thermal stress operated in YIG on YAG.

Many examples are presented by Besser et al.,[96] clearing up
previous inconsistencies in understanding the crazing associated
with these films. Broginski et al.,[98] conclude the mismatch
stresses are of prime importance for YID films on Gd$_{3-x}$Dy$_x$Ga$_5$O$_{12}$
substrates as result of magnetic anisotropy field measurements.
They propose a balance between the thermal expansion stresses and
the misfit dislocations to explain the almost zero stress at large
values of x. As described earlier Carruthers[97] has extended these concepts to a partial relaxation.

In summary, thermal and misfit contributions seem sufficient
to explain epitaxial stresses for the cases in the present literature.

In order to account for the bending that deposited epitaxial films undergo after separation form the substrate, Abrahams et al., (111) had suggested an oriented array of inclined dislocations. For films for $GaAs_{1-x}$ on GaAs, they found a correlation between the number of inclined dislocations, the lattice mismatch, and the amount of bending. An alternative explanation of this well-known bending phenomenon was attributed to the misfit dislocations near the interface. As a stress gradient through the thickness of the film is needed to account for the curling, the criticism of Abrahams et al., that misfit dislocations shoud apply only to 100 Å layers is not valid.

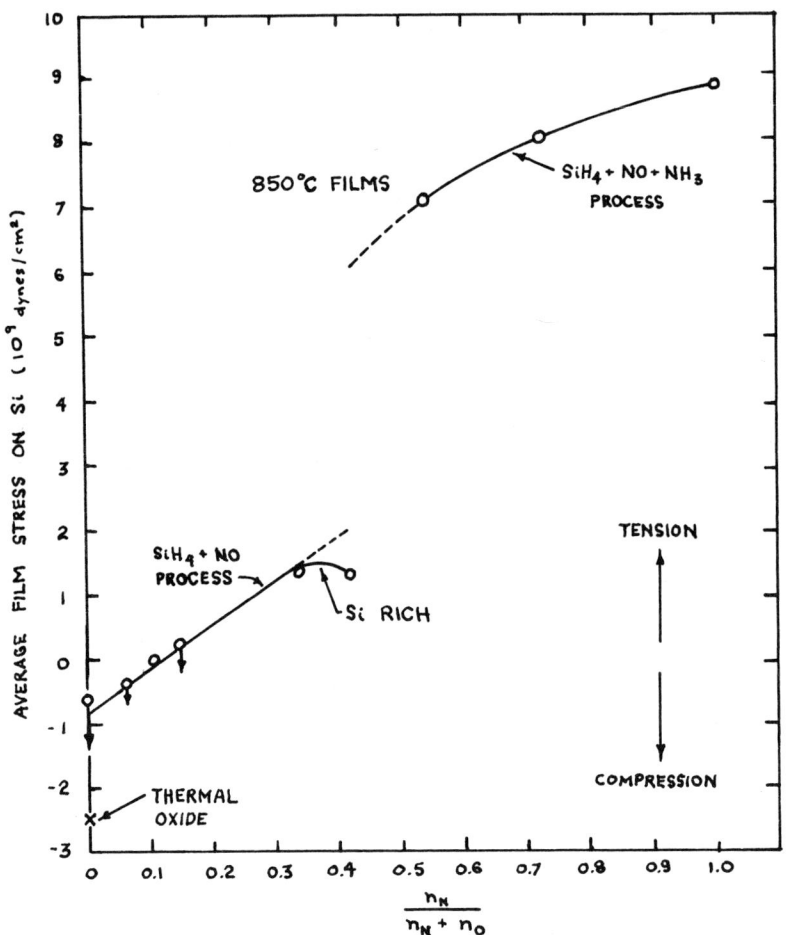

Fig. 31 Average stress in silicon oxynitride films as a function of nitride composition. From Rand and Roberts.(130)

We consider now the last category, namely, films used for insulating, passivating, or isolation purposes. This area is, of course, of extreme importance to the technology of thin films. We regard the electrical effects of such films beyond the scope of this article and content ourselves with some general considerations concerning the mechanical properties. There is a review by Pliskin et al. (126) on thin glass films and Dell'oca et al.(84) consider anodic layers. As the glass film are generally applied on heated substrates, and glass as a material is stonger under compression, one desires the total stress in the film to be compressive at the operating temperature. Thermally grown SiO_2 has perhaps received the most attention. But here the substrate is at high temperatures during the formation of the film and there is only modest control over the resultant compressive stress ($\sim 3 \times 10^9$ dyne/cm^2) as the temperature range to form the oxide is limited, and wet or dry oxidation makes only a small difference.(51, 127). Most authors report no thickness dependence in the range of 0.5 - 2 µm. However, a rapidly increasing compression at small thicknesses is found by Lin and Pugacz-Moraszkiewiez(53) in the range of 0.1-0.3µm . Thermal mismatch is recognized as the origin for thermally grown oxides, but intrinsic mechanisms are reported for CVD SiO_2 films deposited at somewhat lower temperatures(148) when the total stress is about 2×10^9 dyne/cm^2 tension. Post deposition heat treatments change the stress back to compression. A tensile intrinsic stress is also found in the case of thermal decomposition of tetraethylorosilicates.(149)

More recently Si_3N_4(128,129) has been used as a passivation layer for Si. In this case an extremely large intrisic stress of $> 1 \times 10^{10}$ dyne/cm^2 tension is found coupled with thermal stresses of perhaps 10% of the total. The stress does not change with time. Rand and Roberts(130) report amorphous silicon oxynitride films that range in tension of Si_3N_4 as the nitride concentration increases as shown in Fig. 31. These large stresses in both SiO_2 and Si_3N_4 can damage the substrate as we shall see in the next sections.

Although the intrinsic stress rapidly decreases with increasing substrate temperature in metals, we see that substantial intrinsic stresses are present for CVD insulation films.

We try to summarize the literature for non-metallic films as follows :

1. Although thermal contributions may be large, intrinsic stresses are well documented in both polycrystalline and epitaxial films.

2. Real stress contributions have been seen during the isolated island stage of growth, but these contributions should no

longer dominate at average film thicknesses of perhaps two hundred angstroms and are often masked by poor thermal control in the experiment.

3. There is a tendency for stress values to be tensile in polycrystalline and independent of thickness when the deposition parameters are well defined. This suggests that the major contribution to the intrinsic stress in thick films is a volume effect and not an interfacial one. Further, for pure films the stress should be capable of an explanation by similar models as those developed for metals.

4. Compressional contributions are associated with impurities, often oxygen, incorporated in the structure by design or default. These may take place during deposition, or afterward by diffusion. Non-stoichiometric films may also show compression.

5. The decrease in the average stress as the thickness increases may result from a relief process or a change in the intrinsic stress mechanism. The present measurements are not sufficiently complete to generalize, but on energetic grounds a relief mechanism must ultimately operate. Cracking in film or damage to the substrate may result.

6. Both thermal and misfit contributions operate in epitaxial films. The intrinsic contributions are amenable to calculation, and may be either a uniform elastic strain or localized at the substrate interface.

6. CONTROL OF STRESSES, INCLUDING RELAXATION EFFECTS

One of the objectives of understanding the intrinsic stresses in films is to be able to control the stresses and hence the properties for a given application. As we have seen, considerable progress has been made in the case of epitaxial systems. Recently for the case of metals stress control has come about by the use of bias sputtering. Blachman,[150,151] in the case of molybdenum and aluminium, has found a correlation between the stress resistivity and trapped argon content. With increasing negative bias substantial amounts of gas are entrapped which reduce the tension in the film or may even change it to compression. On the other hand, Sun et al.,[167] for the case of sputtered tungsten, find a sizeable compression which is not correlated with the argon concentration In a following paper[209] they find a strong correlation with final grain size in the W films. The structure and stress modifications by ion bombardment during deposition have been studied by Maddox and his co-workers.[168,169,170] Stuart[154] has used the

MECHANICAL PROPERTIES OF NON-METALLIC THIN FILMS

cantilever method to measure the stress in a number of metals deposited by low pressure triode sputtering. Date for the non-metals is not so common, but it appears that one can change the bias to substantially modify the stresses even in these systems. We also call attention to the UV irradiation effects mentioned earlier[210].

The use of elevated substrate temperatures to reduce the intrinsic stress, among other things, has been done for a long time. In addition it is sometimes possible to provide a thermal contribution which gives cancellation with intrinsic stress mechanism by the proper choice of substrate.

Post-deposition treatment may reduce the stress, as it has been known for a some in films of metals deposited at lower substrate temperatures that stress relief mechanisms operate upon heating the substrate to elevated temperatures.[171]. Resistivity and stress changes have been correlated and activation energy measurements made[172] which indicate that very likely a simple defect

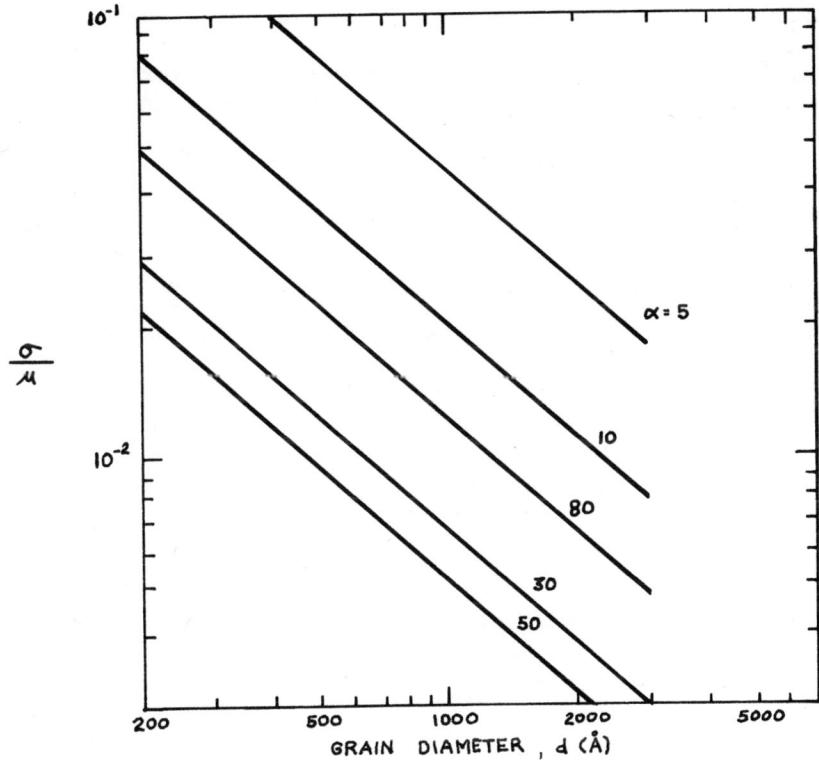

Fig. 32 Critical stress to operate a dislocation source as a function of grain diameter[21] for several inclinations. After Chaudhari.

motion at lower temperatures is followed by recrystallization. Such irreversible changes upon initial heating are usually followed by reversible behavior upon temperatures cycling at lower temperatures. Chaudhari[21] has considered the mechanisms of stress relief in polycrystalline films. Dislocation sources within a grain at a grain boundary or at surfaces are not likely mechanisms because the stress required to operate such dislocation sources is larger than the usual intrinsic stresses within a film. To a first approximation the stress necessary to operate sources varies inversely with the diameter of the grain as shown in Fig. 32. A similar result is observed for sources at a grain boundary edge, but since the intrinsic stresses are the order of 10^{-2} of the shear modulus, these processes are not normally favored. Kinetics equation for the exponential relaxation of the stress on a grain boundary sliding model were also derived by Chaudhary. At most grain boundaries in a film are perpendicular to a free surface under a planar biaxial stress the shear stress acting on the boundary is small and hence, the plastic strain contribution from grain boundaries sliding is expected to be small. The annealing kinetics have been derived for diffusional creep for the cases of volume diffusion and grain boundary diffusion predominating. In this case the thickness of the film matters and of course the activation energies for the two hypotheses are different. Although these equations were derived for the case of a uniform stress, it is pointed out that stress gradients also provide an additional driving force. If a stressed film is covered by a diffusion barrier containing a hole, the film relieves its elastic strains by a flow of matter between the surface and interior of the film. Mass flow out of the film under a compressive stress and into a film under tensile stress, leaving a hillock or depression. Pennabacker[173] has also considered that compressive stresses lead to hillock growth. Chaudhari[174] has recently calculated the hillock density and growth kinetics on the basis of a local relaxation of a compressive stress.

For the case of growing oxide films, Stringer[82] has considered additional mechanisms for plastic flow. The effect of stoichiometry on the oxide plasicity and the reasons for suggsting plastic flow takes place during oxidation are discussed. In addition, vacancy injection and dislocation generation at the metal substrate are examined. A more direct measurement of stress enhanced diffusion in thin films was given by Gangulee[175] in which Be atoms diffuse along grain boundaries of Al films and along the interface between the film and oxidized silicon substrate at a considerably enhanced rate when a tensile stress is applied to the plane of the film.

Diffusion may also be a source of strain. Takai and Francombe (176) have measured a cantilevered deflection during the interdiffusion of gold and aluminium both an increase in stress corresponding to the growth of intermetallic compound faces and stress relief were noted. Tu et al.,[137] have studied the Ni-Si system and Lau and Sun[177] the interdiffusion of Ti-Pd-Au Films.

Similar cases of enhanced migration are commonly found in the recrystallization of amorphous layers of silicon which are produced by ion bombardment of an aluminium coated silicon substrate. Silicon is transported to the surface of the thin aluminium surface according to Hart, et al.[178,179] Additional low temperature rapid migrations have also been found, and this subject has been treated in a recent conference,[180] in which both review papers and research reports will be found.

Helium implanted erbium films examined by Blewer and Maurin (181) have shown dimensional expansion at the surface and microscopic bubbles formed at the surface by release of helium by the films. Additional dilation data is found in the semiconductor literature.

Stress annealing data for non-metallic films is not common. In the case of silicon monoxide Priest and Caswell[146] showed that no stress change took place with a low temperature anneal in vacuum. But as soon as the pressures of oxygen or water vapor exceeded about 10^{-3} Pa, rapid compressive changes took place. This is, of course, related to the degree of oxidation as was discussed earlier and we would anticipate that the changes often found when optical films are exposed to the atmosphere have a related origin. Mattox and Kominiak[168] annealed bias sputtered Corning 1720 glass films at low temperatures. The initial compressive stress became larger, and at temperatures above about 350° the films showed a total tension. This irreversible behavior was again traced to an oxygen deficiency.

At the present times, it is well documented that grain boundary diffusion plays an almost dominant role in small-grained films, even though both lattice and diffusion down dislocation pipes may take place in special circumstances of large grain size, or extreme dislocation concentrations. Nevertheless, the detailed kinetics, especially in the case of impurity diffusion, are the subject of present investigations. The resultant stress relaxation are less well understood, and even the detailed mechanisms are still in doubt for metals,[3] and almost unknown for non-metals.

7. MODULI, FAILURE MODES AND ADHESION

We should comment on the values of elastic modulii for deposited films. Although occasionnally low values are found, as reviewed by Hoffman,[4] most films when carrefully measured show the expected values of the modulii. In addition to direct deflection measurements, resonance techniques[108,182,183] have been used. Spinner [184,185] used a dynamic technique to measure the modulii of many glasses as a function of temperature. Uozumi et al.,[186] have used the pulse-echo technique of sound velocity to determine the modulii. In view of the present interest in amorphous solids we call attention to the work by Chen and Wang, and emphasized by Berry and Pritchet[182] there is a growing body of evidence that Young's modulus for materials in the amorphous state may be characteristically 20-40 % below the value in the crystalline state. Merz et al [211] found a 45% decrease in the shear constant of $Sm_2 Co_{17}$ sputtered alloy, but only a 7% decrease in the bulk modulus compared to the crystalline phase. These measurements support the ideas of Weaire et al [212] that microscopic internal movements take place in amorphous materials under shear. We would expect similar changes in the amorphous layer induced in silicon by ion implantation. [188]. Internal friction measurements have given information as to loss mechanisms and thermal constants[182,186].

Bunshah,[189] has reviewed the properties of evaporated thick films and correlated the mechanical properties of thick, primarly metallic, films were reported in the Conferences on Structure/Property relationships in Thick Films and Bulk Coating.[190]

Kinosita[9] has reported unexpected microhardness measurements for LiF thin films with a Vicker-type indenter, and Winter[191] has developed apparatus for similar measurements using both and indenter and impact from spherical projectiles.

We consider now an outline of the failure modes, which may range from a mechanical distorsion of an optical surface,[152] through drastically changed properties, especially near the interface as indicated in several examples. We concentrate here on the obvious mechanical failures of cracking in the film or substrate and buckling if the film has a large compressive stress.

The importance of the adhesion in preventing failure is well known, although quantitative treatments are difficult to find. Hunt and Gale[192] suggested a model of the adherence to stresses in plastic film, in which the lift-off stress normal to the plane of the film is proportional to the gradient of the tension in the film substrate interface. The importance of the edges of the films or other defects which may be present, is well known and can be explained by the non-uniform stresses near the edges of the film as we mentioned earlier. Plassa[194] has also studied the elastic instability of germanium films as they buckle form a mica substrate. Random wrinkles are found with an isotropic intrinsic stress although oriented wrinkles were found earlier by Yelon and Voegeli [194] in the case of epitaxial films and a suggestion for the sinusoidal wrinkle pattern in terms of a column instability has been made by Plassa and Chopra[11]. As the stress is generally a volume effect, the total force that must be supported across the interface increases as the film becomes thicker. This leads to a critical failure thickness which is a function of the stress in the film and the adhesion to the substrate. The buckling, of course, results from a film under high compression whereas a cracking will be found for films under sufficiently high tension that the fracture stress is exceeded.

Angle of incidence effects may also produce oriented buckled patterns observed in ZnS by Behrndt[195] which were eliminated by rotating substrates, and we discussed earlier the extreme dependence in the SiO system. Recent emphasis on deposition techniques other than evaporation have eliminated many of these problems.

Dislocation generation and even cracking as a result of the residual elastic strains and epitaxial films is well documented. In the galium arsenide photodiode heteroepitaxy lattice paremeter mismatch had to be graded at a slow rate to avoid excessive dislocation generation. We referred earlier to the stress by refringence and cracking in the magnetic bubble materials. Carruthers[97] has treated the deformation and creep as well as estimating the property changes. It was known that thin dielectric films of the order of 1 µm may accomodate the high stresses normally found, thicker films are found to crack. Matthews and Klokholm[196] have considered the fracture of these brittle films under the influence of the misfit stress. The cracks that form in the garnet films are perpendicular to the film plane. If such cracks are to propagate through the film, which is rigidly bonded to its substrate and strained tension, then, the film thickness mus exceed the Griffith crack length. If the stress exceeds the fracture strength of the crystal then, a spontaneous fracture will take place, whereas, for a low stress the film may strain elastically and be stable. These results are summarized in Fig. 33 as a function of the mistfit. Indeed, the presence of cracks in a film indicates that the density of misfit dislocations lies below the optimum value and indicates there are difficulties associated with the generation of these dislocations.

SiO_2 or Si_3N_4 films are known to develop dislocations or cracks in the underlying silicon substrates. Kato et al., Westdorp and Schwuttke,[128] and Tamura and Sunami[129] have considered this problem in detail. The internal stresses were measured, isothermal annealing treatments were carried out, and the density of dislocations introduced during the high temperature anneal were measured. Dislocations in the garnet system have been studied in a series of papers by Matthews and Klockholm.[197]

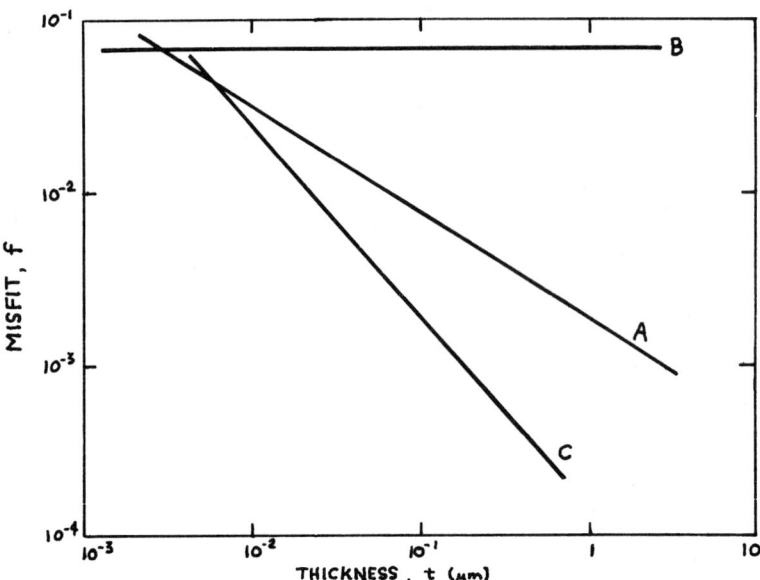

Fig. 33. Crack generation in brittle films as a function of misfit and thickness. Above B film spontaneously cracks, below C it is elastically strained and above A cracks will propagate. After Matthews and Klockholm.(196)

We finally consider adhesion. Much has been said about adhesion in a qualitative way, but to date, the field still suffers from good measurement techniques. Campbell[7] has reviewed this area well and annual conferences of the subject have been published[198] in Aspects of Adhesion. Techniques in cleaning, depositing thin reactive metal layers to aid bonding to glass, interdiffusion, and grading the composition all have been used to enhance the adhesion. Nevertheless, a detailed understanding of even the origin of the forces is obscure, although Van de Vaals coupling is suggested as the most important contributor with band structure effects being important in clean metal surfaces.[199,200]

The stress distributions in an adhesive layer have been calculated by Harrison and Harrison.[201]

We leave the discussion of the most recent measurement techniques to the recent reviews by Chapman[202] and Kendall[203] and the forthcoming review by Weaver[204] and mention here only deposition techniques which should prove useful as well as a brief report of the surface analysis techniques that may provide some of the quantitative information needed for an understanding.

It has long been known that sputtered films generally have better adherence to a substrate than to evaporated ones. The higher energy of the arriving atoms allows a small penetration and / or a cleaning of the surface. The ion plating technique as pioneered by Mattox carries this process even further by subjecting the substrate to a flux of high energy ions before and during the film deposition. From the point of view of adhesion, several benefits may be obtained. The surface will be sputter cleaned and maintained clean until the film begins to form. The high energy flux to the substrate surface provides a high effective surface temperature, enhancing diffusion and chemical reaction. The high defect concentration also provides a physical admixing of the film and substrate material. According to Mattox[205] this process gives rise to sizeable gas incorporation as the negative bias is increased, the gas entrapment decreases again at biases above a few 100 volts, presumably because the high temperature of the deposit allows the gas to diffuse away. Densities of as much as a factor of 2 lower than bulk densities are observed, and the growth morthology is decidedly influenced. As far as the stress is concerned an increase in intrinsic compressive film stress is generally found with increasing negative bias, for metal films.

More recently experimental techniques of AES have been applied to the adhesion question. Staoddart et al.,[206] have studied the effect of the glow discharge on adhesion. Although no surface electrical or topographical effects were seen, gross contamination as well as gas sorption were felt to be important. Houston and Bland[207] found that the discharge current could act as a process control variable to indicate a clean cathode surface, if proper care was taken. Westwood and Bennewitz[208] in reactively sputtered PTO films noted that the presence of oxygen was necessary for good adhesion.

It is premature to generalize from the results but it appears as though certain impurities in small concentrations at the interface are beneficial and, in fact, even needed for good adhesion. Sundahl[193] found the presence of Cu and Si on the surface of high purity Al_2O_3 as well as a small grain size in the substrate were important factors. With the increased availability of surface analysis techniques we expect significant progress in understanding adhesion.

ACKNOLEDGEMENTS

I would like to dedicate this paper to my students, both past and present, who have taught me so much about the mechanical properties. I also wish to thank P. Chaudhari, D. M. Hoffman, A. Kinbara, K. Kinosita, E. Klokholm , S. Mader, E. Ritter, and W. D. Westwood for their helpful discussions and information. Mrs. M. Young and R. Wentz have my gratitude for their faithful efforts in production of the manuscript. The research at Case Western Reserve University was supported by the U. S. Atomic Energy Commission.

REFERENCES

1. See for example : E. Klokholm, J. Vac. Sci. Tech. $\underline{8}$, 148 (1971).
2. M. H. Francombe, A. J. Noreika, W. J. Takei, and S. Y. Wu, J. Vac. Sci. Tech. $\underline{11}$, 130 (1974).
3. A. Gangulee, Acta Met. $\underline{22}$, 177 (1974).
4. R. W. Hoffman, in "Physics of Thin Films" (G. Hass and R. E. Thun, eds.), Vol. 3. Academic Press, New York (1966).
5. R. W. Hoffman in "Thin Films", H. G. F. Wilsdorf, ed. A. S. M. (1964).
6. D. S. Campbell, in " Basic Problems in Thin Film Physics" (R. Niedermayer and H. Mayer, eds), p. 223. Vandenhoeck and Ruprecht, Göttingen (1966).
7. D. S. Campbell, in "Handbook of Thin Film Technology", (Maissel and Glang, eds.), p. 12-1. McGraw-Hill, New York (1970).
8. W. Buckel, J. Vac. Sci. Tech. $\underline{6}$, 606 (1969).
9. K. Kinosita, Thin Solid Films $\underline{12}$, 17 (1972).
10. R. J. Scheuerman, in "Symposium on Deposited Dielectric Thin Films", (F. Vratny, ed.), p. 561. Electrochemical Society, New York (1969).
11. See, for example : K. L. Chopra, "Thin film Phenomena", McGraw-Hill, New York (1969) and, L. Maissel and M. H. Francombe, "Introduction to Thin Film Physics". Gordon and Breach, New York (1973).
12. J. F. Nye, "Physical Properties of Crystals", Oxford Univ. Press, London (1957).
13. C. S. Smith, in "Solid State Physics", Vol. 6, F. Seitz and D. Turnbull, eds. Academic Press, New York (1958).

14. R. W. Vook and F. Witt, J. Appl. Phys. 36, 2169 (1965).
15. J. Turley and G. Sines, J. Appl. Phys. 41, 3722 (1970).
16. T. Y. Thomas, Proc. Nat. Acad. Sci. (U.S.) 55, 235 (1966).
17. A. Brenner and S. Senderhoff, J. Res. Natl. Bur. Stand. 42, 89 (1949).
18. F. A. Doljack, AEC Tech. Rept. 76, Case Western Reserve University, Cleveland, Ohio (1971).
19. E. Klockholm, in "X-ray Diffraction and Stress in Thin Films Symposium", IBM Thomas J. Watson Research Center, Yorktown Heights, New York (March 1969).
20. G. C. Stoney, Proc. Roy. Soc. (London) A32, 172 (1909).
21. P. Chaudhari, IBM J. Res. Develop. 13, 197 (1969).
22. H. J. Oel and V. D. Frechette, J. Am. Ceram. Soc. 50, 542 (1967).
23. J. Aleck, J. Appl. Mech. 16, 118 (1949).
24. K. Haruta and W. J. Spencer, J. Appl. Phys. 37, 2232 (1966).
25. R. W. Hoffman in "The use of Thin Films in Physical Investigations", J. C. Anderson, ed. Academic Press, New York (1966).
26. R. W. Hoffman in "Measurement Techniques for Thin Films" Schwartz and Schwartz, eds. Electrochemical Society (1967).
27. D. G. Ashwell and E. D. Greenwood, Engineering 170, 51 (1970).
28. D. G. Bellow, G. Ford, and J. S. Kennedy, Exp. Mech. 227 (1965).
29. R. E. Rottmayer and R. W. Hoffman, J. Vac. Sci. Tech. 7, 461 (1970).
30. R. W. Springer, AEC Tech. Rept. 79, Case Western Reserve University, Cleveland, Ohio (1972).
31. J. D. Finegan and R. W. Hoffman, Trans. 8th Vacuum Symposium, p. 935. Pergamon Press (1962).
32. F. A. Doljack and R. W. Hoffman, Thin Solid Films 12, 71 (1972).
33. P. H. Robinson and D. J. Dumin, J. Electrochem. Soc. 115, 75 (1968).
34. P. M. Alexander, AEC Tech. Rept. 85, Case Western Reserve University, Cleveland, Ohio (1974).
35. K. Maki and K. Kinosita, Reported in K. Kinosita Proc. Second Colloquium on Thin Films, p. 31 (1967).
36. E. Yoda, J. Appl. Phys. Japan 8, 1355 (1969).

37. E. Yoda, J. Appl. Phys. Japan 8, 1355 (1969).
38. Y. Namba, Oyo Buturi 38, 411 (1969).
39. R. E. Rottmayer, AEC Tech. Rept. 64, Case Western Reserve University, Cleveland, Ohio (1970).
40. R. W. Springer and R. W. Hoffman, J. Vac. Sci. Tech. 10, 238 (1973).
41. R. E. Rottmayer and R. W. Hoffman, J. Vac. Sci. Tech. 8, 152 (1971).
42. A. E. Ennos, Appl. Optics 5, 51 (1966).
43. A. Taloni and D. Hanneman, Surface Sci. 8, 323 (1967).
44. A. G. Blachman, Metallurg. Trans. 2, 699 (1971).
45. K. A. Haines and B. P. Hildebrand, Appl. Optics 5, 595 (1966).
46. R. J. Magill and T. Young. J. Vac. Sci. Tech. 4, 47 (1967).
47. R. Glang, R. A. Holmwood, and R. L. Rosenfeld, Rev. Sci. Instr. 36, 7 (1965).
48. N. N. Axelrod and H. J. Levinstein, unpublished.
49. F. P. Chiang, C. S. Faber, and F. Y. Wang, J. Appl. Phys. 42, 1422 (1971).
50. J. W. Beams, in "Structure and Properties of Thin Films", C. A. Neogebauer, J. D. Newkirk, and D. A. Vermilyea, eds. p. 183. Wiley, New York (1959).
51. R. J. Jaccodine and W. A. Schlegel, J. Appl. Phys. 37, 2429 (1966).
52. C. H. Lane, IEEE Trans. Electron. Dev. ED-15, 998 (1968).
53. S.C.H. Lin and I. Pugacz-Muraszkiewicz, J. Appl. Phys. 43, 119 (1972).
54. C. W. Wilmsen, E. G. Thompson, and G. H. Meissner, IEEE Trans. Electron. Dev. ED-19, 122 (1972).
55. P. M. Schaible and R. Glang "Symposium on Deposited Dielectric Thin Films (F. Vratny, ed.), p. 577, Electrochemical Society, New York (1969).
56. G. Zosi, Z. Angew. Phys. 24, 322 (1968).
57. G. W. Bush and H. J. Read, J. Electrochem. Soc. 111, 289 (1964).
58. A. Taylor, "X-ray Metallography", Wiley, New York (1961).
59. T. I. Kamins and E. S. Meieran, J. Appl. Phys. 44, 5065 (1973).
60. B. D. Cullity, J. Appl. Phys. 35, 1915 (1964).

61. B. Borie, Acta Cryst. 13, 542 (1960).
62. B. Borie, C. J. Sparcks, and J. V. Cathcart, Acta Met. 10, 691 (1962).
63. C. B. McDowell and T. C. Pilkington, J. Appl. Phys. 42, 2958 (1971).
64. A. Authier, "Proc. of the 15th Annual Conf. on Application of X-ray Analysis" (J. B. Newkirk and G. Mallet, eds.), (1967).
65. E. S. Meieran and I. A. Blech, J. Appl. Phys. 36, 3162 (1965).
66. I. A. Blech and E. S. Meieran, Appl. Phys. Letters 9, 245 (1966).
67. G. H. Schwuttke and J. K. Howard, J. Appl. Phys. 39, 1581 (1968).
68. P. Penning and D. Polder, Philips Res. Rept. 16, 419 (1961).
69. R. Zeyfang, J. Appl. Phys. 42, 1182 (1971).
70. G. A. Rozgonyi and T. J. Ciesielka, Rev. Sci. Instr. 44, 1053 (1973).
71. E. P. EerNisse, J. Appl. Phys. 43, 1330 (1972).
72. E. P. EerNisse, J. Appl. Phys. 44, 4482 (1973).
73. F. K. Reinhart and R. A. Logan, J. Appl. Phys. 44, 3171 (1973).
74. R. W. Vook and F. Witt, J. Vac. Sci. Tech., 2 49 (1965).
75. R. Zeyfang, Solid State Electron. 14, 1035 (1971).
76. R. Feder and T. B. Light, J. Appl. Phys. 43, 3114 (1972).
77. H. Schlötterer, Solid State Electron. 11, 947 (1968).
78. A. J. Hughes and A. C. Thorsen, J. Appl. Phys. 44, 2304 (1973).
79. A. C. Thorsen and A. J. Hughes, Appl. Phys. Letters 21, 579 (1972).
80. A. J. Hughes (unpublished).
81. Y. Nakajima and K. Kinosita, Thin Solid Films 5, R5 (1970).
82. J. Stringer, Corrosion Sci. 10, 513 (1970).
83. F. N. Rhines and J. S. Wolf. Reported in Ref. 82
84. C. J. Dell'Oca, D. L. Pulfrey, and L. Young, in "Physics of Thin Films" (M. Francombe and R. W. Hoffman, eds.)., Vol. 6. Academic Press, New York (1971).

85. L. Young, "Anodic Oxide Films", Academic Press, New York (1961).

86. D. A. Vermilyea, in "Advances in Electrochemistry", Vol. 3. Wiley (Interscience), New York (1963).

87. J. H. Van der Merwe, in "Single Crystal Films", (M. Francombe and H. sato, eds.), Pergamon Press, Oxford (1964).

88. J. H. Van der Merwe and N. G. Van der Berg, Surface Sci. 32, 1 (1972).

89. J. H. Van der Merwe, Proc. 6th International Vacuum Congress (to be published).

90. J. W. Matthews, in "Physics of Thin Films" (G. Hass and R. E. Thun, eds.), Vol. 4. Academic Press, New York (1967).

91. J. W. Matthews, S. Mader, and T. B. Light, J. Appl. Phys. 41, 3800 (1970).

92. W. A. Jesser and D. Kuhlmann-Wilsdorf, J. Appl. Phys. 38, 5128 (1967).

93. W. A. Jesser and D. Kuhlmann-Wilsdorf, Phys. Stat. Sol. 19 95 (1967).

94. K. Yagi, K. Takayanagi, K. Kobayashi, and G. Honjo, J. Cryst. Growth 9, 84 (1971).

95. R. Carpenter and D. S. Campbell, J. Mater. Sci. 2, 173 (1967).

96. P. J. Besser, J. E. Mee, P. E. Elkins, and D. M. Heinz, Mat. Res. Bull. 6, 1111 (1971).

97. J. R. Carruthers, J. Cryst. Growth 16, 45 (1972).

98. A. J. Braginski, T. R. Oeffinger, W. E. Kramer, D. K. McLain, and W. J. Takei, IEEE Trans. Mag. 8, 300 (1972).

99. N. Cabrera, Surface Sci. 2, 320 (1964).

100. C. Herring, "Structure and Properties of Crystal Surfaces", (R. Gomer and C. S. Smith, eds.), Univ. of Chicago Press Chicago, Illinois (1953)

101. R. G. Lindford and L. A. Mitchell, Surface Sci. 27, 142 (1971).

102. H. J. Wasserman and J. S. Vermaak, Surface Sci. 32, 168 (1972).

103. R. Ritter, in "Physics of Thin Films" (G. Hass, M. Francombe and R.W. Hoffman, eds) Vol. 8 Academic Press, N.Y. (1975)

104. P. A. Young, Thin Solid Films 6, 423 (1970).

105. E. Klokholm and R. Berry, J. Electrochem. Soc. 115, 823 (1968).

106. S. Mader, J. Vac. Sci. Rech. 11, 131 (1974).
107. P. Chaudhari, J. Vac. Sci. Tech. 9, 520 (1972).
108. J. D. Wilcox, D. S. Campbell, and J. C. Anderson, Thin Solid Films 3, 13 (1969).
109. A. Yelon, J. R. Asik, and R. W. Hoffman, J. Appl. Phys? 33, 949 (1962).
110. D. O. Smith, M. S. Cohen, and G. P. Weiss, J. Appl. Phys. 31, 1775 (1960).
111. M. S. Abrahams, L. R. Weisberg, and J. J. Tietjen, J. Appl. Phys. 40, 3754 (1969).
112. K. Saito, R. O. Bozkurt, and T. Mura, J. Appl. Phys. 43, 182 (1972).
113. J. Reisenfeld and R. W. Hoffman, AEC Tech. Rept. 39 Case Institute of Technology, Cleveland, Ohio (1965).
114. A. F. Turner, "Thick Thin Films", Bausch & Bomb Tech. Rept. (1951).
115. O. S. Heavens and S. D. Smith, J. Opt. Soc. Am. 47, 469 (1957).
116. H. Blackburn and D. S. Campbell, in "Trans. 8th Nat. Vacuum Symp." p. 943. Pergamon Press, Oxford (1961).
117. K. Kinosita, K. Nakamizo, K. Maki, K. Onuki, and K. Takenchi, J. Appl. Phys. Japan 4, (Suppl. 1), 340 (1965).
118. A. Barna, P. B. Barna, J. F. Pocza, and I. Pozsgai, Thin Solid Films 5, 201 (1970).
119. A. Preisinger and H. K. Pulker, private communication (to be published).
120. H. K. Pulker and E. Jung, Thin Solid Films 9, 57 (1971).
121. J. R. Priest, H. L. Caswell, and Y. Budo, J. Appl. Phys. 34, 347 (1963).
122. A. E. Hill and G. R. Hoffman, Brit. J. Appl. Phys. 18, 13 (1967).
123. D. J. Dumin, J. Appl. Phys. 36, 2700 (1965).
124. C. Y. Ang and H. M. Manasevit, Solid State Electron. 8 994 (1965).
125. R. Zeyfang, J. Appl. Phys. 41, 3718 (1970).
126. W. A. Pliskin, in "Physics of Thin Films" (G. Hass and R. E. Thun, eds.), Vol. 4. Academic Press, New York (1969)
127. M. V. Whelan, A. H. Goemans, and L.M.C. Goossens, Appl. Phys. Letters 10, 262 (1967).

128. W. A. Westdorp and G. H. Schwuttke, in "Symposium on Deposited Dielectric Thin Films" (F. Vratny, ed.), p. 546. Electrochemical Society, New York (1969).

129. M. Tamura and H. Sunami, J. Appl. Phys. Japan 11, 1097 (1972).

130. M. J. Rand and J. F. Roberts, J. Electrochem. Soc.

131. H. Blackburn and D. S. Campbell, Phil. Mag. 8, 823 (1963).

132. Y. Doi, Machine Test. Lab. Rept. (Japan) 27, 44 (1958).

133. D. M. Mattox and G. J. Kominiak, J. Electrochem. Soc. 120 1535 (1973).

134. D. S. Campbell, "Trans. 9th National Vacuum Symp." p. 29. The Macmillan Co., New York (1962).

135. A. Catlin and W. P. Walker, J. Appl. Phys. 31, 2135 (1960).

136. H. Schroder and G. M. Schmidt, Z. Angew. Phys. 18, 124 (1964).

137. J. Reisenfeld and R. W. Hoffman, AEC Tech. Rept. 39, Case Institute of Technology, Cleveland, Ohio (1965).

138. J. D. Wilcock, Ph.D Thesis, Imperial College, London (1967).

139. L. Holland, T. Putner, and R. Ball, Brit. J. Appl. Phys. 11, 167 (1960).

140. M. A. Novice, Brit. J. Appl. Phys. 13, 561 (1962).

141. M. A. Novice, Vacuum 14, 385 (1964).

142. J. Priest, H. L. Caswell, and Y. Budo, "Tran. 9th Nat'l Vacuum Symp.", p. 121. The Macmillan Co., New York (1962).

143. J. D. Finegan and R. W. Hoffman, AEC Tech. Rept. 15. Case Institute of Technology (1961).

144. J. Priest, Rev. Sci. Instr. 32, 1349 (1961).

145. Y. Budo and J. Priest, Solid-State Electron. 6, 159 (1963).

146. J. Priest and H. L. Caswell, "Trans. 8th Nat'l Vacuum Symp.". p. 947. Pergamon Press, Oxford (1961).

147. R. J. Scheuerman, J. Vac. Sci. Tech. 7, 143 (1970).

148. H. Sunami, Y. Itoh, and K. Sato, J. Appl. Phys. 41, 5115 (1970).

149. J. A. Aboaf, J. Electrochem. Soc. 116, 1732 (1969).

150. A. G. Blachman, Mett. Trans. 2, 699 (1971).

151. A. G. Blachman, J. Vac. Sci. Tech. 10 299 (1973).

152. R. J. Scheuerman, J. Vac. Sci. Tech. 6, 145 (1969).

153. W. Heitman, Appl. Optics 10, 2685 (1971)
154. P. R. Stuart, Vacuum 19, 507 (1969).
155. D. A. Vermilyea, J. Electrochem. Soc. 110, 345 (1963).
156. S. S. Lau and R. H. Mills, Phys. Stat. Sol. 17, 609 (1973).
157. L. Gmita and E. Teneseu, Rev. Roumaine Phys. 12, 79 (1967).
158. C. Drum and M. J. Rand, J. Appl. Phys. 39, 4458 (1968).
159. J. C. Grosspreutz, Surface Sci. 8, 173 (1967).
160. R. Lathlaen and D. H. Diehl, J. Electrochem. Soc. 116, 620 (1969).
161. R. Carpenter and D. S. Campbell, J. Mater. Sci. 4, 526 (1969).
162. C. Mai, S. Audision, and R. Riviere, C.R. Acad. Sci. Ser. B 269, 1185 (1969).
163. C. M. Drum, R. N. Tauber, J. D. Ashner, P. F. Schmidt, ABS Spring Mtg., Electrochemical Soc. Washington, D. C. (1971).
164. F. B. Micheletti and S. H. McFarlane, ABS Spring Mtg., Electrochemical Soc., Washington, D.C. (1971).
165. A. I. Vousi and L. P. Strakhov, Fiz. Tverd. Tela. 12, 3319 (1970).
166. M. Plassa, Thin Solid Films, 3, 305 (1969).
167. R. C. Sun, T. C. Tisone, and P. D. Cruzan, J. Appl. Phys. 44, 1009 (1973).
168. D. M. Mattox and G. J. Kominiak, J. Vac. Sci. Tech. 9, 528 (1972).
169. D. M. Mattox, 6th International Vacuum Congress, Kyoto 1974 (to be published).
170. R. D. Bland, G. J. Kominiak and D. M. Mattox, J. Vac. Sci. Tech. 11, 671 (1974).
171. H. S. Story and R. W. Hoffman, Proc. Phys. Soc. B 70 950 (1957).
172. N. S. Rasor and R. W. Hoffman, Phys. Rev. 98, 1555 (1955).
173. W. B. Pennebacker, J. Appl. Phys. 40, 394 (1969).
174. P. Chaudhari. Submitted to J. Appl. Phys.
175. A. Gangulee, Phil. Mag., 17, 865 (1970).
176. W. J. Takei and M. H. Francombe, Solid State Electronics 11, 205 (1968).

177. S. S. Lau and R. C. Sun, Thin Solid Films, 10, 273 (1972).
178. R. R. Hart, D. H. Lee and O. J. Marsh, Appl. Phys. Letters 20, 76 (1972).
179. D. H. Lee, R. R. Hart and O. J. Marsh, Appl. Phys. letters 20, 73 (1972).
180. International Conference on Low Temperature Diffusion and Applications to Thin Films. A. Gangulee, P.S. Ho, and K. N. Tu, Yorktown Hts., (1974). To be published in Thin Solid Films.
181. R. S. Blewer and J. K. Maurin, J. Nuc. Mat., 44, 260 (1972).
182. B. S. Berry and W. C. Pritchet, J. Appl. Phys. 44, 3122 (1973).
183. H. Biehl and H. H. Mende, J. Phys. Chem. Solids, 35, 37 (1974).
184. S. Spinner, J. Am. Ceram. Soc., 37, 229 (1954).
185. S. Spinner, J. Am. Ceram. Soc., 38, 113 (1955).
186. K. Uozumi, T. Nakada and A. Kinbara, Thin Solid Films, 12, 67 (1972).
187. H. S. Chen and T. T. Wang, J. Appl. Phys. 41, 5338 (1970).
188. S. I. Tan, B. B. Berry, and L. Crowder, Appl. Phys. Letters 20, 88 (1972).
189. R. F. Bunshah, J. Vac. Sci. Tech., 11, 633 (1974).
190. Conference on Structure/Property Relationships in Thick Films/ and Bulk Coatings, J. Vac. Sci. Tech. 11 (1974).
191. R. E. Winter, Thin Solid Films, 12, 81 (1972).
192. R. A. Hunt and B. Gale, J. Phys. D Appl. Phys., 5, 359 (1972).
193. R. C. Sundahl, J. Vac. Sci. Tech., 9, 181 (1972).
194. A. Yelon and O. Voegeli, in Single Crystal Films, M. Francombe and H. Sato, eds., p. 321. Pergamon Press, Oxford (1964).
195. K. H. Behrndt, J. Vac. Sci. Tech., 2, 63 (1965).
196. J. W. Matthews and E. Klokholm, Mat. Tes. Bull., 7, 213 (1972)
197. J. W. Matthews, E. Klokholm, and T. S. Plasket, IBM J. Res. Develop., 17, 426 (1973).
198. Aspect of Adhesion, D. J. Alner, ed. U. of London Press, London (Multiple Volumes).

199. H. Krupp, Adv. Colloid Interface Sci., 1, 111 (1967).
200. J. Ferrante and J. R. Smith, Surface Sci., 38, 77 (1973).
201. N. L. Harrison and W. J. Harrison, J. Adhesion, 3, 195 (1972).
202. B. N. Chapman, J. Vac. Sci. Tech. 11, 106 (1974).
203. K. Kendall, J. Adhesion, 5, 179 (1973).
204. C. Weaver, to be published.
205. D. M. Mattox, J. Vac. Sci. Tech., 10, 47 (1973).
206. C.T.H. Stoddart, D. R. Clarke, and C. T. Robbie, J. Adhesion 2, 270 (1970).
207. J. E. Houston and R. D. Bland, J. Appl. Phys., 44, 2504 (1973).
208. W. D. Westwood and C. D. Bennewitz, Private Communication (To be published).
209. R. C. Sun, T. C. Tisone and P. D. Cruzan, Private communication. (Submitted for publication in J. Appl. Phys.).
210. I. J. Hodgkinson and A. R. Walker, Thin Solid Films 17, 185 (1973).
211. M. D. Merz, R. P. Allen and S. D. Dahlgren, J. Appl. Phys. 45, 4126 (1974).
212. D. Weaire, M. F. Ashby, J. Logan and M. J. Weins, Acta Metall. 19, 779 (1971).

THIN FILMS IN OPTICS

G. Baldini and L. Rigaldi

Universita degli Studi, Sassari and Gruppo Nazionale
di Struttura della Materia del C.N.R
Via Celoria, 16 Milano Italy

1. INTRODUCTION

In this paper we show how the optical response of a dielectric thin film, or stack of thin films (multilayer), can be evaluated. This subject has been studied by several people, particularly in the last decade, since the use of thin films in optics has many and varied applications. From the old antireflection film coating we go to high reflectance mirrors, beam splitters, interference filters, polarizers etc. Furthermore in recent ye ars, with the wide development of the lasers, there has been a great effort on a new field which is known under the title of "integrated optics". Its main goals are the design and construction of compact optical systems having the properties of some logical circuits that, until now, have been made with the technology of the integrated electronics. Examples of applications of integrated optics are : optical modulators, rotators, waveguides, etc., which are useful for telecommunications and computers. The field is very vast as one can imagine from the examples listed above and therfore it is not possible to cover it in this paper . Therefore we have confined ou r selves to the foundations of light propagation in stratified media with particular attention to transverse and longitudinal propagation. These two aspects have been considered at first for the single film which will be described in the first part whereas multilayer systems will be considered in the second part. Application of several of these concepts can be found in the literature quoted here and, as far as the problem of filters and polarizers is concerned, in the paper of Prof. Pelletier.

2. OPTICAL PROPERTIES OF A SINGLE THIN FILM

In this paper we will consider the propagation of light waves in a thin film which may define as a plane layer with indefinite extension in the x, y plane and finite thickness W along z or, better, with a thickness of the order of the wavelength of light. A substrate is also required for the practical purpose of supporting the film. Its thickness is taken as semiinfinite.

Different arrangements of a thin film can be considered as far as light propagation is concerned. Here we will examine two cases :

a) The indices of refraction of the film and the surrounding media are such that total reflection does not occur (filters, antireflection coating, etc.).

b) Total reflection occurs inside the film (waveguides).

A third case c) may also considered for the sake of completeness but since it is not of interest in the applications which we describe in these papers we will say only that the light may be also totally reflected from the film as shown in Fig. 1 which illustrates cases a), b) and c).

2.1 Reflectance and Transmittance of a Single Film

We examine here the propagation of light when internal reflection does not occur. In Fig. 2 we show the light path assuming that the incident beams is coming from the n_0 side.

In fig. 2 is represented a thin film where plane waves (with amplitudes A and B) are propagating in the directions given by the unit vectors $\tilde{u}$ and $\tilde{v}$. For $z > W$ waves propagate in the medium with refraction index n_2 whereas for $z < 0$ the medium is described by n_0. We define : $k = 2\pi/\lambda$, in vacuum ; $k_0 = k\, n_0$; $k_1 = (\nu - i\varkappa)$; $k_2 = k\, n_2$; waves B in the direction of z (B $\quad \tilde{u}j$; waves A in the direction of $-z$ (A $\quad \tilde{v}j$).

<u>Medium 0</u> (we use vector notation for the amplitude $\tilde{A}$ and $\tilde{B}$ since later we will specify polarization). We have omitted the factor $e^{-i\omega t}$ which can be included in the amplitude

$$\vec{E}_o = \vec{B}_o \exp i(k_o \vec{u}_o \cdot \vec{r}) + \vec{A}_o \exp i(k_o \vec{v}_o \cdot \vec{r})$$
$$\vec{H}_o = \vec{u}_o \times n_o \vec{B}_o \exp(ik_o \vec{u}_o \cdot \vec{r}) + \vec{v}_o \times n_o \vec{A}_o \exp(ik_o \vec{v}_o) \cdot \vec{r}) \tag{1}$$

THIN FILMS IN OPTICS

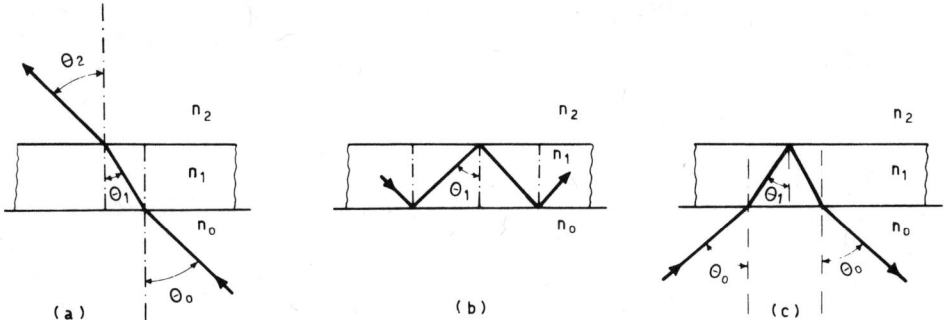

Fig. 1 Three different arrangements for ligth propagation using thin films. Multiple reflections have been neglected in (a) and (b)

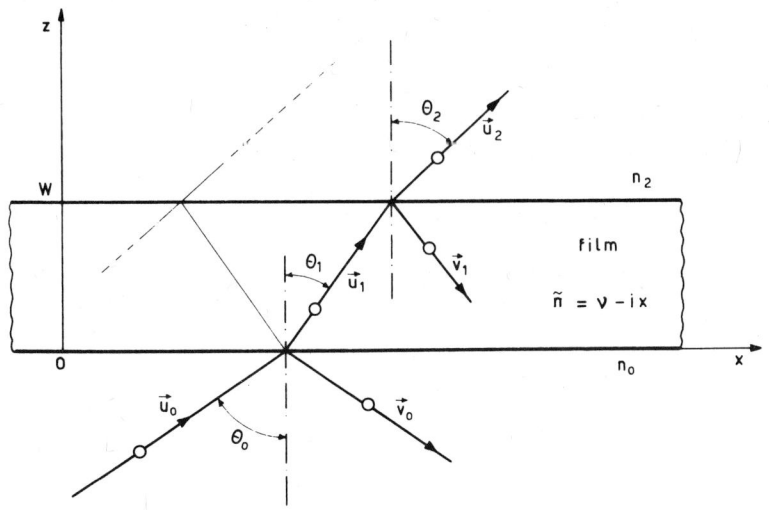

Fig. 2 Film employed in the case (a) of Fig. 1

Medium 1

$$\vec{E}_1 = \vec{B}_1 \exp(ik_1\vec{u}_1\cdot\vec{r}) + \vec{A}_1 \exp(ik_1\vec{v}_1\cdot\vec{r})$$
$$\vec{H}_1 = \vec{u}_1 X n_1 \vec{B}_1 \exp(ik_1\vec{u}_1\cdot\vec{r}) + \vec{v}_1 X n_1 \vec{A}_1 \exp(ik_1\vec{v}_1\cdot\vec{r})$$
(2)

Medium 2 (waves $\vec{v}$ are missing)

$$\vec{E}_2 = \vec{B}_2 \exp(ik_2\vec{u}_2\cdot\vec{r})$$
$$\vec{H}_2 = \vec{u}_2 X n_2 \vec{B}_2 \exp(ik_2\vec{u}_2\cdot\vec{r})$$
(3)

Boundary conditions

$z = 0$

$\hat{z} \times \vec{E}_o = \hat{z} \times \vec{E}_1$

$\hat{z} \times \vec{H}_o = \hat{z} \times \vec{H}_1$

$z = W$ (4)

$\hat{z} \times \vec{E}_1 = z \times \vec{E}_2$

$\hat{z} \times \vec{H}_1 = z \times \vec{H}_2$

We are considering here TE waves ($\vec{E}\perp$ plane of incidence) which are also called s waves. TM waves or p polarization is analogous.

$$\hat{z}\cdot\vec{A}_o = \hat{z}\vec{B}_o = \hat{z}\cdot\vec{A}_1 = \hat{z}\cdot\vec{B}_1 = \hat{z}\vec{B}_2 = 0$$

By using this result and by means of the following definitions:

$$\beta = kn_o\sin\theta_o = k\nu\sin\theta_1 = kn_2\sin\theta_2$$
$$b_o = kn_o\cos\theta_o \; ; \; b_2 = kn_2\cos\theta_2$$
$$b_1 = k(\nu-i\varkappa)\cos\theta_1$$
(5)

we obtain the following boundary conditions:

$$B_o + A_o = B_1 + A_1$$
$$(A_o - B_o)b_o = (A_1 - B_1)b_1$$
$$B_1 \exp(ib_1W) + A_1 \exp(\bar{i}b_1W) = B_2 \exp(ib_2W)$$
$$[B_1 \exp(ib_1W) - A_1 \exp(-ib_1W)] \; b_1 = b_2 B_2 \exp(ib_2W)$$
(6)

THIN FILMS IN OPTICS

Let us recall that the A's and B's may be complex since b_1 is in several cases complex.

From the boundary conditions we obtain the amplitude ratios:

$$\frac{B_2}{B_o} = t \exp(i\delta) = \frac{4\alpha_2 \exp(-ib_2 W)}{(1+\alpha_1)(1+\alpha_2)\exp(-ib_1 W)-(1-\alpha_1)(1-\alpha_2)\exp(ib_1 W)} \quad (7)$$

$$\frac{A_o}{B_o} = r \exp(i\gamma) = \frac{(1-\alpha_1)(1+\alpha_2)\exp(-ib_1 W)-(1+\alpha_1)(1-\alpha_2)\exp(ib_1 W)}{(1+\alpha_1)(1+\alpha_2)\exp(-ib_1 W)-(1-\alpha_1)(1-\alpha_2)\exp(ib_1 W)}$$

where $\alpha_1 = \dfrac{b_1}{b_o}$, $\alpha_2 = \dfrac{b_1}{b_2}$,

γ and δ are the phase shifts of the transmitted and reflected beams.

In order to determine reflectance and transmittance it is required to evaluate the time average of the normal component of the Poynting's vector

$$R = \frac{A_o A_o^*}{B_o B_o} \quad ; \quad T = \frac{B_2 B_2^* b_2}{B_o B_o b_1} \quad (8)$$

Since this procedure is tedious and lengthy we refer to the literature[1]. We note that T is independent of the propagation direction whereas the value of R depends upon it. We may take advantage of this fact in order to determine the optical constants of a film.[2] As an example of the above treatment let us see the expression of T and R for a thin film of real index $\nu \equiv n_1$ at normal incidence.

$$R_1 = \frac{(n_1+n_o)(n_1+n_2)-4n_1 n_o n_2-(n_1-n_o)(n_1-n_2)\cos\phi}{(n_1+n_o)(n_1+n_2)+4n_1 n_o n_2-(n_1-n_o)(n_1-n_2)\cos\phi} = R_2$$

$$(9)$$

$$T = \frac{8n_1 n_o n_2}{(n_1+n_o)(n_1+n_2)+4n_1 n_o n_2-(n_1-n_o)(n_1-n_2)\cos\phi}$$

$$\phi = 2kn_1 W$$

We easily see that R has maxima and minima when $\cos\phi = \pm 1$,

$\cos\phi = 1 \rightarrow \phi = 2\pi m \rightarrow \lambda = \dfrac{2n_1 W}{m}$ $m = 0, 1, 2$

$$\cos \phi = -1 \rightarrow \phi = \pi(2m+1) \rightarrow \lambda = \frac{4n_1 W}{2m+1}$$

As an example of the optical properties of a thin film we shaw in Fig. 3 the normal reflectance of two films of index $n_1 = 1.22$ and $n_1 = 2.0$ deposited onto a substrate with $n_2 = 1.5$.

From the above treatment we see that the optical behaviour of a thin film is completely described when its complex index or, if one prefers, its optical constants are known and its thickness is given. It may occur however that both thickness and optical constants are not known, in which case one should obtain them from experimental data such as reflectance and tranmittance. This problem, and its solutions, are summarized in the following section.

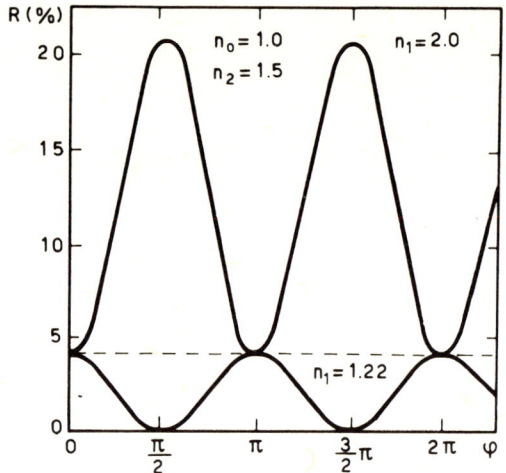

Fig. 3   Reflectance at normal incidence of two films with indices $n_1 = 1.22$ and $n_1 = 2.00$

2.2 Determination of the Optical Constants

The optical constants of thin films can be determined essentially by two methods : (a) polarimetric ; (b) photometric. The polarimetric method employs the ratio of the p- and s-polarized components of the reflected beam, $r_p/r_s = \tan\psi$ and the phase difference of the same components $\delta_p - \delta_s = \Delta$. These two quantities are usually obtained from the parameters that describe the elliptically polarized ligth reflected by the film[3]. If ψ_o and Δ_o, the parameters of the bare substrate, are also known form previous measurements, we may write

$$\Delta - \Delta_o = f(\nu, \chi, W, \lambda)$$
$$\psi - \psi_o = f(\nu, \chi, W, \lambda)$$
(10)

where the two functions of ν, refractive index and χ extinction coefficient, can be written for each wavelength λ, whereas the thickness W of the film under investigation is unknown. The procedure is rather complicated because it is necessary to employ sets of curves that are functions of Δ and ψ and to use successive approximations to find ν, χ and W. Furthermore, polarizers and compensators must be employed. If W is known, as sometimes happens, system (10) can be solved since λ is also known and therefore the two unknowns are ν and χ.

In order to avoid the above limitations one can consider photometric measurements of reflection and transmission at normal incidence. With this procedure, polarizers are not required and the scattered light, particularly important at short wavelengths, will be less than at non-normal incidence.

Existing spectrophotometric methods are based on knowledge of the film thickness but we wish to extend them to absorbing films whose thickness cannot be measured directly. As a consequence, the three unknowns ν, χ and W require the use of at least three independent equations. To this purpose, we observe that three independent measurements can be performed on a thin film supported by a semi-infinite transparent substrate, T, R_1 and R_2. The expression for T, R_1 and R_2 are those, eq(8), given in the previous section where the unknowns are in general ν, χ and W.

We omit from the present review the Kramers-Kronig analysis because it requires measurements over a wide-range of wavelengths. The problem of the determination of optical constants has been discussed by several authors and the interested reader can find severals articles in Optics Journals.

3. TOTAL INTERNAL REFLECTION

Let us consider now the case in which total reflection occurs inside the film. Here we do not examine the way light has entered the film. We can assume that plane waves are reflected back and forth in the film whereas the two media 2 and 0 do not carry any plane wave. The scheme of this case is given in Fig. 4.

We consider here only TE waves since the TM waves can be treated in a completely analogous way. The three refraction indices n_0, n_1 and n_2 are all real. It is also understood that the wave fronts extend from $-\infty$ to $+\infty$ in the y direction so that $\frac{\partial}{\partial y} = 0$
Since we are dealing with TE waves it follows that $E_x = E_z = H_y = 0$ and then we are left with :

$$E_y, \quad H_x = \frac{i}{k} \frac{\partial E_y}{\partial z}, \quad H_z = -\frac{i}{k} \frac{\partial E_y}{\partial x} \qquad (11)$$

where the last two expression stem from Maxwell's equations.

Now we give the fields in the three media.

Field in medium 1 (film)

$$\begin{aligned}
E_y^{(1)} &= A_1 \exp i(\beta x - b_1 z) + B_1 \exp i(\beta x + b_1 z) \\
H_x^{(1)} &= \frac{i}{k} \frac{\partial E_y}{\partial z} = \frac{b_1}{k} [A_1 \exp i(\beta x - bz) - B_1 \exp(\beta x + b_1 z)] \\
H_z^{(1)} &= \frac{i}{k} \frac{\partial E_y}{\partial x} = \frac{\beta}{k} [A_1 \exp i(\beta x - bz) + B_1 \exp(\beta x + b_1 z)]
\end{aligned} \qquad (12)$$

Field in medium 0 (substrate)

The field in this medium can be obtained by recalling that in order to satisfy internal reflection at the substrate-film boundary the inequality

$$\sin \theta_1 > n_0/n_1 \quad \text{must hold and from this :}$$

$$b_0 = \pm ik\sqrt{n_1^2 \sin^2\theta_1 - n_0^2}$$

We choose the plus sign before the root since the field in medium 0 must be finite.

We then write

$$b_0 = ip_0$$

THIN FILMS IN OPTICS

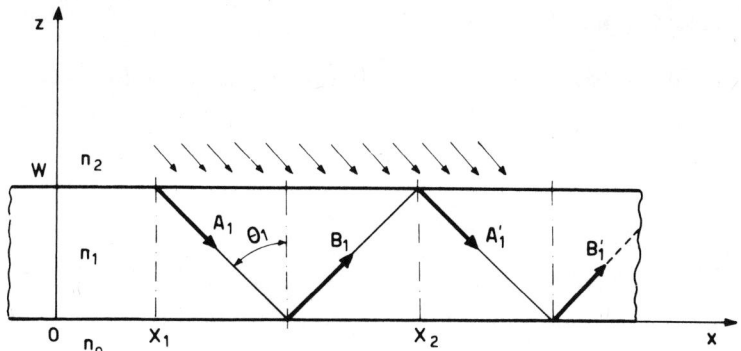

Fig. 4 Film employed in the total internal reflection mode. Light is fed into the film from above

$$E_y^{(o)} = C_o\, e^{p_o z}\, \exp i(\beta x)$$
$$H_x^{(o)} = \frac{i p_o}{k}\, C_o e^{p_o z} \exp i(\beta x) \quad (13)$$
$$H_z^{(o)} = \frac{\beta}{k}\, C_o\, e^{p_o z}\, \exp i(\beta x)$$

Field in medium 2 (air)

With arguments similar to those above we get, with $b_2 = i p_2$,

$$E_y^{(2)} = D_2\, e^{-p_2(z-W)}\, \exp(\beta x)$$
$$H^{(2)} = -\frac{i p_2}{k}\, D_2\, e^{-p_2(z-W)}\, \exp i(\beta x) \quad (14)$$
$$H_z^{(2)} = \frac{\beta}{k}\, D_2\, e^{-p_2(z-W)}\, \exp i(\beta x)$$

Fields in media 0 and 2 are called "evanescent fields". They do not represent free radiation[4]

Boundary conditions

Let us determine the values of the amplitudes B_1, C_o, D_o by employing the boundary conditions at $z = 0$ and $z = W$. For $z = 0$, i.e.

$$E_y^{(1)} = E_y^{(o)}$$
$$H^{(1)} = H^{(o)}$$

which amonts to

$$A_1 + B_1 = C_o$$
$$b_1(A_1 - B_1) = ip_o C_o$$

(15)

With simple passages we get

$$B_1 = A_1 \, e^{-i2\phi_{10}}$$

where

$$\phi_{10} = \tan^{-1} \frac{P_o}{b_1}$$

(16)

which is the well known phase loss in total internal reflection. In the same way at the surface $z = W$ we have:

$$B_1 = A_1 \, e^{i2\phi_{12}}$$

where

$$\phi_{12} = \tan^{-1} \frac{P_2}{b_1}$$

(17)

The range of ϕ_{10} and ϕ_{12} is

$$0 \leq \phi_{10} < \frac{\pi}{2}$$

$$0 \leq \phi_{12} < \frac{\pi}{2}$$

The lower limit corresponds to the critical angle whereas $\frac{\pi}{2}$ corresponds to grazing incidence.

We summarize the above results by saying that in total internal reflection the ray undergoes a phase change $-2\phi_{10}$ at $z = 0$ and $-2\phi_{12}$ at $z = W$.

3.1 Mode Equation

We recall now that the beam A entering the film has a cross

section in the x direction which is much larger than the thickness W of the film. Let us assume that waves A are entering the film at $z = W$ and in particular let us consider the point $x = x_2$, $z = W$. In order to add strength to the waves A in the film the phase of wave A_1' at (x_2,W) must be identical modulo $2m\pi$ to the phase of the wave entering the film. Let us assume that at (x_1,W) both the entering wave and the A_1 wave have phase zero. At point (x_2,W) the phase of the A_1' wave will be $\beta(x_2-x_1)+2b_1W-2\phi_{10}-2\phi_{12}$. The phase of the wave entering the film at (x_2,W) will be $\beta(x_2 - x_1)$.

The above argument leads to the following "mode equation":

$$2b_1W - 2\phi_{10} - 2\phi_{12} = 2m\pi \tag{18}$$

Let us discuss the meaning of the mode equation. Assuming that $n_0 > n_2$ (which requires $n_1 > n_0 > n_2$) we can easily see that in order to have a waveguide mode the following condition must be met:

$$kn_0 < \beta < kn_1$$

When $\beta = kn_1$, then $\theta_1 = \pi/2$, $\phi_{10} = \phi_{12} = \pi/2$, and $b_1 = 0$. It follows also that $W \to \infty$. This is not surprising since grazing incidence is equivalent to direct propagation along x and therefore the boundaries of the film supporting this wave must be at infinite separation. When $\beta = kn_0$ then $\theta_1 = \theta_{1c}$ (critical angle between media 0 and 1), $p_0 = 0$ and $\phi_{10} = 0$. The mode equation (18) gives the minimum thickness in order that the waveguide can support a mode of order m.

$$W_{min} = \frac{1}{k}[m\pi + \tan^{-1}(\frac{n_0^2 - n_2^2}{n_1^2 - n_0^2})^{1/2}] \cdot (n_1^2 - n_0^2)^{-1/2} \tag{19}$$

In the general case when β has intermediate values, between those considered above the physical meaning of the different modes of propagation can be grasped more firmly by writing eq. (18) in the following way

$$W = W_{10} + W_{12} + mW_1 \tag{18'}$$

Where

$W_{10} = \phi_{10}/b_1$,

$W_{12} = \phi_{12}/b_1$,

$W_1 = \pi/b_1$

This is equivalent to say that the film thickness for given β and

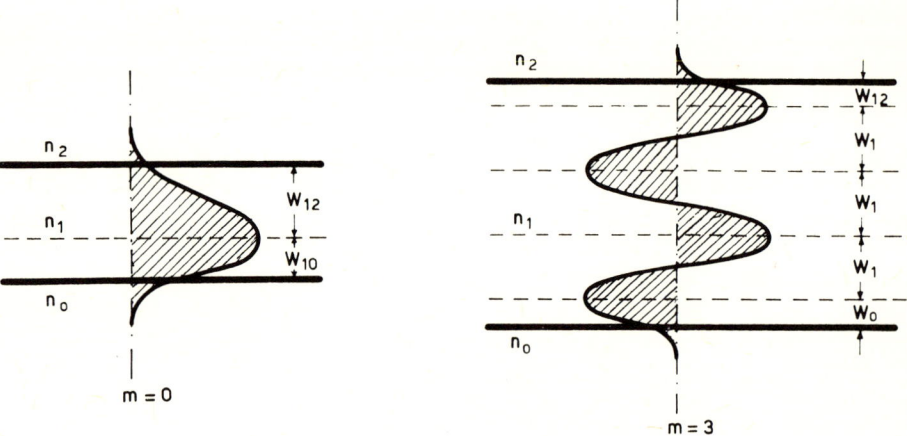

Fig. 5   Examples of thin waveguides showing the electric field amplitude for different propagation modes.

m is that of the m=0 mode plus the quantity mW_1.

Examples are given in Fig. 5.

When n_0, n_1 and n_2 are known we can easily estimate both β/k and W for a given angle of incidence. The curves $W = W(\beta/k)$ with m as a parameter are called "characteristics of the waveguide". From the value of β/k we obtain the wave velocity in the propagation direction : $V_p = C(k/\beta)$, where C is the velocity of light.

In Fig. 6 we report such curves for a GaAs film deposited on an Irtran II substrate, according to Tien and Ulrich[5]. We note that W increases with β/k and with m as expected from the mode equation. The curves of Fig. 6 show also the validity of eq. (18')

3.2 Power Carried in a Waveguide

Let us evaluate the power carried by a dielectric waveguide when the waves are TE. This can be obtained by integrating from $z = -\infty$ to $z = +\infty$ the x component of the Poynting's vector $(c/8\pi)\mathrm{Re}(E_y H_z)$ related to the field which results from the A_j and B_j waves.[4] We note that the field penetrate both media 0 and 2. The electric field in the thin film is given (neglecting $e^{-i\omega t}$) by

$$E_y^{(1)} = A_1 \exp i(\beta x - b_1 z) + B_1 \exp i(\beta x + b_1 z) \tag{20}$$

In order to simplify the interference between waves we translate the x axis in such a way that E_y is maximum or, in other words, to a point were waves A_1 and B_1 are in phase.[5] We have then :

$$E_y^{(1)} = A_1 \exp [i(\beta x - b_1 z)] + A_1 \exp [i(\beta x + b_1 z)] =$$

$$= 2A_1 \cos(b_1 z) \exp(i\beta x) \tag{21}$$

From eq.(10) we get

$$H_z^{(1)} = 2A_1 n_1 \sin\theta_1 \cos(b_1 z) \exp(i\beta y) \tag{22}$$

and then

$$S_x^{(1)} = \frac{c}{2\pi} A_1^2 n_1 \sin\theta_1 \cos^2(b_1 z) \tag{23}$$

which gives for the power carried within the film

$$P_x^{(1)} = \int_{-W_0}^{W_2} \frac{c}{8\pi} \mathrm{Re}(E_y H_z) dz = \frac{c}{2\pi} A_1^2 n_1 \sin\theta_1 \{\frac{1}{2b_1}[\sin(b_1 z)\cos(b_1 z) + b_1 z] \}_{W_0}^{W_2} \tag{24}$$

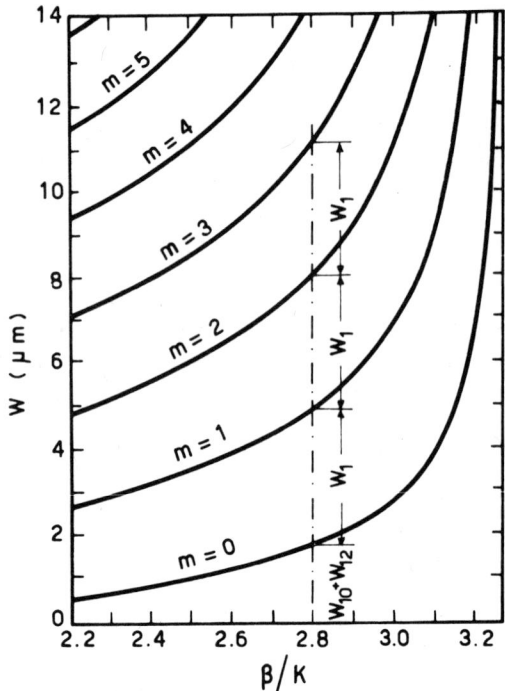

Fig. 6 Characteristics of a thin film waveguide : thickness vs. $n_1 \sin \theta_1$ according to Tien and Ulrich.[5]

In Fig. 7 we have shown the values of $E_y^{(1)}$ vs. z, with the power flow along x.

We have seen that at the boundaries the waves undergo a phase loss so that the fields at $z = W_o$ and $z = W_2$ must be $\pm A \cos \phi_{10}$ and $\pm A \cos \phi_{12}$, respectively, with A a constant, i.e., the value of the field at $z = 0$. If we put $b_1 W_2 = \phi_{12}$ then the field at $z = W_2$ is $A \cos \phi_{12}$. By putting $b_1 W_o = \phi_{10} + m\pi$ we get for $z = -W_o$ the field $+ A \cos \phi_{10}$ if m is even and $- A \cos \phi_{10}$ if m is odd.

THIN FILMS IN OPTICS

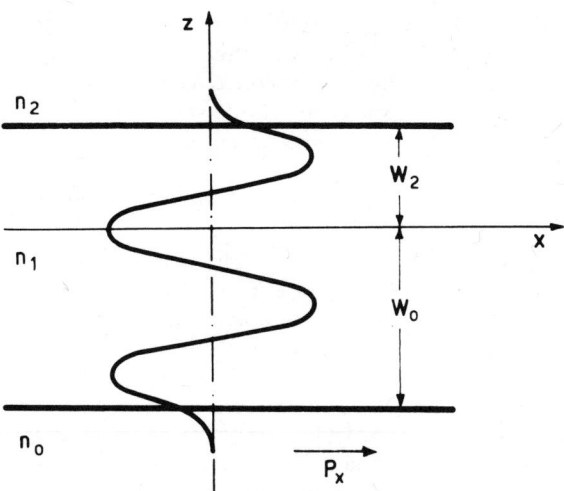

Fig. 7 The diagram shows the choice of the x axis along z. (See text).

It is straightforward to obtain :

$$P_x^{(1)} = \frac{c}{4\pi} A_1^2 n_1 \sin\theta_1 [W + \frac{\sin^2\phi_{10}}{p_0} + \frac{\sin^2\phi_{12}}{p_2}] \qquad (25)$$

By integrating the Poynting's vector from $z = W_2$ to $z = +\infty$ we get the power carried by the air side

$$P_x^{(2)} = \int_{W_2}^{+\infty} S_x^{(2)} dz = \frac{c}{4\pi} A_1^2 n_1 \sin\theta_1 \frac{\cos^2\phi_{12}}{p_2} \qquad (26)$$

and similarly for the substrate side

$$P_x^{(o)} = \int_{-\infty}^{-W_o} S_x^{(o)} dz = \frac{c}{4\pi} A_1^2 n_1 \sin\theta_1 \frac{\cos^2\phi_{10}}{p_0} \qquad (27)$$

The total power is then

$$P_x = P_x^{(o)} + P_x^{(1)} + P_x^{(2)} = \frac{c}{4\pi} A_1^2 n_1 \sin\theta_1 [W + \frac{1}{p_0} + \frac{1}{p_2}] \qquad (28)$$

The quantity $(c/4\pi)A_1^2 n_1 \sin\theta_1$ is the Poynting's vector in the x direction due to the superposition of the A_1 and B_1 waves. The factor $W + (1/P_0)+(1/P_2)$ is the equivalent thickness W_{eq} of the waveguide where the power is confined. Note that $W_{eq} > W$ since the fields extend also into the substrate and into the air.

3.3 Experimental Setups for Coupling Waveguides to External Beams

In order to couple the light to the waveguide some device is necessary since otherwise it is not possible to satisfy the critical angle condition. The most common way of introduction of the radiation into the waveguide is by means of the evanescent field at the base of a prism or by means of a diffraction grating.[5,6]. In the case of the prism coupler it is necessary that the index n_3 of the prism be larger than n_1, the index of the film guide, if all possible modes have to be excited. We have then $n_3 > n_1 > n_0 > n_2$. In Fig. 8 is shown the scheme of a prism coupler.

The evanescent field is created at the gap below the base of the prism and if the gap is a fraction of a wavelength there will be wave propagation in the film. When $kn_3 \sin\theta_3 = \omega/v_p$, where v_p is the phase velocity of one of the modes in the film, the coupling becomes efficient and the power of the laser beam is transferred to the film via the prism.

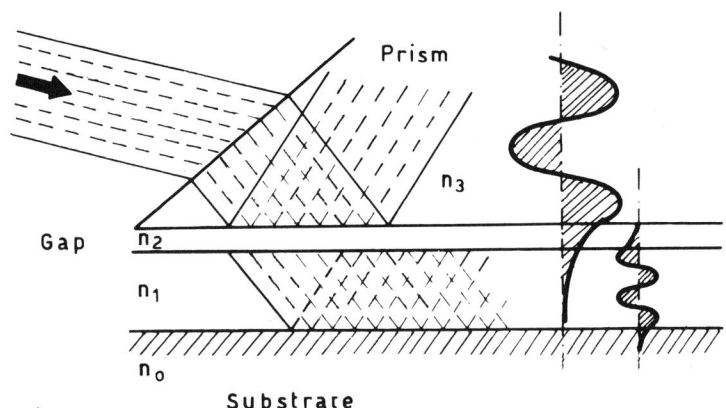

Fig. 8 Description of the prism coupler for feeding light into a thin waveguide.

The length of the coupling region between film and prism cannot be arbitrary since the power flow may change direction and leave the film entering the prism. Obviously a similar prism can be used for extracting power from a film waveguide.

Coupling constant not too far from unity have been obtained (88% Ulrich[7]).

Another coupler is obtained from a grating deposited directly onto the film waveguide. The coupling occurs between the horizontal component of the wavevector in the outer beam and the propagation constant β of one of the allowed modes. It can be shown that the condition for coupling is $c/v = \sin\theta + m(\lambda_o/d)$ where c is the velocity of the guided wave, in the order of the mode, θ the angle of incidence and d the grating constant[8].

Other couplers have been examined and among them we mention the tapered film which allows light to be both fed into the waveguide and to be extracted from it[9].

Goos-Haenchen effect. So far we have considered a simple zig-zag model with phase losses at the two boundaries where total reflection occurs. By doing so we have neglected the contribution of the Goos-Haenchen effect which amounts to a logitudinal shift of the totally reflected ray. In a recent paper[10] it is claimed that although the approach used in these papers is valid for obtaining the mode equation, and the corresponding phase velocities, it fails to predict the correct energy exchanges (among modes e between adjacent guides) and the group velocity. For a treatment of these problems the reader is referred to the paper by Kogelnik and Weber[10].

4. THIN FILM MULTILAYERS

We are going now to study the optical behaviour of a multilayer stack built from several thin films. In order to illustrate the basic concepts which are related to the use of 2×2 matrices as proposed by Herpin and Abelès[11] we consider normal incidence. Oblique incidence can be easily described by means of the so called equivalent indices.[12]. In Fig. 9 is reported the scheme of the multilayer. We omit the term $e^{-i\omega t}$ from the equation and recall that $N_j = \nu_j - \chi_j$, in general. In layer j the electric and magnetic fields result from the superposition of a progressive wave A_j and a regressive wave B_j are given by

$$E(z) = A_j \exp[i(kN_j z - \alpha_j)] + B_j \exp[i(-kN_j z - \beta_j)] \quad (29)$$
$$H(z) = N_j\{A_j \exp[i(kN_j z - \alpha_j)] - B_j \exp[i(-kN_j z - \beta_j)]\}$$

It is obvious that the amplitudes A_j and B_j and the phase constants

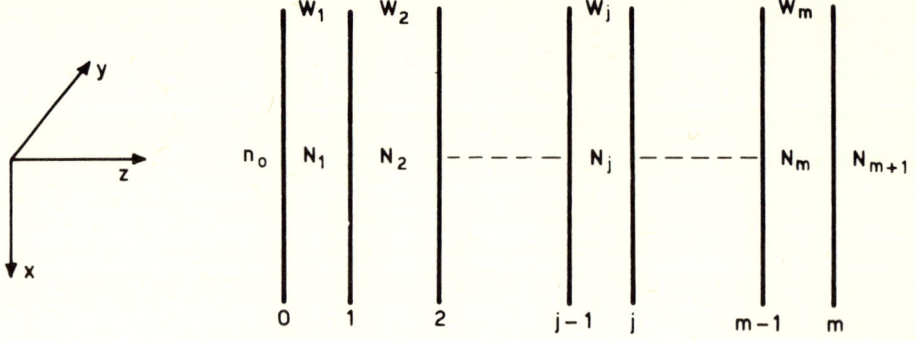

Fig. 9   Scheme of the multilayer considered here

α_j and β_j are not arbitrary but can be determined by means of the boundary conditions. Before employing the boundary conditions let us define the following quantities according to Fig. 10 and using the following notation :

$$E^{(t)}_{j-} = A_j \exp i(\alpha_j - \frac{2\pi}{\lambda} N_j z_j)$$

$$E^{(r)}_{j-} = B_j \exp i(\beta_j + \frac{2\pi}{\lambda} N_j z_j)$$

$$E^{(t)}_{(j-1)+} = A_j \exp i(\alpha_j - \frac{2\pi}{\lambda} N_j z_{j-1})$$

$$E^{(r)}_{(j-1)+} = B_j \exp i(\beta_j + \frac{2\pi}{\lambda} N_j z_{j-1})$$

(30)

From eqs. (30) we easily get

$$E^{(t)}_{(j-1)+} = E^{(t)}_{j-} \exp(i\phi_j)$$

$$E^{(r)}_{(j-1)+} = E^{(r)}_{j-} \exp(-i\phi_j)$$

(31)

with $\phi_j = \frac{2\pi}{\lambda} N_j (z_j - z_{j-1})$

THIN FILMS IN OPTICS

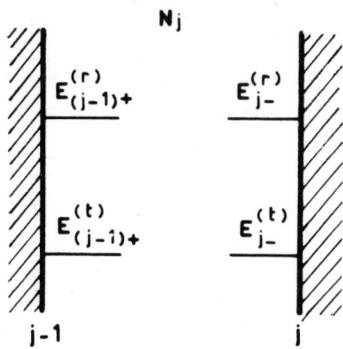

Fig. 10   Light rays at the boundaries of the j-th layer.

By applying the boundary conditions at $z = z_j$ we get

$$E^{(t)}_{j-} + E^{(r)}_{j-} = E^{(t)}_{j+} + E^{(r)}_{j+}$$
$$N_j[E^{(t)}_{j-} - E^{(r)}_{j-}] = N_{j+1}[E^{(t)}_{j+} - E^{(r)}_{j+}] \quad (32)$$

We note that eqs. (32) represent a homogeneous system of $2(m+1)$ linear equations, with $2(m+1)$ unknowns. We have also $E^{(r)}_m = 0$, i.e. there is no reflected component from the right hand side of the last surface and we may put $E^{(t)}_{m+} = 1$ as reference. This reduces the unknowns to $2(m+1)$ and therefore system (32) can be solved. By writing eqs. (30) and (31) for the j-1 interface and recalling eqs 32 we have the recurrence relations for the amplitudes of the transmitted and reflected fields,

$$E^{(t)}_{(j-1)-} = \tfrac{1}{2}(1+N_j/N_{j-1}) \, E^{(t)}_{j-} \exp(i\phi_j) + \tfrac{1}{2}(1-N_j/N_{j-1}) E^{(r)}_{j-} \exp(-i\phi_j)$$
$$(33)$$
$$E^{(r)}_{(j-1)-} = \tfrac{1}{2}(1-N_j/N_{j-1}) E^{(t)}_{j-} \exp(i\phi_j) + \tfrac{1}{2}(1+N_j/N_{j-1}) E^{(r)}_{j-} \exp(-i\phi_j)$$

In particular recalling that $E^{(t)}_{m+} = 1$ and $E^{(r)}_{m+} = 0$ we have

$$E^{(t)}_{m-} = \tfrac{1}{2}(1+N_{m+1}/N_m) \quad ; \quad E^{(r)}_{m-} = \tfrac{1}{2}(1-N_{m+1}/N_m) \quad (34)$$

and from this by the recurrence eqs. 33 we can determine the reflectance and transmittance of the whole multilayer.

4.1 Matrix Technique

All calculations on multilayers (here we are considering isotropic media) can be more easily done by employing the 2×2 matrices an originally proposed by Herpin[11]. Later Abelès showed that, for a general multilayer where the only significant variable is z, the fields at $z = z_1$ and $z = z_2$ can be written as

$$\begin{pmatrix} E(z_1) \\ H(z_1) \end{pmatrix} = \begin{pmatrix} a_{11} & a_{12} \\ a_{21} & a_{22} \end{pmatrix} \cdot \begin{pmatrix} E(z_2) \\ H(z_2) \end{pmatrix} \quad (35)$$

where the a_{ij} element depend only upon the optical constants and thickness of the system between planes $z = z_1$ and $z = z_2$. We note also that the recursion eqs (33) are a linear transformation and can then be written in the following way:

$$\begin{pmatrix} E^{(t)}_{(j-i)-} \\ E^{(r)}_{(j-i)-} \end{pmatrix} = \begin{pmatrix} \dfrac{2N_{j-1}}{N_{j-1}+N_j} \exp(i\phi_j) & \dfrac{N_{j-1}-N_j}{2N_{j-1}} \exp(-i\phi_j) \\ \dfrac{N_{j-1}-N_j}{2N_{j-1}} \exp(i\phi_j) & \dfrac{2N_{j-1}}{N_{j-1}+N_j} \exp(-i\phi_j) \end{pmatrix} \begin{pmatrix} E^{(t)}_{j-} \\ E^{(r)}_{j-} \end{pmatrix} \quad (36)$$

where the 2×2 matrix, from now on called $[N_j]$, depends upon thickness W_j and indices N_{j-1} and N_j. Futhermore the use of matrices becomes more meaningful if we consider the total amplitudes of the fields $E(z)$ and $H(z)$. Eqs. (29) can now be written also as

$$\begin{pmatrix} E_j \\ H_j \end{pmatrix} = \begin{pmatrix} 1 & 1 \\ N_j & -N_J \end{pmatrix} \cdot \begin{pmatrix} E^{(t)}_{j-} \\ E^{(r)}_{j-} \end{pmatrix} \quad (37)$$

having defined $E_j = E^{(r)}_{j\pm} + E^{(t)}_{j\pm}$

$$H_j = H^{(r)}_{j\pm} + H^{(t)}_{j\pm} \quad (38)$$

It is simple to show that we can also write

$$\begin{pmatrix} E_{j-1} \\ H_{j-1} \end{pmatrix} = \begin{bmatrix} S_{j-1} \end{bmatrix} \begin{bmatrix} N_j \end{bmatrix} \begin{bmatrix} j \end{bmatrix}^{-1} \cdot \begin{pmatrix} E_j \\ H_j \end{pmatrix} = \begin{bmatrix} M_j \end{bmatrix} \cdot \begin{pmatrix} E_j \\ H_j \end{pmatrix}$$

where $[N_j]$ is the 2×2 matrix of eq. (36) and $[S_j]$ is the 2×2 matrix of eq. (37) and therefore $[M_j]$ becomes

$$[M_j] = \begin{pmatrix} \cos\phi_j & \dfrac{i}{N_j}\sin\phi_j \\ iN_j\sin\phi_j & \cos\phi_j \end{pmatrix} \qquad (40)$$

It should be noted that the index N_j can be replaced by the so called equivalent index when considering non normal incidence.(12) $N_j \rightarrow N_j/\cos\theta_j$ for p polarization and $N_j \rightarrow N_j\cos\theta_j$ for s polarization. We have seen than that by means of eq. (39), with definition (40), the field at two boundaries are easily related to each other. Expressed through E and H the initial conditions become

$$E_m = E_{m+}^{(t)} = 1, \quad H_m = N_{m+1} \quad E_{m+}^{(t)} = N_{m+1}$$

Then by applying the recurrence eq. (39) we get

$$\begin{pmatrix} E_o \\ H_o \end{pmatrix} = \Pi_j \begin{pmatrix} M_j \end{pmatrix} \begin{pmatrix} 1 \\ N_{m+1} \end{pmatrix} = \begin{pmatrix} M \end{pmatrix} \begin{pmatrix} 1 \\ N_{m+1} \end{pmatrix} \qquad (41)$$

It is now possible to evaluate R and T for any multilayer by means of the matrix elements of $[M]$ by recalling that

$$\begin{pmatrix} 1 & 1 \\ n_o & -n_o \end{pmatrix} \cdot \begin{pmatrix} E_o^{(t)} \\ E_o^{(r)} \end{pmatrix} = \begin{pmatrix} m_{11} & m_{12} \\ m_{21} & m_{22} \end{pmatrix} \begin{pmatrix} E_m \\ H_m \end{pmatrix} \qquad (42)$$

which, when solved with respect to $E_{o-}^{(t)}$ and $E_{o-}^{(r)}$ becomes

$$\begin{pmatrix} E_{o-}^{(t)} \\ E_{o-}^{(r)} \end{pmatrix} = \begin{pmatrix} \tfrac{1}{2}(m_{11}+m_{21}/n_o) & \tfrac{1}{2}(m_{12}+m_{22}/n_o) \\ \tfrac{1}{2}(m_{11}-m_{21}/n_o) & \tfrac{1}{2}(m_{12}-m_{22}/n_o) \end{pmatrix} \cdot \begin{pmatrix} 1 \\ N_{m+1} \end{pmatrix} \qquad (43)$$

This leads to the following expressions for reflectance and transmittance

$$R = \left| \frac{m_{11} - m_{21}/n_o + N_{m+1} m_{12} - N_{m+1} m_{22}/n_o}{m_{11} + m_{21}/n_o + N_{m+1} m_{12} + N_{m+1} m_{22}/n_o} \right|^2 \quad (44)$$

$$T = \frac{\text{Re}(N_{m+1})}{n_o} \frac{4}{|m_{11} + m_{21}/n_o + N_{m+1} m_{22}/n_o|^2} \quad (45)$$

In concluding this section we note that a multilayer is described by a matrix $[M]$ whose determinant is unity. The matrix $[M]$ is the ordered product of individual $[M_j]$ matrices which have real elements along the main diagonal and purely imaginary elements on the other diagonal when the layer is transparent. If the film is absorbing then the elements are complex. It should also be noted that the multilayer is fully described by the 2×2 characteristic $[M]$ independently of the media which surround it (the same can be said of any portion of the multilayer). It can be shown that any multilayer is equivalent to a single layer if the stack is symmetric or to a double layer otherwise. In all cases, however, one should expect strong dispersion and this is what makes the multilayers so useful for many applications.

4.2 Some Examples

In order to show the advantages of the matrix approach let us examine two simple examples which refer to real refractive indices. More detailed applications to multilayer systems will be illustrated by E. Pelletier in his paper[13].

Example (a). Let us consider a single selfsupporting film whose optical thickness is $nW = p(\lambda/2)$ with p integer. Its characteristic matrix M is

$$M = (-1)^p \begin{bmatrix} 1 & 0 \\ 0 & 1 \end{bmatrix}$$

Using this matrix it is shown by means of eq. 44 that $R = 0$ and $T = 1$ which is the expected result for a film known to leave unaffected a light beam of wavelength λ.

Example (b). We examine a dielectric Fabry-Perot filter made of a number 2p of layers with indices n_L and n_H, alternaively, which we indicate with LHLH...LH substrate = (LH)P substrate. If the individual layers have optical thickness $\lambda_0/4$ their matrices are the following:

$$\underset{\sim}{L} = \begin{pmatrix} 0 & i/n_L \\ in_L & 0 \end{pmatrix}, \qquad \underset{\sim}{H} = \begin{pmatrix} 0 & i/n_H \\ in_H & 0 \end{pmatrix}$$

It should be noted that the $\underset{\sim}{L} \cdot \underset{\sim}{H}$ product is a diagonal matrix with real elements

$$L \cdot H = \begin{pmatrix} -n_H/n_L & 0 \\ 0 & -n_L/n_H \end{pmatrix}$$

For a multilayer with p pairs the reflectance, according to eq. 44, then becomes:

$$R_{\lambda_0} = \left\{ \frac{(n_H/n_L)^{2p} - n_S/n_0}{(n_H/n_L)^{2p} + n_S/n_0} \right\}^2$$

where n_0 is the index of the free side medium (air) and n_S is the index of the substrate. As can easily be observed $R \to 1$ at $\lambda = \lambda_0$ when $n_H > n_L$, and p becomes very large. This system is used for several applications such as narrow band filters, laser cavity mirrors, etc. As a final comment to this section we may add that the matrix approach is very practical since it is very suitable for computer design of optical components as will be shown by the paper of Pelletier[13].

5. GENERAL FORMULATION FOR ANISOTROPIC SYSTEMS

In the previous sections we have considered only isotropic media when studying light propagation in thin films. However a great deal of interest has arisen for anisotropic materials in view of their applications to optics and because of their applications to thin film devices (modulators, rotators, etc.). A recent approach to the description of the optical properties of anisotropic multilayers has been developped by Berreman[14] and by Vassell[15] who have employed a 4×4 matrice technique (suggested originally by other authors) which is a generalization on the 2×2 matrix technique described above. Here we are only going to outline this method, referring the reader to the original papers for more details. The method associates with each interface of the multilayer a 4-vector whose components are the tangential projections of the electric and magnetic vectors. The 4-vectors belonging to two different layers can be related (as it has been done previously for the tangential components) of E and H in section 4. for isotropic media) and a 4×4 transfer matrix is established. The transfer matrix of a multilayer results then from the ordered product of individual matrices in a manner analogous to that of isotropic media.

5.1 4 × 4 Matrix Formalism

According to the consideration above, Maxwell's equations may be written in the following matrix formalism

$$\begin{pmatrix} 0 & 0 & 0 & 0 & -\partial/\partial z & \partial/\partial y \\ 0 & 0 & 0 & \partial/\partial z & 0 & -\partial/\partial x \\ 0 & 0 & 0 & -\partial/\partial y & \partial/\partial x & 0 \\ 0 & \partial/\partial z & -\partial/\partial y & 0 & 0 & 0 \\ -\partial/\partial z & 0 & \partial/\partial x & 0 & 0 & 0 \\ \partial/\partial y & -\partial/\partial x & 0 & 0 & 0 & 0 \end{pmatrix} \begin{pmatrix} E_x \\ E_y \\ E_z \\ H_x \\ H_y \\ H_z \end{pmatrix} = \frac{1}{c}\frac{\partial}{\partial t} \begin{pmatrix} D_x \\ D_y \\ D_z \\ B_x \\ B_y \\ B_z \end{pmatrix} \quad (46)$$

This equation may be condensed by the following expression

$$\underset{\sim\sim}{RG} = \frac{1}{c}\frac{\partial}{\partial t} \underset{\sim}{C} \quad (47)$$

where the R matrix is made of two 3×3 matrices with all zeros (2nd and 4th quadrants) and by the 3×3 matrices $\underset{\sim}{R_1}$ a,d $-\underset{\sim}{R_1}$:

$$\underset{\sim}{R} = \begin{pmatrix} 0 & R_1 \\ -R_1 & 0 \end{pmatrix}$$

with $\underset{\sim}{R_1}$ = curl.

Here we are interested in a stack of thin layers at oblique incidence and therefore the propagation vector along x, $\beta = k n_i \sin\theta_i$, is a constant whereas there is no y component. Then $\underset{\sim}{R_1}$ becomes

$$\underset{\sim}{R_1} = \begin{pmatrix} 0 & -\frac{\partial}{\partial z} & 0 \\ \frac{\partial}{\partial z} & 0 & -i\beta \\ 0 & i\beta & 0 \end{pmatrix} \quad (48)$$

Continuing our analysis, which holds for the remainder of these papert, we may write

$$\begin{pmatrix} \underset{\sim}{E} \\ \underset{\sim}{H} \end{pmatrix} = \underset{\sim}{\Gamma}(z) \exp[i(\beta x - \omega t)] \tag{49}$$

$$\begin{pmatrix} \underset{\sim}{D} \\ \underset{\sim}{B} \end{pmatrix} = \underset{\sim}{\Lambda}(z) \exp[i(\beta x - \omega t)]$$

where $\Gamma(z)$ and $\Lambda(z)$ represent the amplitudes of the fields E, H, D B along the direction of the normal to the multilayer. As a consequence of this we may rewrite eq.(47) which becomes :

$$\underset{\sim\sim}{R}\underset{\sim}{\Gamma} = -ik\underset{\sim}{\Lambda} \tag{50}$$

When neglecting non linear effects we may find a linear relation between $\underset{\sim}{\Lambda}$ and $\underset{\sim}{\Gamma}$

$$\underset{\sim}{\Lambda} = \underset{\sim\sim}{M}\underset{\sim}{\Gamma} \tag{51}$$

with the 6×6 matrix M :

$$\underset{\sim}{M} = \begin{pmatrix} \underset{\sim}{\varepsilon} & \underset{\sim}{\rho} \\ \underset{\sim}{\rho'} & \underset{\sim}{\mu} \end{pmatrix}$$

where ε, ρ, ρ', and μ are 3×3 matrices. If we neglect optical rotation, then $\rho = \rho' = 0$ and we are left with the 3×3 dielectric tensor ε and the permeability tensor μ. From eqs.(50) and (51) we get

$$\underset{\sim}{R}\underset{\sim}{\Gamma} = -ik\underset{\sim\sim}{M}\underset{\sim}{\Gamma} \tag{52}$$

Now we apply definitions (48) to (52) and it is found that the 3rd and 6th components of (52) are linear algebraic equations in the six components and the remaining four linear differential equations of the first order can be written for four components. Following reference[14] we eliminate Γ_3 and Γ_6 and one obtains

$$-\frac{i}{k}\frac{\partial}{\partial z}\underset{\sim}{P}\cdot\begin{pmatrix} E_x \\ E_y \\ H_x \\ H_y \end{pmatrix} = \underset{\sim}{S}\cdot\begin{pmatrix} E_x \\ E_y \\ H_x \\ H_y \end{pmatrix} \tag{53}$$

where

$$P = \begin{pmatrix} 0 & 0 & 0 & 1 \\ 0 & 0 & -1 & 0 \\ 0 & -1 & 0 & 0 \\ 1 & 0 & 0 & 0 \end{pmatrix} \quad (54)$$

and S is expressed through M.[14]. Finally with the aid of the permutation matrix Q :

$$Q = \begin{pmatrix} 1 & 0 & 0 & 0 \\ 0 & 0 & 1 & 0 \\ 0 & 0 & 0 & 1 \\ 0 & 1 & 0 & 0 \end{pmatrix}$$

which gives

$$\begin{pmatrix} E_x \\ E_y \\ H_x \\ H_y \end{pmatrix} = Q \psi \quad (55)$$

and

$$\psi = \begin{pmatrix} E_x \\ H_y \\ E_y \\ H_x \end{pmatrix}$$

eq. (53) may be written as

$$\frac{\partial \psi}{\partial z} = ik\Delta\psi \quad (56)$$

where $\Delta = Q^{-1}PSQ$

It can be verified that if $M_{ij} = M_{ji}$ and $\rho = \rho' = 0$ (non optically active media) then also $S_{ij} = S_{ji}$. We cannot give more details here about the matrix Δ and refer the reader to the literature[14,15]

Let us now assume that eq.(56) has the formal solution

$$\psi(z) = \exp[ik(z-z')\Delta] \cdot \psi(z') \quad (57)$$

where the matrix representation of the exponential operator is the transfer matrix $U(z-z')$ of the layer of the thickness $z-z'$. The inverse of this operator is the characteristic matrix C known in the literature.

$$\underset{\sim}{C}(z-z') = \underset{\sim}{U}(z'-z)$$

For the evaluation of $U(z'-z)$ we must refer again to the literature.

Let us now consider a multilayer such as that of Fig. 9. The profile of the field component parallel to the z plane inside the jth layer ($z_{j-1} \leq z \leq z_j$) can be written as

$$\psi_j(z) = U(z-z_{j-1})\psi(z_{j-1}) \qquad (58)$$

With this formalism we generate the 4×4 transfer matrix T for the multilayer

$$T = U_1 \cdot U_2 \cdots U_m \qquad (59)$$

with $U_j = U(z_j - z_{j-1})$

At this stage if the optical constants of the materials employed are known, we are able to evaluate the properties of any stack of anisotropic layers by means of the mere use of the product of the U matrices. The problem can be easily handled by a computer and therefore one can design systems for different applications. A few examples are found in the paper by Vassell[15]. It is shown, for instance, that by straightforward application of the 4×4 matrices the waveguiding condition for a uniaxial layer are easily obtained. Other examples are found in the papaer by Berreman[14].

REFERENCES

1. L.N. Hadley and D.M. Dennison, J. Opt. Soc. Am. **37**, 451 (1947).

2. G. Baldini and L. Rigaldi, J. Opt. Soc. Am. **69**, 495 (1970).

3. Sec. e.g., Meyer, Z. Physik **168**, 169 (1962).

4. M. Born and E. Wolf, Principles of Optics, Pergamon Press

5. P.K. Tien and R. Ulrich, J. Opt. Soc. Am. **60**, 1325 (1970).

6. P.K. Tien and R.J. Martin, Appl. Phys. Lett. **14**, 291 (1969).

7. R. Ulrich, J. Opt. Soc. Am. **63**, 1419 (1973).

8. M.L. Daiess and L. Kuhn, Appl. Phys. Lett. 16, 523 (1970).
    H. Kogelnik and T. Sosnowski, Bell Syst. Tech. J. 49, 1602 (1970).

9. P.K. Tien and R.J. Martin, Appl. Phys. Lett. 18, 398 (1971).

10. H. Kogelnik and H.P. Weber, J. Opt. Soc. Am. 64, 174 (1974).

11. A. Herpin, Compt. Rend. Acad. Sci., 225, 182 (1947).
    F. Abelès, Ann. Phys. (Paris), 5, 596 and 706 (1950).

12. J. MacDonald, Metal Dielectric Multilayers. A Hilger, London (1971).

13. E. Pelletier, Paper in this Book

14. D.W. Berreman, J. Opt. Soc. Am. 62, 502 (1972).

15. M.O. Vassell, J. Opt. Soc. Am. 64, 166 (1974).

RADIATION EFFECTS IN THIN FILMS

A. Holmes-Siedle

J.J. Thomson Physical Laboratory

University of Reading, Reading RG 2AF, England

1. INTRODUCTION

The term "radiation effects" is usually used to describe the secondary phenomena which occur after a high-energy photon or particle has passed through a solid lattice and disturbed it by transferring momentum to some atoms or by raising some atoms to electronically excited states. While these primary transfers of energy are predictable and roughly similar in form for all elements and compounds, the sequelae depend very strongly on the chemical and physical structure of the solid. For thin films, we should thus observe virtually all the effects expected in the corresponding bulk material, to which will be added a range of effects characteristic of the thin-film state such as surface effects, non-stoichiometry, high impurity levels, strain and disorder. Also, since thin films are frequently used in electronic devices, we must interest ourselves in effects associated with drift or displacement of electrons, holes or ions. This variety makes the investigation of radiation effects in thin films a challenging task especially since the nature of thin films eliminates several of the conventional investigative tools used for bulk crystals. However, because of its technological importance, the field continues to develop.

We will first discuss some radiation effects which have proved to be dominant in thin films and will go on to describe some of the techniques used to study these effects and some important cases of radiation environment which may cause degradation of thin films. We will then describe some particular thin-film devices and other interesting interactions of radiation with thin foils or films.

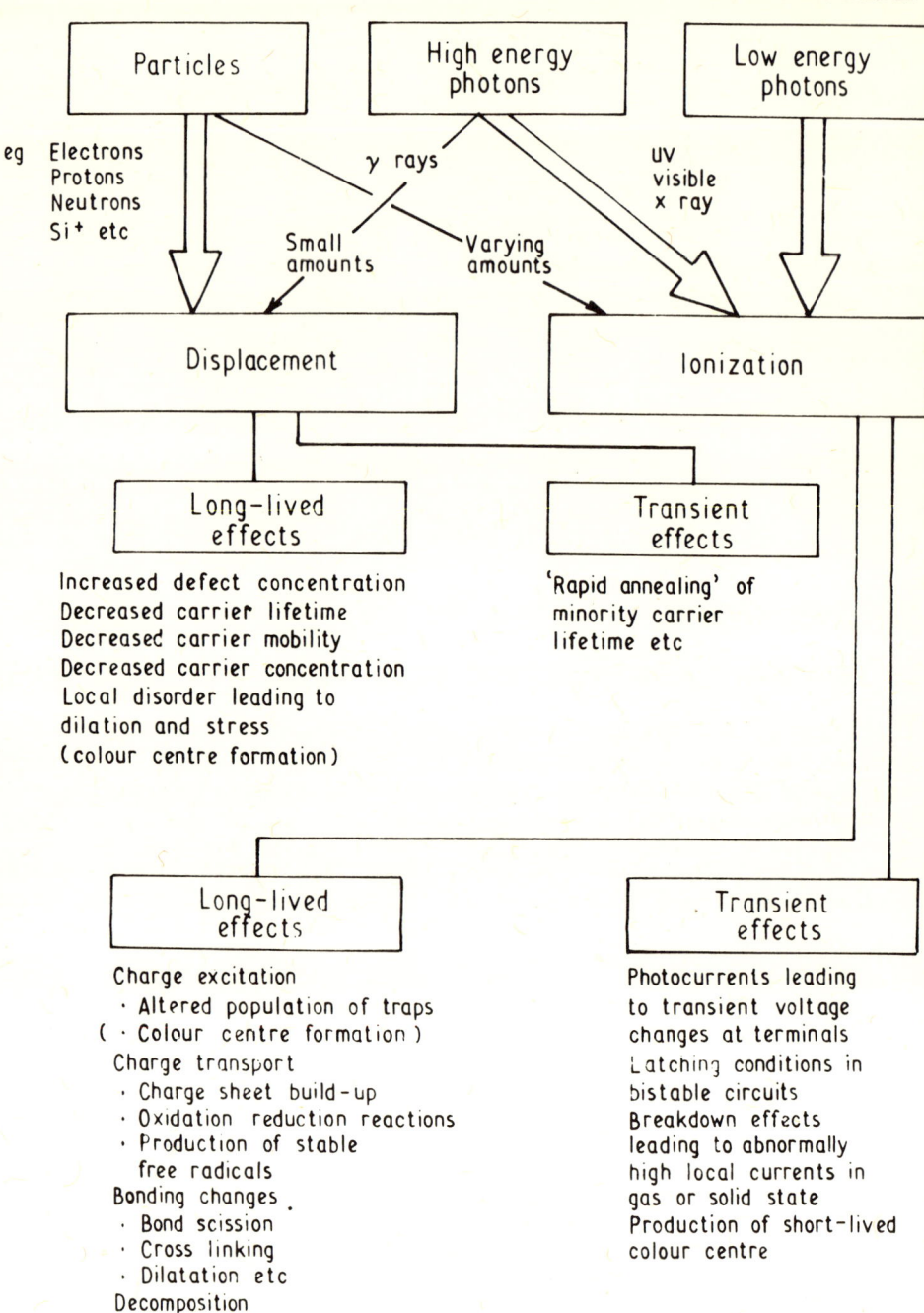

Fig. 1   A classification of radiation effects

2. RADIATION EFFECTS AND POINT DEFECTS

Fig. 1 gives a summary of the wide range of effects which can follow a primary displacement or ionisation in a solid [1]. The distinction made here between high-energy and low-energy photons concerns whether the photon can transfer sufficient energy to the solid to produce atomic displacements, for example by Compton scattering. This requires a photon energy in the MeV range. At lower energies, there are still two large and important groups of phenomena which can occur, namely transient and permanent ionisation effects.

Point defects are involved in two ways. Firstly, the primary vacancies and interstitial atoms may complex with impurities or aggregate into groups and these structures may then be stable for long periods at room temperature ; secondly, point defects formed in a material when it is prepared may interact with the carriers produced by irradiation and change their charge state. Since thin films commonly contain high concentrations of defects formed during preparation, ionization effects will figure quite strongly in our discussion. By the same token, radiation-induced displacement may figure less strongly as a technological problem, in a system which starts its life with a large number of defects although displacement may be interesting scientifically, if only as a method for the controlled change of point defect concentration in films. An example of the latter method is ion implantation ; the range of a kilovolt ion is of the order of hundreds of nanometres ; we can make many uses of this similarity between particle range and film depth.

Other chapters have described the nature of the amorphous state and some defects observed in semiconductor and dielectric films. In this chapter, we will attempt to enlarge on the nature of point defects which exhibit strong electronic activity. Because of the experimental difficulties, a sound knowledge of the point defect structures present in thin films has not yet been built up. We are only at the stage of looking, in the thin films, for analogues of those point defects which have been observed in the corresponding bulk material. The attempt is worthwhile, however, because it is important to control trapping in electronic thin films and success in this will be more likely if we understand the structure of the traps.

3. RADIATION ENVIRONMENTS

While there is not as yet as good collection of data on radiation environments and their relative capacities for damage, a detailed description of all those of interest is outside the scope of this review. There are some partial accounts in the specialist literature[1][2][3][4][5][6][7]. However, in order to determine the precise physical problems which should concern us in the present study, we have to define broadly the types and magnitudes of radiation flux which our materials are likely to encounter and to group or classify them for severity of effect. We thus divide the levels of damage into three groups "ordinary" "high" and "extra high" and, in Tables 1 and 2 the levels of severity are indicated in these terms. The dose ranges associated with each group, expressed in units of "damage equivalent 1-MeV neutrons cm^{-2}" and in "rads" are defined at the bottom of Table 2. Usually, different types of effect have to be considered in each range, defect-controlled electrical properties being the first to appear, with optical and mechanical effects appearing later. The relative magnitudes of the ionizing dose in rads and the degree of atomic displacement in "damage-equivalent" units (see below) will vary with the energy and type of particle. In Table 1, we attempt to show the common associations between certain radiation sources and certain sub-systems, such as space radiation with thermal control surfaces or reactors with control electronics. The diversity of materials and radiation types is obvious.

We can express the integrated damage levels received in terms of two damage parameters - displacement and ionization effect. For displacement effects we can crudely express the damage inflicted by the various particles (electrons, protons, fusion neutrons etc.) in terms of a 'damage-equivalent fluence". It can be said that for many electrical degradation effects in semiconductors, the amount of degradation per particle changes strongly with particle energy and type. For example 14 MeV neutrons generated by a deuterium-tritium reaction cause about three times the degradation produced by a 1 MeV neutron in a silicon junction device. The damage-equivalent fluence for 10^{24} cm^{-2} of fusion neutrons is thus about 3×10^{24} cm^{-2} (1 MeV-equivalent neutrons). We could equally choose another neutron energy for expressing our damage-equivalence or even, with limitations, another particle (space radiation calculations often express the total effect of a broad spectrum of particles in "damage-equivalent, normally incident 1 MeV electron per cm^{-2}"[2][8] In Table 2, the 1 MeV neutron is used for comparison. Also, in this table, a more fundamental equivalence is given, namely "displacements per atom". This self-explanatory term simply indicates the number of primary atomic displacements which have occurred. The damage-equivalent fluence, on the other hand, equates the ultimate effect of that displacement on a measurable property of the material (resistivity etc.) The latter depends on the course of many secondary events (e.g. the number of primary point defects generated along the track of the primary knock-on atom and the subsequent interaction of such point defects with each other or with impurities Thus, the units have different uses but both serve to indicate the different physical problems we face in the three fluence ranges.

TABLE 1 - RADIATION SOURCES, LEVELS AND EFFECTS (from Holmes-Siedle, 1974£)

	Atomic Batteries and space radiation	Ion and electron beam devices	Nuclear explosions	Nuclear Reactors	
				Controls	Internal parts
Class of Levels †	Ordinary	High	Ordinary	High	Extra High
Typical 1-MeV neutron Equivalent	$< 10^{15}$ ncm^{-2}	10^{18}	10^{15}	$10^{16}-10^{18}$	$10^{20}-10^{25}$
Particle	e, p, α	e, X-ray, ions	n, γ, X-ray	n, γ	n, γ, ions
Radiation Characteristics	Low Rates Long Times	High surface doses Long Times	V. High rates V. short times	Medium Rate Long times	High Rate Long Times
Problems	Circuits and materials very intolerant of failure Power Limited	Active materials exposed to high light and radiation levels	High reliability and ruggedness required. Power limited	Shielding may be limited. Control circuits must not drift	High temperatures Material strength deteriorates
Systems Involved	Controls Electronics. Imaging Systems. Scientific Instruments Thermal Control	Ion-Beam processing Systems Electron-Beam Devices	Data Links Guidance Communications	Control Electronics. Sensors Power Regulation	Electrodes Control Rods Reaction Vessel Insulation

† For definition, see Table 2

TABLE 2 – TYPICAL RADIATION ENVIRONMENTS (from Holmes-Siedle 1974)

System	Class of Radiation Level†	Largest Expected Levels Displacement Damage (Equivalent 1-MeV neutrons cm^{-2})	Displacements per atom	Ionisation (Rads)	Types of device or material exposed
Nuclear Fission Reactors					
Power	Extra High				
Propulsion	Extra High				
–Core	High	10^{23}	200	10^{14}	Cladding, Rods, etc.
–Engine	High	10^{20}	0.2	10^{11}	as above
–Electronics	High	10^{19}	0.02	10^{10}	Moving Parts
		10^{16}	2×10^{-5}	10^{7}	Semiconductors Etc.
Nuclear Fusion Reactors					
Containment Vessel	Extra High	10^{24}	2000	10^{14}	First-Wall Structure (possibly insulator over metal) Vacuum seals and Valves
Magnetic Coils	High	10^{18}	2×10^{-3}	10^{9}	Conductors and Superconductors Cable Insulators
Control Electronics	High	10^{16}	2×10^{-5}	10^{7}	Semiconductors and Dielectric devices Optics and Sensors
Electron-Beam Processing	High	10^{13}	2×10^{-8}	10^{8}	Paints, Plastics, Semiconductors

		$\phi_n(n\,cm^{-2})$ (1)		D(rads) (2)	
Ion-Beam Processing	High	10^{18}		Variable with Z of ion	10^{14}	Semiconductors
Nuclear Weapons	Ordinary	10^{15}		2×10^{-6}	10^{6}	Semiconductors, Thermal-Control Coatings, Sensors
Space Radiation						
Jupiter Belts	"	10^{12}		2×10^{-9}	10^{7}	Semiconductors
Earth Belts	"	10^{13}		2×10^{-8}	10^{7}	Thermal-Control Sensors
Solar Flares	"	10^{12}		2×10^{-9}	10^{3}	
Sterlization	"	$< 10^{13}$		$< 2 \times 10^{-8}$	10^{6}	Foods, surgical materials
Radiosotope Generators	"	$< 10^{3}$		$< 2 \times 10^{-18}$	10^{5}	As for Space Radiation
X-Ray Sources	"	0		0	10^{4}	Imaging Devices etc.

†Definition of Radiation Levels

Level	$\phi_n(n\,cm^{-2})$ (1)	D(rads) (2)
EXTRA HIGH	$10^{20}-10^{24}$	$10^{11}-10^{15}$
HIGH	$10^{15}-10^{20}$	$10^{6}-10^{11}$
ORDINARY	Below 10^{15}	Below 10^{6}

(1) Damage-equivalent 1-MeV neutrons
(2) Associated ionisation level varies, but typically of the order of $\phi_n \times 10^{-9}$ (rads)

Ionization effects are equally important in device degradation and optical effects. Here, we must express the total amount of electron-hole pair production produced by the particles or photons in a convenient unit. Units of absorbed energy are frequently used, the rad being equal to 100 erg absorbed per gram.

4. SOME EXAMPLES OF POINT DEFECTS AND TRAPPING IN SOME THIN FILM SYSTEMS

4.1 Introduction

Examples of a well-characterised set of deep traps which act as recombination centres in a semiconductor are the radiation-induced A and E centres in silicon which are vacancy -impurity complexes produced by bombardment[9], the iron and copper centres, which often occur in silicon ingots as grown and the gold centre, which can be introduced by diffusion[10]. Minority-carrier lifetime is short in thin-film silicon (even in epitaxial silicon on sapphire) but, it has not yet been determined whether any of the above traps are the cause of the low lifetime. In amorphous silicon films, no discrete levels appear but rather a continuum of states which completely alter the band structure and transport properties of the semiconductor. Several trapping energy levels found in SiO_2 and Al_2O_3 films as grown have been identified by electrical methods [11] but no structure has yet been assigned to them. The SiO_2 system is, in fact, one of the better experimental vehicles for attempting to make such assignments and we will be discussing its structure in several places in this presentation as a model for understanding point defect in films. Table 3 describes a set of point defects in crystalline silica. Despite our present ignorance concerning the defect structure of thin films, this pair of materials, silicon and silicon dioxide provide about the best hopes for an early advance in understanding such defects for several reasons :
(1) a very large amount of knowledge has been accumulated on the properties of the bulk materials
(2) Both materials have been deposited in well-characterised thin films
(3) Each material has been produced in a stable, amorphous thin film form.

4.2 Observation of Point Defects in Thin-Film Insulators

We will discuss here some electrical effects of point defects in insulators and some optical observations which may throw light on the structure of these traps. Cathodoluminescence studies[12] have shown the presence of three discrete defect levels in thermally grown SiO_2, these are possibly connected with oxygen vacancies

TABLE 3 Colour centres in quartz and α-SiO$_2$

Centre	Possible structures	E_{max}	Trap type
A_1, A_2	associated with Al	1.9-2.6	Hole
B_1, B_2	doubly-unsatisfied Si	4.1-5.2	Electron
E_2, G	singly-unsatisfied Si	5.4	Electron
C, E_1	O vacancy, two non-equivalent Si	5.8	Electron
H	void ? (sputtered SiO2)	6.55	Unknown
D	unknown	7.2	Unknown
E	interstitial	7.6	Hole

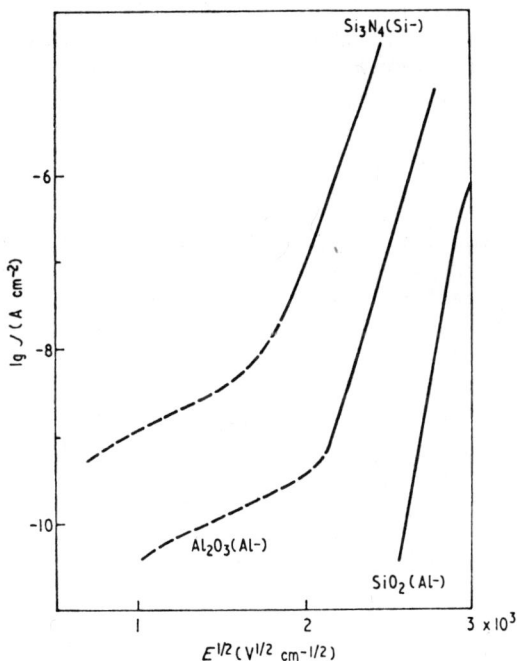

Fig. 2 Field dependence of current in Si$_3$N$_4$, Al$_2$O$_3$ and SiO$_2$. Dotted part of Si$_3$N$_4$ and Al$_2$O$_3$ curves were probably measured during initial transcient.

or impurities, luminescence occurs at 2.8, 3.3 and 4.3 eV. Photodepopulation studies have indicated an electron trap level about 2.1 eV below the conduction band. As will be discussed later, a deep hole-trap with a high trapping cross-section exists in thermally-grown SiO_2 but attempts to characterise it have so far been unsuccessful. In silicon monoxide films Walley and Jonscher [13] have identified a high concentration of traps in the region of 0.2 eV by temperature dependence studies on the I-V characteristic of the film. These levels allow strong Poole-Frenkel conduction, which is not possible in silicon dioxide because of the lower density and greater depth of electron traps in that material. Similarly high Poole-Frenkel conduction occurs in pyrolytic silicon nitride films [14]. Most aluminium oxide films show higher conduction than thermally grown silicon dioxide but the process still appears to be barrier-limited [15] I-V curves for the three dielectric films mentioned above are compared in Fig. 2. As an indication of the sensitivity of thin films to the conditions of preparation, it should be noted that amorphous pyrolytic and sputtered Al_2O_3 are found to have higher conduction than polycrystalline pyrolytic films [16] [17], probably due to Poole-Frenkel conduction. Several different defect levels are detectable even in polycrystalline Al_2O_3 films by measuring the optical release of electrons and holes. Harari and Royce [18] find four trap levels for each carrier in the 2-5 eV range. It is interesting that some of these energies correspond to optical absorption bands in bulk sapphire samples which have been irradiated with neutrons.

These results all suggest that insulator thin films, although frequently deposited or growing in amorphous form, contain reasonably well-ordered structures with at least short range order and a continuous, near stoichiometric network with very few dangling bonds by comparison with the chalcogenide glasses. Thus, the band gap is maintained and likewise the very high resistivity typical of the bulk material. However, this structure also implies that the intrinsic defects which are present will act as deep-lying traps for electrons and holes.

Another disadvantage is that unlike the amorphous chalcogenides, such a network of atoms is strongly affected by impurities which disturb the short range order (as with Na^+ in a silicate glass). Thus, impurity atoms may also act as electrically and optically-active defect levels. Such is the case with the hydrogen atom in Fig. 3, which shows models for defects in an amorphous SiO_2 film [19].

A few observations of transmission and reflectance of thin oxide films have been made in order to try and identify the electrically-active defects. While a strong defect absorption has been observed in one form of evaporated Al_2O_3 film [20] experimental difficulties have so far prevented any correlation of these with the

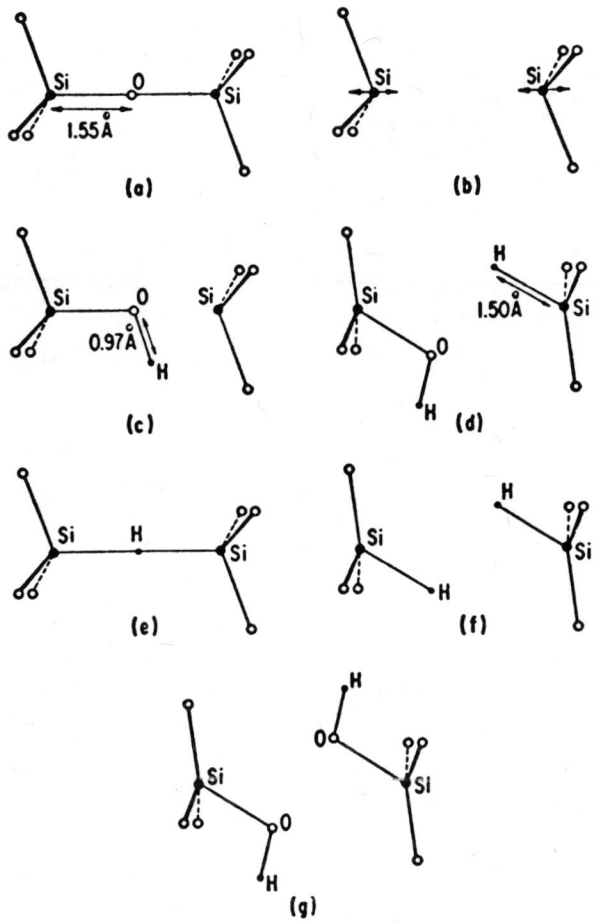

Fig. 3 Atomic configuration in SiO$_2$ in the vicinity of (a) a normal oxygen site (b) an oxygen vacancy and (c-g) various centers containing hydrogen. The plane of the paper is the (110) plane in which we have assumed the defect atoms to lie. The x and y axes are (111) and (112) respectively. The arrows in (b) show the extent of the Si displacements.

electrically-active defects in pyrolytic films. Hickmott[21] has observed defect bands in sputtered silicon dioxide, some of which correspond with those in bombarded bulk SiO_2.

4.3 Interface States

A form of defect which is often important in electronic thin films is the interface state ; when two solids are in intimate contact, it is virtually certain that not all atoms at the interface will be satisfied chemically. The free bonds thus constitute a thin sheet of defects which, while they may exert a strong field effect on the surrounding material, may be impossible to detect by conventional methods of measuring defects (optical, ESR etc.) The important case of the $Si-SiO_2$ interface is described later but significant numbers may also exist between all kinds of deposited layers, a particular example being the layered dielectric of the MNOS memory device[11].

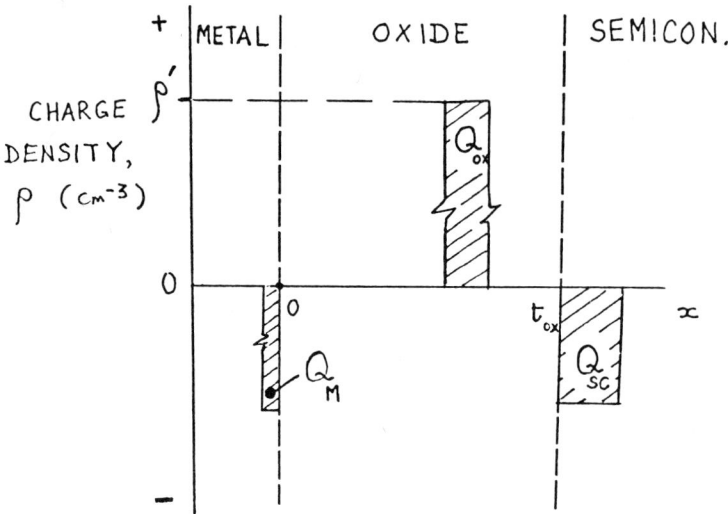

Fig. 4 Image charge in metal and semiconductor produced by a slab of charge trapped in the insulating oxide.

5. THE METAL-INSULATOR-SEMICONDUCTOR SYSTEM

Since irradiation radically affects the charge distribution in MIS devices, we will give a brief résumé of the effects which charges in the insulator have on the semiconductor and metal layer. A simple sheet of charge placed in the insulator as in Fig. 4 produces image charges in the semiconductor space-charge region and the metal. The relative amounts of this image charge in metal and insulator depend on the position of this slab, but charge neutrality demands that the sum of the image charges equal the insulator charge, thus,

$$Q_{ox} - Q_{sc} - Q_m = 0$$

using the notation given in the figure. The same formula applies if, as usually occurs, a charge sheet is present at the insulator/semiconductor interface in the form of "fast states" or if the slab is replaced by some uneven distribution from metal to semiconductor (see, for example [22])

A particularly useful electrical measurement can indicate the magnitude of one of the image charge sheets. This is the high-frequency capacitance-voltage measurement, in which the modulation of MIS capacitance vs. gate bias is measured, as shown in Fig. 5.

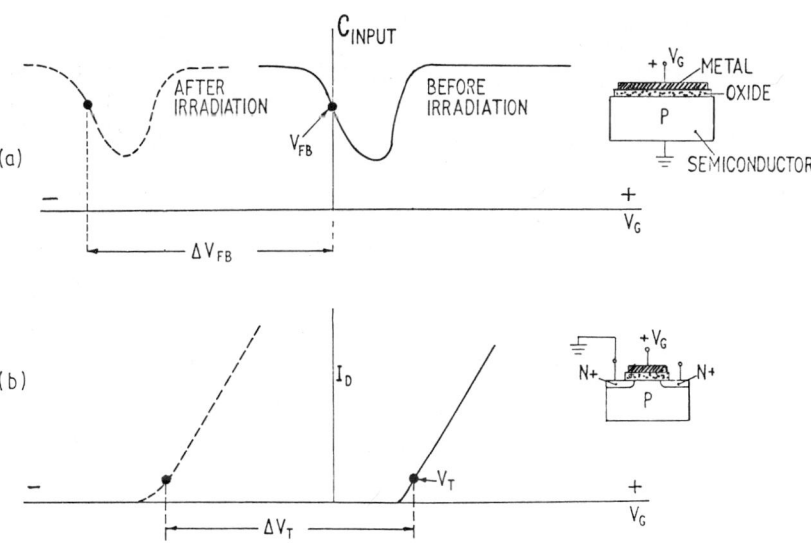

Fig. 5 Change in operating region of MIS device produced by ionizing radiation.

In the simplest case, the shift of the inflection point is a direct measure of the image charge in the semiconductor. Measurements are relatively easily and rapidly carried out (23) and it has been shown that they can yield interesting and meaningful results if the proper experimental conditions are established and if care is taken in their interpretation. For the purposes of this measurement, we can divide the charge-carrying states into two groups :

1) <u>Interface states</u> which are stationary electronic states located right at the plane separating the semiconductor from the insulator. Interface states are analogous to the fast surface states mentioned above and can exchange charge with the semiconductor. Because the insulator has a wide forbidden gap, the energy levels of interface states can lie either within or outside the forbidden gap of the semiconductor. This will determine whether or not they change their charge state when a field is applied between metal and semiconductor.

2) <u>Oxide states</u> bulk defect states, forming a "space-charge region" in the insulator film and not in communication whith the semiconductor surface. There are two forms of this space charge, namely, mobile and immobile. The immobile species is mainly held in traps that are part of the defect structure of the insulator ; the mobile charge is mostly due to ions that are capable of migrating through the insulator, especially during conditions of high field and elevated temperature.

The effect of a buildup of oxide charge such as Q_{ox} is parallel translation of C-V and I_D-V_G characteristics along the voltage axis, each by the same amount, as shown in Fig. 5. We can express this by saying that V_{FB}, the gate voltage at which the energy bands of the silicon are not bent at the surface (flat-band condition) shifts by the same amount as V_T, the voltage "threshold" at which the conductivity of the semiconductor channel region begins to increase, (V_T is usually defined as the gate voltage producing a given low channel current, say I_D = 10 µA). In the case shown in Fig. 5, that of an n-channel transistor which is normally "off" at V_G = 0, the shift in V_T shown causes the transistor instead to be "on" at this voltage. Obviously, such a shift could cause a circuit containing this transistor to malfunction. We shall discuss later some particular cases of degradation of circuit properties produced by irradiation. While the buildup of oxide charge alone leads to exactly equal shifts in V_T and V_{FB}, this is not so when new interface states are introduced. This leads to the <u>distortions</u> of both C-V and I_D-V_G characteristics(24).

While the foregoing account demonstrates that the capacitance-voltage method is one of the more useful techniques to be used in understanding charge buildup in the MIS system, it must be remembered that the precise location of the charge within the insulator is not indicated, only the value of its image in the semiconductor. The distribution of charge density versus depth in the oxide is usually non-uniform and other techniques must be used to determine this profile. The shift in V_{FB} due to any buildup of charge can be represented as

$$\Delta V_{FB} = \frac{1}{C_{ox}} \int_0^{t_{ox}} \frac{x}{t_{ox}} \cdot \rho dx$$

This means that the nearer x is to its limiting value, t_{ox}, the larger the value of Δ_{FB}. The maximum is when the sheet is concentrated next to the semiconductor.

Two other simple cases will be constantly referred to in later sections, namely sheets of uniform density (ρ') and finite thickness (d) at the two extremities of the oxide, i.e. up against the metal or silicon electrodes.

The expressions for these cases are :

(a) Sheet next to silicon (case for positive V_I)

$$\Delta V_{FB} = - \frac{\rho' d^2}{C_{ox} t_{ox}} \;;$$

(b) Sheet next to metal (case for negative V_I)

$$\Delta V_{FB} = - \frac{\rho'}{C_{ox}} \cdot \frac{d^2}{2 t_{ox}} \left(\frac{2 t_{ox}}{d} - 1\right) \;;$$

when d is small, this approaches

$$\Delta V_{FB} = - \frac{\rho'}{C_{ox}} \cdot d = \frac{Q_{ox}}{C_{ox}} = \frac{qN}{C_{ox}}$$

$d \ll t_{ox}$

Note that since a thicker oxide has a higher value of C_{ox}, the same charge sheet next to the silicon has a larger effect on V_{FB} in a thicker oxide. A convenient formula for this dependence is

$$\Delta V_{FB} = \frac{\Delta N_t t_{ox} (\mu m)}{2.1 \times 10^{10}} \;; \qquad d \ll t_{ox}$$

where ΔN_t is the area density (cm^{-2}) of holes trapped in the thin charge sheet next to the silicon (not the number of traps available). Thus, a typical oxide of t_{ox} = 0.12 µm will yield a shift of -1 volt for a charge sheet consisting of 1.76×10^{11} holes per cm^2.

5.2 Basic Features of Radiation Sensitivity

Fig. 6 shows a model for the processes which occur in an MIS system when it is exposed to high-energy irradiation. Holes are trapped in the region near the silicon when the field is as shown, presumably because electrons are largely swept out of this area[25]. The result of the buildup of charge is the shift of electrical characteristics shown in Fig. 5[26]. Despite a large amount of investigation, the precise details of the process are still not clear. For example the hole trap has not been identified and some debate still exists as to the charge transport, mechanisms involved [27] [28][29][30]. Not pictured here is the gradual generation of new Si/SiO$_2$ interface states under irradiation which also occurs[31].

5.3 Techniques for Investigation

After some early attempts to use empirical methods to correct the tendency of thermally-grown silicon dioxide films to build up space-charge under irradiation (see section 8.3) more fundamental investigations have been started using physical techniques such as photoconductivity, optical spectroscopy and ion bombardment. Two processes in thin film MIS insulators have to be understood thoroughly before a proper model for space-charge buildup can be developed. The first is charge transport by electrons, holes and possibly both polarities of ion. The second is the trapping of carriers which, since the ultimate object is to control and modify this trapping, further requires a detailed knowledge of trap structure. As will be seen, the MIS structure is very well adapted to transport studies, since high fields can easily be obtained and transit times are very short. On the other hand, it is very poorly adapted for conventional studies of defect structure, such as electron-spin resonance, diffraction and spectroscopy, since films of less than 1 micrometre, unless they contain very high concentrations of defects, will not contain an absolute number of defects sufficient for detection by such techniques. Thus, for example, colour-centre spectra have been observed in sputtered silicon dioxide films[21] and heavily ion-bombarded films [32] but not in simple thermally-oxidised films. The defect concentration in the first two cases must be greater than 10^{19} cm^{-3} - an unusually high density compared with those normally used in colour-centre studies (say 10^{17}cm^{-3}) Some of the methods and their results will be described.

Williams[33], Goodman[34] Snow et al[35] and Powell[36] have characterised the barriers between silicon dioxide, silicon and various electrode metals, using the threshold wavelength for pho-

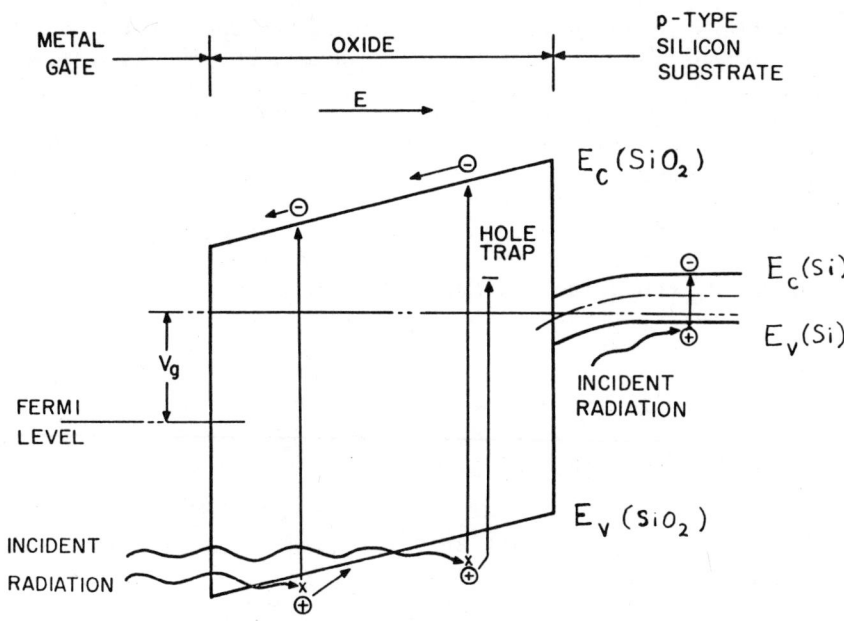

Fig. 6 Mechanism of Space Charge Trapping in Single Oxide Film (after Gwyn[29]).

toemission as a measure of the energy. These studies thereby established the first estimates of the band-gap energy for thermal silicon dioxide, since the sum of the barrier energy for holes and that for electron emission from the same material should equal the band-to-band transition necessary for electron-hole conduction. Photoinjection currents due to electrons could be measured with ease, while those due to holes were at the limit of detection. Two important facts concerning transport could be elicited. The mobility of electrons was surprisingly high for a material presumed to be amorphous (about 30 cm^2 $volt^{-1}sec^{-1}$ as compared with 1 cm^2 $volt^{-1}sec^{-1}$ for amorphous germanium) ; the mobility of holes was vanishingly small, and certainly several orders of magnitude lower than that for electrons.

Another form of injection into MIS insulators is by production

of hot carriers in the semiconductor which can then drift into the
insulator. Nicollian, using a high-frequency, 50-volt bias across
the semiconductor space-charge region, could produce avalanche conditions which led to electron injection but no appreciable injection
of holes into the oxide. However, Werwey[37], using a DC avalanche
at a junction edge, claims to have also detected hole currents in
oxide films. In all these forms of injection, charging effects are
observed after currents have flowed for some time, which implies
that the oxide contains a limited number of traps for electrons and
holes. However, these studies have not proceeded far enough to give
any clues as to the structure of the defects involved, especially
in the case of the hard-to-obtain holes. In the case of the electron traps only, the energy levels of the traps have been estimated, using a measurement of the threshold of light energy at which
the trapped electrons can be de-trapped to produce a photo-depopulation current. The trap depths found in this way vary from 0.4 to
2 eV [33].

Experiments by Zaininger[38], Harari and Royce[18] and Emms,
Holmes-Siedle, Groombridge and Bosnell[30] have also shown that
UV light can be used to annihilate, or "photo-anneal", the radiation-induced positive space-charge. Photoinjection has been used
in another way by Peel and co-workers[39] to observe a barrier
lowering in silicon dioxide films after the buildup of a radiation-induced positive charge.

Since the process which creates space-charge in high-energy
electron or gamma ray irradiation is though to involve electron-hole pair creation, it is of major interest to discover what happens when light with only just enough energy to create the pair is
used, rather than particles with many thousands of times the energy, as with the former type of high-energy irradiation.

Other basic information on oxide structure can also be obtained by such means. For example, it is a well-known technique to use
measurements of the threshold of intrinsic photo-conductivity to
determine the true optical band-gap. This experiment has been performed recently with silicon dioxide films. Di Stefano and Eastman
[40] found that the onset of conduction began sharply at slightly
below 9.0 eV. Powell [41] has confirmed this finding and it is confirmed indirectly by measurments of external photoemission currents
vs. wavelength in very thin oxide films, by Di Stefano and Eastman
[40]. The best figure for conductivity band-gap of amorphous, thermally-grown silicon dioxide films is 8.9 eV. It is interesting that
there seems to be little difference between the optical properties
of thermally-grown silicon dioxide films and those of bulk amorphous
silicon dioxide, such as synthetic fused silica [42]. Furthermore,
the optical properties of the latter two materials has been shown
to exhibit surprisingly little difference from crystalline silicon

dioxide[43]. The importance of this general finding lies in its bearing on the trapping model which we adopt for silicon dioxide films. Firstly, we should not assume that, as for many amorphous compounds, a high density of localised states exists near the conductivity edge ; secondly, we can perhaps borrow more freely than in many other cases on the models used to explain the defect structure of bulk silicas. Powell and Derbenwick[42], Holmes-Siedle and Groombridge[30] and Emms et al[30] have demonstrated the use of irradiation with band-gap light to produce positive space-charge buildup in silicon dioxide films.

Normally, the trapped space-charge in irradiated MOS capacitor samples is extremely stable at room temperature under various ambient and bias conditions, showing a reduction in oxide charge of not more than a few percent within a period of several months. However, annealing of an irradiated MOS sample free of interface states in an inert gas at elevated temperatures results in an approximately parallel positive shift of the C-V characteristic of the sample, indicating a reduction of the positive charge in the oxide. The major changes take place between 150°C and 300°C and by 400°C all of the radiation effects in the oxide appear to have been annealed out[63][35]. Danchenko and co-workers have drawn two conclusions from a careful study of the temperature dependence of annealing of the positive space-charge. The activation energy is in the region of 1.5 eV and seems to correspond to the tunnelling of electrons from the silicon interface into the space-charge region over a barrier of this height.

6. RADIATION SENSITIVITY IN SEMICONDUCTOR FILMS

The majority of the semiconductor devices used today employ the p-n junction in silicon as the basic operating element. However, the fact that most of the operating principles involved in p-n juntion devices require high crystallinity, low defect concentration, low doping levels and surface passivity make silicon devices difficult to employ in a high energy radiation environment. Because of this fact, and because of the desire to cut cost, much work has been done recently towards the development of thin-film semiconductor devices in which the key subelements operate satisfactorily in thin film form, despite the presence of high defect density (polycrystalline films) or absence of crystalline order. Typical products of this research are the cadmium selenide thin-film solar cell and the amorphous chalcogenide switching device. Before these developments, other semiconductor films were already in use, mainly in the field of large area photosensors (such as the selenium films in vidicon TV camera tubes). Table 4 shows a survey list of the most common types of semiconductor device and notes which forms are usually made from bulk crystalline semiconductor and which from thin semiconductor films [1]. 'x' signs indicate

TABLE 4 : DISPLACEMENT EFFECTS IN SEMICONDUCTOR DEVICES : PRIMARY AND SECONDARY FAILURE MODES (Adapted from Holmes-Siedle[1])

	1	2	3	4	5	6	7
Device Function	Usual form of semiconductor		Lifetime Reduction	Carrier Removal	Mobility Decrease	Trapping	Other
	Bulk	Thin Film	key to Symbols XX = Primary Failure Mode X = Secondary Failure Mode				
p-n JUNCTION DEVICES							
(a) Low Reverse Fields							
Bipolar transistors & SCR	o	–	XX	X	–	–	–
MIS (MOS) Field-Effect Transistor	o	o	–	XX	X	–	–
Variable Threshold Trans.	o	o	–	XX	X	–	–
Junction FET	o	–	–	XX	X	–	–
Rectifying/Blocking Diode	o	o	XX	X	–	–	–
Tunnel Diode	o	–	–	–	–	–	X
Schottky Barrier Diode	o	o	XX	X	–	–	–
Junction Photosensor	o	o	XX	X	–	X	–
Optoisolator	o	o	XX	X	–	X	–
Junction Electroluminescent Diode	o	–	–	X	–	XX	–
MIS Electroluminescent Diode	o	o	–	X	–	XX	–
Solar Cell	o	o	XX	X	–	–	–
(b) Avalanche Devices							
Zener and IMPATT diode	o	–	X	XX	–	–	–
Surface-Controlled Avalanche Diode	o	–	–	XX	–	–	X
(c) Other							
Charge-coupled Device	o	–	–	XX	–	X	–
Hall-Effect Device	o	–	–	X	XX	–	–
OTHER DEVICE							
Transferred Electron Device	o	–	–	X	–	XX	–
Photoconductive photosensor	o	o	–	–	X	X	–
Storage Photosensor	–	o	–	–	X	X	–
Mechanical Tranducer	o	o	–	XX	X	–	–
Ovonic Threshold Switch	–	o	–	–	–	X	XX
Ovonic Memory Cell	–	o	–	–	–	X	XX
Amorphous Tunnel Triode	–	o	–	–	–	X	XX
Photostructural Switch	–	o	–	–	–	X	XX
Cold-Cathode Electron Emitter	o	o	–	X	XX	–	–

whether the basic operating principle of the device is likely to be affected by displacement damage in the semiconductor material and the particular effect in the material which leads to degradation of device action. With increasing particle fluence, the earliest affected are those devices which require a high minority carrier lifetime, such as solar cells, bipolar transistors and silicon controlled rectifiers. Field-effect transistor are affected only at much higher fluence levels when resistivity and carrier mobility begin to change. Also, a few devices are sensitive to the slow emptying of shallow defect trapping levels which may produce an inconvenient 'tail' on the falling edge of a square electrical pulse emitted by the device. This would increase the dark current of an imaging device, for example. Only at much higher fluence levels would "other" properties like the electromechanical constant, utilized in a mechanical transducer, or the hopping conductivity of an amorphous semiconductor, be affected significantly. Thus, the reader can choose a semiconductor device principle which, ideally, should be radiation tolerant by noting whether the "x" signs cluster to the left or the right hand side of the table.

It is often found that, if a satisfactory device can be fabricated from thin films, the effect of radiation on the device action is less, presumably because the device has been designed to tolerate the considerable concentration of defects introduced in the fabrication. The solar cell is an example ; thin-film cadmium sulphide cells are very radiation-tolerant compared with silicon cells but cannot be fabricated with such high initial conversion efficiencies as silicon cells.

The ionization effects of most interest within semiconductor materials are non-destructive and consist of the photocurrents produced when carriers are generated, especially if generated near p-n juntion field regions. Diffusion, drift and recombination are the important controlling parameters. Material parameters such as trap concentration radically affect recombination and drift mobility. Even in devices of simple geometry, models for these processes are still difficult to construct because of the very complex gradations of impurity concentration present in a junction device. Thin-film p-n junctions will respond less strongly than diffused p-n junctions of the same area because the collection volume for carriers in the thin film will be much smaller. Against this must be weighed the fact that the high concentration of defects present may lead to high photoconductivity. However, on the whole it can be said that polycrystalline thin film semiconductors should show a lower overall response to irradiation than bulk semiconductors because of size effects and recombination effects. The case of the amorphous film can only be properly discussed with specific device principles in view (see next section).

7. SOME IMPORTANT RADIATION EFFECTS IN THIN FILM DEVICES AND MATERIALS

A few examples are given here of the impact of radiation effects on the working of several important thin-film systems. The first two are not strictly thin-film devices but contain a thin film layer with a critical function as a part of a crystalline semiconductor device.

A typical circuit configuration for MOS devices is shown in Fig. 7(a) namely the complementary-symmetry MOS inverter (a "CMOS logic gate") and the geometry of an integrated form of this logical inverter shown in Fig. 7(b). The semiconductor is crystalline silicon. The voltages and fields across the gate insulators are shown for the two logic states of this digital device. There may be many thousands of such pairs interconnected on a single integrated circuit chip, only a very few being accessible via the external connections. The effect of irradiation on the switching characteristics of the device is shown in Fig. 8 a-b. It can be seen that, under irradiation, not only do the input-output voltage characteristics change but the current drain (and hence power consumed) can also increase or decrease, depending on the logic state while under irradiation. Initially, degradation is gradual as the speed of operation of the circuit and sensitivity to spurious input signals ("noise immunity") deteriorate ; finally, a condition is reached in which the circuit completely ceases to perform its logical function. Examination of the circuit will show that the function of the n-channel device is the one most affected by a given degree of space charge buildup. It is obvious that, for a large-scale logic circuit, a careful circuit analysis is required before one can predict the effect of the individual changes described on the overall operation of the circuit. (See for example Dennehy et al[44]). The "duty cycle" of the device while being irradiated must also be considered ; very often, the logical input signal levels are alternating between "1" and "0" (say between + 5 V and 0 V) while irradiation is proceeding ; in this case, some "radiation-induced annealing" can occur during the time when each gate insulator is under zero field [45] and the net shifts of each transistor lie between the "100% off" and "100% on" values.

Burghard and Gwyn[46] have made a careful comparison of some circuits with precisely the same logical function, made by different manufacturers to the same electrical specification. These differences stem almost entirely from differences in fabrication of the gate insulator. As the next section will indicate, it is not yet clear why a material like silicon dioxide which, when thermally grown, has good stoichiometry, low impurity content and few dangling bonds, should show this great sensitivity to the fabrication technique and research is still proceeding on the topic.

RADIATION EFFECTS IN THIN FILMS

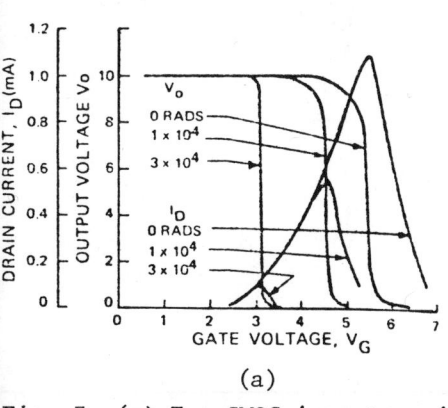
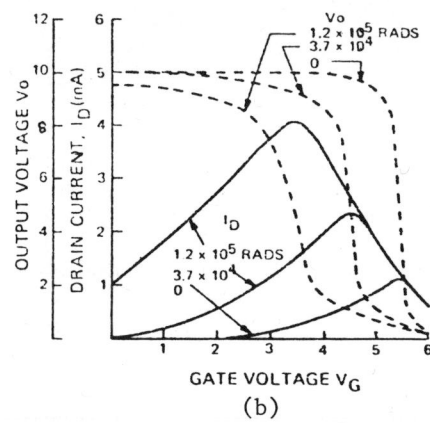

Fig. 7 (a) Two CMOS inverters in cascade
(b) Representation of First Inverter Pair in CMOS Network in (a). Physical Structure, Bias conditions and electric Fields in Insulator.
(from Poch and Holmes Siedle[47])

Planar bipolar transistors employ thermally-grown SiO_2 films for a different but critical purpose, namely to prevent leakage and surface recombination effects where the junctions meet the surface. The oxide is affected by radiation in approximately the same manner as described for MIS devices, despite the radically different field conditions. Since silicon bipolar transistors are used extensively in space and missiles, considerable effort has been spent on analyzing the problem introduced at the junction by the accumulation of positive charge in the overlaying oxide. Fig. 9 shows the regions of oxide important for degradation of gain. While little headway has been made in finding passivation processes which will reduce the effect in junction devices, a clearer understanding is emerging of the damage mechanisms involved, and a method has been developed for distinguishing between and predicting the two kinds of damage effects possible in bipolar transistors, namely, damage to the crystalline silicon lattice, which reduces minority carrier lifetime, and changes in the oxide layer, which increases the surface recombination velocity. In studying the

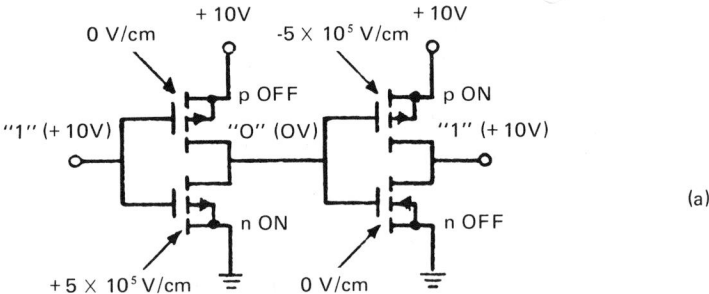
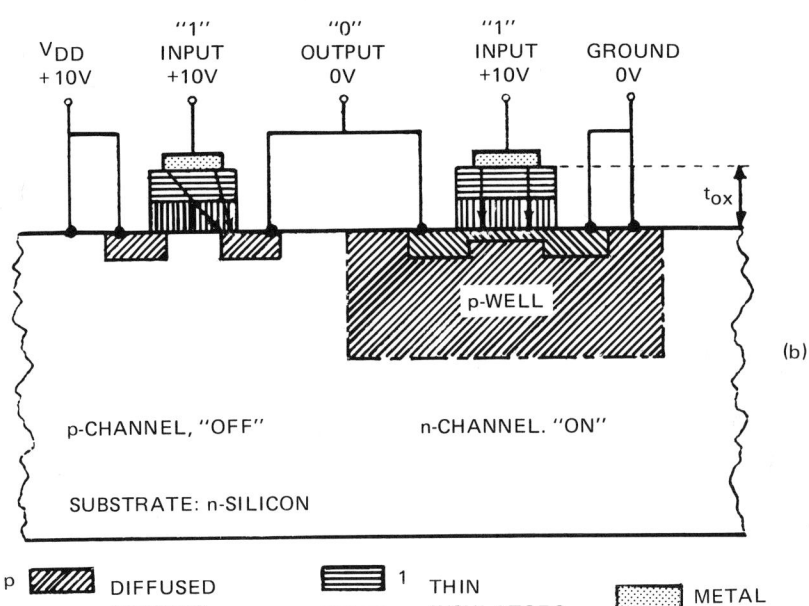

Fig. 8 (a) Change in CMOS inverter Transfer Characteristic. Gate bias during irradiation -10 Volts for the p-channel, 0 Volts for the n-channel; V_{DD} is 10 Volts.
(b) Change in CMOS inverter Transfer characteristic. Gate bias during irradiation 0 Volts for the p-channel and +10 Volts for the n-channel; V_{DD} is 10 Volts.
(from Poch and Holmes-Siedle[47]).

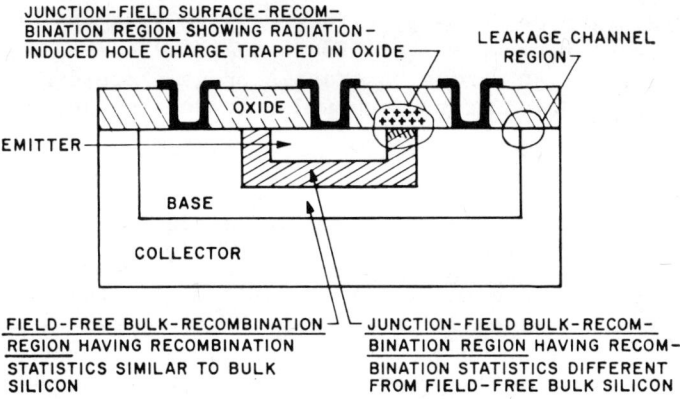

Fig. 9 Cross section of bipolar transistor showing region involved in gain loss due to the generation of oxide change or interface states (from Holmes-Siedle and Zaininger[26]).

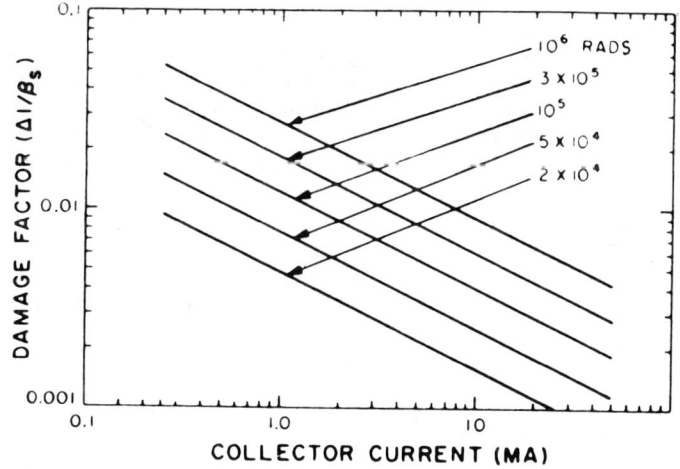

Fig. 10 Form of an engineering specification for the worst case of surface damage factor $\Delta(1/\beta)_s$ as a function of collector current in 2N916 transistors of a given manufacturer actually a specification of 1 percent probability of exceeding the given level of beta change (from Holmes-Siedle and Zaininger [47]).

effect of radiation on transistor forward gain, the author and co-workers (see for example $^{(45)}$) found that the combined effect of bulk and ionization damage on transistor gain could be described as a combination of surface and bulk damage factors as follows

$$\Delta \frac{1}{\beta} \text{ (Total)} = \Delta \frac{1}{\beta} \text{ (Bulk)} + \Delta \frac{1}{\beta} \text{ (Surface)}$$

where β is the forward gain of the transistor. In other words, when damage was expressed in this form, the two types of damage were additive. While the bulk factor could be predicted fairly well by means of an analytical expression, the value of the surface damage factor could not be predicted for a given transistor structure. The effect arose from the oxide and, apparently, was determined in magnitude by several uncontrolled variables in oxide processing which gave great scatter of gain effects from batch to batch of devices or even within a batch.

Units with identical electrical specifications but from different manufacturers (i.e., different processing) vary even more radically than the samples shown. Thus, in order to qualify devices for use in a radiation environment radiation tests are required for every type of device used, these results then being converted by some valid technique into predictions of performance degradation in a given operational situation. For such analysis, the surface damage factor $\Delta(1/\beta)_s$ is a useful tool, especially, since it is found that the damage is very dependent on emitter injection level. The dependence is regular and can sometimes be cautiously extrapolated. Also, the radiation-induced loss of gain, analyzed statistically using $\Delta(1/\beta)_s$ as a measure of damage, was found to have a frequency distribution fitting the normal, bell-shaped curve. Thus, a statistical prediction of the probability of meeting a particular level of degradation in a given number of devices can be developed in a form which can be used by circuit engineers who must provide sufficient allowances in their designs for transistor degradation. Shown in Fig. 10 for the 2N916 transistor are anticipated values of the surface damage factor as a function of collector current at five different radiation dose levels. Knowing the expected dose and collector current, a worst case anticipated value of $\Delta(1/\beta)$ and hence end-of-mission beta can be calculated simply and routinely : This is an example of the practical application of research on thin dielectric film structures under irradiation. Even before a means of suppressing the effects of radiation has been found, an understanding of the effects can be used to develop predictive formulae and preventive measures (see for example $^{(45)}$).

An early application for glassy semiconductors was as thermometers in a nuclear reactor core[48]. Thus, for certain devices the theoretical prediction seems to hold true-namely that the glassy semiconductors in bulk form maintain their properties well above the level at which the transport properties of silicon or gallium arsenide begin to change. The threshold and memory devices have also been tested in reactors and under pulsed X rays. X ray dose rates as high as 1.8×10^{11} rads^{-1} and reactor neutron fluences of 1.2×10^{17} n cm^{-2} did not change the switching characteristics of the thin-film "STAG" materials with graphite electrodes[49]. Smith and his co-workers (1972) tested a more advanced memory device and found no loss of function after reactor irradiation to 10^{16} n cm^{-2} accompanied by 3×10^7 rad of gamma irradiation. These results are generally confirmed by the work of Nicolaides and Doremus[50] and of Flanagan and Wyatt[51] on similar chalcogenide devices. Thus, the particular charge injection processes used in the monostable and bistable chalcogenide switches also appear not to be affected by either transient ionization or the impact of neutrons on the network. However, Chen and Bailey[52] believe that some memory devices will be affected at a fluence about 10^{18} n cm^{-2}, since they have found that a bulk sample of crystalline "memory" material, $Ge_{0.08}As_{0.50}Te_{0.42}$, can be rendered amorphous by a fluence of 3×10^{18} n cm^{-2} in a reactor. Bistable devices made from such material hold information in the form of thin crystalline filaments running between the electrodes[53]. Even partial disordering of these filaments could reduce or extinguish the high conductivity of these paths and hence erase the memory effect. Hench et al[54] find that the dielectric loss in V_2O_5-P_2O_5 glasses is altered after 10^{17} cm^{-2} reactor neutrons while Adler et al[55] find that the resistivity of amorphous silicon films at a fluence of 8×10^{18} cm^{-2} reactor neutrons decreases by a factor of 4. This change is not negligible but is much less than would occur with crystalline silicon and, moreover, would probably be tolerated as a change in the "off" resistance of a monostable device. However, it should be borne in mind that the switching device principles described represent only two of many possible principles which can be developed in future from the amorphous semiconductor field and are somewhat limited in their applications. For example, device principles in which local application of energy produces changes in structure will be much more susceptible to the intense local disturbance produced by heavy particle bombardment. Thus most memory principles may be less intrinsically radiation resistant than monostable principles. Clearly we will have to be cautious of all lightly triggered memory principles, employing what Grigorivici[56] calls "less drastic structural changes than crystallization" such as polarization or minor changes in short-range order. Olley and Yoffe[57] have demonstrated that ion bombardment can introduce strong damage effects in chalcogenides at low temperature, creating local concentrations of "dangling bonds" which are not present in the original evaporated amorphous film. Thus it should be expected that, since many different forms of disorder are possible in the glassy semiconductors, particle irradiation can still effect strong changes in devices formed from them. It is seen that statements, often made, that disordered materials cannot be further damaged is unfounded as a generalization.

The use of ion implantation to modify or create thin surface layers is only starting to be investigated for a few cases. While ion beams have been widely used to place chemical dopant atoms at a desired depth in semiconductors and other materials[58], the accompanying heavy displacement damage has been exploited in a few cases. For example, optical waveguides have been formed in silica surfaces by creating displacement damage in the surface with a beam of ions[59]. Also, one approach to producing radiation-tolerance in MOS devices is to ion-bombard the top half of the oxide, producing electron trapping centres[31][30].

A few cases of radiation effects in protective coatings should be noted. In space vehicles, evaporated, grown or deposited dielectric films are frequently used to maintain the desired radiative equilibrium, between the vehicle, the sun and deep space. For this, a high emissivity in the IR and a low absorptivity in the visible is required. For long missions near the Van Allen particle trapping regions, high fluences of low-energy protons and electrons are accumulated by the outer surfaces of a spacecraft, leading to colour-centre formation ; most space vehicles have thus suffered gradual increases in temperature as the solar absorptivity of the thermal-control surfaces has increased [3]. Some nuclear fuel granules are encapsulated in pyrolytically-deposited ceramic coating to prevent the creation of radioactive dust. These may rupture through accumulated radiation damage, accompanied by helium gas generation from implanted alpha-particles and transmutation of elements.

Finally, some useful scientific information can be obtained from observing microscopically the effects of particles in thin foils. Lead iodide foils are used for studying nuclear fission recoil processes, since the tracks formed are visible[60]. In many materials, the region around the track of a heavy ion becomes more susceptible to etching because of heavy ionisation or displacement damage and the track can be made visible by this etching. The same phenomenon has been observed in thermal silicon dioxide films in which the etch rate can be significantly increased by kilovolt electron bombardment. The phenomenon can be used to define patterns ; electron bombardment is also observed to produce local densification and stress relief in the film[61].

Another interesting case of visualisation of defects is that of the cubic voids observed in alkali halide foils when irradiated for some length of time in the electron microscope beam. The normal production of colour centres by ionisation must occur and be followed by the agglomeration of vacancies[62].

8. CONCLUSIONS

This brief survey has shown that several of the radiation effects already observed in the bulk state of matter are expressed in modified and striking forms in the thin-film state. Such effects are : charge trapping, leading to space-charge buildup ; mobility changes and photo-conductivity ; densification, leading to changes in etch rates and built-in strain ; amorphisation and/or changes in resistivity which may inhibit the processes essential for the action of a device. One effect which is significant only in the thin film state is the radiation-induced generation of interface states.

The primitive state of understanding of the changes which irradiation produces in thin films reflects the general state of understanding of thin films on the atomic scale. However, the deleterious effects of radiation are sufficiently important, especially in dielectrics, that work will continue to improve this knowledge. Also, since other thin films are tolerant to degradation, thin film devices may be developed further for use in radiation environments. Finally, the technique of ion bombardment is so well suited for probing and modifying thin films that the study of the interaction of thin films with radiation is sure to continue for some time for this reason alone.

REFERENCES

1. A.G. Holmes-Siedle, Reports of Progress in Physics, $\underline{37}$ (6), 669 (1974).

2. W.L. Brown, J.D. Gabbe and W. Rosenzweig, Bell Syst. Tech. $\underline{42}$ 1505 (1963).

3. W.S. West, W.J. Poch, A.G. Holmes-Siedle and D. Carroll NASA TR-R-371 (Washington NASA 1971).

4. A.G. Holmes-Siedle, Nature $\underline{251}$, 191 (1974).

5. A. Charlesby, Atomic Radiation and Polymers (Oxford Pergamon) p. 52 (1960).

6. S. Glasstone, The Effects of Nuclear Weapons (US Atomic Energy Commission : Washington DC 1962).

7. G.M. McCracken and S. Blow, UKAEA Culham Report No. CLM-R120 (1972).

8. W.J. Poch and A.G. Holmes-Siedle IEEE Trans. Nucl. Sci. $\underline{NS\ 15}$ (6) 213 (1968).

9. J.W. Corbett, "Electron Radiation Damage in Semiconductors and Metals "New York, Academic Press, (1966).

10. S. Braun and H.G. Grimmeiss, J. Appl. Phys. 45, 2658 (1974).

11. P. Balk, "Solid-State Devices, 1973" Conference Series n°19 (Institute of Physics, London) (1974).

12. P.J. Mitchell and DeNure, Solid State Electronics 16, 825 (1973).

13. P.A. Walley and A.K. Jonscher, Thin Solid Films 1, 367 (1967).

14. S.M. Sze, J. Appl. Phys. 38, 2951 (1967).

15. R.H. Walden, J. Appl. Phys. 43, 1178 (1972).

16. T. Tsujide and K. Iida, Japan J. Appl. Phys. 11, 600 (1972).

17. C.A.T. Salama, J. Electrochem. Soc. 117, 913 (1970).

18. E. Harari and B.S.H. Royce. IEEE Trans. Nucl. Sci. NS-20 (6) 280 (1973).

19. A.J. Bennett, and L.M. Roth, J. Phys. Chem. Solids 32, 1251 (1971).

20. E. Harari and A.G. Holmes-Siedle, Bull. Am. Phys. Soc., Ser II, 16, 500 (1971).

21. T.W. Hickmott, J. Appl. Phys. 42 (6), 2543 (1971).

22. A.S. Grove, "Physics and Technology of Semiconductor Devices" Wiley, N.Y. (1967).

23. K.H. Zaininger, RCA Review 27, 341 (1966).

24. K.H. Zaininger, IEEE Trans. Nucl. Sci. NS-13 (6) 237 (1966).

25. P.J. Mitchell, IEEE Trans. Electron. D. v ED-14 (11) 764 (1967).

26. A.G. Holmes-Siedle and K.H. Zaininger IEEE Trans. on Reliability, 17, 34 (1968).

27. B. Andre, J. Buxo, D. Esteve and H. Martinot. Solid State Electronics 12, 123 (1969).

28. H.L. Hugues, R.D. Baxter and B. Phillips, IEEE Trans. Nucl. Sci. NS-19 (6) 256 (1972).

29. C.W. Gwyn, J. App. Phys. __40__ (12), 4886 (1969).

30. C. Emms, A.G. Holmes-Siedle, I. Groombridge and J.R. Bosnell, IEEE Trans. Nucl. Sci., to be published.

31. H.L. Hughes, Proc. IEEE Reliability Physics. Las Vegas (March 1971).

32. T.J. Russell, T. Pandolfi and B.S.H. Royce, Colour Centre Conference, Japan (1974).

33. R. Williams, Phys. Rev. __140__ (2A), 569 (1965).

34. A. Goodman, Phys. Rev. __152__ (2), 780 (1966).

35. E.H. Snow, A.S. Grove and D.J. Fitzgerald, Proc. IEEE, __55__ 1168 (1967).

36. R.J. Powell, J. Appl. Phys. __41__, (6), 2424 (1970).

37. J.F. Verwey, Appl. Phys. Lett. __43__, 2273 (1972).

38. K.H. Zaininger, Appl. Phys. Letters __8__, 140 (1966).

39. J.L. Peel, R.A. Kjar and R.C. Eden, Appl. Phys. Lett. __17__, 3 (1970).

40. T.H. Di Stefano and D.E. Eastman, Solid State Commun. __9__, 2259 (1971).

41. R.J. Powell, (Private Communication).

42. R.J. Powell and Derbenwick, IEEE Trans. Nucl. Sci. __NS-18__ (6) 99 (1970).

43. H.R. Philipp, J. Phys. Chem. Solids __32__, 1935 (1971).

44. W.J. Dennehy, A.G. Holmes-Siedle and K.H. Zaininger, RCA Review __30__, (4) 668 (1969).

45. A.G. Holmes-Siedle and W.J. Poch, Brit. Interplan. Soc. __24__ 273 (1971).

46. R.A. Burghard and C.W. Gwyn, IEEE Trans. Nucl. Sci. __NS-20__ (6) 300 (1973).

47. W.J. Poch and A.G. A.G. Holmes-Siedle IEEE Trans. Nucl. Sci. __NS-17__ (6) 33 (1970).

48. J.T. Edmond, J.C. Male and P.F. Chester, J. Phys. E. Sci. Instrum. 1 , 373 (1968).

49. S.R. Ovshinsky, E.J. Evans, D.C. Nelson and H. Fritzsche IEEE Trans. Nucl. Sci. NS-15 (6), 311 (1968).

50. R.G. Nicolaides and L.W. Doremus, J. Noncryst. Solids 8-10 (1972).

51. T.M. Flannagan and M.E. Wyatt, J. Noncryst. Solids 2 , 229 (1970).

52. C.W. Chen and D.M. Bailey , Radiation Effects 10, 65 (1971).

53. J. Bosnell and C.B. Thomas, Solid. St. Electron. 15, 1261 (1972).

54. L.L. Hench, J. Noncryst. Solids 2, 229 (1970).

55. D. Adler, H.K. Bowen, L.P.C. Ferrard, D.D. Merchant, R.N. Singh and J.A. Sauvage J. Noncryst. Solids 8-10, 844 (1972).

56. R. Grogorivici, Thin Solid Films, 12, 153 (1972).

57. J.A. Olley and A.D. Yoffe, J. Noncryst. Solids 8-10, 850 (1972).

58. G. Dearnaley, J.H. Freemann, R.S. Nelson and J. Stephen, Ion Implantation (Amsterdam, North Holland) (1973).

59. R.D. Standley, W.M. Gibson and J.W. Rogers, Appl. Optics 11 (6) 1313 (1972).

60. Chadderton, "Fission Damage in Crystals" (Methuen, London) (1969).

61. R.A. Sigsbee and R.H. Wilson, Appl. Phys. Lett. 23 (10) 541 (1973).

62. D.J. Stirland, "The use of Thin Films in Scientific Investigations" (J.C. Anderson Ed. Academic Press. N.Y.), p. 172 (1966).

63. V. Danchenko, U.D. Desai and S.S. Brashears, J. Appl. Phys. 39, 2417 (1968).

Applications

THIN FILM APPLICATIONS IN MICROELECTRONICS

V. Le Goascoz

L.E.T.I. Centre d'Etudes Nucléaires de Grenoble

B.P. 85, Centre de Tri, 38041 Grenoble, FRANCE

1. INTRODUCTION

The purpose of microelectronics is to associate together several physical phenomena to realize a function, a logical or analogical function. Let me mention for example, diode phenomena, injection of current between each side of a barrier, field effect in the bulk of a semiconductor (the field effect transistor), field effect near by surfaces of a semiconductor (the metal oxyde semiconductor field effect transistor), conduction in insulating layers etc... The well known bipolar transistor is an example of such device as well as the MOS field effect transistor which are examples of combination of several and different physical phenomena to obtain a particular device.

In the former case, the bipolar device, main physical phenomena are bulk phenomena. All parameters we need to realize a device or to compute a circuit are bulk parameters and in this case thin film insulating layers are not of main importance. In the other hand, in the later case with MOS devices, surface phenomena are very important; physical parameters are surface parameters and we need thin film layers to realize such a devices not only from the point of view of physics but also for the technological aspect.

For these different reasons, we have chosen MOS Structure to illustrate the use of non metallic thin films in microelectronics. But we cannot forget other important and numerous devices which use thin films at any time of the making. These layers can be only a step in the technological process but they can be an active area of the device. Let me mention some devices as microwave device family, piezoelectric devices used in conjunction with MOS circuit,

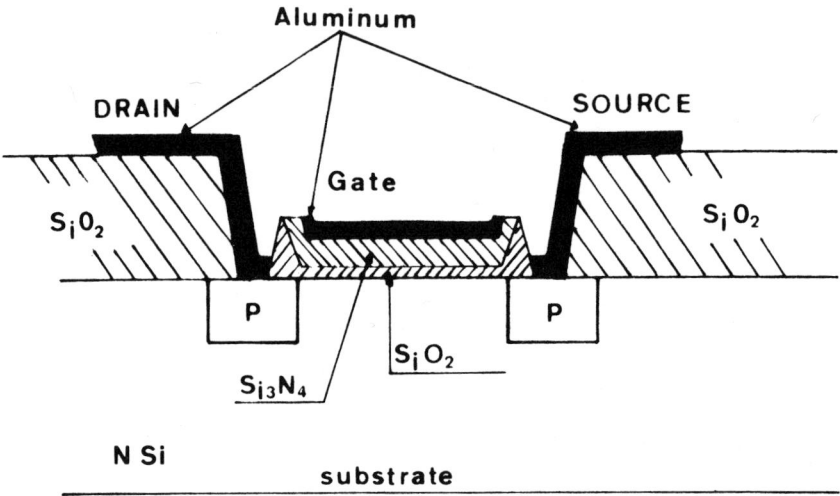

Fig. 1   Schematic diagram of a MOS transistor.

Josephson devices, photodetector devices, light emitting diodes etc...

Technological aspect is a very important one in the use of thin insulating films. In this case these thin films do not serve in the active part of the device and in many case these layers are removed from the device when this one is completely achieved. Such layers are important especially by their mechanical and chemical properties and not at all for their electrical characteristics.

Electrical characteristics can be ranged in insulating properties and conducting properties. In the first case insulating properties are used in a device when a field effect is needed to control, for instance, a voltage level inside the device. These layers can be used in an active part of a device but they only participate in the device for their insulating properties. Therefore the pain property we are searching in such a layer is insulating property. It is difficult to obtain good layers and the most important parameters we must control in the making of these layers are stability and reproductibility of layers.

We can also use these thin films in an active part of a device for their own conduction properties. Conduction is very weak and the film can be considered as an insulator. Such a property is used in MNOS memory devices where differential conduction between two layers is used to inject charges inside insulators, and to trap carriers at the interface between insulators. Reproducibility of layers and a good knowledge of physical aspects of conduction are the more important features to be considered in this case.

To appreciate importance of thin insulating film in microelectronic it is necessary to draw up list of main steps of technology used to realize MOS Field effect transitors and large scale integration circuits. Before technology, a short review on MOS transistor is necessary. The Fig. 1 shows diagrams of a MOS Transistor.

A MOS transistor is a surface device. A silicon surface potential controlled by a insulated Gate, allows current to flow at the silicon surface between two diodes drain and source. The former with a reverse bias and the later with a forward bias. If the silicon surface is not inverted by the action of gate potential, no current can flow. In the case of N silicon substrate if negative voltage above treshold voltage is put on the gate, silicon surface is inverted and a current can flow between drain and source in the channel. Inversion of the channel is obtained by surface field supplied by the gate electrode. This electrode is completely insulated from the channel by an insulator or combination of insulators as can be seen on the Fig. 1. The action of the field decreases with distance and to allow a good control of surface potential it is necessary to use

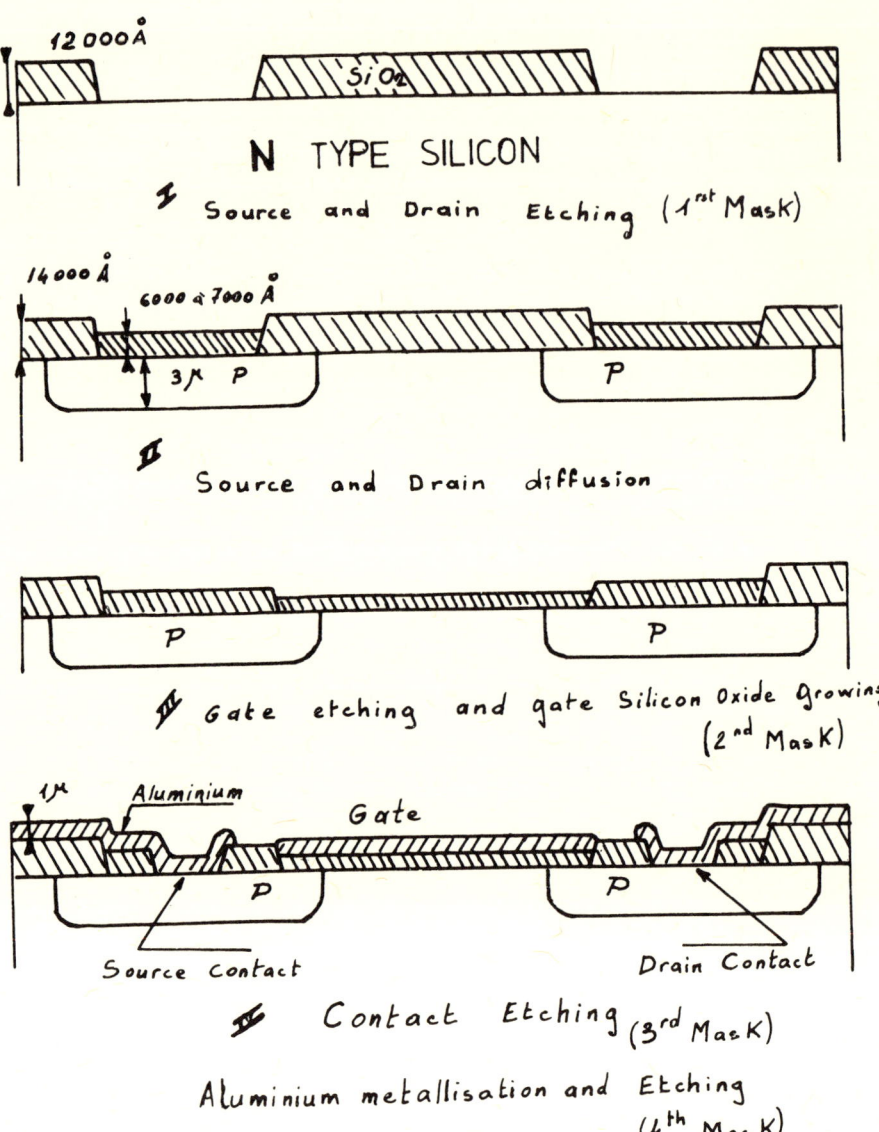

Fig. 2 Main different steps of the bulk MOS technology.

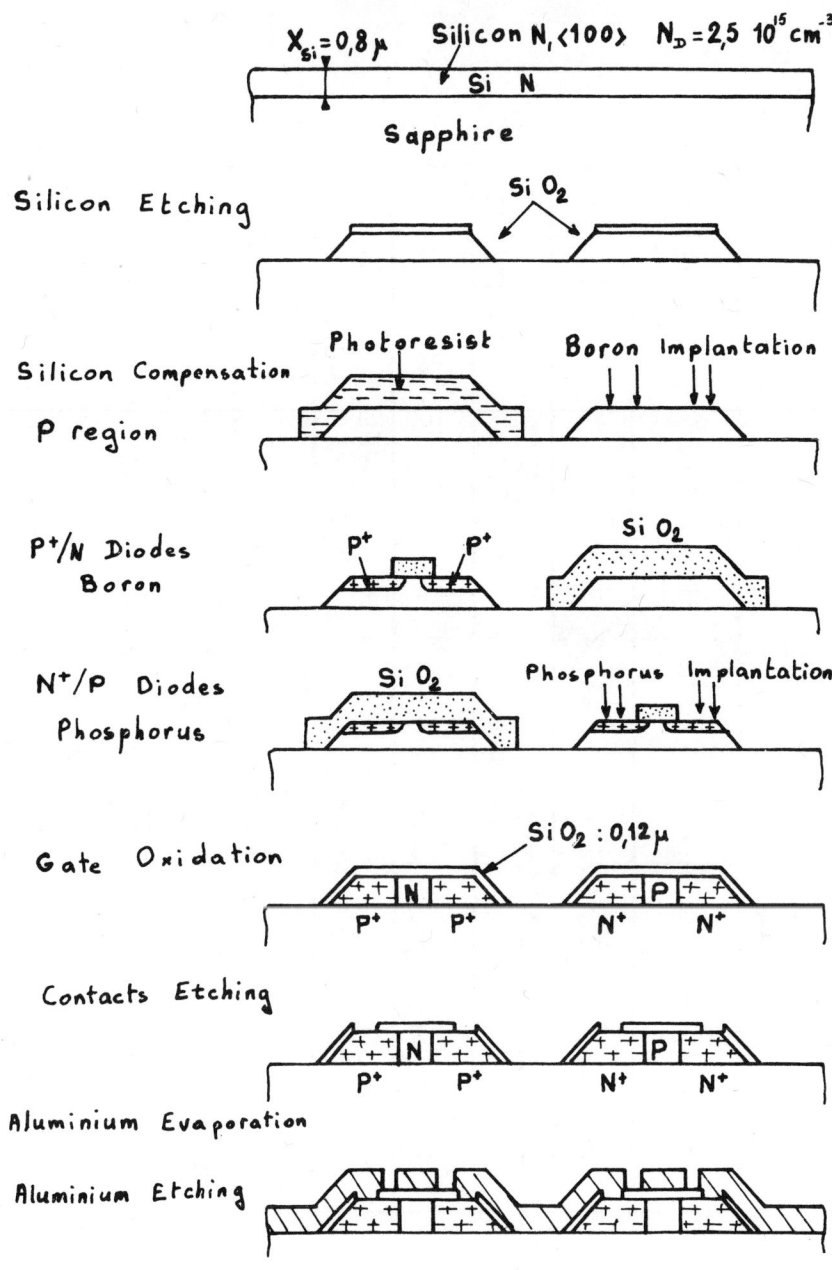

Fig. 3 The different steps of the silicon on sapphire CMOS technology.

TECHNOLOGY	NUMBER OF PHOTOMASKS	
STANDARD P SOSMOS	5	
DEPLETION – ENHANCEMENT P CHANNEL	6	SELECTIVE ION IMPLANTATION
C.MOS / SOS (DEPLETION – N CHANNEL)	6	N^+ DIFFUSION
C.MOS / SOS (ENHANCEMENT)	7	P WELL N^+ DIFFUSION
C.MOS / SOS (DEEP DEPLETION)	6	N^+ DIFFUSION

Fig. 4   Comparison of the complexities of different silicon on sapphire technologies.

THIN FILM APPLICATIONS IN MICROELECTRONICS 423

insulating thin layers in the range of 1000 Å. In the other hand, it is necessary to take contact on diode to apply voltages. Aluminium contact pads allowing device mounting in a package is deposited on a field oxide with thickness in the range of 1.2 µ.

To realize how it is possible to use non metallic thin film in microelectronic, two reviews on MOS bulk technology and on CMOS on sapphire technology will be done. The Fig. 2 shows main steps of MOS process on bulk silicon. We can observe that two kinds of thin film material is used in such case. Organic layers as photoresist which allow to describe a circuit pattern and inorganic materials as silicon oxide, silicon nitride or aluminium. The range of thickness can be from 1000 Å to 15 000 Å.

The first step of the process consist in the cleaning of a N type silicon wafer and then in oxidization of silicon to obtain a SiO_2 12 000 Å thickness layer. This layer is called field oxide or thick oxide, it is used in the process but do not participate in the device function. By means of photoresist pattern we open in the field oxide windows allowing boron diffusion inside silicon. This step is the making of source drains diodes. The next step consists in opening of gate location, in cleaning of silicon surface and in oxidizing of silicon for instance with dry or wet oxygen at a temperature near by 1000°C. SiO_2 thickness obtained is around 1200 Å. Then we must open holes trough field oxide to take contacts on drain source diodes. The last step is aluminium deposit and etching. Aluminium thickness is in the range of 15000 Å. With this technology, only bulk silicon wafer is not a thin film.

On the Fig. 3, we can see the most sophisticated CMOS technology : the SOS CMOS technolgy : the silicon on sapphire complementary MOS technology. In this device, all the layers are thin films, even silicon thickness is in the range of 0.8 µm. The most important layer finds in the process is photoresist layer used as mask. This CMOS Technology is the more complicated and use 6 or 7 photomasks. The Fig. 4 shows complexity of different technology using silicon on sapphire substrate ; and Fig. 5 explains main advantages of one of them, the SOS MOS depletion enhencement technology.

2. DIELECTRICS IN THE MOS TECHNOLOGY

Several dielectric layers can be used to achieve a MOS device. Each step of the technology needs one or several insulating layers to process the device. For example to etch silicon nitride we must use a SiO_2 mask because adhesion of photoresist on silicon nitride is not good and the SiO_2 mask is obtained by a photoresist process. We shall concentrate now on thin insulating films extensively used in MOS technology. Several other dielectric layers exist and can be used but only in the research field because all their properties are not known, and especially their stability with the time.

DIELECTRIC ISOLATION

- DECREASE OF PARASITIC CAPACITANCES
- NO PROBLEM OF FIELD OXIDE THRESHOLD
- NO BULK EFFECT ON THRESHOLD VOLTAGES
- NO PUNCH THROUGH EFFECT
- HIGHER TECHNOLOGICAL YIELD

DEPLETION ENHANCEMENT

- A SINGLE SUPPLY VOLTAGE
- SMALL SENSITIVITY TO SUPPLY VOLTAGE SHIFT
- LOW POWER CONSUMPTION
- TTL COMPATIBILITY
- WELL SUITED FOR ANALOG DEVICES
- HIGH DENSITY OF INTEGRATION

Fig. 5. Advantages of the SOS MOS depletion enhancement technology

2.1 - Dielectrics :

- SiO_2 thermally grown : Thermally grown SiO_2 can be obtained by different ways according to properties and thickness to reach. The steam oxygen growth is used for thick layers because growth rate is very important (90 Å/minute with 1150°C). 12000 Å field oxide in bulk MOS Technology is achieved with steam oxide.

The wet and dry oxygen growth is used for thin layer and especially for gate layers. The growth rate is slow and can be for example 40 Å/minute with dry oxygen. Typically SiO_2 thickness of 1200 Å is used to realize gate oxide.

- SiO_2 deposited with low temperature chemical vapour deposition : In this case, SiO_2 layer is used as field oxide layer in silicon on sapphire MOS technology and as a passivating layer at the end of the process. To prevent holes and defects across the layer, we must deposit more tahn 5000 Å of SiO_2.

- Si_3N_4 by chemical vapour deposition is used as a gate insulating layer over thin SiO_2 or as a passivating layer to prevent pollution by moistures and against radiation damages. In the former case thickness is in the range of 500 Å and in the later case in the range of 5000 Å.

- Silicon : It is of two types : epi silicon CVD deposited or on insulating substrate (sapphire) ,or polycrystalline silicon CVD deposited over an oxide as a metal gate instead of aluminium. In the former case, epi silicon thickness can be in the range of one micron in the SOS technology and up to ten microns in the bulk technology. In the later case polycrystalline thickness is in the range of 5000 Å up to one micron.

2.2 - Main properties :

- Mechanical properties
- insulating support for metal stripe line connecting between them each part of a circuit.
- mask against species diffusion such as boron, phosphorous or arsenic as well as against ion implantation.
- protective layers against pollutions, radiation damage, short circuits.

- Chemical properties : The main chemical property of these insulating thin films used in the MOS technology is the specific

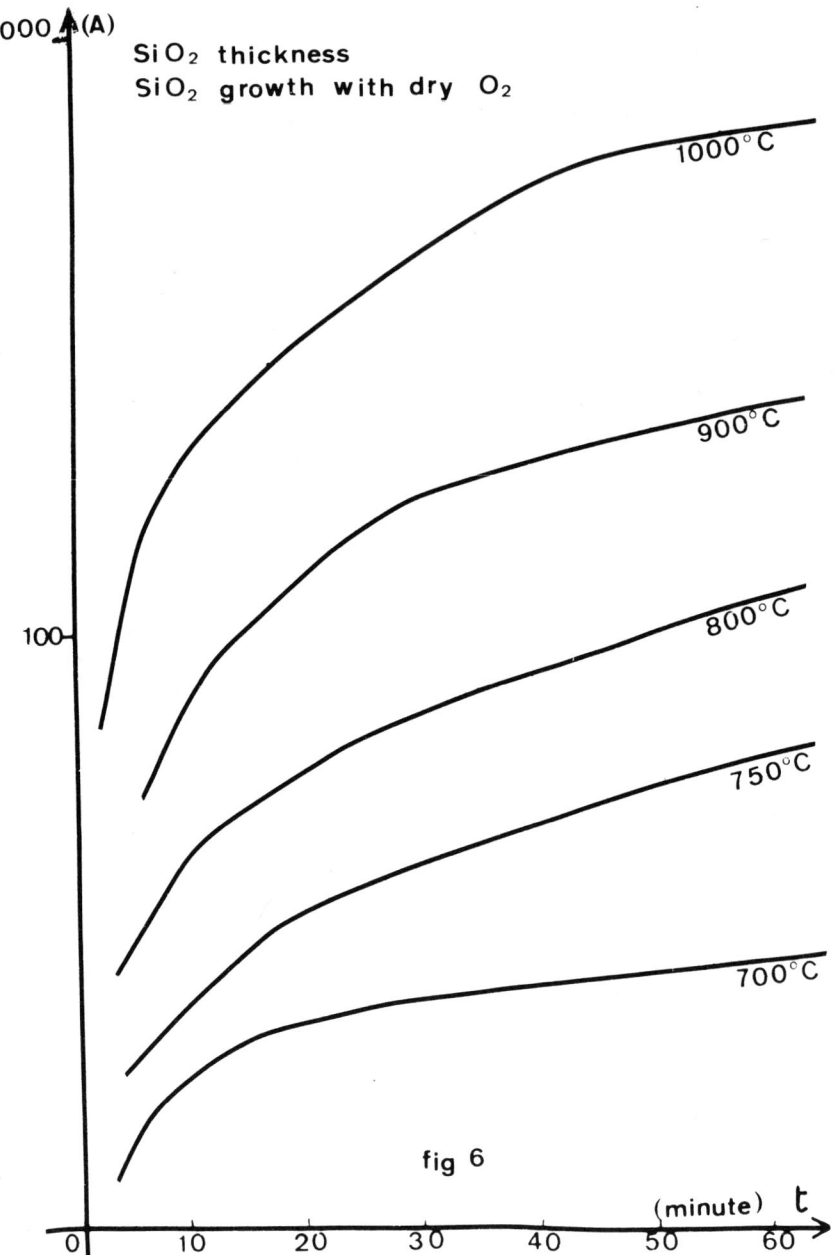

Fig. 6. SiO$_2$ growth curves for dry oxygen thermal oxidation.

etching of these layers versus several acid or basic solutions. For example silicon is etched by a solution of KOH or by a mixture of NO_3H, FH, CH_3COOH; SiO_2 is etched by FH (10% in water); Si_3N_4 is etched by PO_4H_3 (180°C) and lightly by FH (10% in water) and Al is etched by PO_4H_3 (room temperature). Sometimes we can combine the etch rate of two layers to obtain bent etching.

- <u>Electrical properties</u> : The insultating property is the main feature of these layers and it is extensively used in the technology and in the devices. In the MOS devices, gate is insulated from silicon surface by a dielectric and surface potential is controlled by the electric field induced by gate potential. No current flows across gate in the common MOS device.

The insulating properties of non metallic thin films is used to allow aluminium stripe lines to run over the circuit without short circuits. In this case insulation is achieved by the SiO_2 field oxide.

Conducting properties of these insulators are used in some particular devices such as MNOS memory device and MOS Tunnel devices.

- <u>Silicon dioxide material</u> : SiO_2 is the more important dielectric used in semiconductor process. Thermally grown SiO_2 is well known. Numerous workers have contributed to the knowledge of silicon-silicon dioxide interface[1][2]. This interface is very important for the quality of the device, and specially for obtaining high transconductance values of the FET MOS. SiO_2 can be obtained by different ways for the different uses. Quality of SiO_2 depends on the making, and wanted thickness range indicates the kind of oxidation process, that is the technology. Very thick layers are obtained by steam oxidation and thin layers by wet or dry oxidation.

Fig. 6 shows SiO_2 growth curves with dry oxygen and for different oxidation temperatures. Temperature must be choosen to obtain an oxidation time suitable for a good oxidation reprocibility. We give some examples of oxidation time, temperature, and process.

Thickness   1000 Å with wet O_2 temperature 1000°C time 12'
Thickness   100 Å with dry O_2 temperature 800°C time 60'
Thickness   30 Å with dry O_2 temperature 750°C time 20'
Thickness   20 Å with dry O_2 temperature 700°C time 13'
Thickness   17 Å with dry O_2 temperature 680°C time 10'
Thickness   12000 Å with steam O_2 temperature 1150°C time 140'

Main problem with SiO_2 layer is oxide charges introduced in the layer in the thermal oxidation. These charges can be reduced by high temperature annealing in nitrogen after oxidation, or after metallisation by low temperature (400°C, 500°C) annealing in nitrogen or water vapour.

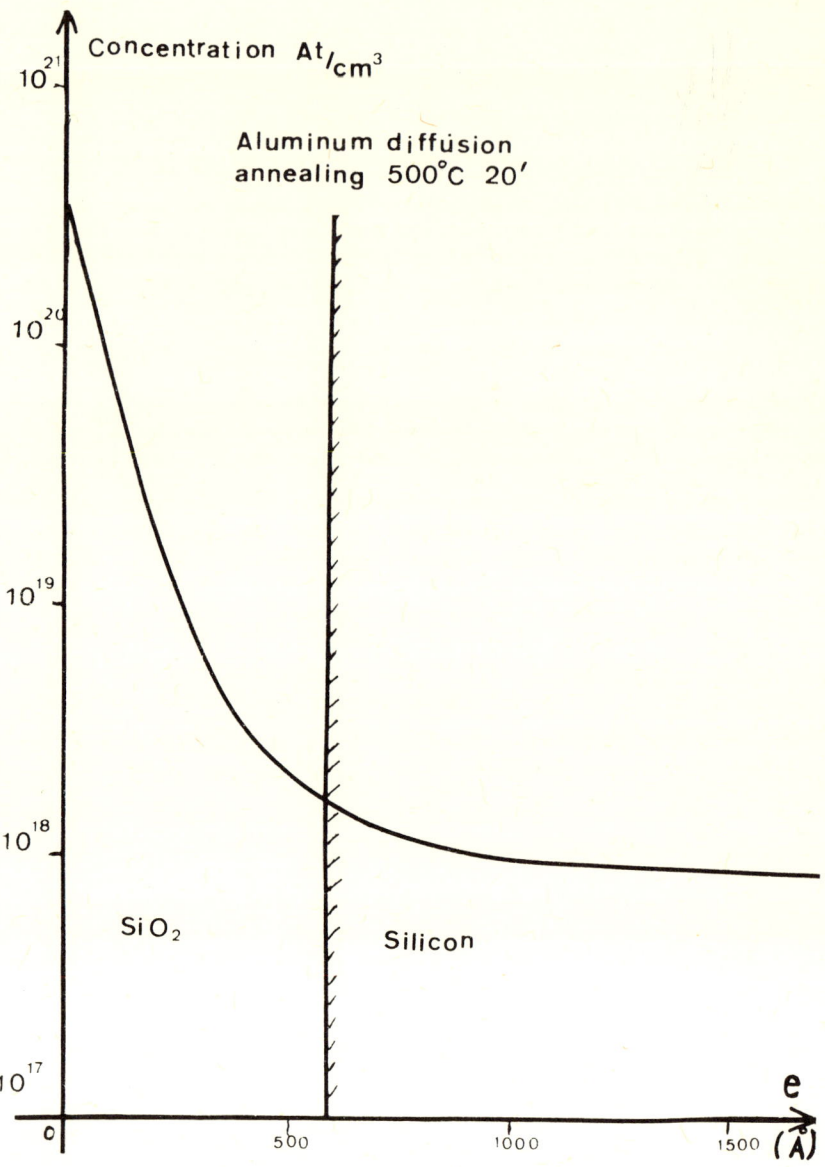

Fig. 7. Aluminium diffusion profile in SiO$_2$-Si system as obtained from ion microprobe measurements.

Other problems can appear in the SiO_2 layers and the more important is the presence in the layer of some species like Na^+ or H^+. These ionized species are very mobile in the electric field and they can drift toward silicon surface or gate electrode involving a shift of threshold voltage.

We have seen to reduce surface states, low temperature annealing after metallisation gives good results and significant decrease of positive oxide charges. However it is necessary to be careful with annealing.

We show in Fig. 7 aluminium profile in SiO_2 after annealing obtained with ion microprobe. We can see diffusion of aluminium inside SiO_2. With a 600 Å SiO_2 layer we obtained 2.10^{18} At/cm^3 of aluminium at the SiO_2-Si interface which is very important. With gate oxide layer of 1200 Å, surface concentration is very weak, but in the case of future technology for instance MICROMOS technology where gate thickness is in the range of 100 Å, these results must be taken into account to prevent short circuits across gate. In the case of MICROMOS technology, either we must reduce annealing temperature or we must use another metal as molybdenum or polycrystalline silicon.

Presence of H^+ species in SiO_2 layers can be reduced by suitable annealing at high temperature with nitrogen whereas presence of Na^+ species is very difficult to avoid;to minimize pollution by Na^+ we must use very pure water and to decrease influence of Na^+ ions we can prepare SiO_2 layer with O_2-HCl mixture. Cl^- ions have the property to trap Na^+ ions and to increase breakdown strength[5].

Now we can realize very good SiO_2 layer in wet or steam oxygen without Na^+ pollution. We generate steam not by boiling water but H_2O synthesis with very pure O_2 and H_2 in a reactor. Results obtained with this process are similar to results achieved with dry oxygen. We have some number of surface state, similar breakdown strength, no shift of threshold voltage under field and temperature. However the microbreakdowns observed with dry oxygen grown layers are not seen with these layers.

Another point of main importance is the capability of dry oxygen grown SiO_2 to trap holes. Even when composition is stochiometric, SiO_2 may exhibit hole trapping due to non-bonding p-valence band levels. The origin of this seems to be the presence of O^{--} ions[6]. In the other hand oxide grown with water does not exhibit this feature and unlike dry oxide, seems to trap electrons[7].

The creation of fast states and positive oxide charges under positive and negative bias (slow trapping instability) has been further investigated in several studies[8, 9, 5]. At the present time, the best process to decrease and sometime to avoid Na$^+$ drift of threshold voltage is to use phosphosilicate glass passivation. Addition of P_2O_5 inside SiO_2 layer allows gettering of Na$^+$ by phosphosilicate glass. With these layers we obtain an increase of dielectric constant from 3.82 to 4.1[3], but dielectric strengh is as good as with as-grown SiO_2 films.

With the silicon on sapphire technology we must use deposited SiO_2 to obtain thick insulating layer allowing passage of metal stripe lines. Deposited SiO_2 is achieved at low temperature (450°C) with O_2, SiH_4 reactions. Quality and density of these layers are not the same than thermally grown SiO_2, but despite of some defects as holes across layers and fast etch rate, bulk technology and SOS MOS technology use extensively deposited SiO_2 as field oxide layer, mask oxide, passivating layers. Quality can be improve by particular annealing in order to obtain densification of the layer. Quality of deposited SiO_2 after densification is quite similar with thermally grown SiO_2 (etch rate has almost the same value).

- Silicon nitride material

Silicon nitride has been one of the several materials introduced, eight years ago in the MOS technology with great interest. However problems connected with the use of deposited Si_3N_4 arised rapidly. Silicon nitride deposited directly on silicon exhibit very important shift of flat band voltage due to injection of charges directly from silicon into trap levels inside Si_3N_4. To avoid this effect it is necessary to put between silicon surface and silicon nitride layer, a thermally grown SiO_2 layer to prevent charges injection into silicon nitride traps. These so called MNOS structures are similar to MOS structures and we can realize MOS integrated circuits using this double layer as insulating layer of the gate. However great care must be taken with this technology and especially with SiO_2 layer quality to prevent threshold instability due to hole injection in SiO_2 layer. These holes choosed up by SiO_2-Si_3N_4 interface and threshold voltage shift can be very important.
However this charge injection and the shift of the flat band voltage have been used to realize a memory, the so called MI_1I2S memory.
In this case, SiO_2 layer is very thin and we favoured electron injection across SiO_2 layer to obtain flat band voltage shift. In the last part we shall speak about these devices.

Deposited silicon nitride layer can be achieved by different methods but the most widely used is the chemical vapour deposition of Si_3N_4 after reaction in vapour phase of SiH_4 (or $SiCl_4$) with NH_3 with N_2 or H_2 gas carrier. Deposit temperature can be in the range of 650°C to 1000°C. Quality and characteristics depend on both temperature and SiH_4 over NH_3 ratio. The index of refraction increases when temperature decreases and decreases when SiH_4/NH_3 decreases. Values can be 1.98 at 850°C and $SiH_4/NH_3=1/100$ and 2.03 at 770° and $SiH_4/NH_3=1/70$. In the same way the dielectric constant depends

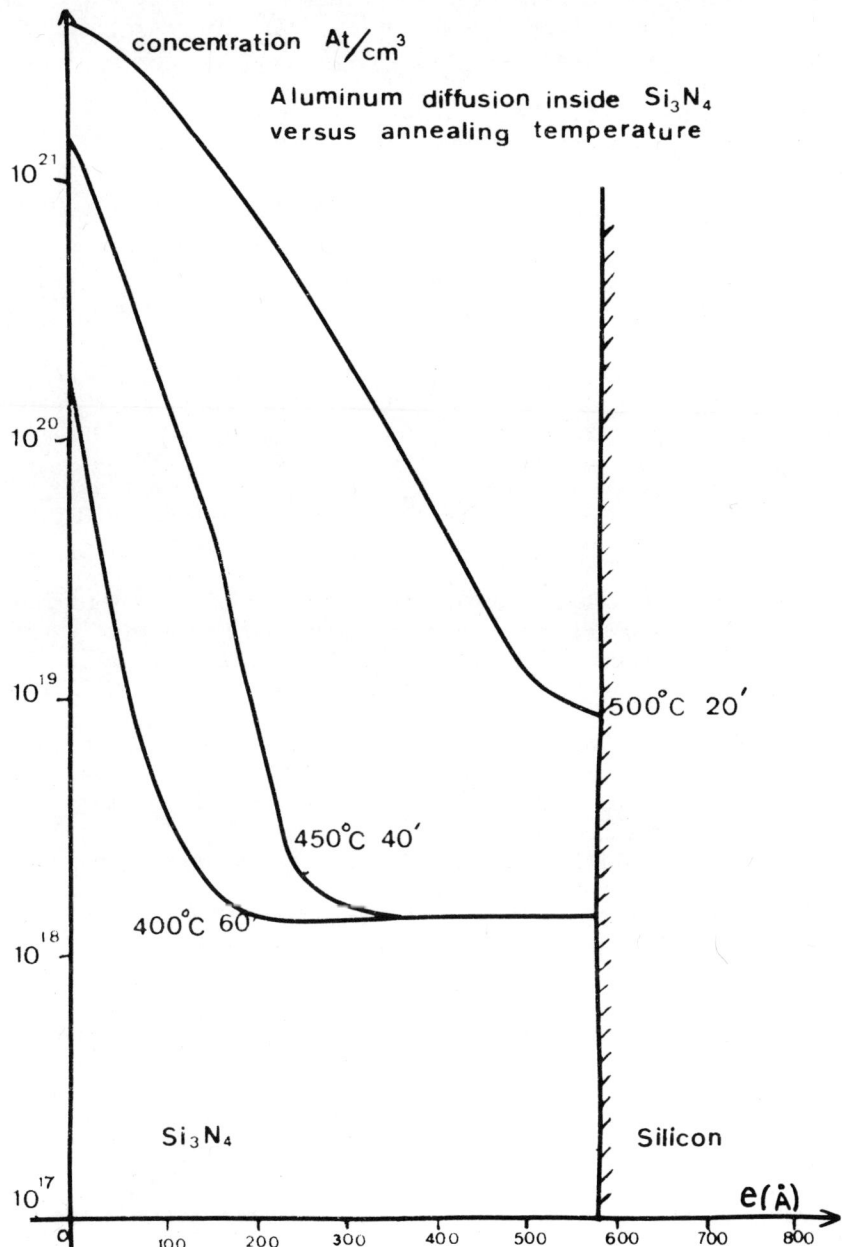

Fig. 8. Aluminium diffusion profile in Si_3N_4-Si system as obtained from ion microprobe measurements.

on the technological parameters. We observe a decrease when temperature decreases and an increase when SiH_4 over NH_3 ratio increases. The values are in the range of 6.5 to 7.4[10,11,12].

In the other hand, the dielectric strength is high and quite similar to SiO_2 dielectric strength (10^7 V/cm)[13,10]. The dielectric strength is better with H_2 deposited than with N_2 deposited silicon nitride and with a very low SiH_4 over NH_3 ratio, the layer has the lowest density of pinholes[12].

From the point of view of the conduction across the layer which is of great interest for the stability of MNOS structures and for MI_1I_2S memory devices, the current transport across silicon nitride is bulk controlled. The conduction due to Poole-Frenkel effect is the most probable[13] at high field. As we have seen with previous parameters, silicon nitride conductivity is strongly dependent on technological parameters. The conductivity increases when temperature decreases and when SiH_4 over NH_3 ratio increases[14]. We can minimize the conductivity by deposition between 800°C and 900°C and with SiH_4 over NH_3 ratio in the range of 1/100.

For devices as MI_1I_2S memory transistors, it is very important to obtain silicon nitride layer with poor conductivity to avoid charge injection into the silicon nitride at the metal silicon nitride interface. According to technological process, pollution by different species can appear at this interface, lowering barrier and allowing easy injection from the metal. Pollution is mainly due to aluminium diffusion inside silicon nitride during annealing. In the case of aluminium diffusion, we can see on Fig. 8 aluminium profile obtained with ionic microprobe in silicon nitride layer for different annealing temperatures. As we have seen previously for silicon dioxide, even at low temperature (400°C) an important part of metal can diffuse inside the insulator and for 500°C this diffusion can reach silicon interface with a doping level closed to $10^{19} At/cm^3$.

What is the impact of these results on physics and on thin film technology ? For layer ticker than 1000 Å, aluminium does not reach silicon interface and no effect can be noted. But with MI_1I_2S memory devices, where conductivity is of main importance to exhibit memory phenomena, this diffusion is of main importance. Effectively aluminium diffusion changes the conductivity of the layers. Fig. 9 shows current versus voltage curves for two devices. The former is annealed at 400°C and the later at 500°C. We observe the important changes of conductivity. The other curves are related to different measuring temperatures. These results show how conduction phenomena and injection phenomena can be modified by technological process.

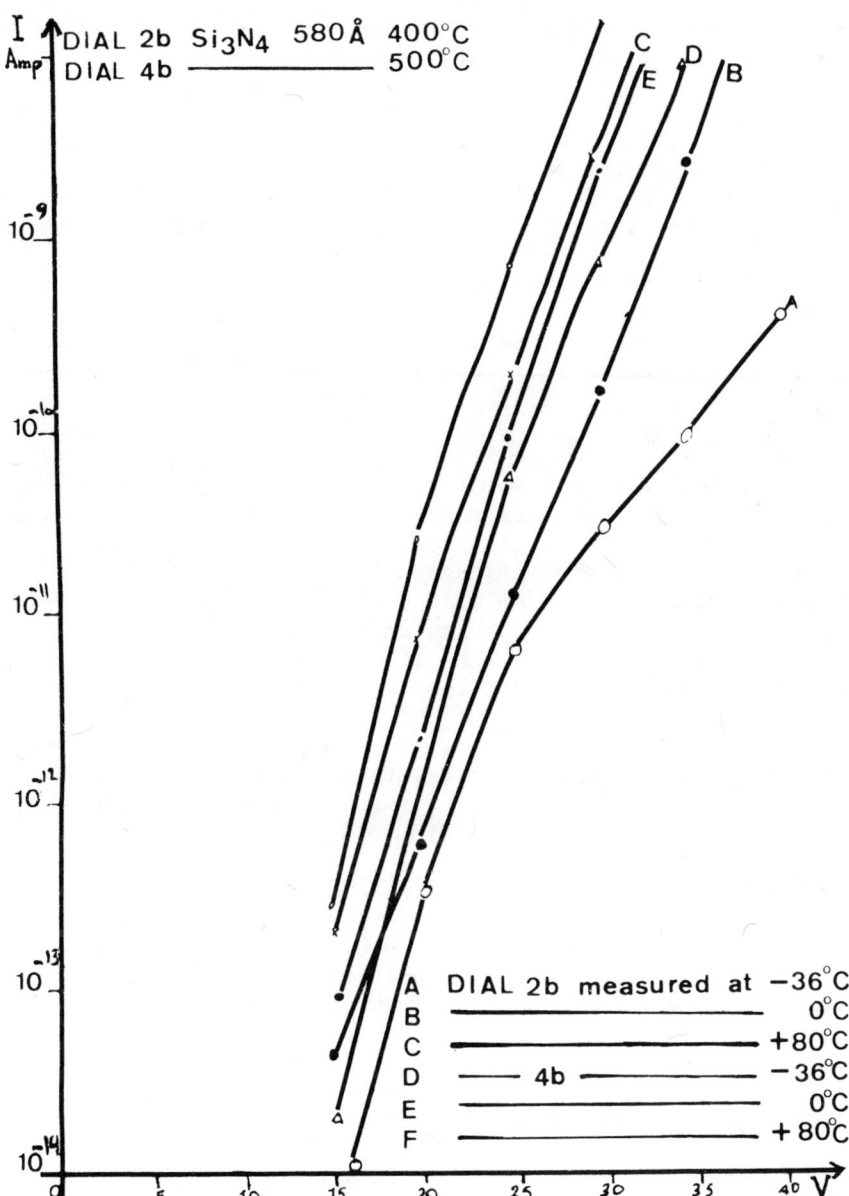

Fig. 9. I(V) characteristics of Si_3N_4 layers for different annealing temperatures.

3. THIN FILM APPLICATIONS IN DEVICES

Up to now we have shown why we need thin insulating films in microelectronic and what kind of dielectric films are used in the technology.

This part will be devoted to the use of these thin films in devices. We have seen that non metallic thin films have several characteristics : chemical characteristics, mechanical characteristics and at last electrical characteristics. These later characteristics are extensively used in microelectronic devices. To illustrate the use of these layers we have choosen three kinds of devices MOS tunnel diode and FAMOS device to illustrate insulating property and MI_1I_2S device to show how it is possible to use very weak conductivity in insulating film to obtain memory phenomena.

3.1 - MOS tunnel diode

In this device, the insulating property of a very thin SiO_2 layer (< 50 Å) and the tunnel conductivity across this layer are used in the same time. Fig. 10 shows the structure : metal, SiO_2, N type silicon. If we apply a negative bias to the metal versus the silicon bulk, the silicon will be in depletion. Minority carrier coming from silicon bulk toward surface are drained off by tunnel current I_{vm}. Therefore silicon surface cannot be in inversion because carriers generated in silicon are not sufficient to accumulate at the interface[15, 16].

We note that the aluminium Fermi level is just facing the forbidden silicon band gap, therefore, no current can flow this way.

If we generate numerous carriers in silicon, for example by light illumination, excess carriers will be at the silicon surface and the surface potential will increase. Therefore metal Fermi level will be facing the silicon conduction band and an important current will flow between aluminium and silicon. In figure 10a we can see MOS tunnel structure in the deep depletion mode. Minority carrier generation is not sufficient to invert silicon surface and to allow metal Fermi level to rise up to the silicon conduction band. In this case no I_{mc} current can flow from metal to silicon. In figure 10b numerous minority carriers are available, due to light generation. In this case, minority tunnel current I_{vm} across SiO_2 from silicon valence band is less important than optical generation current. Therefore minority carrier accumulation occurs at the interface, involving metal Fermi level rising up to the silicon conduction band. The position of metal Fermi level determines current flow and this position is due to silicon surface potential depending on optically generated minority carriers.

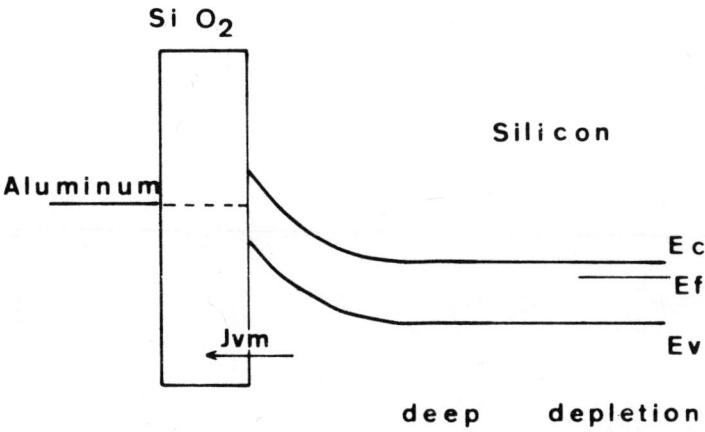

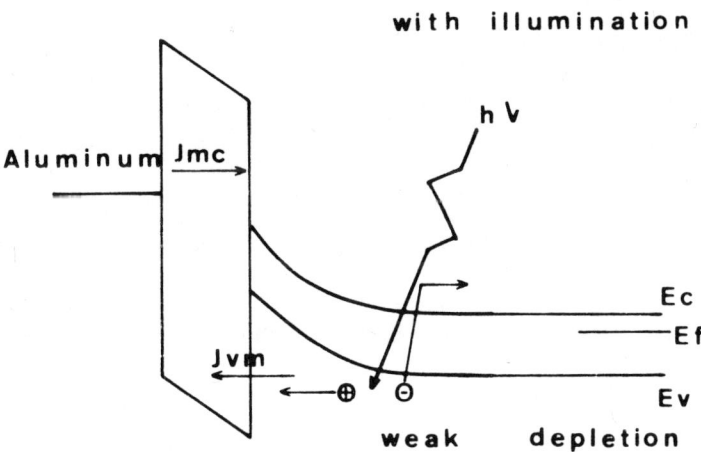

Fig. 10. MOS tunnel diode structure with and without light illumination.

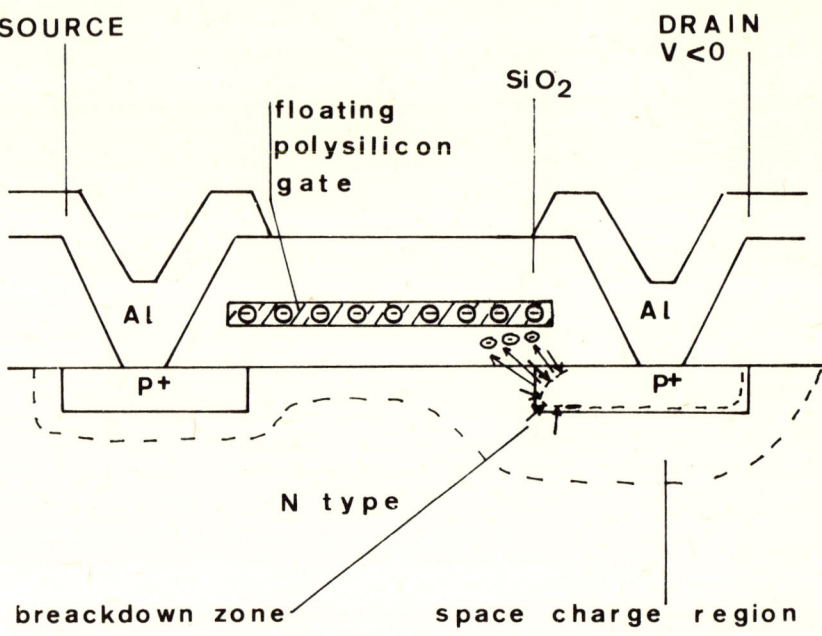

Fig. 11. Vertical section of a FAMOS structure under bias.

This device has an internal gain. We have measured values in the range of 30. The thickness of SiO$_2$ layer must be less than 50 Å to allow significant current to flow by tunnel mode. The main interests of these structures are a very good sensitivity of the device to low lighting and a low value at the current flowing across the device. Then performances of these devices are strongly reduced by interface states at the SiO$_2$-silicon interface. Improvement of technology is still necessary to obtain better results.

These devices can be used as single low level photodetector as well as in photodetector matrix. This structure can also be used for tunnel spectrometry by scanning the metal Fermi level over the band gap to determine interface state location. At last, instead of generate minority carriers by optical means, it is possible to supply minority carriers by peripherical diode.

3.2 - FAMOS devices

Injection of carriers into SiO$_2$ layer of a MOS capacitor or transistor from an avalanche diode has been previously reported by many workers[17,18,19,20]. In these cases charge injection was produced by silicon surface breakdown due to large ac signal on the metal gate. This injection produces a shift of the flat band voltage specially with p substrate. Hot carriers can be produced by another way. When we put a diode in breakdown mode hot carriers are present at the silicon surface and can be injected in or across the SiO$_2$ gate. This effect has been extensively used by Frohman-Bentchowsky[21] from Intel to realize a memory device.

Figure 11 shows a vertical section of FAMOS structure under bias (FAMOS = floating gate avalanche injection MOS). This device is similar to MOS transistor with the difference that the metal gate is completely insulated by a SiO$_2$ layer underlaying the gate. This floating gate is made of polycrystalline silicon. When sufficient voltage is applied to the drain diode to cause breakdown, then hot carriers are injected across gate insulator and are collected by the gate metal. With N type substrate, electrons are injected in the gate insulator and the gate metal is charged negatively by electrons. If the number of charges is sufficient, the field produced can invert the silicon surface and current can flow from drain to source. At the present time, we have no means to discharge electrically this structure and we need U.V. light or X ray to excite carriers and to discharge insulated metal gate. Retention time of charges on insulated metal gate can be important (up to 10 years).

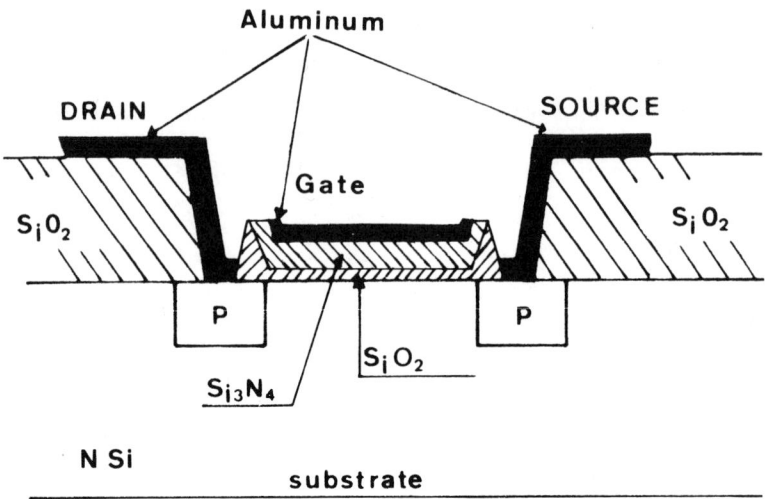

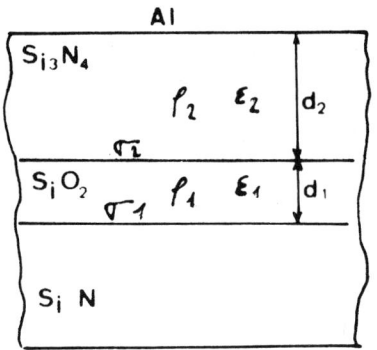

Fig. 12. Vertical section of a MI_1I_2S transistor.

THIN FILM APPLICATIONS IN MICROELECTRONICS

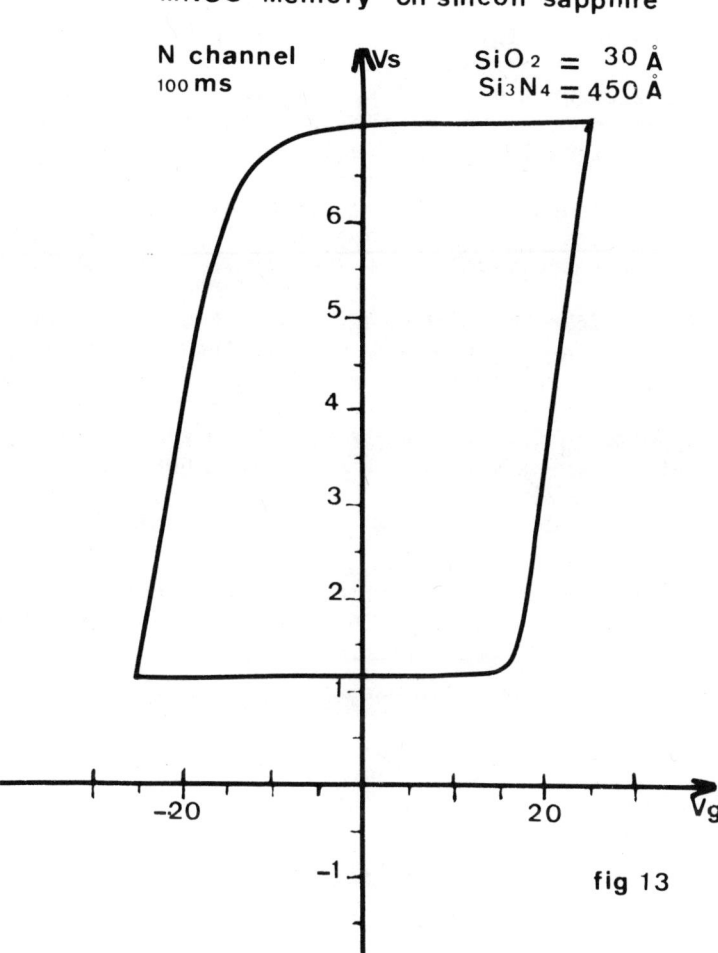

Fig. 13.   Memory cycle obtained with MNOS device in the SOS technology.

3.3. - MI_1I_2S memory devices

MI_1I_2S device [22] is a particular device because it uses the insulating properties of silicon dioxide and silicon nitride films to control silicon surface potential and in the same time it uses conduction properties of these two materials to change the threshold voltage of the transistor. Change of threshold voltage is achieved by charge injection inside the gate insulator.

Figure 12 shows a vertical section of a MI_1I_2S transistor. The gate insulator consists of two insulators. A very thin SiO_2 layer is closed to silicon and a silicon nitride film is deposited over this first layer. The thickness of the SiO_2 layer can be in the range of 30 Å and the thickness of Si_3N_4 is in the range of 500 Å. The conductivity of SiO_2 and Si_3N_4 layer is related to current I_{ox} and I_n respectively, flowing across layers under bias.

For SiO_2 thickness in the range of 30 Å to 50 Å, the conduction mechanism is a Fowler-Nordheim tunneling and for silicon nitride layer, the Poole-Frenkel effect is the most important at high field. When applying a voltage V_g to the MI_1I_2S structure, currents I_{ox} and I_n flow across layer. The magnitude of these currents is determined by electric field distribution :

$$V_g = E_1 d_1 + E_2 d_2$$

Fields and currents are not identical in each layer, and charges σ_2 accumulate at the $SiO_2-Si_3N_4$ interface. The continuity of current at the interface gives relations :

$$\frac{d\sigma_2}{dt} = I_{ox} - I_n$$

$$\varepsilon_1 E_{ox} - \varepsilon_2 E_n = \sigma_2$$

Accumulation of charges at the $SiO_2-Si_3N_4$ interface gives a shift of the flat band voltage :

$$\Delta V_{FB} = -\sigma_2 \frac{d_2}{\varepsilon}$$

Figure 13 shows an example of memory cycle obtained with N channel MI_1I_2S device. Technological threshold is + 1.2 volts ; with a positive pulse (+25 V, 100 ms) we reach a threshold of + 7 volts and then with a negative pulse we go back to technological threshol The thickness of the SiO_2 layer is very important in these devices because injection and rate retention time depend strongly of SiO_2 thickness.

THIN FILM APPLICATIONS IN MICROELECTRONICS   441

With thickness in the range of 30 Å the injection time is about 100 ms and with 17 Å the injection time can be 100 ns. In the former case, conduction across SiO_2 is due to a Fowler-Nordheim process whereas in the later case direct tunneling from silicon to silicon nitride traps allows fast injection. Simultaneously the retention time varies. With thick SiO_2 layers, retention times of more than ten years can be expected, whereas with very thin layers we have only retention times of few weeks.

The main difficulty encountered with these devices is the need to apply positive voltage to inject charges and then negative voltages to discharge the structure. Because the need of these two voltages, it is difficult to realize on bulk silicon a complex device using MI_1I_2S components. With the use of silicon on sapphire technology, these difficulties are removed since each transistor can be insulated from the others by silicon etching. This technology is now in development in several laboratories.

4.   CONCLUSION

This paper presents a brief survey of some thin insulating films used in microelectronic. Our discussion was only concerned by insulators extensively used in microelectronic industry. However in research field, numerous materials are in use to study their properties. These layers as those used for example in Josephson devices will be extensively used in few years but at the present time technology and properties are not well known.

Advanced devices in microelectronic are generally obtained with well known materials except when a new material, with its particular property, is an important and active part of the device. It is the case of MI_1I_2S devices where very thin SiO_2 layers was necessary to obtain memory phenomena.

We have seen that several kinds of non metallic thin films are used in microelectronic, either organic or inorganic material. With the last and most sophisticated technology, the SOS MOS process, all the layers have a thickness less than one micron. In the future new devices are in the spotlight as micro MOS devices where all sizes will be divided by a factor of ten : Channel length in the range of one micron instead of ten micron ; gate thickness in the range of 100 Å instead of 1000 Å. The need of such thin layers indicates that the technological process has to be changed. Etching process of SiO_2 or silicon will be obtained with new methods, for example by ion plasma etching and thin SiO_2 layers will be obtained at lower temperature.

With the technology in use in industry, very large integration circuits are realized. At this time we can have on the same chip more than 10 000 transistors in a circuit. With the new technology in development, complexity of 50 000 transistors and more on the same chip is expected.

REFERENCES

1. A.G. Revesz and K.H. Zaininger, RCA Rev. $\underline{29}$, 22 (1968).

2. A.G. Revesz, K.H. Zaininger and S.A. Evans, J. Phys. Chem. Solids $\underline{28}$, 197 (1967).

3. E.H. Snow and B.E. Deal, J. Electrochem. Soc. $\underline{113}$, 263 (1966).

4. E.H. Snow, A.S. Grove, B.E. Deal and C.T. Sach, J.A.P. $\underline{36}$, 1664.

5. C.M. Osburn and E.J. Weitzman. J. Electrochem. Soc. $\underline{119}$, 603 (1972).

    C.M. Osburn and D.W. Ormond, J. Electrochem. Soc. $\underline{119}$, 597 (1972).

    C.M. Osburn and S.I. Raider. J. Electrochem. Soc., $\underline{120}$, (1973).

    C.M. Osburn and J. Chou, J. Electrochem. Soc., $\underline{120}$, (1973).

6. T.M. Di Stefano Appl. Phys. Lett. $\underline{19}$, 280 (1971).
    J. A. P. $\underline{44}$, 527 (1973).

7. E.H. Nicollian, C.N. Berglund, P.F. Schmidt and J.M. Andrews, J. A. P. $\underline{42}$, 5654. (1971).

8. P. Rossel, H. Martinot and D. Esteve, Solid State Electron. $\underline{13}$, 425 (1970).

9. A. Goetzberger, A.D. Lopez and R.J. Strain, J. Electrochem. Soc. $\underline{120}$, 90 (1973).

10. G.A. Brown, J.C. Robinette and H.G. Carlson, J. Electrochem. Soc. $\underline{115}$, 948 (1968).

11. V.Y. Doo, D.R. Kerr and D.R. Nichols, J. Electrochem. Soc. $\underline{115}$, 61 (1968).

12. M.T. Duffy and W. Kern, RCA Rev. $\underline{31}$, 742. (1970).

13. S.M. Sze, J. Appl. Phys. $\underline{38}$, 2951 (1967).

14. F.A. Sewell, E.T. Lewis and H.A.R. Wegener, Technical Report AFAL-TR-70-148 (1970).

15. R.A. Clarcke and J. Shewchun, Solid State Electron. 14, 957 (1971).

    M.A. Green and J. Shewchun, Solid State Electron. 17, 349 (1974).

16. V. Le Goascoz, J. Borel and A. Payo-Casares, ESDERC, Nottingham (1974).

17. E.H. Nicollian and A. Goetzberger, IEEE Trans. Electron. Devices 15, 686 (1968).

18. E.H. Nicollian, A. Goetzberger and C.R. Berglund, Appl. Phys. Lett. 15, 174 (1969).

19. Poirier and Olivier, Appl. Phys. Lett. 15, 364 (1969).

20. J.F. VERWEY, Appl. Phys. Lett. 15, 270 (1969).

21. D. Frohman-Bentchkowsky, Appl. Phys. Lett. 18, 8 (1971).

22. J.T. Wallmark and J.H. Scott, RCA Rev. 30, 335 (1969).

    E.C. Ross and T. Wallmark, RCA Rev. 30, 366 (1969).

    E.C. Ross, M.A. Goodman and M.T. Duffy, RCA Rev. 31, 467 (1970).

    G. Dorda and M. Pulver, Phys. Status Solidi 1, 71 (1970).

    K.I. Lundstrom and C.M. Svensson, J. Appl. Phys. 43, 5045 (1972); Electron. Lett. 6, 645 (1970).

23. V. Le Goascoz and J. Borel, ESDERC, Munich (1971).

    P. Gentil, V. Le Goascoz and J. Borel. Onde Electrique Nov. (1972).

    J. Borel, E. Mackowiak and V. Le Goascoz, Colloque Mémoires, Paris (1973).

APPLICATION OF THIN NON-METALLIC FILMS IN OPTICS

E. Pelletier

Centre d'Etude des Couches Minces (Associé au C.N.R.S.)

St Jerôme - 130013 MARSEILLE - France

1. INTRODUCTION

The optical properties of a thin film are :
- the reflectance R (or R')
- the transmittance T
- the corresponding phase changes ψ_r (or $\psi_r{}'$), ψ_t
- the absorption is defined as : $A = 1 - R - T$ or $A' = 1 - R' - T$

These measurements can give information about structure and properties of the material of which the layer is composed. A very common application of thin films is to modify the optical properties of a surface, for example to obtain an high reflectance or a low reflectance without modification of the geometrical properties. If the desired result can be obtained with only one layer, then no complex problems of analysis or thechnique need to be faced, however if a single layer does not achieve the desired results, we may be able to conceive a multilayer system which will acheive it. This poses a series of experimental and computational problems which are not all completely resolved. We will discuss these problems in this article.

2. PROBLEMS IN THIN FILM OPTICAL FILTERS

2.1 Multilayer Calculations.

It is easy to compute the optical properties of a single idealized layer of which the index is n (complex) and the thickness is e, deposited on a substrate. The computation of the optical properties of a multilayer stack (design equation : (n_i, e_i) $i = 1, P$, substrate : n_o, incident medium : n_{p+1}) is tedious but it can be easily carried out with a digital computer, even if the n_i values are complex. So that the "analysis" of any complicated design can be obtained in few seconds if we assume perfectly flat, parallel homogeneous layers. However, this is not the main problem encountered in the application of dielectric films. We may have need of a filter with a given profile of R versus wavelength. The main problem is to determine the number of layers P, the indices and the thickness of each one of the component layers to obtain the required properties. We do not have a logical set of rules so that our "synthesis" must proceed by trial and error. The calculations start from a proposed design, and then adjustments may be made until a satisfactory solution is found.

2.2 Problems in the Preparation of Optical Coatings

The calculations to be described permit us to determine the design of a dielectric stack having prescribed optical properties. We have to consider the following parameters :

1) Thickness - The finished product will correspond to the design specifications only if the three following conditions are fulfilled :

a) Uniformity. Control of the uniformity to layer thickness over the area of the substrate is an important factor. A lack of uniformity causes a shift of characteristic wavelength over the surface of the filter. The effects of lack of uniformity are dependent upon the "spatial wavelength" Λ of the defect.
If $\Lambda \ll \lambda$ then we must use "couches de passage" (defect of uniformity of index).
If $\Lambda \gg \lambda$ then we must use calculations carried out with the rules of optical geometry (aberrations).
Between the two domains is a scattering domain.

b) Accuracy. The second factor is control of the overall thickness of each layer. This creates a major practical problem, namely the precise control of the optical thickness of the layer during deposition. The best method is to use a thickness monitoring device which responds directly to changes of optical thickness. Even if the refractive index differs appreciably from the specification, the difference can be partially compensated by an appropriate and automatic adjustment of the geometrical thickness. The optical monitoring of layers of thickness $p\lambda_0/4$ (p integer) is very easy, because the optical properties $T(\lambda_0)$ are very peculiar. However, the monitoring of stacks made of unequal layers is much more difficult.

c) Preparation. The basic manufacturing technique for the construction of thin films filters is that of vacuum deposition. With this process the two above problems can be resolved and the number of available materials is not large. Dielectric materials are of particular interest because of their low absorption.

2) Refractive index - The refractive index depends on the chemical composition of the material. The number of materials found to be suitable for optical coatings is limited, so the number of input values of index should be minimised in the design of filters. This limitation can be partially resolved in two ways :

- The co-evaporation to two dielectric materials mixed in various proportions to provide refractive index values ranging between those of pure materials.

- A mathematical equivalence : A symetrical stack made of two materials n_1, n_2 is equivalent to a single layer of equivalent index n (n_1 < n < n_2)
The judicious design of the thicknesses and of the total thickness gives a stack with the desired values of n and e. This statment is valid at a given wavelength and remains good enough in a broad band if the total thickness is smaller than the wavelength. The main inconvenience of this method is a large increase of the number of layers of the filter.

3) Environmental resistance. Probably the most important aspect of the environmental performance of the filter is its resistance to humidity, but the resistance to other agent such as temperature, shock and corrosive fluids, may all be important. All these difficulties are dependent on the specified application, but we are often thereby confined to a limited set of available materials.

3. MAIN APPLICATIONS : CLASSICAL FILTERS

In this paper it is not possible to give a detailed study of various thin film optical devices. For more details see [1].

1) Antireflection coatings
    - single layer antireflection
    - multilayer antireflection

2) High-reflectance mirror coatings
    - Multilayer dielectric stack $\lambda_o/4$
    - Extending the high reflectance zones

3) Edge filters
    - Symmetrical multilayers and the Herpin index
    - Extending the transmission zone

4) Band pass filters
    - Broad-band pass

- Narrow band filters (Fabry Perot and Double half-wave)

5) Polarizing beam splitter

The main part of these filters are obtained with quarter wave stacks $\lambda_0/4$ made of alternate high index and low index layers. This is for several reasons :
- very little absorption
- the optical properties are stationary in the neighbourhood of λ_0, and of $\theta = 0$, so main part of these filters can be illuminated with a convergent beam.
- optical monitoring is convenient
- the layers deposited can be almost uniform and homogeneous.

4. SYNTHESIS OF MULTI-LAYERS WITH PRESCRIBED OPTICAL PROPERTIES

4.1 The Problem

We can increase the field of application of thin films considerably with stacks made of layers having unequal thicknesses. In order to obtain prescribed optical properties, what are the successive indices and thicknesses making an optimum design ? The synthesis may be performed by successive iterations and the free parameters which are adjusted can be :
- the indices of each layer
- the optical thicknesses

or both together. These specifications can be used to study the different methods of synthesis, which are summarized in the following table.

optical thicknesses $n_i e_i$ / indices n_i	$\lambda_0/4$ stacks	unequal thicknesses
available dielectric materials	classical filters	free parameters e_i
intermediate indices	POHLACK's relations	free parameters n_i and e_i

4.2 Synthesis with Constraint over the Thicknesses

This method involves the case in which the optical thicknesses of the stack are equal to $\lambda_0/4$. For a stack made of P layers we can write :

$$1/T \simeq A_o + A_1\cos \pi \lambda_0/\lambda + \ldots + A_p \cos P \pi \lambda_0/\lambda$$

The identification of this polynomial with an analogous development $a_o, a_1, \ldots a_p$ giving the Fourier coefficients of the prescribed optical properties, can permit us to determine the number of layers. In fact, the application of this method is very limited because the constraints over the optical thicknesses, which must be kept equal, is too restrictive. In addition, the values of the refractive indices found from these equations are between 1 and about 6 and the practical realisation of these designs is impossible exactly in this form. Analogous calculations can be carried out for angles of incidence other than normal : in this case, the effective phase thicknesses are not stationary and the refractive indices should be replaced by the more general optical admittance ($n_i \cos \theta_i$ or $n_i/\cos \theta_i$) according to the polarisation.

4.3 Synthesis without Constraints

This calculation starts from a proposed design and adjustments may be made until a satisfactory solution is found. At the begining, we may define a suitable criterion for comparing together the quality of any one design; so a "merit coefficient", f permits us to compute the distance between curves showing the optical properties. The problem lies in finding the values of the free parameters (that is to say the design) corresponding to the minimum of the distance f.

The vast volume and complexity of these calculations overload the capacity of available computers. The best design found under these conditions is the "nearest" to the starting value. It is completely impossible to study all the designs completely. For example, for the function $f(\ldots n_i \ldots e_i \ldots)$, if P = 10, the calculations must be performed with N values of n and E values of e. The number of designs is $(N.E)^P$. Assume that N = 10, E = 10 and that the computation of f for one wavelength is about 10^{-6} second, then the complete study of the whole solutions is made in 10^{14} second $\simeq 3.10^6$ years ! This is not the only difficulty : it is also not possible to add successive layers to a refined design. The best solution with q layers does not have the same structure as the best solution with p + q layers.

4.4. Synthesis with Available Materials

A method for simplifying the synthesis problem is to reduce

arbitrarily the number of parameters. In addition, to simplify the experimental production of the coatings, only common evaporation materials are used, the only variables in the calculation being the mechanical thicknesses of the films. A wide variety of filtering properties can be obtained by the use of only a limited set of materials. But these restrictions can seriously prevent the determination of an optimum design.

4.5. Some results

The methods of synthesis are not perfect but we are now able to use them to find theoretical designs for numerous problems in the design of filters for normal incidence or oblique incidence We can say that the large part of these problems can be resolved with less than twenty layers, made up of four or five materials. However the methods of calculation are still very empirical. Despite these difficulties, very beautiful results can be shown ; for example in the case of band pass filters in the infra-red. (Dobrowolski). The free parameters are thicknesses and indices.

Another peculiar example concerns the design of a beam splitter for the P polarisation ($\theta = 55°$) for several visible and infrared wavelengths. This design is computed by assuming that the component materials are Zinc Sulfide and cryolite. This result permits the importance of the dispersion of the refractive indices versus the wavelength. This factor cannot be neglected if the optical properties of the design may be determined with good accuracy. It will always be necessary if the allowed tolerances are small.

5. PROBLEMS IN THE PREPARATION OF OPTICAL COATINGS

5.1 Readily available materials

It is sufficient to quote a list of available dielectric materials which are commonly used. The spectral region of transparency and the refractive indices are the two main points. For an accurate synthesis, and above all for precise monitoring, an exact knowledge of the values of the refractive index of a thin film is necessary. Sophisticated methods of determination may be used to obtain this result (for example n and k of ZnS in the visible).

5.2 Preparation of a Layer of Uniform Thickness

The uniformity of a layer depends on the source-to-substrate distance. The layers obtained on a rotating substrate are much more uniform in thickness than on a fixed substrate. With appropriate adjustments, we can experimentally determine the best disposition for a given evaporation geometry. A very good uniformity is of the utmost importance for some kinds of filter such as narrow band pass filters. On the other hand, for achromatic coatings, a perfect uniformity in thickness is not absolutely necessary (However do not forget ψ_r).

5.3 Monitoring of the Deposited Thickness

For more details see(2).
- quartz crystal method
- ionisation gauge
- OPTICAL MONITORING(3)

Thickness monitoring devices respond directly to changes of optical thickness. A difference between the effective and the specified values of refractive index is compensated by an automatic adjustment of the geometrical thickness : the optical thickness is $\lambda_0/4$.

a) Principle: As the thickness of a dielectric film deposited on glass increases, the reflectance or transmittance of the coated surface has a maximum or a minimum value (turning point) at the design wavelength λ_0, whenever the optical thickness of the film passes through a multiple of $\lambda_0/4$.

b) The location of extremums: In order to increase the accuracy of the location of the turning points, two very different methods have been developed :

- Zero of $\partial T/\partial e$. In practice the voltage signal from the detector T, may be recorded and the transmittance is differentiated with respect to time. An extremum is recognized by the zero value of $\partial T/\partial e$ and this is the signal for the end of deposition.

- "Maximètre" of Giacomo-Jacquinot. Essentially, the transmittance (or reflectance) of the coating is differentiated with respect to wavelength. An extremum is recognized by the zero value of $\partial T/\partial \lambda$. It is very important to repeat that the zero of $\partial T/\partial e$ should not be confused with the zero of $\partial T/\partial \lambda$. A complete comparison of the two monitoring processes is necessary.

c) Monitoring of dielectric stacks

- Classical filters: The optical thicknesses of the component layers may be equal to $\lambda_0/4$. The optical monitoring of λ_0 is very easy and a knowledge of the value of the refractive indices is not necessary The whole stack can be monitored on the same substrate and for each lay, we must stop deposition as soon as the zero point of $\partial T/\partial e$ (or $\partial T/\partial \lambda$) is reached.

- Unequal-layer stack: The experimental production of no quarter wave layers requires a "control programme" that must be followed step by step to obtain the required optical properties. If the design is n_1e_1, n_2e_2, n_3e_3, we must first compute the evolution of the spectral profile $T(\lambda)$ during the deposition of layer number one (i.e. the thickness increases from zero to n_1e_1). Now we have to assume that layer number one is <u>perfectly monitored</u>. A new calculation of the optical properties of the stack (glass + n_1e_1 + n_2e_2) gives the wavelengths for direct optical monitoring... This process is continued for each layer ; a trace of the evolution of the spectral profile is made and we can then choose for each one the type of differential and the monitoring wavelength giving the best accuracy (Note that $\partial T/\partial e = 0$ and $\partial T/\partial \lambda = 0$ must not be confused). The values of the refractive indices must be known with a large accuracy in order to perform these calculations.

5.4 The Importance of the Choice of Monitoring Process

Over all the methods, the ones which depend directly on optical properties are usually preferred over those which are less direct.

a) Direct and indirect monitoring: The optical monitoring of the whole stack on only one monitoring plate can be difficult to perform. It seems easier to use several monitoring plates. Under these conditions, we monitor only a small number of component layers on each monitoring plate, the work piece being as close as possible to the monitoring plate and receiving the whole stack (indirect monitoring). Unfortunately the best accuracy is obtained when the monitoring is performed on the whole piece itself (direct monitoring) because the thickness deposited at the same time on the two different plates are never exactly equal. (The angular distribution of the evaporant beam from crucibles varies with time). Sometimes the work piece cannot be used to perform the monitoring (metallic substrate, geometrical size, etc.) and the next best method is to use only one monitoring plate.

b) Relation between production errors and the monitoring process used: Sensitivity is not the lone criterion in a comparison of optical monitoring processes. We can demonstrate the need for a "stable" monitoring process in which the cumulative effect of errors in successive layers is considerably reduced. Under these conditions, whatever small errors occur in the early layers of the stack can be corrected. This notion will be developed in a particular example : take the filter : glass H L H L H L H L 2H L H L H L H L H air with the notations : H high index layer of optical thickness $\Lambda/4$ (zinc sulfide)
L low index layer of optical thickness $\Lambda/4$ (cryolite) $\Lambda = 5461$ Å.
A 10% error in the optical thickness of the spacer layer results in a filter of design :

glass H L H L H L H L (2.2H) L H L H L H L H air of which the peak transmission is in error by about 220 Å. However, this error will have no marked effect on the optical properties of the completed filter if direct optical monitoring of $(\partial T/\partial e)_\Lambda = 0$ is used. For layer number 9 $\partial T_9/\partial e_9 \neq 0$ but the next layer is then corrected such that $\partial T_{10}/\partial e_{10} = 0$ (different from $\Lambda/4$) thus obtaining a filter of design : H L H L H L H L (2.2H)(0.66 L) H L H L H L H air. The error is thereby perfectly corrected and the filter obtained is centred to better than 0.1 Å. Under these conditions, the usefulness of a "stable" monitoring process is obvious.

5. CONDITIONS UNDER WHICH FILTERS MAY BE USED AND MANUFACTURING TOLERANCES

5.1 Performance Specification

To simplify as far as possible the problems of synthesis, the calculation of the optical properties are made for only one single angle of incidence Θ_0. It is important to verify that the optical properties do remain stationary in the neighbourhood of Θ_0. In an optical arrangement the energy carried by a beam of light is directly proportional to the aperture. The performance of a filter is therefore dependent on the conditions under which it is used. In this respect the case of narrow band interference filters is particularly relevant and we give one single example of this.

Consider a filter of design M7 8L M7 illuminated by a parallel beam of light having the following performance :

Maximum transmittance : 0.86, Peak wavelength : 5623 Å, Bandwidth $\Delta\lambda = 14$ Å. If the filter is illuminated by a beam of aperture f/2.8 the performance will be only : $T_m = 0.31$ $\lambda_m = 5608$ Å $\Delta\lambda = 47$ Å. As soon as the aperture of the beam is greater than f/10, a considerable degradation of the performance of a narrow band filter may be observed. In addition, in the calculation of the optical properties of a filter the component layers are assumed to have plane parallel surfaces. In fact, the layers have microdefects which are important in the case of narrow band interference filters. These cause both a reduction of peak transmission accompanied by a broadening of the pass band and a loss of luminous energy by scattering.

Thus to compare theoretical and experimental performances it is best to illuminate only a small area of the filter.

Furthermore, the peak wavelength of a filter depends on the ambient temperature, the variation being of the order of 0.2 Å per degree C.

The conditions under which narrow band interference filters should be used are relatively restricted. This is not usually the case for other types of filter although the possibility of such restrictions should always be borne in mind.

5.2 Improvment of Manufacturing Precision

Taking account of all the above elementary precautions, the next step is the comparison of calculated and experimental performance. If there is considerable difference between these it is of course necessary to investigate the causes so that the manufacturing process may be improved.

a) Defects in the monitoring system - During deposition, the observed values of T_i (or $\partial T_i/\partial \lambda)_{\Lambda_i}$ will be in good agreement with predicted values if and only if the following conditions are fulfilled :

- the indices of refraction of the deposited layers are exactly known
- each of the layers in the stack has exactly the required thickness.

Computer simulation has been found to be a most powerful method for the analysis of possible defects in the monitoring method and has also yielded much information on the magnitude of refractive index and thickness errors which may be expected in practice.

Let us consider for example a filter which consists of only three layers. The required thicknesses are e_1, e_2, e_3. If we use refractive indices n_1, n_2, n_3, we find that the correct wavelengths for use with the Maximètre are Λ_1, Λ_2, Λ_3. Now let us suppose that in practice the actual values of refractive index are slightly different, say, N_1, N_2, N_3, with $N_1 \simeq n_1$, $N_2 \simeq n_2$, $N_3 \simeq n_3$. If the layers of the completed filter had nevertheless the correct values e_1, e_2, e_3, then the spectral profile of the filter, $T(N_i e_i)$ would be very nearly the same as the calculated profile $T(n_i e_i)$. However, the errors in the values of refractive index affect the accuracy of the monitoring process and therefore, the thicknesses of the deposited layers. Using control wavelength, Λ_1, Λ_2, Λ_3, perfect monitoring, that is to say monitoring in which there are no other errors, would give thicknesses E_1, E_2, E_3, such that $(\partial T_i/\partial \lambda)_{\Lambda_i}=0$ i.e. E_1 is such that for the combination (glass + $N_1 E_1$) we have $(\partial T_1/\partial \lambda) = 0$ for the wavelength Λ_1. E_2 is such that for (glass + $N_1 E_1 + N_2 E_2$) we have $(\partial T_2/\partial \lambda)_{\Lambda_2} = 0$ and so on. The filter obtained under these conditions can have a sepctral profile $T(N_i E_i)$ very different from that expected, $T(n_i e_i)$. Not only are these errors in the values of refractive index but much more important there are considerable errors in the layer thicknesses.

Computer simulation is an excellent tool in the determination of precise values of refractive index. It permit one to take account of errors which are introduced when deposition is not terminated exactly at the correct point when a zero of $\partial T/\partial \lambda$ is observed. It has been used to demonstrate the importance of choosing a "stable" monitoring process in which the cumulative effect of errors in the thickness of layers already deposited on the thicknesses of succeeding layers is not at all catastrophic. Under these conditions the manufactured filters will have acceptable performance regardless of those errors which may exist in the thicknesses of the individual layers. For the best monitoring procedure it is necessary to choose for each layer that monitoring method (e.g. $\partial T/\partial \lambda$ or $\partial T/\partial e$) which is optimal, taking account of both the sensitivity and the "stability" of all possibilities.

b) Study of a completed experimental filter - It is intersting to determine the errors which have been made in an experimental filter in order to avoid making the same error in subsequent filters. In order to do that it is necessary to calculate the thicknesses which have actually been deposited. This is a synthesis problem : what are the thicknesses of the component layers knowing the spectral profile of the filter ? Without additional information this problem cannot be simply resolved and synthesis programmes can give several solutions. Computer simulation of monitoring together with the study of measurements obtained during deposition of the filter are therefore most useful. Under these conditions the synthesis problem can be easily resolved if the number of unknown parameters is small. If, during the deposition, a rapid measurement of the filter could be made with good precision, it would be possible to check step by step, the performance of the stack and to make slight adjustments of the control programme which could minimise the effects of any error on the monitoring of the following layers.

In addition, synthesis calculations can be performed actually during deposition to calculate the optimum thicknesses for the subsequent layers of the filter taking account of the errors which have already been made. All this can be obtained with a real time computer. This method is difficult to put into operation. However it does represent a necessary step if complex filters are to be made successfully in industry.

c) Spectral profile - Among the earliest control techniques, the observation with the naked eye of the colour of the coated glass (in reflection) was already a global evaluation of spectral profile. This method should be perfected so that it can be used with any kind of stack. With the help of a rapid scanning monochromator one could measure over a large spectral range the transmittance (or reflectance) of a stack during deposition. In this way one would be replacing a very precise measurement at one or two wavelengths with a less precise control of the global spectral response. The usefulness of this method is especially clear for filters which are to be used over a large spectral range. Because of its complexity, this method is not in current use.

5.3 Manufacturing Tolerances

Taking account of the required optical properties (and of the tolerances in these optical properties) it is intersting to be able to put a figure on the accuracy necessary in obtaining the correct values of both refractive index and thickness for each layer during the production of a filter. This calculation is far from being evident. Nevertheless it is necessary to complete our synthesis methods so that we can give to the manufacturer designs which can be easily produced. The synthesis problem must therefore be enlarged by the introduction of this "new dimension". In this way one would have the maximum chance of success for industrial production.

At the start of the section on synthesis we defined a "coefficient of merit" representing the distance between the ideal optical properties and those of the filter calculated at each iteration.

$$f = || T_{ideal} - T(e_i n_i) ||$$

For $T_{perfect}$ the best solution obtained, $f = \varepsilon \simeq 0$

This same definition can be used to express the distance between the optical properties of the perfect filter and those of the filter which is actually produced

$$f = || T_{perfect} - T_{exp} ||$$

If the parameters $(...e_i,.....n_i, ...)$ are independent, the production tolerances will be large if the partial derivatives

$$\partial f/\partial e_i \simeq 0 \quad ; \quad \partial f/\partial n_i \simeq 0$$

The minimum $f(....n_i....e_i....)$ is large and deep. In general these solutions are those which have the best chance of being found by our synthesis methods.

In fact, to obtain good precision it is necessary to use direct monitoring method. But in this case the errors in the optical thicknesses of successive layers are not independent. We have just seen that they perturb the control of subsequent layers (unless one has a real time computer). Optical monitoring has the advantage of measuring the optical thickness $n_i e_i$: an error in n_i is partially compensated by an error in e_i. But the control of the following layers risks being seriously perturbed. Thus the tolerances necessary for the ith layer can be very tight even if the derivative $\partial f/\partial e_i = 0$; they depend on the monitoring method chosen (optical monitoring, direct or semi-direct, detection of the zeros of $\partial T/\partial e$ or of $\partial T/\partial \lambda$, ... choice of the wavelength for each layer ....) and on the errors which have occured in earlier layers. Further, the accuracy of the measurement of the optical thicknesses depends on the monitoring method chosen for each layer. Numerical calculation of manufacturing tolerances can only be carried out in several particular cases. It is quite impossible, at the present stage of development, to give general conclusions.

6. CONCLUSION

To day, in the industrial sphere, an extensive range of fil - ters with very diverse optical properties is produced. Without doubt, in the years to come, one will see spectacular progress in this field. This will certainly pose a whole series of completely new problems. The development of filter design synthesis must take account of the problem of manufacture. Only a limited number of materials are available and also the conditions which are necessary to obtain resistant layers (heated substrates) are difficult to reconcile with accurate monitoring. The indices of refraction of component layers can vary easily by several percents as a function of parameters which are difficult to control : pressure, temperature, and so on. Mixed layers are a fortiori more difficult to produce with precision. The control of optical thickness becomes difficult when the required accuracy is of order of 1%. In any case we can never avoid some imprecision. In addition, absorption, structural irregularities are never strictly zero. The manufactured product can differ considerably from that projected. In such cases one must simply look for the causes and it is thus that one makes progress.

ACKNOWLEDGMENTS

The author wish to thank Dr A. Holmes Siedle and Dr H.A. MacLeod for their help in the translation of this work.

REFERENCES

1. H.A. MacLeod "Thin Film optical filters " Adam Hilger Ltd London (1969).

2. "Compte rendu des journées de contrôle continu de déposition sous vide" Genève 21-22 juin 1971

3. "Optical Monitoring" Le Vide 1, 157 (1972).

SOME APPLICATIONS OF NON-METALLIC THIN FILMS

M. H. Francombe

Westinghouse Research Laboratories

Pittsburgh, Pennsylvania 15235   (U.S.A)

1. INTRODUCTION

Thin films of elemental semiconductors such as Si, and of passive dielectrics, such as SiO_2 and Si_3N_4, have already found widespread applications in solid state component technology and in high-density microelectronics. As preparative techniques develop, films of more complex semiconducting and insulating materials are receiving increased study and are gradually being incorporated in a variety of novel devices, both passive and active. Many of these devices cannot yet be considered as part of the standard range of microelectronic components, and in fact their hybrid organization, and final structural form and function are still the subject of speculation and experimental development. The numerous innovations occuring in thin film devices are amply illustrated by recent work in fields such as microwave diodes and transistors, thin film transistors, infrared detectors, magnetic bubble memories, photovoltaic solar cells, microwave acoustics, capacitors and solid state imaging and display systems.

In this lecture we shall consider some examples of recent applications of thin films, in which much of materials and fabrication technology is still undergoing research and development.
The special features of thin film preparative approaches being used for these devices will be discussed briefly and the problems for optimizing film quality and the device structural form in relation to its required function will be outlined. There are, in fact, numerous examples of applications in various stages of development from which we could choose, but time and space does not permit us to cover more than a representative selection. In the following sections we shall discuss thin film aspects in the development of the GaAs field

effect transistor, infrared detectors magnetic bubble and ferroelectric memories, capacitors and some magneto-optic and electro-optic structures suitable for use in integrated optic elements, displays or imaging systems. For a more complete review of work on these and other film applications the interested reader is referred to journals and texts dealing with thin film and solid state device topics [e.g., References (1)-(3)] and to special transactions of the IEEE covering microwave devices, solid state imaging and microwave acoustics.

2. THE GaAs FIELD-EFFECT TRANSISTOR

2.1 General Considerations

Historically, thin films have played an important role in the development of transistors. This is especially true of the thin film transistor (TFT) of the Weimer type [4] which utilizes a thin, polycrystalline high-mobility semiconductor film, usually of CdS or Te, together with an insulated metal gate in a field-effect structure. It also is true of conventional bipolar and field-effect transistors based upon epitaxial Ge, Si or GaAs structures. As the need for higher frequency signal generation and amplification at increased power levels has grown, special requirements for shallow diffused regions or for thin epitaxial films (especially in Si and GaAs) have emerged. For bipolar transistors operating at GHz frequencies, base regions of submicron thicknesses are required, and defect-free material relatively free of traps is mandatory in order to keep the transit time for minority carriers short.

To achieve high-frequency and high-power operation, semiconductors possessing high mobility and large energy gap are preferred, and thus recent work in this field has been directed towards compound semiconductors such as GaAs or InP.[5] Unfortunately, despite improvements in crystal growth techniques it has proven very difficult to reduce the traps in such materials to densities small compared with the minority carrier density, and as a result, operating frequencies in minority carrier devices have been disappointingly low. Three-terminal devices of the FET type, in which the current flow may be due to either minority or majority carriers, possess several attractive advantages. In these devices current flow between the source and drain contacts (see Fig. 1) is modulated by a voltage applied to the gate. The active region of the device is a thin film ($\sim$ 1 micron) which is electrically isolated from its substrate either by making the substrate from insulating material or by using material of opposite electrical type.

APPLICATIONS OF NON-METALLIC THIN FILMS 461

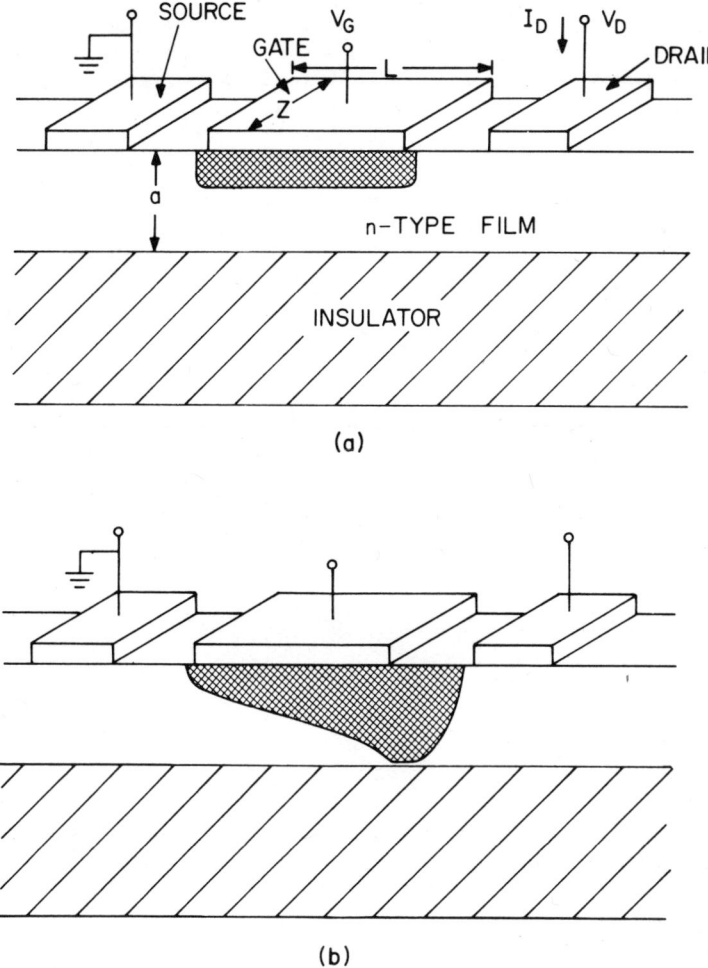

Fig. 1 Simplified view of a junction FET mounted on an insulating substrate :   (a) depletion layer profile for operation in the linear region ;   (b) depletion layer profile for operation in saturation.

   In silicon integrated circuits used for memory or logic applications, FET devices with insulated gates, MOSFET's, built either on silicon substrates of opposite conductivity type or on sapphire, find widespread use. In these FET's the source and drain electrodes constitute junctions (e.g., with p-type regions diffused in an n-type layer) and the device is made to conduct by applying a gate potential of sign and magnitude suitable to introduce an inversion region or channel under the gate insulator. The channel conductance between source and drain then occurs at the silicon surface via transport of minority carriers. In the junction-, or Schottky-barrier FET, the source and drain electrodes constitute ohmic contacts to the active film region, and with no gate potential applied the channel region conducts via transport of majority carriers. As reverse bias is applied to the gate junction or Schottky barrier the channel is depleted of majority carriers, and with increasing gate bias the conducting region is progressively constricted and eventually "pinched-off" [Fig. 1(b)].

In majority carrier devices of the junction FET or Schottky-barrier FET type reasonably high levels of defects and traps can be tolerated. However, it is desirable to use semiconductors with high carrier mobility and saturated drift velocity. Also, for high-temperature or high-power operation both the energy gap and thermal conductivity should be high. With the exception of the thermal conductivity, which is about three times that of Si, bulk n-type GaAs satisfies these requirements admirably. The problem in making an FET device is to duplicate the desirable bulk GaAs properties in a thin electrically isolated thin film.

2.2. Electrical Properties

The requirements placed upon the fabrication technology for a GaAs microwave FET may best be appreciated by considering the factors influencing the low- and high-frequency operation of the device. Following the treatment given by Hower et al. [6] for a junction device as depicted in Fig. 1, we may use the gradual-channel solution approach of Shockley to describe the form of the I-V characteristics for the FET. For a small potential difference between source and drain the depletion layer will extend uniformly under the biased gate [Fig. 1(a)]. However, as the drain current, I_D, is increased, a significant potential drop builds up along the channel and the depletion layer varies in thickness [Fig. 1(b)]. The gradual channel solution can be used to predict the I-V characteristics if the drain-to-gate bias does not exceed $U_0 = qN_D a^2/2K\varepsilon_0$, a quantity termed the pinch-off bias, where N_D is the film doping, a is the film thickness, q is 1.602×10^{-19} C, ε_0 is 8.85×10^{-14} F/cm, and K is the relative dielectric constant ($\simeq 12$). The resultant I-V characteristics are shown in Fig. 2, and are obtained by plotting the normalized current I_D/I_0, derived from a function which is the result of integrating Ohm's law over the entire length of the channel. The current is plotted as a function of normalized drain voltage V_D/U_0 for different values of normalized gate voltage, $\eta = (V_B - V_G)/U_0$ (where V_G is the gate voltage and V_B the junction built-in-voltage). The normalizing current, I_0, is related to the device parameters by

$$I_0 = G_0 U_0 / 3 \tag{1}$$

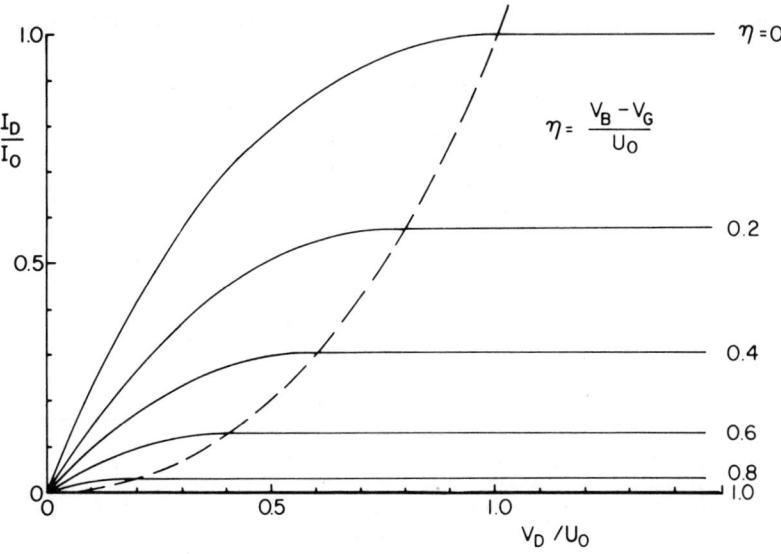

Fig. 2 Common source current-voltage characteristics of an FET based on the gradual channel solution. Dashed line indicates the onset of saturation for drain-to-gate bias sufficient to cause pinch-off.

where

$$G_o = 1/R_o = q\mu_o N_D a Z/L \ ; \qquad (2)$$

μ_o is the mobility and Z/L the geometrical width-to-length ratio ; G_o is the unmodulated conductance of the material under the gate.

If the gate voltage is kept constant and the drain voltage increased, a point is reached when the potential drop from drain to gate is U_o. This condition is termed "saturation" and occurs along the line where

$$V_D/U_o = V_{D_{sat}}/U_o = 1 - \eta = 1 + [(V_G - V_B)/U_o] \qquad (3)$$

plotted as the dashed line in Fig. 2. The drain current in saturation varies only with gate voltage according to the relation

$$I_{D_{sat}}/I_o = 1 - 3\eta + 2\eta^{3/2} \qquad (4)$$

When $\eta = 1$ the drain current is reduced to zero and the entire channel is in the pinched-off condition. The corresponding value of "pinch-off" voltage is $V_p = U_o - V_B$.

Two small-signal quantities of interest are the source-drain conductance in the linear range

$$g_{ds_o} = \partial I_D/\partial V_D = G_o(1 - \sqrt{\eta}) \tag{5}$$

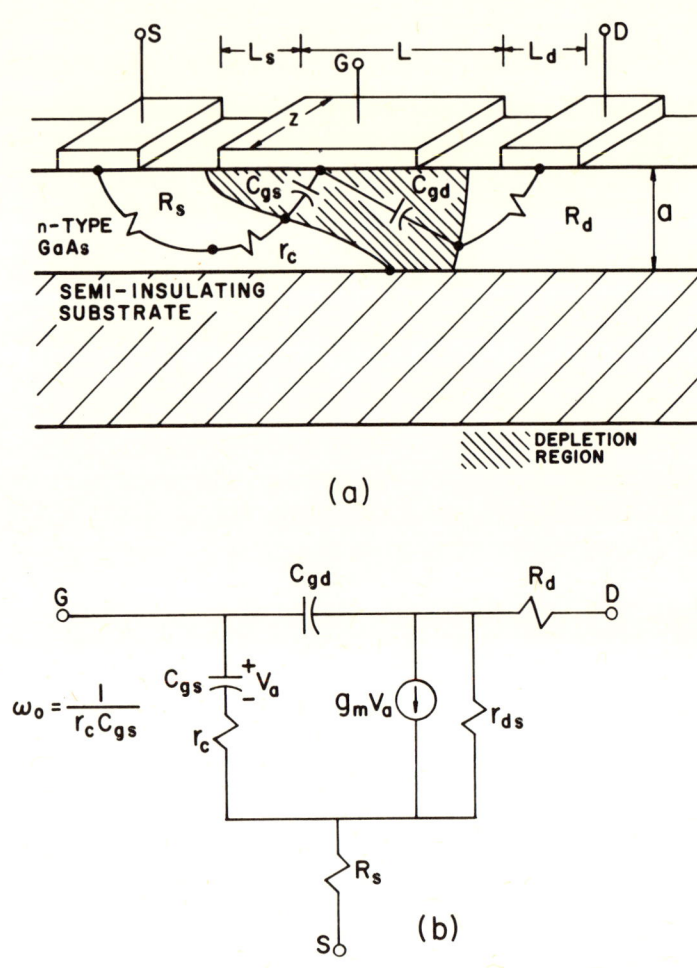

Fig. 3 (a) Active portion of the FET in saturation. (b) Circuit model of the FET.

and the transconductance in saturation. This transconductance is defined as $g_m = \partial I_D/\partial V_D$, and for $V_D \gg V_{D_{sat}}$ is also equal to $G_o(1 - \sqrt{\eta})$.

With reference to the above relationships and to the circuit parameters shown in Fig. 3 we can describe some of the quantities used to define the frequency limits of performance. Figure 3(a) depicts the neutral and depleted regions of the device for terminal bias values corresponding to the conditions of saturation. Figure 3(b) shows the corresponding lumped-element circuit model where the R_S and R_D are the source and drain extrinsic resistances and include the effect of spacings L_S and L_D.

An internal cut-off frequency for the device can be defined in terms of the time for the gate capacitance, C_{gs}, to be changed through the "channel resistance," r_c. The corresponding radian frequency is $\omega_o = 1/r_c C_{gs}$. If the effect of the parasitic resistive element, R_s, is considered, the frequency is modified to a new value,

$$\omega_1 = \omega_o (G_s + g_m)/(G_s + g_c) \tag{6}$$

where $G_s = 1/R_s$ and $g_c = 1/r_c$. For the ideal case of $R_s = R_D = 0$ it may be shown that $g_c = 3 g_m$ and the cut-off frequency is given approximately as

$$\omega_o = 3 g_m/C_{gs}. \tag{7}$$

Other high-frequency performance criteria often quoted are the unilateral gain, U, and the maximum frequency of oscillation, f_{max}. The gain may be expressed as

$$U = \frac{g_m'}{4 g_{ds} g_c'} \frac{1}{x^2} \tag{8}$$

where the prime denotes the inclusion of the effect of the parasitic element R_s and $x = \omega/\omega_1$; ω is the operating frequency. The unilateral gain decreases as the square of the frequency (f^2), i.e., by 6 dB/octave (see Fig. 4). Setting $U = 1$ gives the maximum frequency of oscillation.

$$f_{max} = \frac{\omega_1}{4\pi} \left(\frac{g_m'}{g_c'}\right)^{1/2} \left(\frac{g_m'}{g_{ds}}\right)^{1/2}. \tag{9}$$

To obtain high values of U and f_{max}, devices with small gate lengths (and hence small C_{gs}) and also low values of parasitic resistance are required.

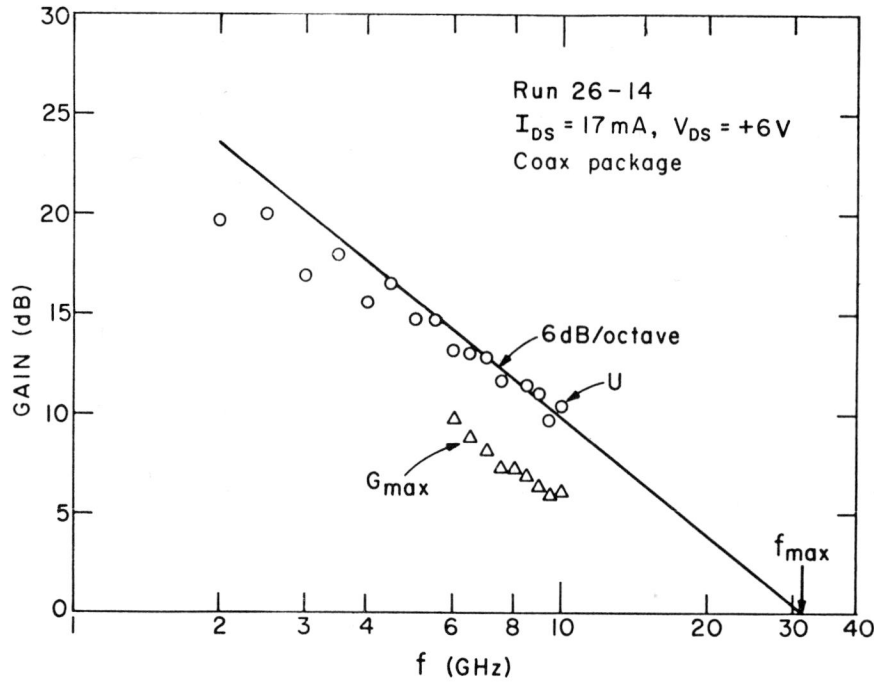

Fig. 4 Unilateral gain, U, and maximum available gain G_{max} for GaAs FET, "TFE-3" geometry in coaxial package.

2.3 Device Fabrication Aspects

Turning now to the aspects of FET fabrication, we discuss briefly some thin film problems vital to the optimization of device parameters. The GaAs epitaxial layer usually is grown on a semi-insulating GaAs substrate by the well-known open tube vapor transport technique developed by Knight et al.[7] using high purity $AsCl_3$, Ga and H_2. Using this method, epitaxial films with net donor concentrations in the $10^{14}/cm^3$ range with Hall mobilities of 8300-8400 cm^2/V-sec at 300°K have been achieved. Films grown by liquid-phase epitaxy show even higher mobilities but are difficult to produce with the thickness uniformity and surface smoothness needed for FET fabrication. Accurate control of GaAs film thickness and of epitaxial quality are important. The thickness,

APPLICATIONS OF NON-METALLIC THIN FILMS

in relation to doping concentration, must be tailored so that pinch-off can be achieved at gate voltages not exceeding breakdown for the Schottky barrier. This breakdown voltage decreases with increasing carrier concentration of the GaAs and is lowered locally by the presence of defects in the epitaxial film.

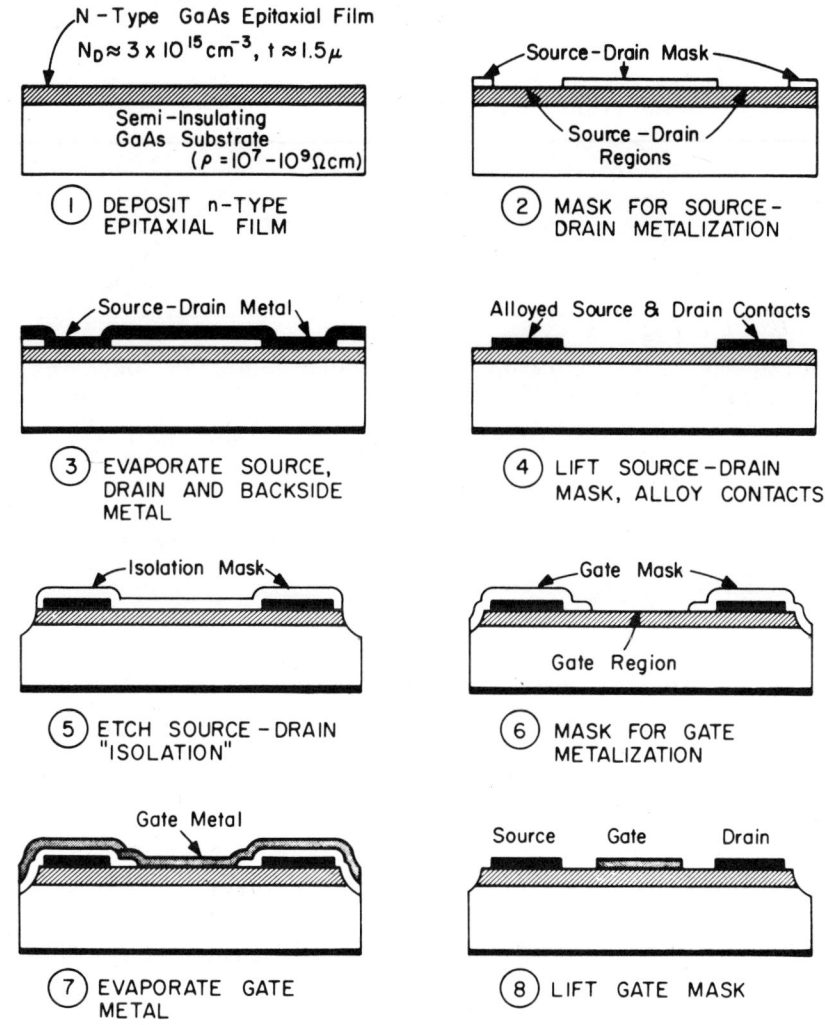

Fig. 5 Process steps for fabrication of a Schottky-barrier-gate GaAs FET.

Figure 5 illustrates schematically the main processing steps involved in fabricating a GaAs FET device. Assuming a gate length of L = 5 µm, the epitaxial thickness is choosen to be 1.5 µm. If a pinch-off voltage of 6V is desired, then the donor concentration, $N_D = 3 \times 10^{15}/cm^3$. The process steps in Fig. 5 are self-explanatory and simply depict the use of photolithographic masking steps to incorporate ohmic source and drain contacts and Schottky barrier gate in the device. The ohmic contacts are produced by evaporating a mixed metal film, consisting typically of Au-Ge (95/5) or Ag-In-Ge (90/5/5). Alloy formation with the GaAs surface, leading to the production of ohmic source and drain contacts, is obtained by annealing in an inert gas atmosphere at temperatures ranging (depending upon the composition used) from 450 to 600°C. The final steps outlined in Fig. 5 illustrate the formation of the Schottky barrier gate contact (usually aluminium) by evaporation. The process steps are subject to many variations and these involve changes both in the methods of forming the contacts and in the geometry of the device. Thus, for the Schottky-barrier contact other metals or alloys, such as Pt-Ni, have been used, while to obtain reduced contact resistance a surface n^+ layer is produced (e.g., by ion implantation) interposed between the GaAs layer and the metal alloy contacts. Accurate registration of the gate relative to the source and drain contacts is difficult to achieve with small spacing using the two-mask procedure shown in Fig. 5. To obtain better control of registration, schemes such as those using self-alignment of the gate in a gap defined by the inner edges of the source and drain contacts have been proposed. Figure 6 shows a photograph of a final device illustrating the geometry of the contact pads chosen to optimize high-frequency performance.

3. INFRARED DETECTORS

Considerable work has been done over the past three decades in the field of infrared sensitive, photoconductive films, especially those based upon the lead salts PbS, PbSe and PbTe. More recently, the need in military applications for infrared detectors of high sensitivity and low noise to be used for surveillance and imaging has prompted increased research on single-crystal materials, in many cases in thin film form. The response characteristics of such detectors are specially tailored to fit certain "spectral windows," i.e., infrared wavelength ranges within which attenuation of radiation in the atmosphere is relatively small. Two such windows lie in the ranges 3-5 µm and 3-14 µm.

Infrared detectors are chosen on the basis of their energy gaps, E_G, and reference to the semiconductor and infrared detector literature shows that the pure compounds available do not provide a good match to the spectral windows mentioned above. To obtain a

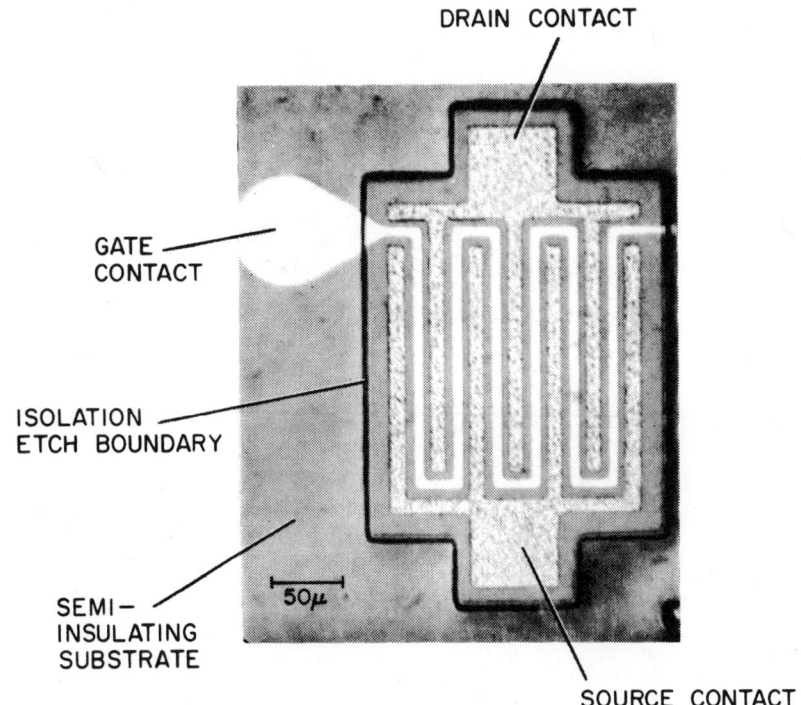

Fig. 6 Surface photograph of Schottky-barrier-gate GaAs FET

good match suitably chosen pure compounds are alloyed together to give pseudo-binary solutions of the type In(As,Sb), (Pb,Sn)Te, (Hg,Cd)Te, Etc. By adjusting the composition, materials with peak response at any desired wavelength can be obtained. Thus, the alloy $Pb_{0.81}Sn_{0.19}Te$ provides a peak response at approximately 11 μm (Fig. 7) while $InAs_{0.9}Sb_{0.1}$ peaks at about 4μm. In most alloy systems of this type the value of energy gap E_G varies roughly linearly with composition. However, the system $InAs_xSb_{1-x}$ is a notable exception (Fig. 8) and displays a shallow minimum in the band gap value at about x = 40%. It is interesting to note that, in principle, detectors matching both of the spectral windows referred to above can be produced from compositions in this single alloy system.

3.1 Detector Parameters

The performance of infrared detectors is characterized by a number of important parameters [8] such as the noise equivalent

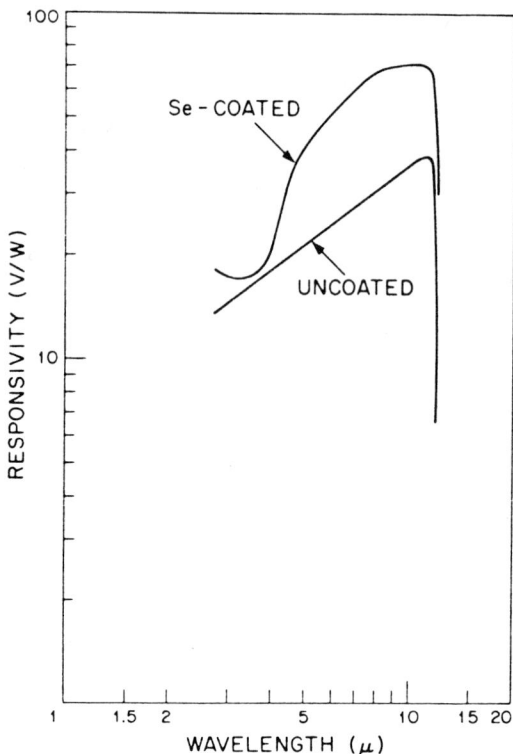

Fig. 7 Responsivity spectra of a $Pb_{0.81}Sn_{0.19}Te$ diode at 77°K before and after application of a selenium antireflection coating.

power, detectivity, responsivity and speed of response. The conditions of measurement for these parameters are standardized in terms of the temperature of the radiating source, the modulating frequency (frequency at which the radiation incident on the detector is chopped) and the amplifier bandwidth. For long IR work a blackbody temperature of 500°K usually is used, while to minimize the noise variations within the frequency interval over which noise is measured a narrow amplifier bandwidth ($\sim$ 5 Hz) is employed. The noise equivalent power (NEP) may be written as follows:

$$\text{NEP } (500°K, 900 \text{ Hz}, 5 \text{ Hz}) = \frac{P_D A}{S/N} \qquad (10)$$

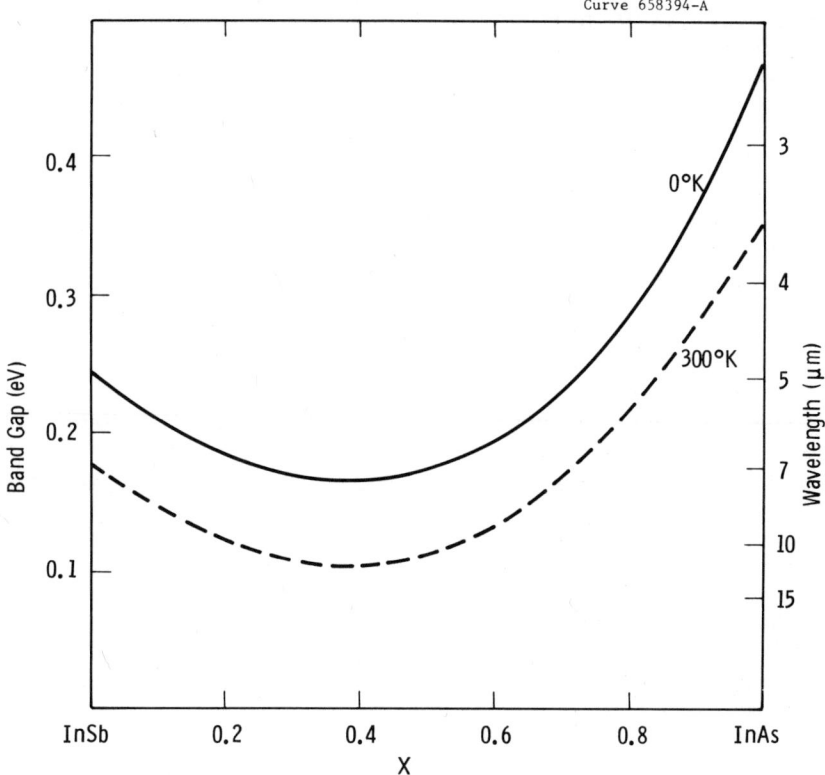

Fig. 8 Variation of bandgap and equivalent cutoff wavelength with composition in system $InAs_xSb_{x-1}$.

The quantities in parentheses refer to the blackbody temperature, modulating frequency and amplifier bandwidth, respectively, S and N represent signal and noise under the conditions of measurements, A is the detector area and P_D the radiant power density reaching the detector from the black body.

The reciprocal of NET is the detectivity, D, and is often normalized to an amplifier bandwidth of 1 Hz and detector area of 1 cm². This yields the parameter D^*, expressed as

$$D^* = \frac{S/N}{P_D}\left(\frac{\Delta f}{A}\right)^{1/2} \tag{11}$$

The normalization is based upon evidence that noise varies as the square root of the amplifier bandwidth and that D varies inversely as the square root of the detector area. Unless otherwise stated D^* is given for a field of view of 2π sr and a background temperature of 300°K.

Thin film detectors will vary in structure depending upon the proposed mode of operation, and before proceeding further with our discussion of detector parameters, it will be useful to describe these modes. In general, three types of structure have been considered, i.e., photoconductors, photovoltaic and MOS photodiode. For operation in the photoconductive mode, semiconductors of high purity possessing low majority carrier concentrations are preferred ($\sim 10^{14}/cm^3$). For use in the photovoltaic or MOS photodiode mode, material requirements are less stringent and, for example, majority carrier densities on each side of a p-n junction should be in the 10^{17}-10^{18} range. The MOS photodiode detector is an interesting special case of the photovoltaic mode and involves the use of an optically transparent electroded capacitor structure on the semiconductor. In the case of n-type InSb this leads to an inverted p-type region adjacent to the oxide interface and hence to an induced p-n junction. Since much of the recent and current research on thin film detectors is aimed towards photovoltaic structures, we may illustrate the other important detector parameters by reference to junction detectors.

A typical junction detector in (Pb,Sn)Te would comprise a bulk p-type crystal with a shallow n-type region formed at the surface by diffusion. Photons absorbed in the surface region generate carriers which diffuse towards the p-n junction. In the open circuit case they produce a change in junction potential as a result of an increase in minority carrier density on both sides of the junction. If the junction is shorted externally a current flows tending to bring the minority carrier distribution back to its equilibrium state. If reverse bias is applied to the junction the photocurrent adds to the normal reverse diode current. For high sensitivity as many as possible of the photogenerated carriers should reach the junction (high efficiency), and the incremental junction resistance should be maximized.

For a diode detector the current I_t can be expressed as the sum of the photocurrent, I_p, and a current, $I(V)$, which flows in the absence of incident radiation as a result of the applied voltage V.

$$I_t = I_p + I(V) \tag{12}$$

In an ideal diode in which all of the current is due to injection

$$I(V) = I_s[\exp(qV/kT) - 1] \tag{13}$$

where I_s is the diode saturation current. The photocurrent I_p may be expressed as

$$I_p = q \eta N \tag{14}$$

where N is the number of incident photons per second and η is the quantum efficiency (i.e., the number of light-generated carriers crossing the junction per incident photon).

For small voltages I(V) can be assumed linear and is expressed as (1/R)V. With the diode operated as a photovoltaic detector, $I_t = o$, then

$$V = R I_p = q \eta NR = (q \eta R/E_\lambda)P_\lambda \tag{15}$$

where V is the open-circuit photovoltage, P_λ the power of incident photons ($P_\lambda = NE_\lambda$), and E_λ is the energy per photon at wavelength, λ.

The voltage responsivity $R_{V,\lambda}$ at wavelength λ can be expressed as

$$R_{V,\lambda} = V/P_\lambda = \eta q R/E_\lambda . \tag{16}$$

The noise voltage of the open circuit junction is given by the Johnson (or thermal) noise of the incremental diode resistance.

$$\bar{V}^2 = 4 kTR \Delta f. \tag{17}$$

If the 1/f noise due to surface and contact contribution, which predominates at low frequencies, is neglected, the noise equivalent power may be derived for a 1 Hz bandwidth by using Eqs. (15) and (17). Then the detectivity for an area A can be expressed as

$$D^*_\lambda = \eta q (AR)^{1/2}/2 E_\lambda (kT)^{1/2}$$

In addition to detectivity and noise parameters another important criterion in the characterization of diode performance is response speed. This can be influenced by both the effective life-time of photoexcited carriers and by the effective time constant of the junction. The second effect depends on the magnitude of diode resistance, R and capacitance, C. In turn, the capacitance is influenced by factors such as carrier concentration in the p and n regions, dielectric constant ε, and magnitude of reverse bias. In general, fast response is favored by short carrier lifetime and a low RC product. However, in adjusting parameters (such as R) for fast response it is important to avoid sacrificing detectivity.

3.2. Preparation and Structure of Thin Film Junction Detectors

The fact that the active region used in an infrared detector is very shallow (a few microns at most) implies that thin film geometries and processing approaches should be ideally suited for such devices. The increasing role of thin films in this area is effectively illustrated in particular by recent work on (Pb,Sn)Te[9,10] and (Hg,Cd)Te[11] detectors. Thin films of these compounds (usually in epitaxial form) have been produced by a variety of techniques, e.g., vacuum sublimation, chemical vapor deposition, liquid phase epitaxy, vacuum evaporation and sputtering. To a large extent the motivation in this work has arisen from a need for large-area detector arrays with uniform characteristics.

Much of the earlier research on (Pb,Sn)Te and (Hg,Cd)Te detectors involved the use of bulk crystals in which junctions were formed by diffusion. More recently, studies by Holloway and coworkers[9] have shown that device-quality epitaxial layers of PbTe and (Pb,Sn)Te could be grown on foreign substrates such as BaF_2, and that photovoltaic diodes could be formed in these either by proton bombardment[9a] (to produce shallow n-type regions in p-type films) or by the application of rectifying metal contact layers[9b] (using for example evaporated films of Pb) to produce a Schottky barrier. Both proton bombardment induced- and Schottky barrier junctions in PbTe epitaxial films have yielded high peak detectivities, with values of D_λ^* (5.5 µm, 600, 1) of about 6×10^{11} cm $Hz^{1/2}$ W^{-1}. However, Schottky junctions on films of the pseudo-binary (Pb,Sn)Te compositions tend to be leaky with rather low resistances.

A desirable aim for both the (Pb,Sn)Te and (Hg,Cd)Te film materials is to be able to fabricate p-n junctions by processes such as diffusion or successive epitaxial growth. Unfortunately, epitaxial films grown on foreign substrates often contain numerous grain boundaries or crystallographic faults, and attempts at diffusion lead to the localized penetration through the entire film thickness. This can be avoided by selecting a substrate such as PbTe or CdTe for film growth which is miscible with the epitaxial composition. The resulting film then possesses essentially a bulk monocrystalline structure. Successive deposition approaches can be used providing that growth conditions can be established for the controlled deposition of p and n layers. Recent work by Krikorian et al. [10] and by Cohen-Solal et al. [11] indicates that this can be achieved through growth of the layers by cathodic sputtering.

In the case of (Pb,Sn)Te the electrical type of both bulk crystals and films may be controlled by adjusting the metal-tellurium stoichiometric balance. This can be illustrated with reference to Fig. 9, which shows a magnified region of the phase diagram in the vicinity of a compound $Pb_{0.8}Sn_{0.2}Te$. Compositions within the enclosed single-phase region which are Te-rich are p-type while those which are metal-rich are n-type. It is clear from Fig. 9 that the

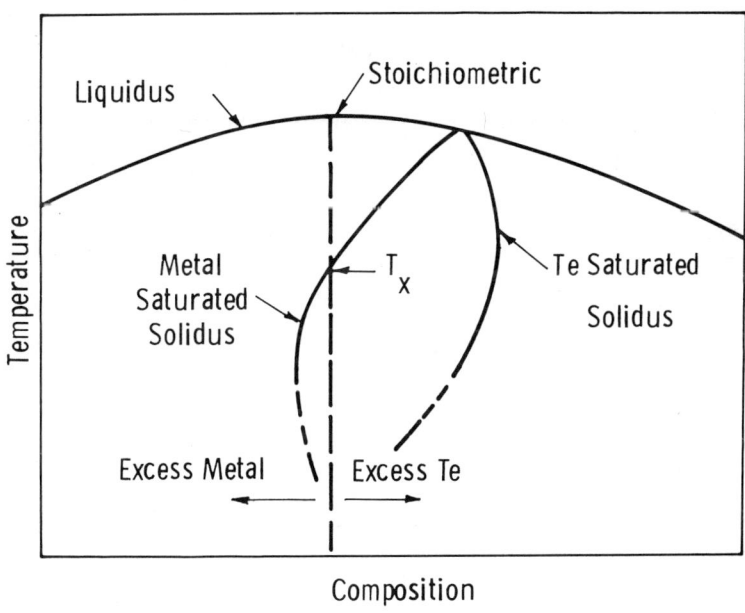

Fig. 9 Magnified view of phase diagram for approximate composition $Pb_{0.8}Sn_{0.2}Te$ in vicinity of stoichiometric compound.

the type and carrier concentration of a (Pb,Sn)Te crystal or film can be adjusted by annealing in either a metal- or tellurium-rich atmosphere at an appropriately chosen temperature. However, in the case of thin films it is virtually impossible to control selectively the type of a surface region in this way since rapid diffusion causes the whole film to convert. The preferred approach is to grow the p and n layers successively from the vapor phase, and the studies of Krikorian et al.[10] indicate that this should be possible simply by adjusting the substrate temperature and deposition rate. As shown in Fig. 10, which is based upon results for epitaxial layers produced by triode sputtering from a target of composition $Pb_{0.8}Sn_{0.2}Te$, low deposition rates and low substrate temperatures favor the growth of n-type films while, high rates

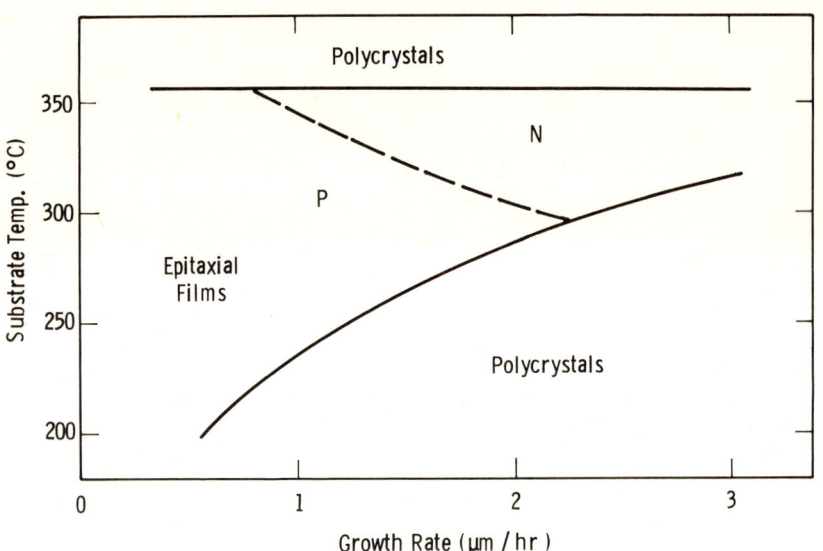

Fig. 10 Epitaxial-polycrystalline phase diagram for $Pb_{.8}Sn_{.2}Te$ films.

and elevated temperatures lead to the formation of p-type films. No data on p-n junctions produced by this approach are yet available.

Cohen Solal et al.[11] have been successful in growing p-n hetero- and homojunctions in the solid solution series $Cd_xHg_{1-x}Te$ by triode sputtering in mercury vapor from alloy cathodes in a vacuum system evacuated by a mercury diffusion pump. Due to the high partial pressure of Hg in the system, loss of mercury from the alloy film at elevated growth temperatures could be avoided and epitaxial layers with low x values were successfully grown on CdTe substrates. The deposited films were made p and n type by impurity doping with metals such as gold which were incorporated in the growing layers by using an auxiliary metal sputtering source. Table 1 shows some typical results obtained for two compositions with peak spectral response occurring at 6.5 and 11.5 µm, respectively.

TABLE 1

PROPERTIES OF FAR I.R SPUTTERED PHOTODIODES AT 77°K

λ_p micron	R_d ohm	η percent	D^* cm. $Hz^{1/2}$. W^{-1}	
6.5	1,000.000	35 - 50	$\lambda_{6.5}$	10^{10}
11.5	1,500.000	55 - 60	$\lambda_{10.6}$	4×10^{10}

4. THIN-FILM CAPACITORS

Thin film insulator materials have found many useful applications in integrated circuits as dielectric layers in devices and as protective or passivating coatings. One of the more obvious advantages derives from the fact that due to their small thickness, high values of capacitance can in principle be achieved in very small volumes. Numerous materials, processed by several preparative techniques, have been examined and some have found application. Those studied include SiO, SiO_2 and Al_2O_3 formed by evaporation, sputtering, chemical vapor deposition and glow-discharge chemical decomposition, anodic oxides such as Ta_2O_5 formed by anodization of sputtered metal films, polymer layers such as parylene or polytetrafluorethylene (PTFE) deposited by uv polymerization or sputtering, and high permittivity mixed oxide dielectrics such as ferroelectric $BaTiO_3$ and $Bi_4Ti_3O_{12}$ grown by flash- or multi-source evaporation or by sputtering.

For thin film capacitors two types of structure usually are sought, both capable of yielding relatively high specific capacitance. One of these utilizes thin amorphous films, for example of SiO or SiO_2, which can be made free from cracks or pinholes, and where a high capacitance (together with reasonably high voltage breakdown) can be obtained by using small thickness. Capacitance values can be further enhanced (to about 0.1 $\mu F/cm^2$) as in the case of tantalum capacitors by using a dielectric such as Ta_2O_5 ($\varepsilon_{Ta_2O_5}$ = 21, cf. ε_{SiO} = 5). This choice would be especially appropriate where the same metal, tantalum, is used for resistors in other parts of the thin film circuit.[12] The second structure utilizes polycrystalline films of a high-permittivity mixed oxide such as $BaTiO_3$. Here, the high permittivity, typical of the bulk dielectric, can only be achieved by developing a relatively well-crystallized film structure, and this in turn requires the use of high growth temperatures. This is illustrated by Fig. 11 which shows the variation of dielectric constant with substrate temperature for flash-evaporated films of $BaTiO_3$ grown at different positions on a resistively heated platinum substrate. Similar results have been obtained for sputtered films of $SrTiO_3$. The resulting film structure contains grain boundaries, and to avoid low-voltage breakdown the films must be made thick (> 1 μm), thus sacrificing capacitance.

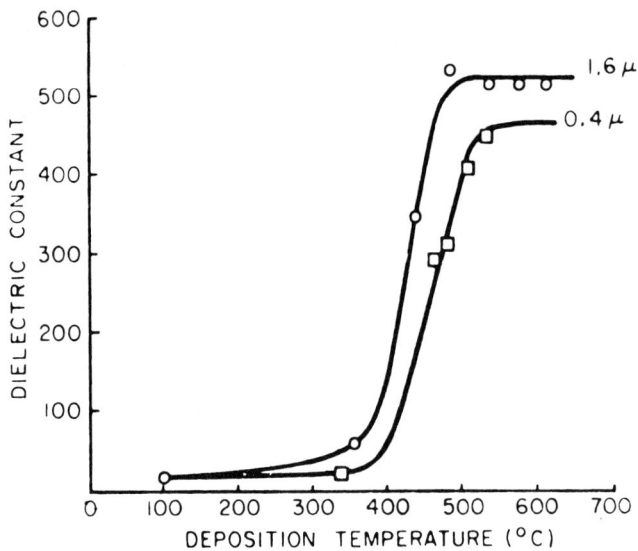

Fig. 11   Dependence of dielectric permittivity on deposition temperature for flash evaporated $BaTiO_3$ films.

Since in monolithic silicon analog integrated circuits there is little need for high capacitance, satisfactory low-value capacitors can be produced from simple dielectrics such as SiO, SiO_2 or anodic Ta_2O_5. In hybrid microelectronics, especially in those circuits operating at lower frequency, chip capacitors are found to be both satisfactory and economical. It appears, therefore, that the main area for technological expansion in the thin film capacitor field will be in high-value discrete capacitors, using thin organic films such as parylene or PTFE, or in microwave integrated circuits used in multi-module systems such as phased-array radars where high-value bypass capacitors are needed. In the latter case capacitances of the order of 1 $\mu F/cm^2$ are desirable together with very low dispersion characteristics.

High specific capacitance values have been obtained with a number of ferroelectric materials in thin film form,[13] as indicated in Table 2. Unfortunately, with some of these materials the dielectric loss is high, and dispersion effects occur at higher frequencies. Figure 12 shows dispersion effects observed in vacuum evaporated $PbTiO_3$ films at frequencies in the kHz range. Presumably, the large changes with frequency in capacitance and dissipation factor arise from chemical inhomogeneity, perhaps due to chemical reduction of the dielectric film surface. Supporting evidence for this model is obtained by assuming a film structure comprising high and low conductivity regions in series. From calculations based on the assumed values of capacitance and resistance shown in Fig. 12 the frequency behaviour found with the $PbTiO_3$ films can be closely duplicated.

TABLE 2

DIELECTRIC PROPERTIES OF HIGH PERMITTIVITY MIXED OXIDE CAPACITOR FILMS PREPARED BY VACUUM DEPOSITION

Material	Method	Deposition Temp. (°C)	t (μm)	C/A (μF/cm^2)	ε	D(%)
$BaTiO_3$	FE	550	0.2	2.25	500	4
$BaTiO_3$	EB	>600	0.46	1.58	820	4.1
$BaTiO_3$	RF	720	1.3	0.49	700	3.0
$PbTiO_3$	EB	700	---	1.81	---	3.7
$SrTiO_3$	RF	500	0.24	0.75	200	1.0
$Bi_4Ti_3O_{12}$	RF	550	0.50	0.4	200	0.2

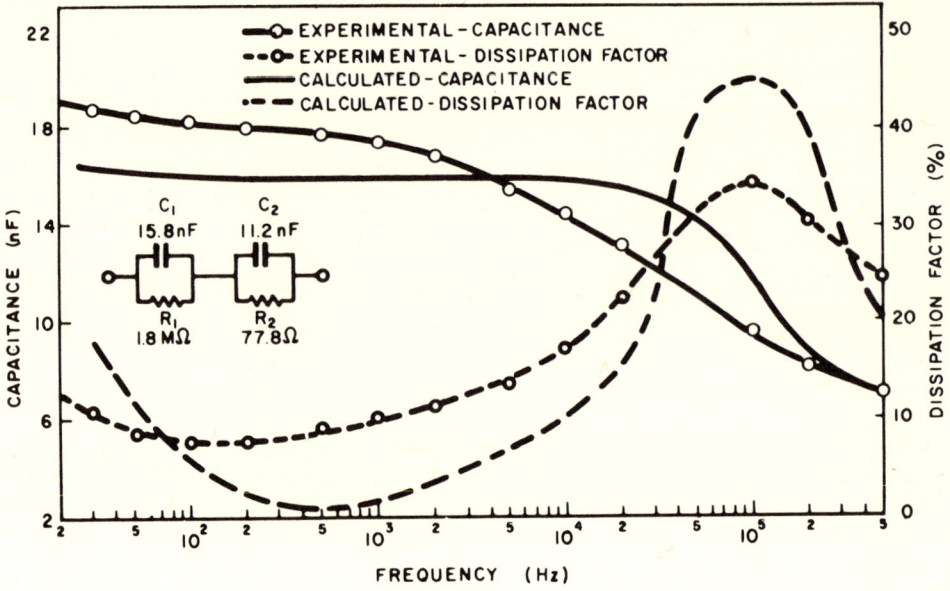

Fig. 12 Equivalent circuit fit for polycrystalline PbTiO$_3$ films

Films of SrTiO$_3$ and Bi$_4$Ti$_3$O$_{12}$ prepared by rf sputtering are found to have somewhat lower loss and better high frequency characteristics, while still offering high specific capacitance. In the case of the SrTiO$_3$ films pulse measurements have shown that no significant loss occurs at frequencies up to 1 GHz, while for sputtered films of Bi$_4$Ti$_3$O$_{12}$ similar data were obtained for frequencies as high as 4 GHz. Both materials would be excellent candidates for bypass capacitors.

5. THIN FILM MEMORIES

Probably the best known application of thin films to memories is that involving the use of thin magnetic permalloy (Ni$_{81}$Fe$_{19}$) films. Considerable research and development effort has been invested in this area over the past 15 years in the hope of developing fast, high-density memories based upon thin film matrices which would out-perform the existing ferrite core memories. While permalloy film memories have found some application in special systems, they have not achieved the multi-megabit scale with acceptable performance justifying replacement of core memories. This was attributable to problems of obtaining reproducibility, uniformity and disturb insensitivity in the permalloy matrices, and also to the fact that more attractive memory structures using high-reliability, stable semiconductor devices and also a highly stable magnetic alternative utilizing single-crystal rather than polycrystalline media have recently become available. In this section we shall discuss briefly the characteristics of such magnetic bubble memories fabricated in single-crytal, magnetic oxide, film media and some more speculative memory work based upon ferroelectric thin film.

5.1 Magnetic Bubble Memories

In magnetic media possessing large values of saturation magnetization M_S (e.g., permalloy) reduction to a thin-film geometry gives rise to a very high demagnetizing field ($4\pi M_S$) perpendicular to the film surface. As a result, the magnetization is constrained by shape anisotropy effects to lie in the plane of the film. However, in media with relatively low M_S values, other competing magnetic anisotropy terms (e.g., crystalline, growth-induced or magnetostrictive effects) may possess energy exceeding that of the shape anisotropy, and allow the formation of a magnetic easy axis perpendicular to the film surface. This situation holds true in the case of many optically transparent magnetic oxides such as the orthoferrites (e.g., $YFeO_3$) or garnets. In the demagnetized condition a thin crystal is divided up into strip domains (Fig. 13), in which the magnetization directions are antiparallel and normal to the large surface. Optical contrast between the domains is obtained by virtue of the Faraday rotation effect.

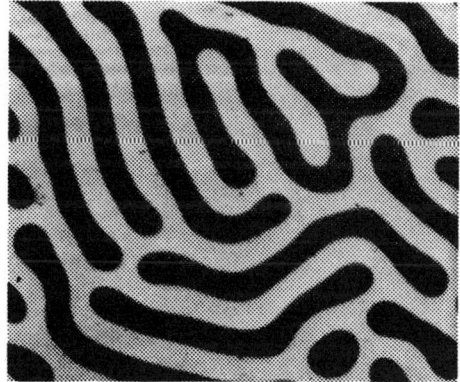 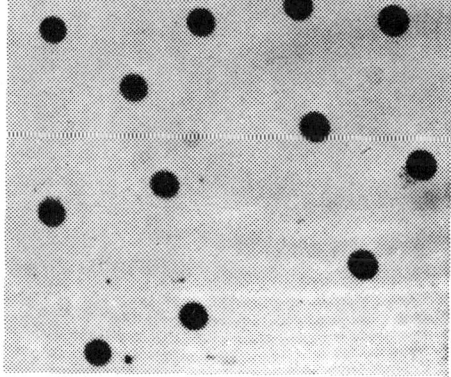

Fig. 13 Strip domains (left) seen by means of Faraday rotation of polarized light represent regions magnetized alternately inward and outward in a 55 μm thick platelet of Ytterbium orthoferrite. Domains are typically 90 μm in width. When a 50 Oe bias field is applied perpendicular to the orthoferrite sheet (right) a cylindrical domain or "bubble" precisely 35 μm in diameter is formed for each properly oriented single-walled strip domain.

In practice, single-crystal magnetic films several microns in thickness, with domain structure of the type shown in Fig. 13, may be grown epitaxially on a supporting substrate of non-magnetic garnet such as $Gd_3Ga_5O_{12}$ using chemical vapor deposition (CVD) or liquid phase epitaxy (LPE) growth techniques. The garnet composition must be optimized to obtain a good lattice and thermal expansion match with the substrate. In the case of the LPE approach, which appears to be the best growth technique evolved to date, films of excellent quality with uniform thickness are currently achieved by spinning the substrate about an axis normal to its plane while suspending it in the molten garnet solution. A typical garnet composition grown by this method, possessing excellent magnetic characteritics for high-speed memory application, is $Y_{1.03}Gd_{1.29}Yb_{0.68}Al_{0.7}Fe_{4.3}O_{12}$.

If a bias field H_A is applied normal to the plane of the magnetic oxide film, those strip domains whose magnetization is opposed to the field shrink in size and eventually change shape to a cylindrical geometry of the type shown in Fig. 13 (right). These cylindrical domains are called bubble domains, and their size and stability are dependent upon relative contributions respectively from the applied field, the domain wall and magnetostatic energies. The theory of the domain structure has been developed by Thiele,[14] while the general technology of magnetic bubble media and their applications to memory and logic devices have been extensively explored by Bobeck [15] and other workers.

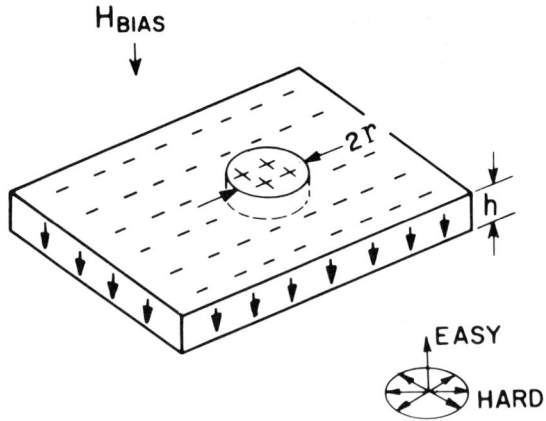

Fig. 14   Geometry of a cylindrical domain in a uniaxial platelet. The domain wall width is assumed much less than 2r.

APPLICATIONS OF NON-METALLIC THIN FILMS

The theory of Thiele leads to the following conclusions : (1) The diameter of the smallest stable domain realizable is approximately 4ℓ. The term ℓ is a material length defined as $\sigma_w/4\pi M_s$. (2) The media thickness h which yields the smallest bubble diameter is $\pi\ell$. (3) A bias field $H_A = (0.3)4\pi M_s$ is needed to support the minimum diameter bubble domain. Figure 14 shows the geometry for a cylindrical bubble domain in a film of thickness h. In Fig. 15 the dependence of domain diameter on thickness is illustrated. The upper curve defines the bubble collapse. Between the curves stable bubbles with diameters inversely proportional to the applied field can exist.

The utility of magnetic bubbles for memory devices arises from the fact that they can be moved in directions parallel to the film surface by the application of a field gradient. Several methods of applying gradients have been proposed. In one, forming the basis of a shift register, an array of photolithographically printed conductors is used to step bubbles at a high data rate by sequentially

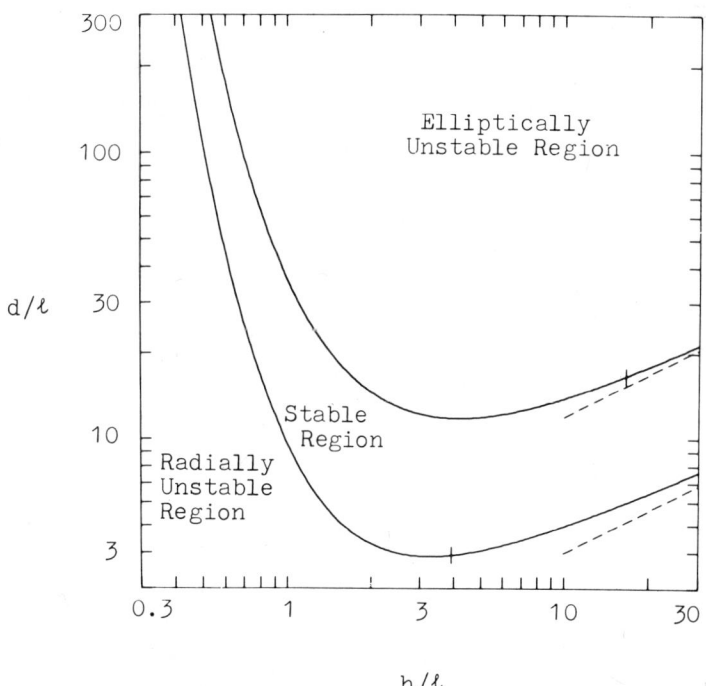

Fig. 15 Dependence of the normalized diameter d/h on the normalized thickness h/ℓ. The upper curve defines the bubble to strip instability and the lower curve bubble collapse.

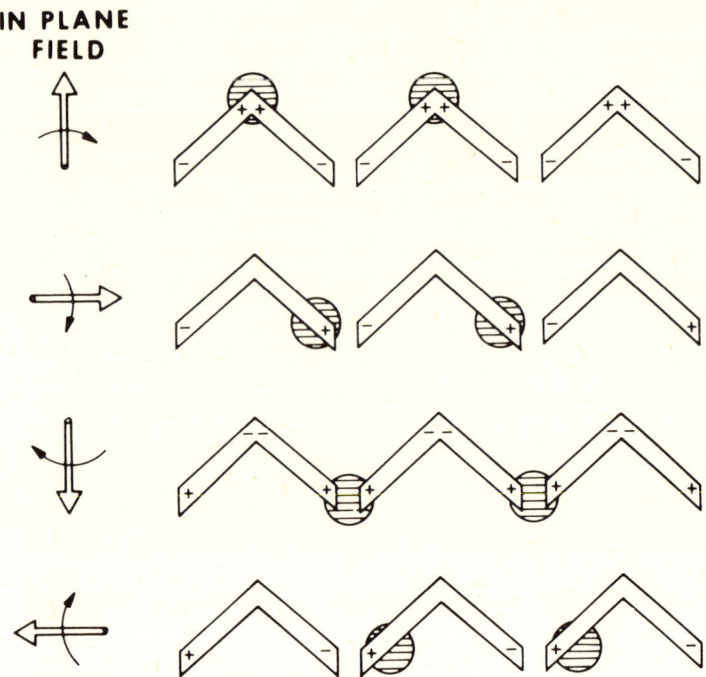

Fig. 16 Step by step bubble propagation in chevron circuit caused by clockwise rotation of in-plane field.

applied current pulses. The bubbles can also be moved using a superimposed array of magnetic permalloy elements. When a field rotates in the plane of these elements, travelling positive and negative magnetic poles are generated at their extremities, as shown for example in Fig. 16. These poles selectively attract and repel the bubble so as to control its motion. Several element geometries have been studied for such permalloy circuits, but the chevron design shown in Fig. 16 is the most widely used.

The generation and detection of bubbles requires specially designed circuit elements. One method of generation is to provide for expansion of a bubble and its separation into subsidiary domains. Several methods of detection have been proposed including Hall effect detectors, magneto-optic detectors employing the Faraday rotation effect and magnetoresistive detection using thin permalloy elements. The magnetoresistive detector seems to be the most generally used, and its sensitivity may be optimized by using a special expander circuit to increase the size of the bubble just prior to detection.

Much research is still in progress with the aim of achieving higher densities and data rates with magnetic bubbles. Several properties must be optimized for this purpose. Among the key parameters are coercivity H_c, which should be lower than 1 Oe, bubble mobility, μ, which should be in excess of 100 cm/sec Oe, and bubble diameter, which should be small, consistent with the highest resolution drive-circuit elements obtainable by lithographic procedures. Since electron beam lithography is capable of submicron resolution, this implies the ability to handle small bubbles in close spaced geometries which should lead to the evolution of very high density data stores. Amorphous alloy media of the Gd-Co and Gd-Fe type have recently been discovered which display bubble domain behavior, with bubbles in the micron size range. These should provide good candidates for high density information processing.

5.2 Ferroelectric Memories

The charge storage properties of ferroelectrics which derive from the P-E hysteresis loop characteristic, offer memory storage potentialities analogous to those available, and fully exploited, with ferrimagnetic metrials. Unfortunately, several practical problems have been encountered over the years which have prevented the realization of those potentialities. In fact, these arise from the intrinsic tendency for ferroelectrics not to possess a well-defined threshold field for switching, a factor resulting in considerable leakage of stored charge under the influence of adjacent disturb fields. In addition, serious difficulties were encountered with control of material parameters and in devising access techniques which could place ferroelectric memories on an economic footing competitive with that for magnetic core systems.

Given the situation that cheap, low-coercivity, square-loop ferroelectrics are available with reproducible properties, the corresponding availability of the same materials in thin film forms immediately offers considerable further advantages. Reduction in material thickness would enable switching voltages to be reduced, while large-area thin-film ferroelectrics would permit the use of high-resolution photolithographic methods in making thin film address conductors and electrodes. A recent evaluation by Chapman[16] and further results published by Mehta[17] indicate that developments with ferroelectric films (specifically with the PZBFN-65 ceramics*) and in the areas of CRT, phosphor, photoconductor and

---

* These are slip-cast foils of nominal composition $Pb_{0.92}Bi_{0.07}La_{0.01}(Fe_{0.405}Nb_{0.325}Zr_{0.27})O_3$. Films of the same composition were also prepared by rf sputtering.

high-resolution optics technology have now reached a level where a new type of fast-access large-capacity ferroelectric memory becomes feasible.

A large storage capacity memory with ultimate capacities in excess of 10^9 bits can in principle be built from an assembly of photoconductor-ferroelectric chips. Each chip in the array is addressed by a flying light spot about 20 μm in diameter, produced by focusing the 10 mil spots on a CRT phosphor screen. The photoconductor provides a light-controlled division of the address voltage applied to the PC/FE sandwich, and essentially supplies an artifical 'threshold' condition for switching. The chip comprises the layered structure—transparent electrode/photoconductor/ferroelectric/electroded substrate—and might be addressed as shown schematically in Fig. 17. In his assessment of feasibility, Chapman assumed a switching voltage for a PZBFN-65 ferroelctric layer 4 μm thick of 20 volts, giving a switching time of 2 μsec. Allowing for a complete cycle time of 2.5 μsec and a total applied voltage pulse of 30 volts, 2/3 of this applied pulse would have to appear across the ferroelectric in 0.5 μsec. Given the specific capacitance for the ferroelectric, time constant and voltage division considerations now lead to values for the light and dark resistivities of the photoconductor. It appears that these photoconductor properties, including the desired rate of photocurrent decay, estimated from the mobility, can be supplied by several available materials including CdS, CdSe and ZnSe. The general conclusion of this study, which is too detailed for reproduction here,

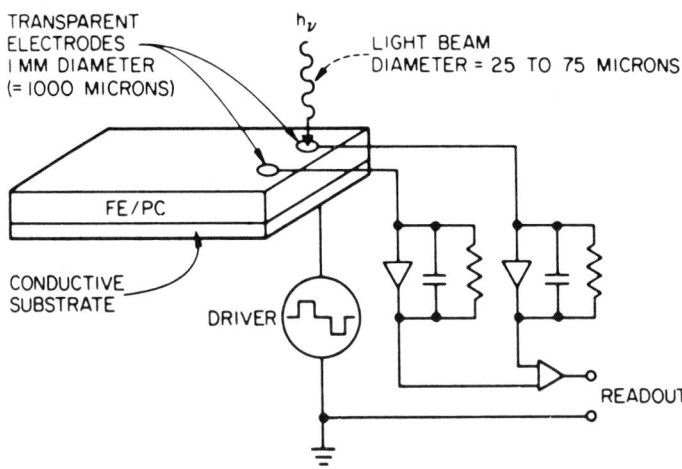

Fig. 17 Test circuit for evaluating storage characteristics of photoconductor-ferroelectric memory chip.

APPLICATIONS OF NON-METALLIC THIN FILMS

was that all aspects of the proposed memory system except the storage medium could be engineered with existing technology.

In a subsequent treatment Mehta[17] carried out experimental tests on a storage medium comprising the PZBFN-65 ferroelectric ceramic at a thickness of 2-2.5 µm in conjuntion with a CdSe photoconductor layer 0.5 µm thick. Light address was accomplished using illumination form a He-Ne laser (6328 Å) covering a spot size of about 25 to 75 µm (Fig. 17). To separate the small signal generated in switching the illuminated spot, it was necessary to compensate for the output produced by partial switching of the non-illuminated area. This was achieved in a test situation by using equivalent transparent electrode areas of 1 mm diameter for the dark and illuminated areas and employing the differential sensing scheme indicated in Fig. 17. The feasibility of addressing the medium using a 40 µm diameter spot focused from a 400 µm spot on a CRT was demonstrated successfully. The ability of the ferroelectric

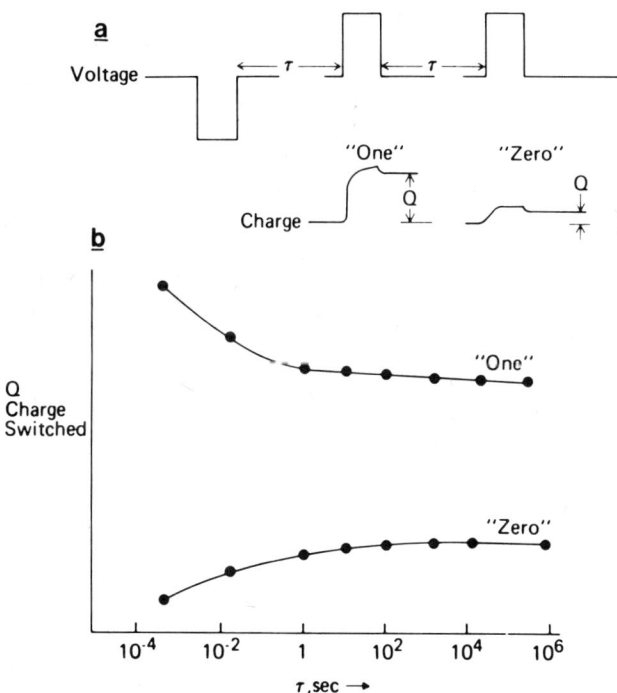

Fig. 18 The ability to store 'one' and 'zero' bits was determined with the indicated voltage pulses. Spontaneous depoling causes the 'one' to decrease and the 'zero' to increase as the time interval between pulses increases.

films to retain their stored charge was also tested under various experimental conditions,[16] and was found to be satisfactory. Typical data for stored 'one' and 'zero' states as a function of the time interval between address pulses are shown in Fig. 18.

These tests revealed certain shortcomings in the CdSe photoconductor and in the CRT mode of address. Low dark resistivity of the CdSe layer caused a considerable amount of charge to be switched in unaddressed spots and made the device unusually disturb sensitive. Also, the low sensitivity of the photoconductor and its long decay time limited cycle times to about 250 μ sec. Variations occurred between the 'one' and 'zero' signal levels, which appeared attributable to non-uniformities in the device or in the CRT light intensity.

6. MAGNETO-OPTIC AND ELECTRO-OPTIC DEVICES

The interaction of light with thin films forms the basis of many useful and commercially profitable communication and display devices. In the previous section we considered examples of such interactions in the field of high-density memories. There are in fact numerous other examples in which the modulation of light by transparent thin film media can be used in integrated optic circuits, as well as magneto, electro- and acousto-optic displays. In some such displays intensity modulation is achieved by variable scattering in a thin liquid-crystal layer, or light is generated at varying levels of intensity by an electroluminescent thin film. Space does not permit us to describe more than two types of devices in this wide class.

Integrated optic circuits offer exciting possibilities in the communication field because of the high speed and density at which information can be transmitted. The four basic elements of such circuits require the means for (a) light generation, (b) light propagation and deflection in thin-film wave guides, (c) light modulation via e.g. magneto- or electro-optic switching effects, and (d) light detection. Research is still progressing with elements for all of these purposes. One obvious candidate for the modulation of light being propagated in a thin film wave guide is the magneto-optic Faraday rotation phenomenon in transparent magnetic films. High-quality garnet films have already been found suitable for use as single-crystal wave guides. It has been shown recently by Tien and his coworkers[18] that by inducing a magnetic anisotropy axis to lie parallel with a garnet film surface a superimposed rf signal can be used to modulate the intensity of light traversing the film. A view of the experimental arrangement is shown in Fig. 19. The garnet film comprises a thin circular disc of composition $Y_3Ga_{1.1}Sc_{0.4}Fe_{3.5}O_{12}$ grown epitaxially on a $Gd_3Ga_5O_{12}$ substrate. Light is cou-

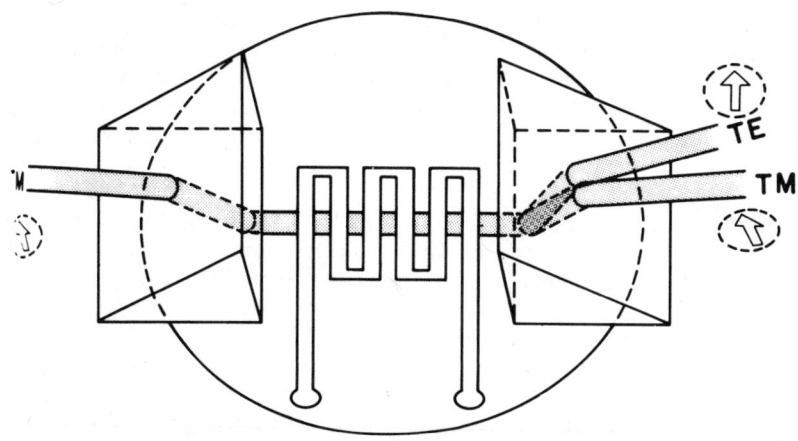

Fig. 19 Experimental arrangement of a magneto-optic switch or modulator.

pled into and out of the film by means of titanium oxide prisms, and the modulation signal is applied with a superimposed serpentine shape thin film coil which is fabricated by standard photolithographic techniques. Typically, the magnetization is constrained by a dc field to lie at 45° to the light propagation direction. The TE (transverse electric) and TM (transverse magnetic) waveguide modes propagating through the film will be angularly split on arriving at the second birefringent prism. The component of magnetization along the propagation direction causes conversion of the TM into the TE mode and thus influences the relative intensities of the split beam components. Superposition of the rf field causes high frequency modulation of this magnetization component, and consequently modulates the intensities of the TM and TE beams. Modulation at frequencies higher than 80 MHz has been obtained with applied fields as small as 0.2 Oe. The frequency appears presently to be limited by the response characteristics of available detectors. This approach offers a novel means of transmitting electrical signals via light beams in thin films.

Modulation of light intensity can also be produced by any of the well-known electro-optic effects, but until recently the only film media available for this purpose were liquid crystals. The phenomenon chosen for intensity modulation will depend upon whether the light is being propagated in a waveguide film structure parallel to the surface as in integrated optics, or through the thickness of the film as in a large-area optical display or memory. In the first case, the path length over which the intensity is modulated can be chosen arbitrarily, depending upon the magnitude of the electro-

optic effect used. In the second, the path length is of necessity
limited to a few microns by practical considerations such as film
growth rate, transparency and adhesion to the substrate. The electro-optic effect should therefore be strong and preferably independent of optical path length. One obvious effect would be based
on extinction of polarized light due to field-induced rotation of
the optic axis of a single-crystal film. Among a small class of
ferroelectric structures currently under investigation, bismuth titanate $Bi_4Ti_3O_{12}$ offers an unusual extinction mode of contrast
which fulfills this need. The polarization axis in $Bi_4Ti_3O_{12}$ lies
in the a-c plane of the monoclinic (pseudo-orthorhombic) structure
at about 4° to the a-axis, giving rise to a small polarization
component (4 µC/cm) along the c-axis and a large component
(50 µC/cm) along the a-axis. Reversal of the c component of P
leads to a rotation of the optical indicatrix major axis (in a
a-c plane) from +25° to -25° relative to the c-axis.[19]. With
the crystal placed between crossed polarizers; near-optimum contrast can be obtained. Unfortunately, the natural growth habit
of $Bi_4Ti_3O_{12}$ (platelets with the c-axis normal to the large face)
makes it difficult to produce large a-c face areas suitable for
the display by mechanical sectioning. A solution to the problem
has been found in the author's laboratory, in which large (approximately 1 cm^2) a-c crystal faces have been grown by epitaxial
deposition of the compound onto single crystal of MgO and $MgAl_2O_4$.
(13,20)

The as-grown epitaxial bismuth titanate films deposited onto
(110) $MgAl_2O_4$ are initially multi-domain, with antiparallelism of
the a and c polarization components. To pole and subsequently
switch a large-area film in situ on the substrate for display purposes, a test structure consisting of an interdigitated (IDT) electrode array incorporated with a 45° electrode scheme was employed.
This structure comprises a set of electrode fingers at the substrate-film interface, followed by a titanate layer 10 µm thick and finally an upper set of electrode fingers connected (in common with
the underlying electrodes) to peripheral contact pads. The principle of operation of the test structure can be explained from the
unit cell enclosed in the dotted lines, as illustrated in Fig. 20.
Each unit cell consists of two outer electrodes with one center
electrode which divides the active area into two regions. To operate the structure, the outer electrodes are grounded. A field
high enough to switch both a and c polarization components is then
applied to all the center electrodes, resulting in a poled structure in the two active regions of each cell. The resulting polarization directions are electrically opposed but the antiparallel
orientations of the indicatrix are optically equivalent (see primed
letters). Switching of a particular cell is effected by addressing
its center electrode with an opposite lower field sufficient to
reverse only the low coercive force c-axis polarization compo-

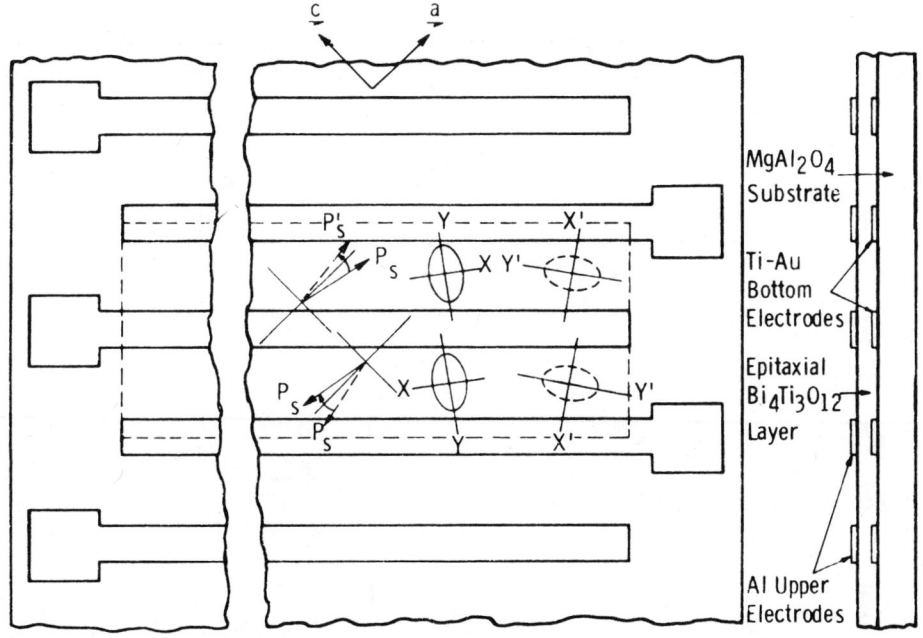

Fig. 20 Schematic diagram of a unit cell of the IDT structure used with a bismuth titanate epitaxial film, showing the tilt of the optical indicatrix caused by switching the c component of P_s.

nents* in each region. This reversal produces the same tilt in both of the optical indicatrices (as shown dashed in Fig. 20) and thus a cooperative change in optical transmission.

To evaluate the contrast obtainable in 10 μm layers, preliminary electro-optic switching experiments were carried out using this test structure, with all the center electrodes connected together. Results of measurements with a photodetector of the contrast ratio, based on the ratios of the light intensity transmitted at different stages of c-axis polarization reversal to the residual intensity at extinction, are shown in Fig. 21. The contrast ratio achieves a maximum of 7/1 at an effective field of about 60 kV/cm, and then gradually decreases with further increase in field. This decrease is due to the progressive reversal of the a-axis compo-

---

*The coercive fields required to switch the c and a components of polarization in these films separately are respectively about 5 and 80 kV/cm.

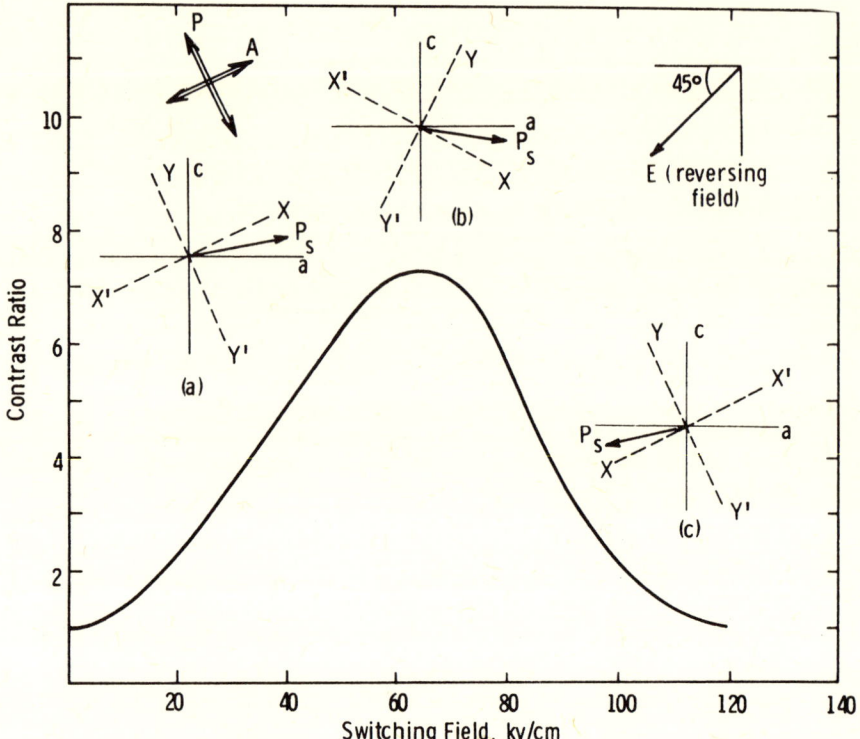

Fig. 21 Contrast ratio vs. switching field plot for 45° address of bismuth titanate films. Inserted also are sketches showing the P_s and optical indicatrix settings, (a) at low reversing field, (b) at the field for maximum contrast, and (c) at high reversing field.

nent of polarization, which tilts the optical indicatrix back into antiparallelism with the original extinction setting. The gradual variation of the contrast ratio with switching field indicates that the epitaxial bismuth titanate film also has a grey scale capability. It can be used for optical display and memory application in optical, electron-beam and matrix addressing modes by special designs of the electrode cell structure incorporated in the film.

REFERENCES

1. "Physics of Thin Films" (eds. G. Hass, M. H. Francombe and R. W. Hoffman), Vol. 1-7, Academic Press, New York.

2. "Application of Thin Films", Proceedings of International Conf. on Thin Films, Venice, May 15-18, 1972 ; Thin Solid Films 13, 1-443 (1972).

3. "Handbook of Thin Film Technology," (eds. L. I. Maissel and R. Glang), McGraw-Hill Book Co., New York (1970).

4. P. K. Weimer, "Physics of Thin Films," (eds. G. Hass and R. E. Thun), Vol. 2, p. 148, Academic Press, New York (1964).

5. "Gallium Arsenide and Related Compounds," Proceeding of the Third International Symp., Aachen, Germany, October, 1970; The Institute of Physics, London (1971).

6. P. L. Hower, W. W. Cooper, B. R. Cairns, R. D. Fairman and D. A. Tremere, in "Semiconductors and Semimetals" (eds. R.K. Willardson and A. C. Beer), Vol. 7, Applications and Devices, Part B, p. 147, Academic Press, New York (1971).

7. J. R. Knight, P. Effer and P. R. Evans, Solid-St. Electron. 8, 178 (1965).

8. "Semiconductors and Semimetals," (eds. R. K. Willardson and A.C. Beer), Vol. 5, "Infrared Detectors," Academic Press, New York (1970).

9. (a) E. M. Logothetis, H. Holloway, A.J. Varga and W.J. Johnson, Appl. Phys. Letters 21, 411 (1972).

   (b) E. M. Logothetis, H. Holloway, A. J. Varga and E. Wilkes, Appl. Phys. Letters 19, 318 (1971).

10. E. Krikorian, R. Longo and M. Crisp, "$Pb_{1-x}Sn_xTe$ Variable Band Gap Alloy System Study," First Interim Report. Contract No. F33615-72-C-1042, May 1972.

11. G. Cohen-Solal, C. Sella, D. Imhoff and A. Zozime, Sixth International Vacuum Congress - Kyoto, March 25-29, 1974. To be published in Japan J. Appl. Phys.

12. "Thin Film Technology," Robert W. Perry, Peter M. Hull and Murray T. Harris, D. Van Nostrand Co., Inc., Princeton, New Jersey (1968).

13. M. H. Francombe, Thin Solid Films 13, 413 (1972).

14. A. A. Thiele, Bell System Tech. J. 48, 3287 (1969).

15. A. H. Bobeck, J. Vac. Sci. Technol. 9, 1145 (1972).

16. D. W. Chapman, J. Vac. Sci. Technol. 9, 425 (1972).

17. R. R. Mehta, J. Appl. Phys. 42, 1842 (1971).

18. P. K. Tien, R. J. Martin, R. Wolfe, R. C. LeCraw and S. L. Blank, Appl. Phys. Letters 21, 394 (1972).

19. S. E. Cummins and L. E. Cross, J. Appl. Phys. 39, 2268 (1968).

20. M. H. Francombe, Ferroelectrics 3, 199 (1972).

PARTICIPANTS

DIRECTOR

C.H.S. DUPUY
Département de Physique des
  Matériaux
Université Claude Bernard -
  Lyon I
43 Bd du 11 novembre 1918
69621 Villeurbanne FRANCE

COMMITTEE:

F. ABELES
Laboratoire d'Optique des
  Solides
Université Paris VI
Tour 13 - 12
4 Place Jussieu
75230 Paris Cedex 05 FRANCE

P. CAMAGNI
Centre Commun des Recherches
  Euratom
Division de Physique
EURATOM   ISPRA (Varese) ITALIE

H. CURIEN
Délégué Général à la Recherche
  Scientifique et Technique
Ministère de l'Industrie et
  de la Recherche
35 Rue St Dominique
75007 Paris   FRANCE

G. DEARNALEY
Theoretical Physics Division
UKAEA Research Group
AERE Harwell
Berkshire ENGLAND

M.H. FRANCOMBE
Westinghouse Research Laboratories
Research and Development Center
Beulah Road
Pittsburg, PA 15235   USA

A.K. JONSCHER
Department of Physics
Chelsea College
University of London
Pulton Place
London SW 6 5PR ENGLAND

R. NIEDERMAYER
Ruhr Universität
Institut für Experimentalphysik
Arbeitsgruppe 4
463 Bochum   GERMANY

LECTURERS:

Prof. D.S. CAMPBELL
Dept. of Electronic and Electri-
  cal Engineering
University of Technology
Loughborough
Leicestershire LE 11 ENGLAND

Dr. D. GREENWOOD
H.H. Wills Physics Laboratory
University of Bristol
Tyndall Avenue
Bristol BS 8   ENGLAND

Dr. R.M. HILL
Dept. of Physics
Chelsea College
University of London
Pulton Place
London SW 6 5PR   ENGLAND

Dr. A.G. HOLMES-SIEDLE
J.J. Thomson Physical Laboratory
Whiteknights
Reading RG6 2AF   ENGLAND

Prof. R. NIEDERMAYER
(cf. Committee)

Prof. H. RAETHER
Universität Hamburg
Institut für Angewandte Physik
2 Hamburg 35
Jungiusstrasse 11   GERMANY

Porf. G. BALDINI
Universita degli Studi di Milano
"Instituto di Scienze fisiche
  Aldo Pontremoli"
20133 Milano
Via Celoria 16   ITALY

Prof. H.K. HENISCH
Materials Research Laboratory
Pennsylvania State University
University Park, PA 16802   USA

Prof. R.W. HOFFMAN
Dept. of Physics
Case Western Reserve University
Cleveland, Ohio 44106   USA

Dr. M.H. FRANCOMBE
(cf. Committee)

Prof. S.J. FONASH
Engineering Science Major
The Pennsylvania State University
231 A Sackett Building
University Park, PA 16802   USA

Dr. A. CACHARD
Département de Physique des
  Matériaux
Université Claude Bernard Lyon I
43 Bd du 11 novembre 1918
69621 Villeurbanne   FRANCE

Dr. E. PELLETIER
Université de Provence
Laboratoire d'Optique
13, rue Henri Poincaré
13397 Marseille Cedex 4   FRANCE

Dr. V. LE GOASCOZ
L.E.T.I.
Centre d'Etudes Nucléaires de
  Grenoble
B.P. 85 Centre de Tri
38041 Grenoble   FRANCE

PARTICIPANTS:

M. ADAMOV
Boris Kidric Institute
Belgrade P.O. Box 522 YOUGOSLAVIE

ARCONADA
Dept. Héliphysique
Université de Provence -
  Centre St Jérôme
13 Rue H. Poincaré
13013 Marseille   FRANCE

J. AVARITSIOTIS
Loughborough University of
  Technology
Physics Dept.
Leicestershire   ENGLAND

PARTICIPANTS

BANGERT
Institut für Physik der Technische Hochschule Wien
1040 Karlsplatz 13 Wien AUSTRIA

B. BANYAI
Institute of Physical Chemistry
Str. Galati 31 Sec. 2
Bucharest ROUMANIE

M. BEN MALEK
Lab. Chimie Nucléaire
Institut de Physique Nucléaire
Université Claude Bernard Lyon I
69621 Villeurbanne FRANCE

T. BOTILA
Institute of Physics
Bodül Pacii
Bucharest ROUMANIE

P. BRICK
Chelsea College
Physics Dept. Lab.
Pulton Place
London SW 6 ENGLAND

J. CARDOSO
Institut für Angewandte Physik
2 Hamburg
36 Jungiusstrasse 11 GERMANY

D. CARLES
Faculté des Sciences de Rouen
76130 Mont St Aignan FRANCE

A.N. CASPERD
Newcastle upon Tyne Polytechnic
   Dept. Physics Research
Elison Building - Elison Place
Newcastle upon Tyne NEI 855
ENGLAND

G. CEASAR
Dept. of Chemistry
University of Rochester
River Station
Rochester, N.Y. USA

M. CELASCO
I.E.N.G.F.
Corso d'Azeguo 42
10125 Torino ITALIA

T. CHAKUPURAKAL
Celestijnenlaan 200 D 3030
   Heverlee
Dept. of Physics
University of Leuven BELGIQUE

A. De CHATEAU THIERRY
I.N.S.T.N.
B.P. 6
91 Gif sur Yvette FRANCE

P. CHEYSSAC
Lab. Electro Optique
Faculté des Sciences
Parc Valrose
06 Nice FRANCE

A. CHOUDRY
Physics Dept.
University of Rhode Island
Kingson, R.I. 02881 USA

T. CLAESON
Dept. of Physics
Chalmers University of Technology
Fack S. 40220 Gothenburg 5
SUEDE

G. COHEN SOLAL
Lab. Physique du Solide
1 Place Briand
92190 Bellevue FRANCE

Le CONTELLEC
C.N.E.T.
Route de Tregastel
22301 Lannion FRANCE

H. COSTEANU
Institute of Physical Chemistry
Str. Galati 31 Sec. 2
Bucharest ROUMANIE

M. CROSET
Lab. Central de Recherches
  Thomson C S F
Corbeville 91 Orsay   FRANCE

B. DELAUNAY
D P H N B C C E N Saclay B.P.2
Gif sur Yvette   FRANCE

S. FEIERABEND
Ruhr Universität Bochum Inst.
  für Experimental Physik
Arbeitsgruppe 4
463 Bochum Post Fach 2148   GERMANY

J.P. FILLARD
Laboratoire de Physique des
  Solides II U.S.T.L.
Place Bataillon
34000 Montpellier   FRANCE

H. FRELLER
SIEMENS AG.Z F A F T E 2 F T 4
85 Nürnberg Katzwangerstrasse
GERMANY

C. FURTADO
Lab. de Physica da Universidade
COIMBRA   PORTUGAL

M. GED
C.N.E.T.
196 Rue de Paris
92220 Bagneux   FRANCE

A. GLADIEUX
I N S T N   B.P. 6
Gif sur Yvette   FRANCE

A. GUSTINETTI
Institute of Physics
University of Pavie
Pavie   ITALY

K. HARTIG
Ruhr Universität Bochum Inst.
  für Experimental Physik
Arbeitsgruppe 4
463 Bochum Post Fach 2148
GERMANY

K. HECQ
Université de Mons
23 Av. Maistrian
7000 Mons   BELGIQUE

C. HOLM
Norvegian Defense Research
  Establishment P.O. Box 25
2007 Kjeller   NORVEGE

K. LYNN
University of Utah
Salt Lake City 2006 MEB
Materials Science and Engineering
Utah 84112   USA

U. KICHKO
Ruhr Universität
Bochum Institut für Experimental
  Physik
Arbeitsgruppe 4
463 Bochum Post Fach 2148
GERMANY

M. KLEITZ
Ecole Electrochimie - Laboratoire
  Cinétique Electrochimie
  Minérale
E.N.S.E.E.G. B.P. 44
38401 St Martin d'Hères   FRANCE

R. KOWALCZYK
03 839 Warszawa Ul. Grochowska
  320
Lab. of Physik Polish Optical
  Work   POLOGNE

C. KUHL (Mrs.)
SIEMENS Zentrale Forschung und
  Entwicklung FL Halbleiter-
  systeme
D 8000 München Post Fack 801709
RFA

B. SAINT MARTIN LAVILLE
Lab. Physique Appliquée -
  Couches Minces
61 Rue A. Camus
68 Mulhouse
FRANCE

PARTICIPANTS

I.L. MAC WALTER
Imperial College of London
Dept. of Electrical Engineering
London SW 7   ENGLAND

MAZETTI
I E N G F
Croso d'Azeguo 42
10125 Torino   ITALIE

V. NAHOO (Mrs.)
Imperial College of London
Dept. of Electrical Engineering
London SW 7   ENGLAND

T. NENADOVIC
Boris Kidric Institute Belgrade
P.O. Box 522
Belgrade   YOUGOSLAVIE

R. NICOLAIDES
Picatinny Arsenal Building 350
Dover, N.J. 07801   USA

G. OLIVE
B.C. Research 3650 Wesbrook CRES
Vancouver BC   VGT 1W5   CANADA

A.E. OWEN
University of Edinburg
School of Science and Engineering
King's Building, Mayfield Rd.
Edinburgh EH9 3JL   ENGLAND

CONTRERAS J. PEDRAZA
Université de Provence
Laboratoire d'Optique
13 Rue H. Poincaré
13397 Marseille Cedex 4   FRANCE

PELOZZI
Laboratory of Special Materials
University of Parme
Parme   ITALIE

A. DE POLIGNAC (Mrs.)
Lab. Electronique
Faculté des Sciences de Reims
Moulin de la Housse
51100 Reims   FRANCE

E. PRAVECZKI
University of Bristol
H.H. Wills Physics Laboratory
Royal Fort Tyndall Avenue
Bristol BS 8 ITL   ENGLAND

PUYCHEVRIER (Mrs.)
C.N.E.T.
196 Rue de Paris
92220 Bagneux   FRANCE

H.F. RAGAIE
LETI CEN Grenoble
B.P. 85 Centre de Tri
38041 Grenoble Cedex 4   FRANCE

H.S. REEHAL
University of Bradford
Physics Dept.
Bradford 7 Yorkshire   ENGLAND

L. RIGALDI
Institute di Fisicia -
   Gruppo Solidi
Via Celoria 16
Milano   ITALIE

A. ROGER
Dept. Physique des Matériaux
Université Claude Bernard LYON I
43 Bd du 11 Novembre 1918
69621 Villeurbanne   FRANCE

N. ROMEO
Institute of Physics
University of Parme
Parme   ITALIE

H.W. RUDOLF
Universität Munchen Lehrstuhl
   Rollwagen 8
Schelling Strasse 4 Munchen
GERMANY

SCHATT SCHNEIDER
Institut für Angewandte Physik
   der Technischen
Hochschule Wien 1040 Karlplatz
13 Wien   AUTRICHE

P.E. SCHMIDT
I.V.I.C.
Apartado 1827 Caracas VENEZUELA

L. SELLA
Lab. Micro Electronica
  OLIVETTI
10015 Ivrea TO   ITALIE

L. DE SHAZER
Solid State Physics
Centre for Laser Studies
University of Southern California
Los Angeles, CA 90007   USA

J. SIEJKA
Groupe de Physique des Solides
  E.N.S. Tour 23
2 Place Jussieu
75221 Paris Cedex 05 FRANCE

W.R. SMITH
Pennsylvania University
University Park, PA 16802   USA

A. STEPANESCU (Mrs.)
I.E.N.G.F.
Corso d'Azeguo 42
10125 Torino   ITALIE

A.C. TSUI
Physics Dept.
University of Bradford
Bradford 7 Yorkshire   ENGLAND

P.C. THACKRAY
348 Benedum Hall
University of Pittsburgh
Pittsburg, PA 15621   USA

B. VIDAL
Lab. d'Optique
Faculté des Sciences de
  St Jerôme
13013 Marseille   FRANCE

D.T. VIGREN
Freie Universität Berlin 33
Institute für Theorestishe Physik
Wes Arnin Allee 3   GERMANY

VUILLERMOZ
C.N.E.T.
Route de Tregastel
22301 Lannion   FRANCE

WAGENDRISTEL
Institut für Angewandte Physik
  der Technischen Hoschschule
Wien 1040 Karlplatz
13 Wien   AUTRICHE

F. WEHKING
Ruhr Universität Bochum
Postfach 2148
Bochum   GERMANY

J. WEINER
Imperial College of London
Dept. of Electrical Engineering
London SW 7   ENGLAND

R. WENZ
Case Western Reserve University
Dept. of Physics
Cleveland, Ohio 44106   USA

Y. YODOGAWA
Dept. of Electrical Engineering
Kyoto Technical University
Matsugasaki Sakyo Kyoto
Japan 606   JAPAN

INDEX

Absorption coefficients, 190
Activation energy, 199, 221
Adhesion of films
  failure of, 342
  preventing film failure, 340
  sputtering and, 343
Alkali halide films, stress in, 331
Alkali halide foils, voids in, 410
Amorphous films
  dielectric polarization, 225
  electronic conduction, 248
  for capacitors, 478
  Poole-Frenkel effect in, 221
  radiation effects, 392
  reflectivity of, 190
Anistropic materials, optical properties, 377
Anodisation, 31
Anticlastic bending, 290
Anti-reflection coatings, 4, 356, 447
Atomic diffusion on surfaces, 51

Band pass filters, 447
Beam splitters, polarizing, 448
Behaviour of thin films, 2
Bending,
  anticlastic and synclastic, 290, 291

Bending (cont'd):
  linear theory, 291
  stresses involved, 280, 318
  origin of, 334
Bipolar transistors, radiation effects in, 405, 407, 408
Bismuth titanate films, optical devices from, 490
Boundary conditions
  optical properties and, 373
  reflection and, 364
Boundary monomer concentration, 56, 67, 69
  evaluation of, 63

CMOS inverters, 404
Cadmium films in memory devices, 486
Cantilever plate method of measuring stress, 289
  experimental results, 295
Capacitance methods of thickness measurement, 174, 180
Capacitors, thin film, 477
Capture rate, 51, 65, 73
Chemical properties of dielectrics, 425
Chemical reduction plating, 40
Clusters
  coalescence, 70, 76
  decay of,
    rate of, 56
    relation to substrate and film, 50
  direct impingement on, 70, 78

Clusters (cont'd):
  growth rate, 72
  model, 59
  shape of, 62
  size of, 66, 85
Cluster scattering, 99, 102, 104
Composition of thin films,
  analysis by nuclear reactions, 148
  analysis of thicker films, 143
  ion beam analysis, 142
    backscattering, 142, 147, 155
    in-depth, 146
    quantitative, 145, 151
    X-ray emission, 154
  MeV energy range ion beam analysis, 142
  photoelectron spectroscopy, 158
  secondary ion mass spectrometry, 155
  X-ray spectroscopy, 157
Concentration profile determination, 146, 152, 153
Conductivity, 189, 203
  injection limited, 211
  space charge limited, 212
Crazing in films, 333
Crucibles for evaporating sources, 16
Crystalline films
  activation energy, 199
  characteristics, 203
  defect density, 221
  dielectric polarization, 225
  electronic transport properties
    doping, 203
  injection limited conduction, 211
  physics of, 215
  reflectivity of, 190
  stress in, 316
    control of, 338
    origins of, 318

Crystalline films (cont'd).
  thermal expansion of, 308
  thermal switching in, 266
  transport processes
    Kubo-Greenwood formula, 196
    mobility, 201

Decay rates of clusters, 56, 63
Definitions of thick and thin films, 1, 6
Deposition of thin films, 3
  chemical methods, 30
    anodisation, 31
    decomposition, 36
    nitriding, 36
    oxidation, 35
    polymerisation, 35
    reduction, 36, 40
    thermal growth, 30
    vapour phase growth, 34
  control of thickness, 11
  electroplating, 37
  evaporation methods, 10, 178
    electron-beam heating, 17
    reactive, 18
    sources, 14
  growth rate, 11
  ion beam, 26, 27
    stress and, 323
  ion plating, 29
  methods, 9
    summary of, 44
  physical methods, 10
  plasma reactions, 28
  polymers, 43
  rate of
    measurement, 178
  solution deposition, 42
  sputtering, 19
    adhesion and, 343
    Getter and bias methods, 24
    glow discharge, 22
    principles, 19
    reactive, 23
    R. F., 26, 27
    stress and, 323
    stress control and, 336
    triode method, 25

INDEX

Deposition of thin films (cont'd):
  stress and, 323, 327, 328, 330, 332
    control of, 336
    distribution of, 280
    temperature stabilization of substrate, 293, 299
  thermal stress and, 306
Dielectrics
  in MOS technology, 423
  properties, 425
  silicon and silicon oxides, 425, 427
Dielectric characterisation, 4
Dielectric constants, 220
Dieletric properties of thin films, 225–252
  boundary conditions and, 231
  dipole-like phenomena, 246, 248, 249
  interface effects, 229, 247
    band bending at, 235
    low frequency capacitance at, 240
  ions as carriers, 233, 234
  Maxwell-Wagner model, 238, 239, 240, 241, 244, 250
  polarization and effective polarization, 225
    characteristics, 246
    effective currents, 231, 232
    general solutions, 232
Dielectric relaxation time, 266
Diffusion source of stress, 338
Diffusion equation, 54
  solutions of, 65
Diffusion potentials, 229
Diode infrared detectors, 472
Diode phenomena, 417
'Dirty' films, 22
Doping, 203

Edge filters, 447
Elastic moduli, 339

Elastic problem, 274
  thermal expansion and, 296
Electric flux configuration, 226
Electrical characteristics of films, 419
Electrical conduction, 3
Electrical properties, 3
  of FET devices, 462
  of MOS devices, 427
Electroforming, 40
Electroless deposition, 40
Electron-beam heating, 17
Electronic transport properties, 236
  activation energy, 199
  continuity, 206
  density of ionised donors, 209
  doping, 203
  excitation, 205, 209
  experimental results, 190
  hopping, 248
  injection limited conduction, 211
  ionic defect hopping, 249
  Kubo-Greenwood formula, 196, 197, 209
  localisation and disorder, 195
  mott hopping, 205
  Poole-Frenkel and Poole behaviour, 208, 213
  single carrier control, 213
  space charge control, 212
  specific processes, 203
  temperature and, 192
  thermopower, 200
Electro-optic devices, 488
Electroplating, 37
Epitaxial films
  in junction detectors, 474
  stresses in, 332
    control of, 336
Evaporation energy, 60

Failure of films, 273
  adhesion and, 340, 342
FAMOS devices, 436, 437
Ferroelectric memories, 485
Field effect transistors, 460, 461, 462

Field effect transistors (cont'd):
  electrical properties, 462
  fabrication of, 466
Filament
  definition, 261
  thermal switching and, 262
Filters, 356, 447
  band pass, 447
  conditions for use, 453
  dielectric stacks, 451
  edge, 447
  manufacturing tolerances, 456
  performance specification, 453
  problems in, 445
Floating gate avalanche injection MOS, 436, 437
Foils, particles in, 410
Friedel sum rule, 99

Gallium-arsenic field effect transistor, 460
Gallium films, stress in, 334
Garnet films, stress in, 341
Gaseous anodisation, 28, 29, 33
Germanium, reflectivity of, 190
Glow discharge condition, 20
Gold films
  growth of, 77
  thickness measurement, 177
Goos-Haenchen effect, 371
Grain boundary diffusion, 339
Grain boundary potential, 320
Grain growth and stress, 319
Graphite films for thermal switching, 255
Growth of thin films, 3, 49–91
  boundary monomer concentration, 56, 67, 69
    evaluation of, 63
  capture rate, 51, 65, 73
  characterisation, 81
  clusters,
    coalescence, 70, 71, 76
    decay rate of, 56

Growth of thin films (cont'd):
  clusters (cont'd):
    direct impingement on, 70, 78
    growth rates, 72
    model, 59
    size, 66, 85
  coalescence of nuclei, 70, 76, 81
  decay rates, 56, 63
  diffusion equation, 54
    solution of, 65
  equations, 70
    computer calculations, 77
    solution of, 73
  evaluation of time constants, 54
  kinetics, 50
  monomer concentration, 78
  prediction of mode of, 85
  rate of twin formation, 69
  rates, 72
    connection with material constants, 50
    evaporation, 10, 11
    liquid phase epitaxy, 42
    of polymer films, 43
    sputtering, 22
    with anodisation, 32
    with vapour phase reactions, 34
  stresses in, 313, 314, 315, 319, 320

Hamiltonian, one electron, 93
Helium implanted erbium films
  stress in, 339
Hydrophilic films, 42

Infrared detectors, 468
  parameters, 469
  structure, 472
Injection limited conduction, 211
Insulation devices, 419, 427
  point defects in, 390
Integrated optics, 355, 488
Interface, 229
  metal-insulator-semiconductor system, 396
  polarization, 229, 235, 240, 247, 250

Interface (cont'd):
  radiation effects in, 394
  stresses at, 287, 288, 304, 309, 321
    impurities and, 343
    origin of, 315, 316
Ion beam analysis
  backscattering, 142, 147, 155
  in-depth, 146
  MeV energy range, composition determination of, 142
  of thick films, 143
  quantitative, 145, 151
Ion beam measurement of film thickness, 177
Ion implantation, 410
Ion induced X-ray emission, 154
Ion mass spectrometry, secondary, 155
Ion plating, 29
Ionic defect hopping, 249
Ionic transport in thin films, 219-224
  drift current, 221
  impurity, 219, 222
  mobility, 222
Ionization gauge ratemeter, 179, 180

Junction detectors, preparation and structure, 474

Kubo-Greenwood formula, 196, 197, 201, 209

Lead salts in infrared detectors, 468
Light
  interaction with thin films, 488
  modulation of intensity, 489
Light waves
  propagation in thin films, 356

Lithium films
  microhardness, 340
  stresses in, 331
Localization and disorder in transport, 195
Logical inverters, 404

Magnetic permalloy films, 480
Magneto-optic devices, 488
Mechanical properties of thin films,
  adhesion, 340
  anisotropic stresses, 292
  edge effects, 285, 303
  elastic problem, thermal expansion and, 296
  for dielectrics, 425
  non-metallic, 273-353
    elasticity, 274
    forces acting on film and substrate, 277, 278
  stresses (see Stresses)
  thermal effects, 294
  thermal expansion, 287
  thermal stresses, 306-312
    deposition and, 306
Memory devices, 409, 440, 461, 480
  magnetic buble, 481
Memory transistors, silicon nitride layers in, 432
Metal films, thickness reproducibility, 176
Metal foil structures for film preparation, 14, 15
Metal-insulator semiconductor system
  hot carriers, 400
  radiation effects, 395
Metal-thin film-metal structure, dielectric properties, 226
Metal-thin film-semiconduction structure
  dielectric properties, 226
Microelectronics
  CMOS technology, 421
  dielectrics, 425
    properties, 425
  MOS technology, 420

Microelectronics (cont'd):
  thin films in, 417-443, (see also Specific devices)
Mirror coatings, high reflectance, 447
Modulators, 377
Molybdenum films, dielectric properties, 245
Monomers
  concentration, 56, 67, 69, 78, 85
    evaluation, 63
  condensation by direct impingement, 70
MOSFETS, 461
MOS depletion enhancement technology, 424
MOS floating gate avalanche injection, 436, 437
MOS technology, 420
  complementary, 421
  dielectrics in, 423
  silicon nitride films in, 430
  silicon on sapphire, 423
MOS tunnel diode, 434
Mott hopping, 205
Multicomponent glasses, 253
  thermal switching in, 265

Naturally occuring films, 2
Non-metallic thin films
  anistropic stresses, 292
  applications of, 459-494, (see also Separate devices)
  edge effects, 285, 303
  elastic problem, 274
  forces acting on, 277, 278
  mechanical properties of, 273-353
  stresses in
    anisotropic, 292
    at interface, 287, 288, 303, 304, 309, 321
    impurities and, 343
    components of, 306
    compression and, 331

Non-metallic thin films (cont'd):
  stresses in (cont'd):
    control of, 336
    deposition and, 323, 327 328, 330, 332
    distribution of, 278, 279, 280
    edge effects, 285, 303, 310
    measurements, 288
    mechanism, 306
    origin, 312
    thermal effects, 294
    thermal expansion in, 287
    thermal stresses in, 306-312
    deposition and, 306
Nuclear reactions, analysis by, 148

Optical coatings
  preparation of, 446, 450
  monitoring, 452
  reactive index, 455
  spectral profiles, 455
Optical constants, deterimination of, 361
Optical filters, 4
Optical properties of thin films, 400, 445
  anisotropic, 377
  multilayer, 371
    matrix technique, 374
  single, 356
Optical thin films, 355-382
  applications, 445-457
    problems, 445
  determination of constants, 361
  stress in, 323
  synthesis of multilayer with prescribed properties, 448
Optic circuits, integrated, 488
Oxidation in preparation, 35
Oxide capacitor films, 477
Oxide films
  as thickness guide, 167
  growth stresses, 313, 314
  space charge trapping, 399
  stress in, control of, 338
  transmission and reflectance, 392

INDEX

Oxygen transport in anodic aluminium oxide, 148, 152

Passivation layers, stresses in, 335
Phase transformations, stresses and, 321
Photoelectron spectroscopy, 158
Photometric determination of optical constants, 361
Photometric methods of film thickness measurement, 168
Photons, passing through solid lattices, 383, 385
Physico-chemical analysis of thin films, 141-161
Plasmons, 138
Polarimetric methods of film thickness measurement, 169
Polarizing beam splitters, 448
Polarization, 225
  characteristics of materials, 246
  dipole-like phenomena, 246, 248, 249
  in optic devices, 490, 491, 492
  interface effects, 229, 235, 240, 247, 250
Polymer films, growth of, 43
Polymerisation in deposition, 35
Poole-Frankel effect, 208, 213, 221, 392
Preparation of thin films, 2, 9
  (see also Deposition)
  methods, 9
Properties of thin films
  electronic transport
    (see Electronic transport properties)
  relation to thickness, 163

Protective coatings, 477
  radiation effects in, 410

Quartz, colour centres in, 391
Quartz crystal measurement systems of thickness, 173

Radiation effects in thin films, 383-414
  basic features of sensitivity, 398
  classification of, 384
  damage equivalent units, 386
  displacement, 386
  environments, 385, 388
  in MOS devices, 404
  in metal-insulator-semiconductor system, 395
  in protective coatings, 410
  interface states, 394
  in transistors, 405, 407, 408
  investigation techniques, 398
  ionization 386, 403
  point defects and, 385, 390
  sources and levels, 387
  surface damage factor, 408
  trapping, 390
Re-crystallised materials, reflectivity of, 190
Reflectance, 445
  in multilayer thin films, 376
Reflection in films, 356
  total internal, 362
Refreactive index of films, 447
Research policy, 5
Rotators, 377

Schottly-barrier-gate FET, 467, 468
Secondary ion beam mass spectroscopy, 155
  measurement of film thickness by, 177
  quantitative, 156
Semiconductors, 3, 459
  activation energy, 199
  amorphous, 96
    surface states, 106, 113
    wave functions, 105, 110

Semiconductors (cont'd):
  cluster scattering, 99, 102, 104
  compound, 460
  covalently bonded, 112
  density of states, 215
  displacement effects in, 402
  electronic states of, 93
    nature of, 105
    one dimensional, 108, 109
  energy band structure, 94, 97, 104, 109, 194
  failure modes, 402
  germanium and silicon polymorphs, 104
  interface with metal, 116
  ionization in, 403
  lifetime, electronic theory, 264
  multiple scattering approach to density of state, 97, 104
  one electron Hamiltonian, 93
  orbitals, 94, 95, 105, 112
  physical properties of, 190, 191
  radiation effects, 385
    trapping, 390
  radiation sensitivity, 401
  relaxation, electron theory for, 266
  stresses in, 305
  stresses build-up in, 304
  surface states, 106, 113
    effects of, 114
    localised, 114
    occupation, 115
  X-ray spectroscopy, 157
Silicon in MOS tunnel diode, 434
Silicon dioxide
  atomic configuration, 393
  colour centres in, 391
  in dielectrics, 427
  trapping, 429
Silicon dioxide films,
  optical properties, 400
  radiation effects, 390
  radiation effects, investigation of, 398

Silicon films
  ion transport in, 222, 223
  stress in, 333, 334, 335, 341
Silicon monoxide films, stress in, 329, 339
Silicon nitride films, in MOS technology, 430
Silicon on sapphire complementary MOS technology, 423
Silicon on sapphire technologies, 421, 422, 423, 430
Silicon oxide
  growth curves, 426
  in thin film capacitors, 477
  measurement of thickness, 174
Silicon semiconductors, radiation sensitivity, 401
Single carrier control, 213
Space charge control, 212
Stresses, 274, 275
  anisotropic, 292
  as interface, 287, 288, 303, 304, 309, 321
    impurities and, 343
  components of, 306
  compression and, 331
  control of, 336
    post-deposition, 337
  deposition, and, 323, 327, 328, 330, 332
  distribution within film, 278, 279, 280
  edge effects, 285, 303, 310
  from bending, 280
  growth, 313
  index to literature, 324
  measurements,
    cantilever plate method, 289
    indirect methods, 306
    X-ray methods, 302
    with measurement of mass change, 305
  mechanisms of, 306
  origin of, 312
    differences of lattice spacing, 312, 317
    diffusion, 338
    grain growth and, 319
    incorporation of atoms, 313
    phase transformation, 321

Stresses (cont'd):
  origin of (cont'd):
    recrystallization
      processes, 318, 339
    voids and dislocation
      arrays, 321
  summary of data of, 322
  thermal, 306-312, 322, 332
    depostion and, 306
Structure of thin films,
    123-140
  amorphous state, 124
  crystalline, 124
    electron diffraction of,
      126, 138
    electron microscopy, 130
    lattice planes, 131, 132
    orientation, 126
    roughness of surface,
      136, 137, 138
    size of crystals, 129
  determination, 123
  distortion of, 195
  geometry of, 123
  ideal characterisitcs, 123
  inelastic scattering, 134
  schematic, 124
  surface, 133, 135
    roughness of, 135
  thickness, 123, 124, 141
Substrate
  bending of, 280, 282, 289
  during deposition, 282
  failures in, 273
  in stress control, 337
  plastic deformation in, 285
  relation to growth rate and
    cluster decay, 50
  stresses, 274, 275, 289
    distribution of, 280
  temperature history of, 294
  temperature stabilization
    during deposition,
    293, 299
  thermal stress and, 306,
    307
Substrate-film composite
  edge effects, 286, 303
  forces acting on, 277, 278

Surface of thin films, 133, 135
  atomic diffusion on, 51
  energy, 60
  plasmons, 138
  roughness of, 135
Switching devices, 409

Talystep, 170
Tantalum for capacitors, 477, 479
Tellurium in junction detectors,
    475
Thermal effects, 294
Thermal expansion, 287
  stress and, 308
Thermal expansion tensor, 308
Thermal stress, 306-312, 322, 332
  deposition and, 306
Thermal switching
  critique of models, 267
  current distribution, 261, 262,
    263
  electronic model, 264, 267
    nature of the ON-state, 264
  'electrothermal', 254
  experimental problems, 269
  'forming', 268
  heating effects, 254
  minimum holding current, 269
  non-thermal (electronic)
    models, 253
  phenomenological
    characteristics, 255
  polarity effects, 268
  primary characteristics, 255,
    256, 257
  relation between ambient
    temperature and voltage, 260
  secondary characteristics, 255,
    258, 259
  switch structure, 255, 256
  thermal model, 261, 267
Thermal theory for uniform
    structures, 261
Thermometers, 408
Thermopower, 200
Thickness of thin films, 123,
    124, 141
  for optical coatings, 446, 450,
    454

Thickness of thin films (cont'd):
  measurement of, 163–185
    capacitance methods, 174, 180
    chemical techniques, 177
    direct methods, 164
    electrical methods, 173, 179
    interference methods, 164
    ionisation gauge systems, 179
    magnetic methods, 177
    mechanical methods, 169, 178
    methods, 163
    micro-balances, 170
    oxide on metal as guide, 167
    photometric methods, 168
    polarimetric methods, 169
    probe systems, 169
    quartz crystal systems, 173, 179
    radioactive techniques, 177
    rate control systems, 178
    resistance methods, 176
    rotating systems, 171
    summary of methods, 181, 182
  optical applications and, 449
  relation to properties, 163
Thiele theory, 483
Threshold switching, 253–272
  thermal models, 253
  thermal theory, 261
Transistors
  bipolar, radiation effects, 405, 407, 408
  field effect, 460, 461, 462
    electrical properties, 462
    fabrication of, 466
  GaAs field-effects, 460
  MOS, 419
    diagram of, 418

Transistors (cont'd):
  memory, 440
    silicon nitride layers in, 432
  radiation effects in, 405, 407, 408
Transport, electronic (see Electronic transport)
  ionic (see Ionic transport)
Transport across films, 4
Transport properties, electronic (see Electronic transport properties)
Twin formation, rate of, 69

Uniform structures
  electronic theory for, 264
  thermal theory for, 261

Waveguides, 366, 367, 410
  characteristics, 368
  coupling to external beams, 370
Wigner delay time, 98, 99

X-ray measurements of stress, 302
X-ray spectroscopy of films, 157

Ytterbium orthoferrate films in magnetic bubble devices, 481

Zinc sulphide films
  deposition and stress, 327
  failure of, 341
  structure of, 329